The Classification of Finite Simple Groups

Mathematical
Surveys
and
Monographs

Volume 172

The Classification of Finite Simple Groups

Groups of Characteristic 2 Type

Michael Aschbacher
Richard Lyons
Stephen D. Smith
Ronald Solomon

American Mathematical Society
Providence, Rhode Island

2010 *Mathematics Subject Classification.* Primary 20D05; Secondary 20C20.

ABSTRACT. We complete an outline, aimed at the non-expert reader, of the original proof of the Classification of the Finite Simple Groups.

The first half of such an outline, namely Volume 1 covering groups of noncharacteristic 2 type, had been published much earlier by Daniel Gorenstein in his very detailed 1983 work [**Gor83**].

Thus the present book, which we regard as "Volume 2" of that project, aims at presenting a reasonably detailed outline of the second half of the Classification: namely the treatment of groups of characteristic 2 type.

Aschbacher was supported in part by NSF DMS 0504852 and subsequent grants.

Lyons was supported in part by NSF DMS 0401132, NSA H98230-07-1-0003, and subsequent grants.

Smith was supported in part by NSA H98230-05-1-0075 and subsequent grants.

Solomon was supported in part by NSF DMS 0400533, NSA H98230-07-1-0014, and subsequent grants.

For additional information and updates on this book, visit
www.ams.org/bookpages/surv-172

Library of Congress Cataloging-in-Publication Data

The classification of finite simple groups : groups of characteristic 2 type / Michael Aschbacher... [et al.].
p. cm. — (Mathematical surveys and monographs ; v. 172)
Includes bibliographical references and index.
ISBN 978-0-8218-5336-8 (alk. paper)
1. Finite simple groups. 2. Representations of groups. I. Aschbacher, Michael, 1944–
QA177.C53 2011
512′.2—dc22

2010048011

10 9 8 7 6 5 4 3 2 1 16 15 14 13 12 11

To the memory of Danny Gorenstein

Contents

Preface

The present book, "The Classification of Finite Simple Groups: Groups of Characteristic 2 Type", completes a project of giving an outline of the proof of the Classification of the Finite Simple Groups (CFSG). The project was begun by Daniel Gorenstein in 1983 with his book [**Gor83**]—which he subtitled "Volume 1: Groups of Noncharacteristic 2 Type". Thus we regard our present discussion of groups of characteristic 2 type as "Volume 2" of that project.

The Classification of the Finite Simple Groups (CFSG) is one of the premier achievements of twentieth century mathematics. The result has a history which, in some sense, goes back to the beginnings of proto-group theory in the late eighteenth century. Many classic problems with a long history are important more for the mathematics they inspire and generate, than because of interesting consequences. This is not true of the Classification, which is an extremely useful result, making possible many modern successes of finite group theory, which have in turn been applied to solve numerous problems in many areas of mathematics.

A theorem of this beauty and consequence deserves and demands a proof accessible to any mathematician with enough background in finite group theory to read the proof. Unfortunately the proof of the Classification is very long and complicated, consisting of thousands of pages, written by hundreds of mathematicians in hundreds of articles published over a period of decades. The only way to make such a proof truly accessible is, with hindsight, to reorganize and rework the mathematics, collect it all in one place, and make the treatment self-contained, except for some carefully written and selected basic references. Such an effort is in progress in the work of Gorenstein, Lyons, and Solomon (GLS) in their series beginning with [**GLS94**], which seeks to produce a second-generation proof of the Classification.

However in the meantime, there should at least be a detailed outline of the existing proof, that gives a global picture of the mathematics involved, and explicitly lists the papers which make up the proof. Even after a second-generation proof is in place, such an outline would have great historical value, and would also provide those group theorists who seek to further simplify the proof with the opportunity to understand the approach and ideas that appear in the proof. That is the goal of this volume: to provide an overview and reader's guide to the huge literature which makes up the original proof of the Classification.

Soon after the apparent completion of the Classification in the early 1980s, Daniel Gorenstein began a project aimed at giving an outline of the original proof. He provided background in a substantial Introduction [**Gor82**], in particular discussing the partition of simple groups into groups of odd characteristic and groups of characteristic 2 type. Then in Volume 1 [**Gor83**] he described the treatment of the groups of odd characteristic in detail. However he did not complete the rest of his project, in part because the proof for groups of characteristic 2 type remained incomplete, specifically that part of the proof treating the quasithin groups undertaken by Mason [**Mas**]. This gap was recently filled by the Aschbacher-Smith classification of the quasithin groups [**AS04b**]. Hence it is now possible to finish Gorenstein's project by outlining the proof for the groups of characteristic 2 type. We accomplish that goal here, adopting his title, and regarding the work as "Volume 2" in the series.

While we recommend that the interested reader consult Gorenstein's books, we also intend that our treatment should be sufficiently self-contained that those works will not be a prerequisite. Therefore in Chapter 1, we supply an overview of the treatment of the groups of odd characteristic, which is much briefer than Gorenstein's detailed treatment.

In fact, throughout our exposition, we will be less detailed than Gorenstein, since we believe that a briefer outline of the main steps will be more accessible and useful to most readers. On the other hand, we are careful to honor the important fundamental goal of explicitly listing those works in the literature which make up the proof that all simple groups of characteristic 2 type are known.

Mathematics, particularly the proof of a complex theorem, is hierarchical. We will list the results on groups of characteristic 2 type at the top of that hierarchy, which we refer to as "level 0" results. These are the papers containing subtheorems whose union affords the classification of the groups of characteristic 2 type. We also discuss the papers at level 1: the principal subsidiary results used in the proofs of subtheorems at level 0. We will *not* usually attempt an analysis through levels 2 and beyond; that is, as a rule we do not discuss those papers used to establish the subsidiary results, and so on, down to first principles and the level of textbooks. But our outline could be used as a starting point for such a deeper analysis of the proof.

Finally we will typically assume that the reader has some familiarity with concepts, terminology, notation, and results from elementary group theory, such as might be standard in a first year graduate algebra course. Beyond that, we will try to give more advanced definitions when they arise in our discussion. In addition we provide in Chapter A of the Appendix a review of some intermediate material on simple groups and their properties. The Index should be helpful when encountering new terminology and notation; normally the index entry given in **boldface** indicates either the definition, or the most fundamental page reference.

Acknowledgments. We would like to thank various colleagues for helpful comments on early stages of this work; especially Rebecca Waldecker. (And thanks as usual to the referee.)

Smith is grateful to All Souls College Oxford for a Visiting Fellowship during Hilary Term 2009.

Background and overview

CHAPTER 0

Introduction

In this introductory chapter we give an overview of the strategy behind the proof of the theorem classifying the finite simple groups, while for the most part attempting to avoid the fine structure of the proof. At various places we make oversimplifications to avoid technicalities; this means that some statements are not quite true in special cases, but hopefully this approach conveys the flavor of the proof more clearly than a more involved, technically correct discussion.

To be a bit more specific, we describe how, for a suitable choice of prime p, each finite simple group G is determined up to isomorphism by its p-local subgroups. This makes it possible to classify the simple groups in terms of their local structure.

We explain why the prime 2 plays a special role and is usually the optimal choice for p. We sketch a proof of the Dichotomy Theorem 0.3.10, which partitions the simple groups into groups of Gorenstein-Walter type and characteristic 2 type, according to two possible 2-local structures. We also describe the partition of each of these two types of simple groups into large and small groups of the given type, leading to a four-part subdivision of the proof of the Classification.

0.1. The Classification Theorem

We begin with a statement of the Classification Theorem:

THEOREM 0.1.1 (Classification Theorem). *Each finite simple group is isomorphic to one of:*

(1) A group of prime order.
(2) An alternating group on a set of order at least 5.
(3) A finite simple group of Lie type.
(4) One of 26 *sporadic simple groups.* □

In Section 0.4 we discuss each of the four classes of groups in more detail. But first we begin with a brief discussion of local group theory, and the role it plays in the proof of the Classification. We assume the reader is familiar with the most basic notation and terminology from finite group theory. For those who are not, some notation and terminology is defined at the start of Section A.1 in Appendix A.

Throughout this chapter, G is a finite group and p is a prime. Further $Syl_p(G)$ denotes the set of all Sylow p-subgroups of G. Pick $S \in Syl_p(G)$, and write $\mathcal{P}$ for the set of nontrivial subgroups of S.

A subgroup of G is called *p-local* if it is the normalizer $N_G(P)$ of some nontrivial p-subgroup P of G. We speak simply of a *local subgroup* if we do not wish to specify the prime p. The study of finite groups from the point of view of their local subgroups is called local group theory.

The local theory is the most important tool in the proof of the Classification of the finite simple groups. In particular the proof of the Classification is based on the following two principles:

Principle I: Recognition via local subgroups. If G is simple and the p-local structure of G is sufficiently rich, then G is determined up to isomorphism by the local subgroups $(N_G(P) : P \in \mathcal{P})$; and even by a subfamily $(N_G(Q) : Q \in \mathcal{Q})$ for some suitable small subset $\mathcal{Q}$ of $\mathcal{P}$.

Principle II: Restricted structure of local subgroups. If G is simple then the structure of the locals $(N_G(P) : P \in \mathcal{P})$ is highly restricted. Indeed up to a suitable notion of isomorphism, there are only a small number of choices for the family $(N_G(\mathcal{P}) : P \in \mathcal{P})$.

To oversimplify a bit, the Classification amounts to translating the two Principles into precise statements, and writing down proofs of those statements, at least in enough special cases to include all simple groups. Once this is accomplished, given a finite simple group G, Principle II tells us that for some prime p, G has the same p-local structure as one of the groups $\overline{G}$ listed in the Classification Theorem. Then Principle I says that G is determined up to isomorphism by its p-locals, so G is isomorphic to $\overline{G}$.

0.2. Principle I: Recognition via local subgroups

To make Principle I more precise, we will use the language of topological combinatorics:

NOTATION 0.2.1 (Graphs and complexes associated to p-subgroups). Given a positive integer k, define $\Lambda_k = \Lambda_k^p(G)$ to be the commuting graph[1] on the set of elementary abelian p-subgroups E of G with p-rank $m_p(E) \geq k$. Define $\Lambda_k^p(G)^\circ$ to be the subgraph of Λ_k consisting of those E with $m_p\big(C_G(E)\big) > k$. Notice that if $m_p(G) \leq 1$, then $\Lambda_1^p(G)$ is totally disconnected, and $\Lambda_1^p(G)^\circ$ is the empty graph. Thus these graphs are of no interest unless $m_p(G) \geq 2$.

For a subset $\mathcal{Q}$ of $\mathcal{P}$, form the corresponding set $\mathcal{N} = \mathcal{N}_\mathcal{Q} := \{N_G(Q) : Q \in \mathcal{Q}\}$ of local subgroups, and regard $\mathcal{N}$ as a poset (partially ordered set), ordered by inclusion.

The corresponding *amalgam* $\mathcal{A}(\mathcal{N})$ (cf. the discussion in the relevant subsection of Section 3.3, e.g. 3.3.14) roughly encodes the elements of members of $\mathcal{N}$ as generators for any abstract group $\overline{G}$ satisfying the relations defining the members of $\mathcal{N}$, plus the relations induced by the inclusions among the members of $\mathcal{N}$.

Also let $\Sigma(\mathcal{N})$ denote the "space of cosets" $\coprod_{N \in \mathcal{N}} G/N$, regarded as a poset via $Nx \leq My$ iff $N \leq M$ with $Nx \subseteq My$. Write $\mathcal{O}(\mathcal{N})$ for the order complex of the poset $\Sigma(\mathcal{N})$. That is, $\mathcal{O}(\mathcal{N})$ is the abstract simplicial complex whose simplices are the chains in $\Sigma(\mathcal{N})$. Observe that G acts as a group of automorphisms of $\Sigma(\mathcal{N})$ and $\mathcal{O}(\mathcal{N})$ via right multiplication. ◊

We remark (cf. the discussion after 0.4.1) that if G is a finite simple group of Lie type defined over a field of characteristic p, and $\mathcal{N}$ is the set of all p-local subgroups of G, then $\mathcal{O}(\mathcal{N})$ is the spherical Tits building associated with G. Further

[1] i.e. where edges are defined by the corresponding subgroups centralizing each other; cf. Definition 1.3.8 for further discussion of commuting graphs. Also the p-rank $m_p(X)$ is the maximal rank of an elementary abelian p-subgroup of X; cf. Definition A.1.6.

if G has BN-rank at least 3, then G is determined as a distinguished subgroup of the automorphism group of $\mathcal{O}(\mathcal{N})$. Thus Principle I holds in this important case.

Recognition via simple connectivity. For the purposes of Principle I, the p-local structure of G is "sufficiently rich" if (again oversimplifying for expository purposes) Λ_1 is connected, and there exists $\mathcal{Q} \subseteq \mathcal{P}$ such that the simplicial complex $\mathcal{O}(\mathcal{N})$ is simply connected. The condition that Λ_1 is connected is equivalent to $G = \langle N_G(P) : P \in \mathcal{P} \rangle$, while $\mathcal{O}(\mathcal{N})$ is connected if and only if $G = \langle \mathcal{N} \rangle$. Thus the condition that Λ_1 is connected is implied by the condition that $\mathcal{O}(\mathcal{N})$ is simply connected, but we include that condition anyway for emphasis. The condition that $\mathcal{O}(\mathcal{N})$ is simply connected implies that G is the free amalgamated product of the amalgam $\mathcal{A}(\mathcal{N})$; that is, that G is the largest group $\overline{G}$ generated by the elements of members of $\mathcal{N}$, subject to defining relations for those members, and relations from the inclusions among members of $\mathcal{N}$. Hence G is indeed determined up to isomorphism by the amalgam $\mathcal{A}(\mathcal{N})$ of local subgroups, so that Principle I holds.

Recognition via a single subgroup (such as an involution centralizer). Now in practice the amalgam $\mathcal{A}(\mathcal{N})$ is a fairly complicated and unwieldy object. Thus we often focus on one carefully chosen local $H := N_G(Q)$, where we have $N_S(Q) \in Syl_p(H)$; and show that $\mathcal{A}(\mathcal{N})$ can be retrieved from H—together with the embedding of H in G imposed by the constraint that H is a local subgroup, and the G-conjugacy pattern of suitable elements of $N_S(Q)$. Hence G also is often determined up to isomorphism by one local H, its embedding in G, and *fusion* (i.e. G-conjugacy) in the Sylow p-subgroup $N_S(Q)$ of H.

Let us look at an example from the early days of the modern effort to classify the simple groups. We can hope to have some freedom in choosing a prime p, subject to the constraint that the p-local structure is rich. In practice the 2-local structure is usually rich, so we can take $p = 2$. Moreover most often it is best to choose $Q = \langle t \rangle$ with t of order 2 (an *involution*), so that $H = N_G(Q)$ is the involution centralizer $C_G(t)$.

This is precisely the approach for studying finite simple groups proposed by Richard Brauer in his 1954 address [**Bra57**] to the International Congress of Mathematicians in Amsterdam, where he suggested that simple groups should be characterized in terms of the centralizers of involutions. Moreover his program was anchored and motivated by his seminal result with Fowler:

THEOREM 0.2.2 (Brauer-Fowler [**BF55**]). *Let H be a finite group. Then there are at most a finite number of finite simple groups G possessing an involution t such that $C_G(t) \cong H$.*

Now an obvious necessary condition for the 2-local structure of a nonabelian simple group G to be rich is that 2 be a prime divisor of the order of G. An old conjecture of Burnside from the beginning of the twentieth century postulates that each nonabelian finite simple group *is* of even order. Some years after Brauer proposed his program, Feit and Thompson proved Burnside's conjecture in their monumental work [**FT63**]. The Feit-Thompson Odd Order Theorem (which we discuss later as 1.2.1), and Thompson's thesis (establishing an old conjecture of Frobenius), were two of the early major papers in local group theory.

The case where $\Lambda_1(G)$ is disconnected. In order to classify finite simple groups in terms of their p-local structure, we also need to prove some results in the case where that structure is not sufficiently rich in the above sense; in particular, we need to first determine those groups G such that p divides the order of G, but $\Lambda_1^p(G)$ is disconnected.

As noted earlier, this is always the case when G has p-rank 1. However, an early theorem of Brauer-Suzuki (see 1.4.2), in conjunction with the Odd Order Theorem, guarantees that every non-abelian simple group G has 2-rank at least 2.

But there are also disconnected examples with 2-rank at least 2: for example if $G = A_5$ (the alternating group of degree 5), then $\Lambda_1(G)$ has five connected components, each of which is a complete graph on three vertices. Furthermore given the knowledge of finite groups supplied by the Classification, it can be checked that, with certain known exceptions such as the example just mentioned, if the p-rank $m_p(G)$ of a simple group G is at least 2, then $\Lambda_1(G)$ is connected. (Cf. [**GLS98**, Sec 7.6].)

But of course we need to *prove* a result of this type—*without assuming the Classification*—as part of a proof of the Classification Theorem via p-local structure. Fortunately such a result was proved for the prime 2 by Suzuki and Bender;

THEOREM 0.2.3 (Bender-Suzuki [**Suz62, Suz64, Ben71**]). *Let G be a finite simple group of even order such that $\Lambda_1^2(G)$ is disconnected. Then $G \cong SL_2(2^n)$, $PSU_3(2^n)$, or one of the simple Suzuki groups $Sz(2^{2n+1})$; i.e., G is a simple group of Lie type, of characteristic 2 and BN-rank 1.*[2]

We will discuss this result in more detail later at 1.3.5. And we will discuss the groups of Lie type further in Section 0.4 in this chapter.

However it should be emphasized that when p is odd, the only known determination of groups G such that $\Lambda_1^p(G)$ is disconnected, depends upon the Classification. In effect this means that in classifying simple groups, one must always keep the prime 2 in the picture. In our discussion of groups of characteristic 2 type, after 2.2.1 and at various later points, we will see that the absence of an analogue of the Bender-Suzuki Theorem 0.2.3 for odd primes leads to great difficulties in the treatment of those groups.

Existence and uniqueness problems. Finally observe that if we are to characterize simple groups in terms of some p-local condition, denoted say by $\mathcal{C}$, then we must address the corresponding problems of existence and uniqueness:

Existence Problem. There exists *at least* one simple group satisfying $\mathcal{C}$.

Uniqueness Problem. There exists *at most* one simple group satisfying $\mathcal{C}$.

Existence. For most of the finite simple groups, the Existence Problem is almost trivial: Namely, with the exception of some of the sporadic groups and groups of BN-rank 1, each finite simple group G is essentially the group of automorphisms of some highly symmetric and natural object X (cf. our discussion in later Section 0.4). Indeed for the alternating group A_n, X is just the set $\{1, 2, \ldots, n\}$; and for the classical linear groups, X is a vector space or a space with a form. Thus the existence of X gives an easy proof of the existence of G. The representation of G on X makes it relatively easy to check that G satisfies a suitable p-local defining condition $\mathcal{C}$.

[2]The groups of rank 1 and even characteristic are often called *Bender groups*.

The representation of G on X is also the primary tool for proving the many properties of G which are used in an inductive context in the proof of the Classification. These properties and others in addition provide the basis for applying the Classification—to prove numerous results both in finite group theory and other areas of mathematics.

Uniqueness. Similarly the representation of G on X is usually the basis for a fairly simple treatment of the Uniqueness Problem. For example, we can choose our set $\mathcal{N}$ of local subgroups so that the complex $\mathcal{O}(\mathcal{N})$ is isomorphic to a geometry visible in X, with the members of $\mathcal{N}$ the stabilizers of simplices in that geometry. The natural description of X and its symmetry makes it relatively easy to prove that $\mathcal{O}(\mathcal{N})$ is simply connected, hence establishing the uniqueness of G.

Unfortunately some of the sporadic groups have no known representation as a group of automorphisms of some nice highly symmetric object. For these groups, either or both of the Existence and Uniqueness Problems can present difficulties.

0.3. Principle II: Restricted structure of local subgroups

As discussed in Section 0.1, the proof of the Classification amounts to translating the two Principles listed there into precise statements, and providing proofs of those statements.

In Section 0.2 we gave a fairly precise translation of Principle I, and some indication of how to implement that Principle. In this section we discuss Principle II—although we won't get very far in restricting locals in simple groups. More significant restrictions must wait until Section 0.5, following our discussion of the finite simple groups in Section 0.4.

The generalized Fitting subgroup. Our eventual objective is to see what distinguishes local subgroups of simple groups, from local subgroups of the general finite group. We first need some tools for discussing the structure of general finite groups. (Cf. Appendix A starting at Definition A.1.17.)

The first and most important tool is the generalized Fitting subgroup $F^*(H)$ of a finite group H. (We write H in place of G, since often we will be considering the situation where G is a finite simple group, and H is some local subgroup of G.) We define $F^*(H)$ precisely in Definition A.1.19; in this paragraph, we give a more informal introduction. First a *component* of H is a subnormal quasisimple subgroup of H. For the purposes of this section, we can think of these components roughly as certain nonabelian simple normal subgroups of H; their product is denoted by $E(H)$, and we can think of this product as roughly being a direct product. Then $F^*(H)$ is the product of $E(H)$ with the classical Fitting subgroup $F(H)$ of H (i.e. the largest normal nilpotent subgroup of H). We write $O_p(H)$ for the largest normal p-subgroup of H, and it is easy to see that $F(H)$ is the direct product of the subgroups $O_p(H)$, as p varies over the prime divisors of the order of H.

The definition of $F^*(H) = E(H)F(H)$ is due to Bender, building on earlier work of Wielandt, and of Gorenstein and Walter. The fundamental property of $F^*(H)$ is:

THEOREM 0.3.1 (Bender). *$C_H(F^*(H)) = Z(F^*(H))$.*

PROOF. See e.g. 31.13 in [**Asc00**] □

Notice for example that $F^*(H) = 1$ implies $H = 1$; more generally, $F^*(H)$ roughly "organizes" the structure of H: For as $F^*(H)$ is normal in H, the conjugation map c takes H into $Aut(F^*(H))$; and we see by the self-centralizing property 0.3.1 above that $\ker(c) = Z(F^*(H))$. The latter group is abelian, and typically we can expect it to be only a small part of H; thus $c(H)$ should be a "nearly" faithful representation of H as a group of automorphisms of $F^*(H)$. Furthermore $F^*(H)$ has a natural product structure—and its automorphisms correspond essentially to those of the factors, along with permutations of isomorphic components. Thus $F^*(H)$ is a relatively uncomplicated group, which controls the structure of the possibly highly complex group H.

The generalized Fitting group in an involution centralizer. Moreover as we continue through this section, we will see that for P a p-subgroup of G, the embedding of $F^*(N_G(P))$ in G is fairly restricted.

Indeed first we will examine two examples of very natural groups G, which produce contrasting structures for $F^*(C_G(t))$ in the centralizer of an involution t. Then later we will see that those are essentially the *only* two possible structures.

EXAMPLE 0.3.2 (Involution centralizers in matrix groups of odd characteristic). First let G be the "nearly-simple" group $GL_n(q)$, i.e. the group of all invertible $n \times n$ matrices with entries from a finite field of order q. In this example we take q to be a power of an *odd* prime.

Then every involution t of G is conjugate to a diagonal matrix with -1 and 1 as the eigenvalues. (The choice of q odd is implicitly used, since we distinguish -1 from 1.) That is, for some $1 \leq k < n$, we have the block-diagonal form:

$$t = \left(\begin{array}{c|c} -I_k & 0 \\ \hline 0 & +I_{n-k} \end{array}\right).$$

By standard linear algebra, the centralizer $H := C_G(t)$ of t preserves the distinct eigenspaces of t. Hence the centralizer also has block-diagonal form, namely a product

$$H = \left(\begin{array}{c|c} GL_k(q) & 0 \\ \hline 0 & GL_{n-k}(q) \end{array}\right) = \{\left(\begin{array}{c|c} A & 0 \\ \hline 0 & B \end{array}\right) : A \in GL_k(q),\ B \in GL_{n-k}(q)\}.$$

Thus $H = L_1 \times L_2$ is the direct product of two nearly-simple factors isomorphic to $GL_k(q)$ and $GL_{n-k}(q)$. More accurately, $E(H)$ is the product of two (quasisimple) components $E_1 \cong SL_k(q)$ and $E_2 \cong SL_{n-k}(q)$.[3] So $F^*(H)$ has some components and $|H : E(H)|$ is small. ◇

Next we consider a related example yielding a strikingly different structure:

EXAMPLE 0.3.3 (Involution centralizers in matrix groups of characteristic 2). This time consider the linear group $G := GL_n(2)$ over the field $\mathbb{F}_2$ of characteristic 2. The diagonal matrix of the previous Example 0.3.2 is now just the identity, since $-1 = 1$ in characteristic 2; so to get an involution, we instead take t to be a transvection (i.e., the matrix $t - I_n$ has rank 1—cf. Section A.6). By conjugating

[3]Even this statement is sometimes inaccurate—for example, quasisimplicity fails if $k = 1$ or $n - k = 1$. But it holds for "typical" values of k (and n and q).

in G, we can arrange for t to have the following convenient block-triangular form:

$$t := \left(\begin{array}{c|c|c} 1 & 0 & 0 \\ \hline 0 & I_{n-2} & 0 \\ \hline 1 & 0 & 1 \end{array}\right).$$

Again let $H := C_G(t)$ be the centralizer in G of t; it has the corresponding block-triangular form

$$H = \left(\begin{array}{c|c|c} 1 & 0 & 0 \\ \hline * & SL_{n-2}(2) & 0 \\ \hline * & * & 1 \end{array}\right),$$

where $*$ indicates arbitrary elements of $\mathbb{F}_2$. Indeed H is even a semidirect product $H = UL$, where

$$U := \left(\begin{array}{c|c|c} 1 & 0 & 0 \\ \hline * & I_{n-2} & 0 \\ \hline * & * & 1 \end{array}\right), \text{ and } L := \left(\begin{array}{c|c|c} 1 & 0 & 0 \\ \hline 0 & SL_{n-2}(2) & 0 \\ \hline 0 & 0 & 1 \end{array}\right).$$

In this example, $U = F(H) = O_2(H)$ is the largest normal nilpotent subgroup of H. Further H has no components at all (in particular the group L is *not* normal). Thus $E(H) = 1$, so that in fact we have $F^*(H) = F(H) = U$; and in particular

$$F^*(H) = O_2(H) \tag{0.3.4}$$

is a 2-group. (This last feature also occurs if we let G be the nearly-simple group $PGL_n(2^m)$ for any $m \geq 1$—and even if we choose t to be *any* involution in G.) Notice furthermore that H is much larger than $F^*(H)$ in this example, so the full structure of H is not so well predicted by the structure of $F^*(H)$. Nevertheless, it is still true that $H/Z\big(F^*(H)\big)$ embeds into $Aut\big(F^*(H)\big)$. Indeed H/U embeds into $GL(\overline{U})$, where $\overline{U}$ is the $\mathbb{F}_2$-vector space $U/\langle t\rangle$. ◇

The Dichotomy Theorem distinguishing odd and even cases. Next, motivated by the two examples of linear groups discussed above, we define two classes of finite groups in terms of the local structure of the groups (rather than in terms of linear representations).

The even case. First, as suggested by $GL_n(2)$ in Example 0.3.3, the p-local subgroups in each simple group G of Lie type over a finite field of characteristic p satisfy condition (0.3.4), except with 2 replaced by p. (Indeed we state this property later as 0.4.1.3.)[4] This suggests the following terminology:

DEFINITION 0.3.5 (Characteristic p type). A finite group G is said to be of *characteristic p type* if $F^*(H) = O_p(H)$ for each p-local subgroup H of G. ◇

The groups in our *even* case are the groups of characteristic 2 type. In particular the groups $GL_n(2)$ in 0.3.3 are of characteristic 2 type, so they are even groups.

---

[4]In some of the more recent literature, our terminology of "characteristic p type" in 0.3.5 is replaced by the alternative terminology *local characteristic* p; and groups H satisfying the condition in (0.3.4) are said to be of *characteristic* p. We have retained the more traditional terminology of 0.3.5, to clearly distinguish this more general class of groups from the class of groups of Lie type in characteristic p.

The odd case. The groups in our "odd case" have involution centralizers similar to those in $GL_n(q)$ with q odd, in Example 0.3.2, in that some centralizer has components. But for our proof below of the Dichotomy Theorem 0.3.10, we will need a certain generalization of the notion of component, which we now introduce after a few preliminaries.

DEFINITION 0.3.6. For p a prime, $O_{p'}(G)$ denotes the largest normal subgroup of G of order coprime to p. And roughly dually, $O^{p'}(G)$ denotes the smallest normal subgroup H of G such that G/H is a p'-group. Equivalently, $O^{p'}(G)$ is the normal subgroup of G generated by all of the p-elements of G. ♢

In our study of the structure of p-local subgroups H of G, the groups $O_{p'}(G)$ and $O_{p'}(H)$ will often loom as possible "obstructions" to various desirable properties.

For example consider the components in $E(\overline{H})$, where $\overline{H} := H/O_{p'}(H)$: Here for $q \neq p$ we have $O_q(\overline{H}) \leq O_{p'}(\overline{H}) = 1$ by definition of $\overline{H}$, so that $F(\overline{H}) = O_p(\overline{H})$; hence $E(\overline{H})$ is the obstruction to the condition $F^*(\overline{H}) = O_p(\overline{H})$ (as in (0.3.4))—*for the quotient* $\overline{H}$ (cf. B.1.9). This motivates the study of:

DEFINITION 0.3.7 (The p-layer and p-components). Define $O_{p',E}(H)$ to be the preimage in H of $E\big(H/O_{p'}(H)\big)$, and set $L_{p'}(H) := O^{p'}\big(O_{p',E}(H)\big)$. We call $L_{p'}(H)$ the *p-layer* of H.

A subnormal subgroup $L = O^{p'}(L)$ of H which surjects onto a component of $E\big(H/O_{p'}(H)\big)$ is called a *p-component* of H. Hence $O_{p'}(L)$, when it is not central in L, is the obstruction to L being an actual component (i.e. quasisimple) of H. ♢

We will see that p-components arise naturally in arguments later in this section,[5] although the p-components in centralizers H in our examples of simple groups G—notably for $G = GL_n(q)$ (q odd) in Example 0.3.2—are actual components. So for expository purposes in this section, where usually $p = 2$, the reader should think of $L_{2'}(H)$ as being roughly the same as $E(H)$. But the statement and proof of the Dichotomy Theorem require the following more general notion:

DEFINITION 0.3.8 (component type). A group G is said to be of *component type* if $L_{2'}\big(C_G(t)\big) \neq 1$ for some involution t in G. ♢

Thus Example 0.3.2 shows that for q odd, $GL_n(q)$ is of component type for all $n \geq 3$—with the exception of $GL_3(3)$ because of the solvability of $SL_2(3)$. Indeed small exceptions such as $GL_2(q)$ and $GL_3(3)$ motivate the following additional definition (which we have named after two of the major contributors to the classification of the corresponding groups):

DEFINITION 0.3.9 (Gorenstein-Walter type). We say that a finite group G is of *Gorenstein-Walter type* (or *GW type*) if either (1) G has 2-rank $m_2(G) \leq 2$, or (2) G is of component type. ♢

The groups in our *odd* case are the groups of Gorenstein-Walter type.

[5]More precisely, it is the *absence* of p-components which makes possible results such as (0.3.21).

The Dichotomy Theorem. Recall that Principle II says that the structure of local subgroups of finite simple groups should be restricted. Our first major result in that spirit shows that the 2-local structure of a simple group satisfies the conditions in either Definition 0.3.5 or Definition 0.3.9, and hence its 2-locals resemble those in Example 0.3.2 or Example 0.3.3:

THEOREM 0.3.10 (Dichotomy Theorem). *Assume that G is of even order and satisfies*
(a) $G = O^2(G)$, and
(b) $O_{2'}(G) = 1$.
Then G is either of Gorenstein-Walter type or of characteristic 2 type.

The Dichotomy Theorem begins to distinguish the 2-local structure of simple groups from that of the general finite group, supplying some progress toward making Principle II precise. Specifically, if G is an arbitrary finite group which is not of component type, and if H is an involution centralizer in G, then by definition H has no 2-components and hence no components, so that $F^*(H) = F(H)$—but a priori, this only says that $F(H)$ is nilpotent, not a 2-group. The contribution of the Dichotomy Theorem is to show, when G is also simple and of 2-rank at least 3, that $F^*(H)$ must in fact be a 2-group (as in (0.3.4)); and indeed this holds for any 2-local subgroup H.

After a brief discussion in Section 0.4 of the finite simple groups, we will further pursue Principle II by giving a somewhat more detailed idea of the structure we expect in 2-locals in simple groups of Gorenstein-Walter type, and in groups of characteristic 2 type.

A proof of the Dichotomy Theorem. But in the remainder of this section, we first sketch a fairly complete proof of the Dichotomy Theorem. Our aims are:

- to demonstrate that the basic ideas underlying the dichotomy are actually fairly elementary (admittedly modulo assuming certain standard results); and
- to introduce, along the way, some further concepts and methods which are of fundamental importance throughout the proof of the Classification Theorem.

Some general preliminaries. In the next few lemmas, we let the superscript $*$ denote images under the natural surjection of G onto $G^* := G/O_{p'}(G)$.

We had mentioned after 0.3.6 that the subgroups $O_{p'}(H)$ of p-local subgroups H of G can be an obstruction to establishing various properties of G. To help overcome this problem, we would like to show that generically the image $O_{p'}(H)^*$ is small (ideally even trivial).

Such analysis uses results on "coprime action", as in Section B.1 of Appendix B; here we mention two useful elementary facts (cf. Lemma B.1.8):

LEMMA 0.3.11 (Coprime Action). *Let P be a p-subgroup of G. Then*
(1) $N_{G^*}(P^*) = N_G(P)^*$.
(2) If X is a p'-subgroup of G normalized by elementary abelian P, then:
$$X = \langle C_X(Q) : Q \leq P \text{ with } P/Q \text{ cyclic}\rangle.$$

PROOF. See e.g. 18.7.4 in [**Asc00**], and 11.13 in [**GLS96**]. □

There is a global-implies-local principle (see also Lemma B.1.4) for the condition in (0.3.4):

LEMMA 0.3.12. *If $F^*(G) = O_p(G)$, then G is of characteristic p type.*

PROOF. See e.g. 31.16 in [**Asc00**]—which is an easy consequence of Bender's self-centralizing property 0.3.1 of $F^*(G)$, and the Thompson $A \times B$-Lemma B.1.1. □

This in turn leads to a similar global-implies-local principle for the triviality of the layer of H; and indeed to a sufficient condition for the triviality of the image $O_{p'}(H)^*$ in G^* of the local obstruction $O_{p'}(H)$:

LEMMA 0.3.13. *Assume that $L_{p'}(G) = 1$. Then for each p-local subgroup H of G, $L_{p'}(H) = 1$ and $O_{p'}(H) \leq O_{p'}(G)$.*

PROOF. Let $H = N_G(P)$ for some nontrivial p-subgroup P of G, so by Coprime Action 0.3.11.1, $N_{G^*}(P^*) = N_G(P)^* = H^*$. The hypothesis $L_{p'}(G) = 1$ implies $F^*(G^*) = O_p(G^*)$ (cf. B.1.9). So also $F^*\big(N_{G^*}(P^*)\big) = O_p\big(N_{G^*}(P^*)\big)$ by Lemma 0.3.12—that is, $F^*(H^*) = O_p(H^*)$. So $F^*\big(O_{p'}(H^*)\big) \leq O_{p'}\big(F^*(H^*)\big) = 1$, and it follows that $O_{p'}(H^*) = 1$ using 0.3.1. Since $O_{p'}(H)^* \leq O_{p'}(H^*)$, we get $O_{p'}(H) \leq O_{p'}(G)$, as required. Indeed then $O_{p'}(H) = O_{p'}(G) \cap H$, so that we in fact have $H^* \cong \overline{H} := H/O_{p'}(H)$; hence $F^*(\overline{H}) = O_p(\overline{H})$, and we get $L_{p'}(H) = 1$ by B.1.9, completing the proof. □

Some basics of signalizer functor theory. One focus of our proof is the function $a \mapsto O_{p'}\big(C_G(a)\big)$, defined on the set of elements a of G of order p. In fact this function is just one example among many in p-local analysis which map p-elements or p-subgroups of G to suitable p'-subgroups.[6] We will explore this important topic in more detail later, notably in Section B.3 of the Appendix. We begin here with the notion of a signalizer functor:

DEFINITION 0.3.14 (signalizer functor). Let A be an elementary abelian p-subgroup of G of rank at least 3.[7] A *signalizer functor on A* is a function θ defined on the nontrivial elements of A (we denote this subset by $A^{\#}$) such that

(S1) For each $a \in A^{\#}$, $\theta(a)$ is a p'-subgroup of $C_G(a)$, normalized by A.

(S2) For all $a, b \in A^{\#}$, $\theta(a) \cap C_G(b) \leq \theta(b)$. ◇

The definition is due to Gorenstein, with simplifications due to Goldschmidt; see e.g. [**GLS96**, Sec 20]. The condition (S2) is usually called *balance*; it reflects a certain naturality (to which we will return just before Definition 0.3.16) of θ with respect to the inclusions of $\langle a \rangle$ and $\langle b \rangle$ in $\langle a, b \rangle$, and indeed in A. (We consider variations on the notion of balance in Section B.3.)

In proving the Dichotomy Theorem, we will see at (0.3.21) that once we assume that G is not of component type, the absence of 2-components in involution centralizers will cause the abovementioned function $a \mapsto O_{p'}\big(C_G(a)\big)$ to satisfy the balance condition (S2), and so to define a signalizer functor. This will then be a key tool in proving G is of characteristic 2 type.

And to make that tool effective, we will need to *assume* the Signalizer Functor Theorem—a nontrivial but fundamental result due to Gorenstein, Goldschmidt,

[6]Such p'-subgroup images were called *signalizers* by Thompson.

[7]It turns out that the concept is not very interesting for A of rank at most 2.

Glauberman, and McBride (see e.g. [**GLS96**, 21.3, 21.4]). One conclusion of the result, usually called *completeness*, shows that the values of θ generate a p'-subgroup $\theta(A)$; and conversely this *single* group determines $\theta(a)$ via $\theta(a) = \theta(A) \cap C_G(a)$—so that the balance condition (S2) in Definition 0.3.14 is satisfied in an obvious way. We will give a more detailed statement in the Appendix as B.3.13; here is a rough statement for our present introductory exposition:

THEOREM 0.3.15 (Signalizer Functor Theorem). *Let G be a finite group, and assume that θ is a signalizer functor on some elementary abelian p-subgroup A of G with $m_p(A) \geq 3$. For $B \leq A$, set $\theta(B) := \langle \theta(b) : b \in B^{\#} \rangle$. Then*

(1) $\theta(A)$ is a p'-group.

(2) For each $a \in A^{\#}$, $\theta(a) = C_{\theta(A)}(a)$.

(3) For each noncyclic subgroup B of A, $\theta(B) = \theta(A)$. □

In fact (3) is a corollary of (1) and (2): We are done if $B = A$, so consider any $a \in A \setminus B$. Then B normalizes $\theta(a)$ by (S1) of Definition 0.3.14; so by Coprime Action 0.3.11.2, $\theta(a)$ is generated by the subgroups $C_{\theta(a)}(C)$, for the maximal subgroups C of the noncyclic group B. And for $b \in C^{\#}$, using (2) we get $C_{\theta(a)}(C) \leq C_{\theta(A)}(b) = \theta(b) \leq \theta(B)$. Thus $\theta(a) \leq \theta(B)$, and varying a we conclude that $\theta(A) = \theta(B)$.

Next let Ω be the category where subgroups of G are objects, and for $P, Q \leq G$, we take $\mathrm{Mor}(P,Q) := \{g \in G : P^g \leq Q\}$; and let Ω_p be the full subcategory of Ω on $\Lambda_2^p(G)^\circ$. We extend the notion of a signalizer functor[8] to obtain contravariant functors $\overline{\theta}$ and $\overline{C} : B \mapsto C_G(B)$ from Ω_p to Ω (where $\overline{\theta}(b) = \theta(b) \cap C_G(B)$ for $b \in B^{\#}$ and $\overline{\theta}(g) = \overline{C}(g) = g^{-1}$) such that the inclusion map is a natural transformation from $\overline{\theta}$ to $\overline{C}$.

DEFINITION 0.3.16. Let $\mathcal{I}(G)$ denote the set of elements $a \in G$ of order p which also satisfy $m_p(C_G(a)) \geq 3$. Define a *G-equivariant signalizer functor on $\mathcal{I}(G)$* to be a map θ from $\mathcal{I}(G)$ into the set of p'-subgroups of G such that

(I1) For each $a \in \mathcal{I}(G)$ and $g \in G$, $\theta(a) \leq C_G(a)$ and $\theta(a^g) = \theta(a)^g$.

(I2) For each pair of commuting elements $a, b \in \mathcal{I}(G)$, $\theta(a) \cap C_G(b) \leq \theta(b)$. ◊

Observe that if A is an elementary abelian p-subgroup of G with $m_p(A) \geq 3$, then $A^{\#} \subseteq \mathcal{I}(G)$. Also (I1) for g in A gives the A-invariance for (S1) of 0.3.14, while (I2) restricted to $A^{\#}$ is just (S2). Thus $\theta|_A$ is a signalizer functor on A, so the signalizer functor theorem allows us to define $\theta(B)$ for any nontrivial subgroup B of A.

In the generic situation, the commuting graph $\Lambda_2^p(G)^\circ$ (from Notation 0.2.1), on suitable subgroups of rank at least 2, is connected. In this case condition 0.3.15.3 says that $\theta(B) = \theta(B')$ for each B, B' in a connected component of that graph:

LEMMA 0.3.17. *Assume that $m_p(G) \geq 3$, with $\Lambda_2^p(G)^\circ$ connected, and that θ is a G-equivariant signalizer functor on $\mathcal{I}(G)$. Then $\langle \theta(a) : a \in \mathcal{I}(G) \rangle$ is a normal p'-subgroup of G. In particular for each $a \in \mathcal{I}(G)$, $\theta(a) \leq O_{p'}(G)$.*

PROOF. For $E \in \Lambda_2^p(G)^\circ$, set $\theta(E) := \langle \theta(e) : e \in E^{\#} \rangle$.

For A an elementary abelian p-subgroup of G with $m_p(A) \geq 3$, we saw after Definition 0.3.16 that $\theta|_A$ is a signalizer functor on A. So by the Signalizer Functor

[8]The argument from here on is a simplified version of the material leading up to Theorem B.3.25 in the Appendix.

Theorem 0.3.15, $\theta(A)$ is a p'-subgroup of G, and $\theta(A) = \theta(B)$ for each noncyclic subgroup B of A.

In particular if A' is another elementary abelian subgroup with $m_p(A') \geq 3$ which contains B, then $\theta(A) = \theta(B) = \theta(A')$. So as $\Lambda_2^p(G)^\circ$ is connected by hypothesis, it follows that for each pair $B, D \in \Lambda_2^p(G)^\circ$, $\theta(B) = \theta(D)$; and so we see that $\theta(B) = \langle \theta(a) : a \in \mathcal{I}(G) \rangle$. Then by the G-equivariance condition (I1) of 0.3.16, for $g \in G$, $\theta(B)^g = \theta(B^g) = \theta(B)$, completing the proof. □

Using $O_{p'}\big(C_G(-)\big)$ as a signalizer functor. We can now complete the proof of the Dichotomy Theorem 0.3.10—by using signalizer functor theory to show that groups which are not of Gorenstein-Walter type are of characteristic 2 type. First for any prime p we have:

LEMMA 0.3.18. *Assume that $O_{p'}(G) = 1$, $m_p(G) \geq 3$, and $\Lambda_2^p(G)^\circ$ is connected. Then either*

(1) $L_{p'}\big(C_G(a)\big) \neq 1$ for some $a \in \mathcal{I}(G)$, or

(2) for each $a \in \mathcal{I}(G)$, $F^*\big(C_G(a)\big) = O_p\big(C_G(a)\big)$.

PROOF. We may assume (1) fails, so

$$L_{p'}\big(C_G(a)\big) = 1 \text{ for all } a \in \mathcal{I}(G). \tag{0.3.19}$$

Then in particular we obtain

$$F^*\Big(C_G(a)/O_{p'}\big(C_G(a)\big)\Big) = O_p\Big(C_G(a)/O_{p'}\big(C_G(a)\big)\Big) \tag{0.3.20}$$

using Appendix Lemma B.1.9. We also get:

(0.3.21) (0.3.19) *implies $O_{p'}\big(C_G(-)\big)$ is a G-equivariant signalizer functor.*

For take commuting $a, b \in \mathcal{I}(G)$. It is elementary that

$$O_{p'}\big(C_G(a)\big) \cap C_G(b) \leq O_{p'}\big(C_G(a) \cap C_G(b)\big);$$

and now in view of (0.3.19), we may apply Lemma 0.3.13 to the right-hand side, with $C_G(a) \cap C_G(b)$ and $C_G(b)$ in the roles of "H" and "G", to get

$$O_{p'}\big(C_G(a) \cap C_G(b)\big) \leq O_{p'}\big(C_G(b)\big).$$

Thus we obtain the balance condition (I2) of Definition 0.3.16; while condition (I1) there is immediate.

Using Lemma 0.3.17 and the hypothesis that $\Lambda_2^p(G)^\circ$ is connected, we conclude that $O_{p'}\big(C_G(a)\big) \leq O_{p'}(G)$. Then as $O_{p'}(G) = 1$ by hypothesis, (0.3.20) becomes $F^*\big(C_G(a)\big) = O_p\big(C_G(a)\big)$, giving conclusion (2). □

For $p = 2$, Lemma 0.3.18 will lead to our desired dichotomy—although now in the generic case we require only that the commuting graph on involutions should be connected:

LEMMA 0.3.22. *Assume that $G = O^2(G)$, $O_{2'}(G) = 1$, $m_2(G) \geq 3$, and $\Lambda_1^2(G)$ is connected. Then G is either of component type, or of characteristic 2 type.*

PROOF. We may assume G is not of component type. Thus $L_{2'}\big(C_G(a)\big) = 1$ for all involutions a in G; so as in (0.3.21), we also get that $O_{2'}\big(C_G(-)\big)$ is a signalizer functor. However, so far we have only a weaker connectivity hypothesis than in Lemma 0.3.18.

Furthermore we may assume that $Z^*(G) = 1$:[9] For otherwise, our hypothesis that $O_{2'}(G) = 1$ implies that $Z(G)$ contains an involution x, so that $G = C_G(x)$; and then by the noncomponent type hypothesis, we have $F^*(G) = O_2(G)$, which means that G is of characteristic 2 type by 0.4.1, and so we would be done.

Now, since $p = 2$, $G = O^2(G)$, $Z^*(G) = 1$, and $m_2(G) \geq 3$, our hypothesis of connectivity for $\Lambda_1^2(G)$ implies the connectivity of $\Lambda_2^2(G)$. The proof is elementary but somewhat lengthy, so we have placed it in the Appendix, as the final statement in B.4.11.[10]

Now note that if B and C are (distinct) commuting 4-groups[11] defining an edge of $\Lambda_2^2(G)$, then BC is elementary abelian of rank at least 3; thus $B, C \in \Lambda_2^2(G)^\circ$, and $BC^\# \subseteq \mathcal{I}(G)$. Hence the connectivity of $\Lambda_2^2(G)$ implies that every 4-subgroup of G lies in $\Lambda_2^2(G)^\circ$, so that $\Lambda_2^2(G)^\circ = \Lambda_2^2(G)$, and thus $\Lambda_2^2(G)^\circ$ is connected. Also since every involution is in some 4-group as $m_2(G) > 1$, it follows that every involution of G is in $\mathcal{I}(G)$. Thus as G is not of component type, 0.3.18 now gives $F^*\big(C_G(a)\big) = O_2\big(C_G(a)\big)$ for all a in $\mathcal{I}(G)$, i.e. for all involutions of G.

But (see B.1.6) this condition implies $F^*(H) = O_2(H)$ for all 2-local subgroups H; that is, G is of characteristic 2 type. □

To include the atypical case (where $\Lambda_1^2(G)$ is disconnected) in our proof, we also *assume* the Bender-Suzuki Theorem 0.2.3, which we can use to show:

THEOREM 0.3.23. *Assume G is a finite group of even order, with $G = O^2(G)$ and $O_{2'}(G) = 1$. Then one of the following holds:*

(1) $m_2(G) \leq 2$.

(2) G is of component type.

(3) G is of characteristic 2 type.

In particular this holds for any finite simple group of even order.

PROOF. We may assume that (1) fails, so that $m_2(G) \geq 3$.

Assume first that $\Lambda_1^2(G)$ is disconnected. Then as $G = O^2(G)$ and $O_{2'}(G) = 1$, the general version of the Bender-Suzuki Theorem (which we have stated for simple groups as 0.2.3) says that G is of Lie type in characteristic 2 and of BN-rank 1. So, as we mentioned before Definition 0.3.5 (and state later in 0.4.1.3), G is of characteristic 2 type, so that (3) holds.

Thus we may now assume that $\Lambda_1^2(G)$ is connected. Then 0.3.22 shows that (2) or (3) holds. □

From Definition 0.3.9, G is of Gorenstein-Walter type if it satisfies (1) or (2) of Theorem 0.3.23, so that theorem is a restatement of the Dichotomy Theorem 0.3.10—and hence completes our proof of 0.3.10.

We will also discuss in more detail the role of the Dichotomy Theorem in the proof of the CFSG, later in remarks leading up to 1.3.10.

[9] We recall that $Z^*(G)$ is the preimage of $Z\big(G/O_{2'}(G)\big)$.

[10] We mention that the argument at B.4.11 also makes use (via the reference to B.4.10) of the $\mathcal{K}$-group Hypothesis 0.5.4, that all proper simple sections of G are known. By contrast, in Section B.5 we will provide a proof of the Dichotomy Theorem (due essentially to Thompson and Gorenstein) which does not appeal to B.4.11, and hence requires no $\mathcal{K}$-group assumptions. It also replaces the hypothesis that $Z^*(G) = 1$ by the hypothesis that G is not of characteristic 2 type—see especially the argument from just before B.5.3 through B.5.4. However, we felt that the shorter proof given here is more suitable for expository purposes.

[11] A (Klein) 4-*group* (or *fours-group*) is an elementary abelian group of order 4, i.e. $\mathbb{Z}_2 \times \mathbb{Z}_2$.

0.4. The finite simple groups

Denote by $\mathcal{K}$ (for "known") the set of (isomorphism classes of) simple groups appearing in the Classification Theorem 0.1.1. In this section, we give a brief description of the groups in $\mathcal{K}$. Further details on the groups can be found in Appendix A.

The groups of prime order in case (1) of the Classification Theorem are the *abelian* simple groups. For each prime p, there exists a unique such group, the cyclic group $\mathbb{Z}_p$ of integers modulo p under the group operation of addition of congruence classes.

Sporadic groups. Each of the groups in cases (1)–(3) of the Classification Theorem is a member of a natural infinite family of simple groups. But the 26 sporadic groups in case (4) live in no such family that we know of; hence the designation of "sporadic group".

The sporadic groups are usually denoted just via the names of their discoverers. Here we will just list the corresponding names of the groups, leaving a discussion of their properties to Appendix A and the references there. The 26 sporadic groups are:

the five *Mathieu groups* M_{11}, M_{12}, M_{22}, M_{23}, and M_{24};
the four *Janko groups* J_1, J_2, J_3, and J_4;
the three *Conway groups* Co_1, Co_2, and Co_3;
the three *Fischer groups* Fi_{22}, Fi_{23}, and Fi'_{24};
the *Higman-Sims group* HS;
the *McLaughlin group* McL;
the *Lyons group* Ly;
the sporadic *Suzuki group* Suz;
the *Held group* He;
the *Rudvalis group* Ru;
the *O'Nan group* $O'N$;
the *Harada-Norton group* HN (or F_5);
the *Thompson group* Th (or F_3);
the *Baby Monster* BM (or F_2); and
the *Monster* M (or F_1).

Natural structures for action of simple groups. The non-sporadic finite simple groups G can be described as (essentially) the group of automorphisms of a suitable highly symmetric natural object X. The representation of G on X allows us to study the group and establish many of its properties.

Alternating groups. For example, consider the alternating group A_n of degree n, which is the normal subgroup of all even permutations in the symmetric group S_n on a set X of order n. The group S_n is the full automorphism group of X in the category of sets, and A_n is of index 2 in S_n. For $n \geq 5$, A_n is simple, and indeed we have $A_n = F^*(S_n)$.

Lie type groups. The groups of Lie type are responsible for most of the complexity and difficulty in the proof of the Classification Theorem. Thus we devote most of the rest of this section to a discussion of those groups (again with more details postponed to Appendix A). In particular, we begin by indicating several versions of a natural structure "X", that can be used to describe the finite simple groups G of Lie type.

In one approach (via the "Chevalley construction") such a G is (essentially) the group of automorphisms of some suitable polynomial function f on[12] a vector space V over some finite field F. Thus the object X defining G is the pair (V, f). More precisely, if $A := Aut(V, f)$ is the group of automorphisms, and $P : GL(V) \to PGL(V)$ is the projection map of $GL(V)$ into the group of automorphisms of the full projective geometry $PG(V)$ on V, then G arises as the generalized Fitting subgroup $F^*(PA)$ of the full automorphism group PA of the subgeometry X of $PG(V)$.

For another approach via algebraic groups: Let $\overline{F}$ denote the algebraic closure of F; and then write $\overline{V} := \overline{F} \otimes_F V$, and $\overline{f} := \overline{F} \otimes_F f$—a polynomial function on $\overline{V}$. Then $\overline{A} := Aut(\overline{V}, \overline{f})$ is an algebraic group, as is $\overline{G} := F^*(P\overline{A})$. And we can instead regard $(\overline{V}, \overline{f})$ as the fundamental structure "X"—since as we vary the particular finite subfield F of $\overline{F}$, the finite groups A and G arise as forms determined by F of the algebraic groups $\overline{A}$ and $\overline{G}$.

Finally in a third approach, the polynomial f induces a subgeometry $\mathcal{B}$ of the projective geometry $PG(V)$, called the *(Tits) building* of G; and we can also take $\mathcal{B}$ for our structure "X"—indeed with the exception of the case of G of Lie rank 1, we have $G = F^*\big(Aut(\mathcal{B})\big)$.

We give a few specific examples of f and X below.

Some facts about groups of Lie type. We next provide a preview of some material from Remark A.3.1 in Appendix A.

The groups of Lie type live in infinite families, indexed by finite fields, and by certain kinds of polynomials.

The *(ordinary) Chevalley groups* (also called *untwisted groups*) are the groups corresponding to the simple Lie algebras over $\mathbb{C}$ or their Dynkin diagrams. These are the groups

$$A_n(q) \cong L_{n+1}(q), B_n(q) \cong P\Omega_{2n+1}(q), C_n(q) \cong PSp_{2n}(q), D_n(q) \cong P\Omega^+_{2n}(q),$$
$$E_6(q), E_7(q), E_8(q), F_4(q), G_2(q).$$

There are also the *twisted groups*

$$^2A_n(q) \cong U_{n+1}(q), {}^2D_n(q) \cong P\Omega^-_{2n}(q), {}^3D_4(q), {}^2E_6(q),$$
$$^2B_2(2^{2m+1}) = Sz(2^{2m+1}), {}^2G_2(3^{2m+1}) = Ree(3^{2m+1}), {}^2F_4(2^{2m+1})'.$$

Some polynomials f and their natural structures X. For example, the *symplectic groups $PSp_m(q)$* and the *orthogonal groups $P\Omega^\epsilon_m(q)$* are the generalized Fitting subgroup of the (projective) group of isometries of a symplectic or orthogonal form f (respectively), on an m-dimensional vector space V over $F := \mathbb{F}_q$. Thus in this case the polynomial f is of degree 2. These groups, together with the linear groups $L_n(q)$ and the unitary groups $U_n(q)$, are the *classical matrix groups*.

As in Example 0.3.2, we will often take as our standard example of a group of Lie type, the general linear group $G := GL(V) = GL_n(q)$ of all invertible linear maps on an n-dimensional vector space over the field $F := \mathbb{F}_q$. Of course G is usually "nearly-simple" rather than actually simple; but we use it for expository purposes, since minor adjustments to our statements about $GL_n(q)$ will hold for the simple section $L_n(q) = PSL_n(q)$. In the case of the linear group, f can be

[12]Sometimes we will of course mean "on $V \times V$", e.g. in the case of an inner product.

taken to be the zero polynomial; the corresponding algebraic group $\overline{G}$ is $GL(\overline{V})$, and the building $\mathcal{B}$ is the full projective geometry $PG(V)$.

Some notions involving linear groups. We regard our group G of Lie type as a linear group on a vector space V over a finite field F of characteristic p. Thus each $g \in G$ is a linear map on V. We call g *semisimple* if g is diagonalizable over the algebraic closure $\overline{F}$, and we call g *unipotent* if all its eigenvalues are 1. As F is finite of characteristic p, the semisimple elements are the p'-elements, and the unipotent elements are the p-elements of G.

A subgroup H of G is *unipotent* if there is a chain (or *flag*):

$$0 = V_0 \leq V_1 \leq \cdots \leq V_m = V,$$

of subspaces such that H induces the identity on each quotient space V_{i+1}/V_i. The unipotent subgroups are in fact precisely the p-subgroups of G.

The *parabolic subgroups* of G are the stabilizers of simplices in the building $\mathcal{B}$ for G. For example in the case of the classical groups, the simplices are flags[13] of subspaces of V which are totally singular with respect to the relevant form f. Given a parabolic P, the *unipotent radical* of P is the largest normal unipotent subgroup $R(P)$ of P. From the previous paragraph, $R(P) = O_p(P)$.

The group G is transitive on the maximal simplices of $\mathcal{B}$, each of which contains exactly one vertex from each G-orbit on vertices. The number of vertices in a maximal simplex is the *Lie rank* or *BN-rank* of G.

Recall that the *Bender groups*, arising as conclusions in the Bender-Suzuki Theorem 0.2.3, are the groups of Lie type over fields of characteristic 2 which have Lie rank 1; these are the groups $L_2(q)$, $U_3(q)$, and $Sz(q) = {}^2B_2(q)$, q a power of 2.

In our standard example $G = GL(V)$ of a group of Lie type, the simplices of $\mathcal{B}$ are just all flags $\mathcal{F} = (V_0 < \cdots < V_m)$, and the unipotent radical $R(P)$ of the parabolic P stabilizing $\mathcal{F}$ is the unipotent subgroup centralizing each quotient space in the flag $\mathcal{F}$. The Lie rank of G is the length $\dim(V) - 1$ of a maximal flag.

Next for r a prime, we wish to examine the r-local structure of G.

p-local structure. Suppose first that $r = p$, the "natural" characteristic of G (i.e., that of the underlying field F). We mentioned above that the p-subgroups of G are the unipotent subgroups, and the p-elements are the unipotent elements.

The following result is fundamental; its core is the Borel-Tits Theorem (see e.g. [**GLS98**, Sec 3.1]). Conclusion (3) had been mentioned earlier.

LEMMA 0.4.1. *Let H be a nontrivial p-subgroup of G of Lie type in characteristic p. Then:*

(1) There exists a proper parabolic subgroup P of G such that $H \leq R(P)$ and $N_G(H) \leq P$.

(2) $F^*(P) = R(P) = O_p(P)$.

(3) G is of characteristic p type.

PROOF. For part (1), see [**GLS98**, 3.1.3.a]; for (2) see [**GLS98**, 2.6.5.e].

By (2) we may apply Lemma 0.3.12 to the parabolic P containing the 2-local H in (1), to conclude that $F^*(H) = O_p(H)$. Thus (3) holds. □

[13] The notion of "flag" is slightly adjusted in the case of orthogonal spaces of plus type and even dimension.

Note that when $G = GL(V)$, (1) and (2) are easy to prove using some elementary linear algebra, along with the facts that H is unipotent and $\mathcal{B} = PG(V)$ is the full projective geometry.

Next let $\mathcal{M}$ be the set of proper parabolics of G containing a Sylow p-subgroup S of G. Then we can choose the collection of Principle I to be the set of unipotent radicals $\mathcal{Q} := \{R(P) : P \in \mathcal{M}\}$. Indeed the set $\mathcal{N}$ of normalizers of members of $\mathcal{Q}$ in Section 0.2 is actually equal to $\mathcal{M}$, and the simplicial complex $\mathcal{O}(\mathcal{M})$ determined by all cosets of the members of $\mathcal{M}$ gives one version of the building $\mathcal{B}$.

Furthermore a fundamental theorem of Tits (cf. [**Tit81**, p. 541]) states that $\mathcal{B}$ is simply connected if (and indeed only if) the Lie rank of G is at least 3. So in that case, G is determined up to isomorphism by the amalgam $\mathcal{A}(\mathcal{M})$, as discussed in Section 0.2 (cf. also 3.3.14).

r-local structure for $r \neq p$. Next we consider the case when the prime r is different from the natural characteristic p of G. Recall that the r-elements of G are semisimple, that is, diagonalizable over the algebraic closure $\overline{F}$ of F.

For expository purposes, consider our standard example $G = GL(V)$, where we have $g \in G$ of order r, and diagonalizable even over F. Then the r-th roots of 1 $\{\lambda_i : 1 \leq i \leq r\}$ are the (potential) eigenvalues of g, and $V = V_1 \oplus \cdots \oplus V_r$, where V_i is the g-eigenspace corresponding to λ_i. Then $C_G(g) = G_1 \times \cdots \times G_r$, where G_i acts faithfully as $GL(V_i)$ on V_i, and centralizes V_j for $j \neq i$. Subject to our convention that any $GL(U)$ is a nearly-simple group, each factor G_i is nearly-simple, and hence $(G_i : 1 \leq i \leq r)$ are essentially components in $C_G(g)$.

This generalizes our earlier Example 0.3.2, when p is odd and $r = 2$. Indeed, with few exceptions, if G is a group of Lie type over a field of odd characteristic p, then G is of component type. Moreover in this case, we have

$$L_{2'}\big(C_G(g)\big) = E\big(C_G(g)\big); \qquad (0.4.2)$$

that is, the 2-components in the definition of component type turn out to be actual components.

In fact, this statement holds for involution centralizers in all finite simple groups; it is essentially the content of Thompson's *B-Conjecture* (cf. 1.6.1). The *B*-Conjecture makes precise the sense in which, for a 2-local H in a finite group X, the image of $O_{2'}(H)$ in $X/O_{2'}(X)$ is small. We will see in the next section that we need a result of this sort to avoid obstructions in analyzing the 2-local structure of a simple group G arising from the subgroups $O_{2'}(H)$. The proof of the B-Conjecture was one of the important steps in the Classification; we will say a little more about it in the next section (with a fuller discussion later in Sections 1.6–1.8).

Of course our summary of the Classification Theorem 0.1.1 can be expanded into a detailed statement listing the individual groups in $\mathcal{K}$, by adding to conclusions (3) and (4) the specific families of groups Lie type and the specific sporadic groups indicated in this section.

0.5. The Classification grid

The Dichotomy Theorem 0.3.10 partitions the finite simple groups into the groups of Gorenstein-Walter type and the groups of characteristic 2 type. As we saw in Section 0.4, the groups of Lie type over fields of characteristic 2 are of characteristic 2 type, while most groups of Lie type over fields of odd characteristic are of Gorenstein-Walter type. (Indeed, the only exceptions are $PSp_4(3)$, $U_4(3)$, and $G_2(3)$.)

A partition by size. Our first objective in this final section of the Introduction is to give a partition of the groups of each type into large and small groups. This provides the Classification grid 0.5.3 below, which partitions the finite simple groups into four classes.

Size for groups of GW type. When G is of Gorenstein-Walter type, we decree that G is *small* if $m_2(G) \leq 2$, and *large* if $m_2(G) \geq 3$. So from Definition 0.3.9, the large groups are of component type.

REMARK 0.5.1 (2-rank and signalizer functors). The primary reason for this choice is technical: As we saw in Section 0.3, for Principle II we must control the subgroups $O_{2'}(H)$, for H a 2-local subgroup of G; and our principal tool for doing so is the Signalizer Functor Theorem 0.3.15, which requires $m_2(G) \geq 3$. ◇

But there is another important reason for this choice, having to do instead with the recognition of G as in Principle I. Namely the large groups should be our *generic* examples (that is, large enough to exhibit "typical" behavior), and to classify the generic groups using the approach described in Section 0.2, we want a suitable complex of 2-local subgroups to be simply connected. Such complexes will tend to have the homotopy type of the *clique complex* $K_1^2(G)$, whose simplices are the cliques of the graph $\Lambda_1^2(G)$. And a necessary condition for $K_1^2(G)$ to be simply connected is that $m_2(G) \geq 3$ (cf. Theorem 2 in [**Asc93**]).

Size for groups of characteristic 2 type. Next suppose G is of characteristic 2 type. In this case the structure of a 2-local H is dominated by the 2-subgroup given by $F^*(H) = O_2(H)$, so it is hard to get a handle on H in terms of inductive knowledge of *simple* groups. For this reason, we would like to transfer attention to a suitable odd prime p, and seek components in centralizers of elements of order p.

However we mentioned in Section 0.2 that no analogue of the Bender-Suzuki Theorem 0.2.3 is available for odd primes, so we must also continue to keep the prime 2 in the picture. In practice this means that the p-local structure of some 2-local H should be rich; and from the discussion above, this means in turn that we want $m_p(H) \geq 3$. This leads to the following definitions, due to Thompson in his work [**Tho68**] on N-groups (roughly, minimal simple groups).

DEFINITION 0.5.2 (2-local p-rank and $e(G)$). Given a finite group G and an odd prime p, define the *2-local p-rank* of G to be

$$m_{2,p}(G) := \max\{m_p(H) : H \text{ is a 2-local subgroup of } G\}.$$

Then define

$$e(G) := \max_{p \text{ odd}} m_{2,p}(G).$$

◇

As in Remark 0.5.1, Principle II and the Signalizer Functor Theorem 0.3.15 (for odd p) suggest that it is natural to define a group G of characteristic 2 type to be *small* if $e(G) \leq 2$, and to define G to be *large* if $e(G) > 2$. (The small groups G with $e(G) \leq 2$ are usually called *quasithin*.)

Moreover, again Principle I also suggests that this is the correct place to make our subdivision, as we now indicate. Recall that by 0.4.1.3, a natural example of a simple group of characteristic 2 type is a group of Lie type over a field of characteristic 2. In such a group G, the parameter $e(G)$ is usually maximized on a diagonal subgroup in a Borel subgroup—and so is a good approximation of the Lie rank l of G. Moreover as we saw in Section 0.4, given a fixed Sylow 2-subgroup S of G, each 2-local of the form $N_G(Q)$, $1 \neq Q \leq S$, is contained in one of the proper parabolics over S. Now if $\mathcal{N}$ is the set of such parabolics, then we saw in Section 0.2 that we want the complex $\mathcal{O}(\mathcal{N})$ to be simply connected. Further we indicated in Section 0.4 that $\mathcal{O}(\mathcal{N})$ is a version of the building $\mathcal{B}$, which by the result of Tits is simply connected iff $l \geq 3$.

REMARK 0.5.3 (The "grid" for the original proof of the CFSG). In summary, we have a four-part subdivision of the Classification:

	GW type	characteristic 2 type
small	2-rank ≤ 2	$e(G) \leq 2$ (quasithin)
large	component type	$e(G) \geq 3$

◇

The standard form approach to components in centralizers. We close this chapter by describing more precisely the sorts of restrictions on 2-locals we aim for in Principle II, when G is a large simple group of Gorenstein-Walter type (that is, of component type). Analogous restrictions will apply to large groups of characteristic 2 type.

Some general background. The proof of the Classification Theorem is inductive, so one considers a counterexample G of minimal order to the Theorem. For any proper subgroup H of G, the simple composition factors of H are then known by induction; so G satisfies:

HYPOTHESIS 0.5.4 ($\mathcal{K}$-Group Hypothesis). *Define a finite group G to be a $\mathcal{K}$-*group *if the composition factors of each proper subgroup of G are in the set $\mathcal{K}$ of known simple groups—that is, in the list of the Classification Theorem 0.1.1. (The term $\mathcal{K}$-*proper *is also used .)* ◇

Thus many of the subtheorems used in the proof of the Classification are theorems about $\mathcal{K}$-groups. One such result is the following theorem of Gorenstein and Walter, which provides an important tool in local group theory:

THEOREM 0.5.5 (Gorenstein-Walter L-Balance Theorem [**GW75**]). *Let G be a finite $\mathcal{K}$-group. Then for each p-subgroup P of G, $L_{p'}(N_G(P)) \leq L_{p'}(G)$. (Indeed if $O_{p'}(G) = 1$, then $L_{p'}(N_G(P))$ centralizes $O_{p'}(N_G(P))$.)*

PROOF. See e.g. 31.17.2 in [**Asc00**]. □

(We discuss various general notions of "balance" in the Appendix starting at (B.3.20).)

Standard form for groups of component type. In our discussion in (0.4.2) we mentioned that one of the important steps in the proof of the Classification was the proof of the B-Conjecture 1.6.1; our discussion below will assume that the B-conjecture has already been established—and we state the result in the form:

If G is a finite group with $O_{2'}(G) = 1$, then for each 2-local subgroup H of G, we have $L_{2'}(H) = E(H)$.

Now suppose that G is our minimal counterexample to the Classification. Let t and s be commuting involutions in G, and assume that we have a component L of $C_G(\langle t, s\rangle)$. (So in particular we are assuming that G is of component type.)

Then L is a component of $C_{C_G(s)}(t)$. As the B-conjecture holds in G, the L-Balance Theorem 0.5.5 says that $L \le L_{2'}(C_G(s)) = E(C_G(s))$. Similarly we get $L \le E(C_G(t))$. This embedding of L in $E(C_G(s))$ and $E(C_G(t))$ leads to a partial ordering defined on the components of involution centralizers; and makes it possible to prove that (generically) if t is an involution in G, and L is a component of $H := C_G(t)$ which is maximal in that ordering, then H is a so-called *standard subgroup* of G.

A standard subgroup satisfies various strong properties (see 1.6.3 for details). In particular, if L is standard then $C_G(L)$ is essentially cyclic, and we will assume $C_G(L)$ *is* cyclic for expository purposes. Thus $H = N_G(L)$, and the kernel of the conjugation map $c : H \to Aut(L)$ is $C_G(L)$ which is cyclic. So the structure of H is controlled by L, and is highly restricted, as desired for Principle II.

Let's consider a variation on Example 0.3.2, now taking G to be the nearly-simple group $SL_n(q)$, with both n and q odd. If we simply order components of involution centralizers by cardinality, then the maximal components correspond to $k = n - 1$ in that Example, and are isomorphic to $L = SL_{n-1}(q)$; they arise when

$$t := \left(\begin{array}{c|c} -I_{n-1} & 0 \\ \hline 0 & +I_1 \end{array}\right).$$

Moreover $H = N_G(L) = C_G(t) \cong GL_{n-1}(q)$, and in fact (when we pass to the quotient $PSL_n(q)$) we see that $C_G(L) = Z(H)$ is the cyclic group of diagonal matrices which are scalar on L.

In this example, assuming $L \cong SL_{n-1}(q)$ is standard should also lead to the identification of G as the simple group $L_n(q)$. More generally, a standard subgroup L from a given family of groups of Lie type of odd characteristic in Section 0.4 should lead to a larger group in the same family. A similar remark will hold for L (and hence G) of alternating type. For large groups of GW type, this is the approach to recognizing the group via its 2-locals, as in Principle I.

An analogue for odd centralizers in characteristic 2 *type.* The analysis in the above example also remains valid if we take G to be $SL_n(2^m)$ for $m > 1$—but now take t instead to be a diagonal element of odd prime order p, which has one eigenspace of dimension $n - 1$: This again yields a component of the form $L \cong SL_{n-1}(2^m)$ in $H = C_G(t)$, with $C_G(L) = Z(H)$ cyclic.

Furthermore we can again hope to produce a standard subgroup in a p-local, as we did above in the centralizer of an involution, since the following generalization of the B-Conjecture in fact holds:

B_p-Theorem: Let G be a finite simple group, p a prime, and g an element of G of prime order p. Then $L_{p'}(C_G(g)) \le E(C_G(g))$.

In fact the only known proof of the B_p-Theorem involves quoting the full Classification Theorem. However, when G is a simple group of characteristic 2 type, and p

is odd with $m_{2,p}(G) \geq 3$, it is possible to prove a weak version of this generalized B-Conjecture. That result,[14] in conjunction with the L-Balance Theorem 0.5.5, still yields an inclusion of a component L of $C_G(\langle t, s\rangle)$ in $E(C_G(s))$ and $E(C_G(t))$, and, as before, this leads in turn to a standard subgroup in some p-local for our odd prime p; see *standard type* at 5.1.4.

For G of characteristic 2 type, usually the standard subgroup L will be of Lie type in characteristic 2; and again Principle I will aim to show that G is a larger group in the same Lie family as L.[15]

Summary. Thus, whenever G is "large", it is possible to find an element t of prime order p (with $p = 2$ when G is of GW type, and p is odd when G is of characteristic 2 type) for which we will obtain $H := C_G(t)$ as a standard subgroup of G; whence the structure of H is controlled by the structure of its component L and is highly restricted, as desired for Principle II. We then use these restrictions on the p-local subgroup H to characterize G—as in Principle I.

This completes our quick overview of the general standard-subgroup strategy underlying the proof of the Classification Theorem. In 1.1.1 in the following chapter, we will supply a more detailed outline of an idealized approach to pinning down the centralizer of some element of prime order.

The remainder of the next chapter will then provide a somewhat more detailed historical discussion of the portion of the proof which treats the simple groups of Gorenstein-Walter type. (A far more detailed presentation of that case is given by Gorenstein in [**Gor83**].) The other chapters of the present volume will provide a rather detailed discussion of the other half of the proof of the Classification, namely the part treating the simple groups of characteristic 2 type.

[14]See (14-5) in [**GL83**]. The related discussion in Section I.18 there uses the language of "$L_{p'}$-balance" rather than L-balance; cf. our exposition after 5.2.4, and after 5.3.4.

[15]A more refined ordering of components is necessary to make this statement accurate.

CHAPTER 1

Overview: The classification of groups of Gorenstein-Walter type

We recall from our discussion of the Dichotomy Theorem 0.3.10 that the proof of the CFSG divides into two major subcases: the classification of the groups of Gorenstein-Walter (*GW*) type, and the classification of the groups of characteristic 2 type. While the present book (which we regard as "Volume 2" in Gorenstein's outline project) focuses on groups of characteristic 2 type, in this chapter we will first give an overview of the classification of simple groups of *GW* type.

That classification was documented in great detail by Daniel Gorenstein in Volume 1 [**Gor83**].[1] (His treatment relied in turn on his earlier "Introduction" volume [**Gor82**].) Here we will give a much shorter outline of the principal steps, together with clarification of certain points, including some updating of citations in the literature. We will indicate periodically where in [**Gor83**] the reader can find more details on particular topics; and usually we will mention explicitly when our discussion deviates significantly from that in Gorenstein.

One goal of our shorter treatment is to isolate, and emphasize, a comparatively small number of key results. So in the course of our development, we identify roughly a dozen major subtheorems—which are at "level 0" in the language of the Preface; that is, their union supplies a proof of the classification of groups of *GW* type. We also discuss a number of level 1 results which are important in the proof of many of the level 0 results. Then after our discussion of the original historical process, we recapitulate those major subtheorems in a summary at the end of the chapter.

The Main Theorem for groups of Gorenstein-Walter type

We recall Definition 0.3.9: a finite simple group G is of *Gorenstein-Walter (GW) type* if one of the following two conditions holds:

(1) G has 2-rank at most 2; or
(2) for some involution $z \in G$, $L_{2'}(C_G(z)) \neq 1$ (i.e., G is of component type).

We indicated before Remark 0.5.1 that condition (1) distinguishes the groups we choose to regard as small; while condition (2) characterizes those groups we regard as large or generic—since it is satisfied in the main target groups, i.e., in almost all groups of Lie type over fields of odd characteristic, as well as in most alternating groups. And indeed these families appear as conclusions (3) and (2) in the following result which classifies groups of *GW* type.

[1]Gorenstein used the term *noncharacteristic* 2 *type* rather than *GW* type.

THEOREM 1.0.1 (Main Theorem for GW type groups). *Let G be a K-proper finite simple group of GW type. Then one of the following conclusions holds:*

(1) G is cyclic of prime order; or

(2) G is an alternating group of degree $n \geq 5$, $n \neq 8$; or

(3) G is a finite simple group of Lie type,[2] defined over a field of odd characteristic, but not isomorphic to $G_2(3)$, $PSp_4(3)$, or $U_4(3)$; or

(4) G is $U_3(4)$; or

(5) G is one of the following 19 sporadic simple groups:

$$M_{11}, M_{12}, J_1, J_2, HS, McL, Ly, Suz, He, Ru,$$

$$O'N, Co_3, Co_1, Fi_{22}, Fi_{23}, Fi'_{24}, HN, BM, M.$$

1.1. A strategy based on components in centralizers

In order to highlight various features common to the treatment of the Odd and Even Cases (recall from our discussion culminating in the Dichotomy Theorem 0.3.10 that these are GW type and characteristic 2 type), we begin with a sketch of several specific steps in the approach, which were only briefly summarized at the end of Section 0.5. In particular, this will give a bit more insight into the role of signalizer functors than might have been apparent from our discussion of the Dichotomy Theorem in Section 0.3.

An idealized overview of the strategy. We begin with a list of steps in the process; afterwards we will provide a brief introduction to each step and explain the terminology involved.

STRATEGY 1.1.1. We fix the prime $p = 2$ in the Odd Case, but take p odd in the Even Case. We let t denote some element of order p in our unknown simple group G. In most cases we proceed as follows:

Step 1: Treat certain cases where uniqueness subgroups arise.
Step 2: Treat some small-rank cases.
Step 3: Show for some prime p and some element t of order p that $C_G(t)$ contains a p-component L.
Step 4: Show for suitably chosen (t, L) that L is quasisimple.
Step 5: Show for a suitably maximal pair (t, L) that L and $C_G(L)$ have a certain "standard form".
Step 6: Use the standard form of L and $C_G(L)$ to determine G. ◇

In later chapters, we will discuss Strategy 1.1.1 in more detail—while also indicating some detours which occurred during its actual implementation. But for the moment, we will give just a brief introductory discussion of the various steps.

Step 1: Treat certain cases where uniqueness subgroups arise. In several of the steps one constructs "uniqueness" subgroups[3] M of G. A uniqueness subgroup is a proper subgroup M of G which contains $N_G(D)$ for each D in some large collection $\mathcal{D}$ of nontrivial p-subgroups of M. Indeed usually M is the unique member of some suitable set of subgroups of G which contains $N_G(D)$ for any given D in $\mathcal{D}$.

[2] The full list of groups arising here can be obtained from Remark A.2.1; note that as G is of odd characteristic, the twisted types ${}^2B_2$ and ${}^2F_4$ are not examples here.

[3] A fuller discussion of uniqueness subgroups begins with Example 1.3.1. Some results which determine such pairs include Bender's Strongly Embedded Theorem 1.3.5, Aschbacher's $C(G, T)$ Theorem B.7.3, and Aschbacher's Uniqueness Case Theorem 8.0.2.

For example a uniqueness subgroup may arise as the normalizer of the completion of a signalizer functor (cf. the proof of Lemma B.4.2); or as the subgroup generated by the local overgroups of a fixed Sylow 2-subgroup; or as the normalizer of the subgroup generated by a set of "neighboring" p-components. One then wishes to determine the cases where M is proper in G, ideally showing that then G possesses a split BN-pair of rank 1 in characteristic p, with M a Borel subgroup of G.

Step 2: Treat some small-rank cases. At Remark 0.5.1 we observed that the construction of signalizer functors requires p-subgroups of rank at least 3. This dictates the boundaries defining small rank—namely those we had indicated in the classification grid 0.5.3:

$$\begin{array}{ll} m_2(G) \leq 2 & \text{in the Odd Case} \\ e(G) \leq 2 & \text{in the Even Case.} \end{array}$$

Hence the work in this step is to determine the groups in those small-rank cases.

Step 3: Show for some prime p and some element t of order p that $C_G(t)$ contains a p-component L. After Step 2, signalizer functors can be defined; then the phrase "signalizer analysis" refers to the detailed study of the properties of the functors. We give a brief[4] overview of some of that analysis. (Recall that we considered a special case for $p = 2$ in our discussion of the Dichotomy Theorem 0.3.10.)

Assume that Step 3 fails, so that $C_G(t)$ contains no p-component for any t of order p. Then, as we saw in (0.3.21), $O_{p'}(C_G(-))$ defines a G-equivariant signalizer functor. If the functor is not trivial, then the normalizer M of the value of the functor on a suitable connected component of the graph $\Lambda_1^p(G)$ (cf. Theorem B.3.25, using the connectedness argument underlying Lemma 0.3.17) satisfies $O_{p'}(M) \neq 1$. But $O_{p'}(G) = 1$ as G is simple, so $M < G$, and hence M is a uniqueness subgroup. Ideally this situation has already been handled in Step 1.

REMARK 1.1.2 (The Gorenstein-Walter Alternative). The general form of the dichotomy at this point—namely either the existence of uniqueness subgroups or the triviality of signalizer functors—is called the *Gorenstein-Walter Alternative* in [**GLS02**, p. 49]. ◊

To be a bit more precise, suppose first that Step 3 fails for $p = 2$. Since we have completed the small 2-rank case in Step 2, the Dichotomy Theorem 0.3.10 shows that we are in the Even Case. Now suppose that Step 3 also fails for all odd primes p. Then the Klinger-Mason Dichotomy (see B.9.1) shows that we are in the small-rank case $e(G) \leq 2$ handled in Step 2. Hence Step 3 must hold for some prime p.

Step 4: Show for suitably chosen (t, L) that L must be quasisimple. When L is large, this step follows from the triviality of a suitably chosen signalizer functor, again using Step 1. But we will see later that when L is small, substantial complications can occur.

Step 5: Show for a suitably maximal pair (t, L) that L and $C_G(L)$ have a certain "standard form". If in Example 0.3.2 we take k minimal, then we obtain a component L which is as large as possible. Gorenstein and Walter exploited this choice; and subsequently Aschbacher axiomatized properties shared by large

[4]Major results implementing this step include Theorems 1.3.10 and 2.2.1.

pairs L, $C_G(L)$ in his notion of *standard form*. His fundamental Standard Component Theorem 1.8.12 essentially implemented in the Odd Case what we are here calling Step 5. A related notion of *standard type* appears in the Even Case, and provides the generic situation in that case. This analogy is the main motivation for our idealized overview in Strategy 1.1.1 of parallels between the Odd and Even Cases.

Step 6: Use the standard form of L and $C_G(L)$ to determine G. While there may be many different groups L to treat, it is usually possible to proceed, in a straightforward and fairly uniform way, from the standard form in Step 5 to the hypotheses of an existing recognition theorem for a subgroup $G_0 = \langle L, J \rangle$, where the p-component J is a "neighbor" of L. Since the generic target groups are those of Lie type, in many cases G_0 can be identified using the Curtis-Tits Theorem A.5.1 and variants. Finally, if $G_0 < G$, one shows that G_0 is a uniqueness subgroup, a case handled in Step 1. So this process completes the identification of G.

The strategy actually used for groups of *GW* type. In the remaining sections of the chapter, we will follow the historical evolution of the proof of the Main Theorem. It may not be immediately evident how the actual proof that developed fits into our six-step strategy above, so at relevant points in our narrative, we will discuss how the actual proof required some adjustments to compensate for complications to our idealized strategy. As a brief preview, we provide below a quick informal summary of the main stages; in particular we supply some names and reference numbers for the major subtheorems which will emerge during the course of the subsequent discussion.

OUTLINE 1.1.3 (Overview of the *GW* type classification).
I. (small case) Groups of 2-rank at most 2
- A. The Odd Order Theorem 1.2.1
  - (Level 1) The Strongly Embedded Theorem 1.3.5 and the Dichotomy Theorem 0.3.10
- B. The Brauer-Suzuki Theorem 1.4.2 and the 2-Rank 2 Theorem 1.4.6
  - (Level 1) The Sectional 2-Rank 4 Theorem 1.5.7 and the 2-Generated Core Theorem 1.5.10

II. (generic case) Groups of component type: $L_{2'}(C_G(z)) \neq 1$ for an involution z
- A. Reduction to groups with a standard component
  - i. The $2A_n$ Theorem 1.7.3 and the Classical Involution Theorem 1.7.5
  - ii. The Unbalanced Group Theorem 1.7.1 and the B-Theorem 1.8.11
  - iii. The Standard Component Theorem 1.8.12
- B. Identifying groups with a particular standard component
  - i. The Odd Standard Component Theorem 1.9.1
  - ii. The Aschbacher-Seitz Reduction Theorem 1.9.3
  - iii. The Even Standard Component Theorem 1.10.1 ◊

So we turn now to the history of the classification of groups of *GW* type.

1.2. The Odd Order Theorem

The proof of the solvability of groups of odd order by Feit and Thompson in [**FT63**] was the spark which ignited the effort that became the CFSG.

From the viewpoint of simple groups of *GW* type, we may state the Feit-Thompson result as giving most of conclusion (1) in the Main Theorem 1.0.1:

THEOREM 1.2.1 (The Odd Order Theorem). *Let G be a finite simple group of 2-rank 0. Then G is cyclic of odd prime order.*

The original proof of Feit and Thompson outlined by Gorenstein in Section 1.2 of [**Gor83**].

The local analysis in that proof was later significantly simplified by Bender and Glauberman in [**BG94**], using Glauberman's ZJ-Theorem [**Gla68**] and Bender's proof of the Feit-Thompson Uniqueness Theorem (which is discussed as Theorem 1.2 in Gorenstein [**Gor83**]). The character theoretic part of the Feit-Thompson proof was simplified by Peterfalvi in [**Pet00**], using ideas of Dade and Sibley. Peterfalvi also simplified Chapter VI of the proof in [**Pet84**] (English version in [**BG94**, App C]).

In view of the Odd Order Theorem 1.2.1, *in the remainder of this chapter we may assume that our simple group G has even order. In particular, G has 2-rank at least 1 and hence contains an involution.*

1.3. (Level 1) The Strongly Embedded Theorem and the Dichotomy Theorem

We continue to describe the history of this part of the CFSG, but our analysis will now deviate somewhat from Gorenstein's outline [**Gor83**]: Namely we regard much of the work involving uniqueness subgroups as being at level 1 rather than level 0. In doing so, we are emphasizing that these results

- are not directly concerned with groups of GW type; and
- are used in important ways in both the Odd and Even Cases of the CFSG.

We can also view this distinction in the context of Strategy 1.1.1: The previous section treated the 2-rank 0 subcase of the Small Odd Case (2-rank at most 2) in Step 2 of the Strategy. Before completing that Step in the following section, we first discuss uniqueness subgroups—corresponding to the exceptional branch in Step 1 of the Strategy.

The Strongly Embedded Theorem as the main 2-Uniqueness Theorem. So in this section, we temporarily widen our focus beyond the groups of GW type, in order to examine the following situation:[5]

EXAMPLE 1.3.1. The term "uniqueness subgroup" refers to a proper subgroup containing many local subgroups. The notion goes back to Thompson's early work.[6] We will illustrate the general concept with an important example which comes up in many subsequent classification problems.

The Odd Order Theorem 1.2.1 tells us that a simple group G always has some 2-local subgroups. Generically these 2-locals lie in many different maximal subgroups of G. So if we can force a reasonably large collection of 2-locals to lie in a *single* subgroup M, we can usually expect to show that M is equal to G. Thus we can hope to determine all pairs $M < G$ with the property that M is the unique maximal subgroup of G containing some large set of 2-locals.

[5]The material in this section is described in Gorenstein's introduction [**Gor82**, Sec. 4.2], rather than in his outline [**Gor83**].

[6]E.g. the Feit-Thompson Uniqueness Theorem mentioned in the previous section. See also the discussion in Gorenstein's introduction around [**Gor82**, Cor 4.18].

We typically refer to such a subgroup M as a *uniqueness subgroup* of G. Later we will encounter various hypotheses defining uniqueness subgroups. But first we discuss a particular class of pairs M,G which arise in various contexts.

We begin with the question: when does the normalizer of a Sylow 2-subgroup lie in a unique maximal subgroup M of G? Consider the important class of simple groups G of Lie type in characteristic 2. From the theory of parabolic subgroups of such groups (cf. A.3.1), the Sylow normalizers are the Borel subgroups B of G, and the BN-rank of G is the number of distinct maximal subgroups containing B.

Thus B lies in a *unique* maximal subgroup of G precisely when G has BN-rank 1. The relevant groups G are $L_2(2^m)$, $Sz(2^{2m-1})$, and $U_3(2^m)$, $m \geq 2$, and M is the Borel subgroup B itself. ♢

The groups of BN-rank 1 in characteristic 2 will figure prominently in this section—as we now focus on an important early result due to Bender.[7] We will begin by discussing several other situations where rank-1 groups arise. Eventually we will see that the hypotheses defining these situations are equivalent, and in one of them it will be clear that the Borel subgroup is a uniqueness subgroup.

Doubly transitive permutation groups. Suzuki's papers [**Suz62, Suz64**] continued the investigation of doubly transitive permutation groups going back to Jordan, Frobenius, and Zassenhaus. The relevant properties are suggested by the structure of $L_2(2^m) \cong SL_2(2^m)$:

EXAMPLE 1.3.2 (Double transitivity of $SL_2(2^m)$). Consider $G := SL_2(2^m)$. This corresponds to the "degenerate" case $n := 2$ from the discussion in the latter part of Example 0.3.3. Rather than considering an involution centralizer, we consider instead a Borel subgroup B, consisting of the lower triangular matrices. We have $B = UH$, where U is the subgroup of all lower unitriangular matrices, and H is the Cartan subgroup given by diagonal matrices:

$$U := \begin{pmatrix} 1 & 0 \\ * & 1 \end{pmatrix}, \text{ and } H := \begin{pmatrix} \alpha & 0 \\ 0 & \alpha^{-1} \end{pmatrix}.$$

Now G acts by conjugation on its Sylow 2-subgroups; and the Sylow 2-subgroup U has stabilizer B. Furthermore U is regular on the remaining Sylow 2-subgroups; so the action is doubly transitive (that is, transitive on ordered pairs of distinct points). ♢

Suzuki abstracted the appropriate permutation-theoretic conditions from these and the other groups of BN-rank 1 in characteristic 2, obtaining the following characterization:

THEOREM 1.3.3. *Let G be a simple 2-transitive permutation group of odd degree, in which the stabilizer B of a point is the semidirect product of a regular normal nilpotent subgroup U with a complement H of odd order. Then $G \cong L_2(2^m)$, $Sz(2^{2m-1})$ or $U_3(2^m)$, $m \geq 2$.*

A revised proof of this result appears in Peterfalvi [**Pet86**].

---

[7] Our wider focus in this section is now apparent, since Bender groups are of characteristic 2 type, so among these groups, only $L_2(4)$ and $U_3(4)$ are of *GW* type, because they are the only such groups of 2-rank at most 2.

Strongly embedded subgroups. In the groups G above, the subgroup U of the point stabilizer B is a Sylow 2-subgroup of G, while the subgroup H of odd order is a 2-point stabilizer—of the form $B \cap B^g$ for $g \in G - B$. (In Example 1.3.2, we could take B^g to be the *upper* triangular matrices.) So it follows that every involution in G fixes exactly one point.

Bender in [**Ben71**] weakened Suzuki's hypotheses considerably—by studying transitive permutation groups of even order in which all involutions have a unique fixed point. This is in turn essentially equivalent[8] to the following more group-theoretic condition on a subgroup M:

DEFINITION 1.3.4 (strongly embedded). Let G be a finite group having a proper subgroup M of even order, such that $M \cap M^g$ is of odd order for all $g \in G - M$. Then we say that M is a *strongly embedded* subgroup of G. ◇

We use this definition of strong embedding (although later we mention some equivalent definitions) in stating the fundamental result of [**Ben71**]—where Bender made use of Suzuki's theorem to obtain the same conclusions starting from the weaker hypothesis of strong embedding. (We already stated an equivalent version of the result at 0.2.3, under the hypothesis that the commuting graph $\Lambda_1^2(G)$ of 0.2.1 is disconnected.)

THEOREM 1.3.5 (Strongly Embedded Theorem, or 2-Uniqueness Theorem). *Let G be a finite simple group with a strongly embedded subgroup.*
Then $G \cong L_2(2^m)$, $Sz(2^{2m-1})$ or $U_3(2^m)$, $m \geq 2$.

This hypothesis is equivalent to the condition $\Gamma_{1,S}(G) < G$ for $S \in Syl_2(G)$ in Remark 1.3.7, and also to $\Lambda_1(G)$ disconnected in Remark 1.3.9.

The nickname "2-Uniqueness Theorem" will seem more appropriate in some of the equivalent formulations mentioned in the theorem, and developed below. These involve general notions which will recur in the analysis in later chapters.

The 1-generated core. We will focus here on the case $k = 1$ (and later on the case $k = 2$) of:

DEFINITION 1.3.6 (The k-generated core). Let k be a positive integer. Assume that X is a finite group, p is a prime, and P is a p-subgroup of X with $m_p(P) \geq k$. Define

$$\Gamma_{k,P}(X) := \langle N_X(Q) : Q \leq P, m_p(Q) \geq k \rangle.$$

When $P \in Syl_p(X)$, we call $\Gamma_{k,P}(X)$ the *k-generated p-core* of X. ◇

REMARK 1.3.7 (Strong embedding and the 1-generated core). It is easy to check (see B.4.7) that a proper subgroup M of G is a strongly embedded subgroup of G if and only if M contains $\Gamma_{1,S}(G)$ for some $S \in Syl_2(G)$.

Note that $\Gamma_{1,S}(G)$ contains the normalizers of all nontrivial subgroups of S, and so is a uniqueness subgroup in the sense of 1.3.1; hence it is reasonable to call the Strongly Embedded Theorem a 2-uniqueness theorem. ◇

---

[8]E.g. if K is the intersection of all the G-conjugates of M satisfying 1.3.4, then $|K|$ is odd, and G/K is a transitive permutation group on the cosets G/M, in which every involution fixes exactly one point.

Commuting graphs. The notion of strong embedding can be phrased from the graph-theoretic point of view introduced in 0.2.1. Again we focus on the case $i = 1$, with the case $i = 2$ figuring prominently later.

DEFINITION 1.3.8 (commuting graphs). Let G be a finite group of even order. For a fixed positive integer i, let $\Lambda_i(G)$ denote the *commuting graph*, whose vertices are all elementary abelian 2-subgroups of G of rank i, and whose edges are given by commuting pairs of such subgroups. ◊

REMARK 1.3.9 (Strong embedding and $\Lambda_1(G)$ disconnected). In the case $i = 1$, the commuting graph $\Lambda_1(G)$ on subgroups of order 2 may be identified with the graph just on the involutions of G, whose edges are pairs $\{x, y\}$ such that $xy = yx$.

Again it is a standard elementary fact (cf. B.4.7) that $\Lambda_1(G)$ is disconnected if and only if $N_G(\Lambda_0)$ is strongly embedded for a connected component Λ_0 of Λ. Then since $N_G(\Lambda_0)$ contains $C_G(x)$ for each $x \in \Gamma_0$, we see that $N_G(\Lambda_0)$ is also a uniqueness subgroup in the sense of 1.3.1. ◊

In the language of Strategy 1.1.1, the Strongly Embedded Theorem 1.3.5 treats the Uniqueness Case (Step 1) for groups of GW type—and this is why we will sometimes be referring to it as *the* 2-Uniqueness Theorem for groups of GW type.

Further there is another related situation which is usually regarded as also part of Step 1 in the treatment of groups of GW type: namely a Preuniqueness Case, which must be reduced essentially to the Uniqueness Case. We will meet this case later at Definition 1.5.4—where we will see that it is in turn described in terms of the commuting graph $\Lambda_2(G)$ on Klein 4-subgroups of G.

The Dichotomy Theorem and its connection with trichotomy. The Strongly Embedded Theorem 1.3.5 was in fact used heavily, throughout the CFSG.

We already encountered one such application, in the way in which connectivity of $\Lambda_1(G)$ entered into our proof of Lemma 0.3.22—which of course was our eventual re-statement of the Dichotomy Theorem 0.3.10. It will be illuminating to pursue the theme of dichotomy a little further to draw some connections with an analogous notion which arises in the Even Case.

We already indicated that the Dichotomy Theorem logically precedes the treatment of groups of GW type—in the sense that it established the main Odd/Even subdivision of the CFSG, in which groups of GW type form the odd branch. Furthermore we emphasized that the only theorems which we had to assume in our proof of that result are the Strongly Embedded Theorem (in the form of 0.2.3), along with the Signalizer Functor Theorem 0.3.15 (which dates from that same early period). So this is a natural place to discuss the dichotomy result within the context of our historical narrative.[9] We recall that the result states:

The Dichotomy Theorem 0.3.10 Every finite simple group is either of Gorenstein-Walter type or of characteristic 2 type (with 2-rank at least 3).

We single out one part of the proof of the Dichotomy Theorem for comparison with the Characteristic 2-Type Case.[10] Namely (as suggested in our discussion of Step 3 of Strategy 1.1.1) the Uniqueness Case arises at least implicitly during the

[9] We have included this result in a section labeled "level 1"—since it is not at level 0 from the point of view of the classification of the groups of GW type in this chapter. However, it certainly *is* a level 0 result from the viewpoint of the entire CFSG.

[10] So this part of our discussion does not appear in Gorenstein [**Gor83**].

proof—in effect giving a third branch. However, that branch remained essentially implicit: for the Strongly Embedded Theorem could be applied during the proof of 0.3.22, and then the groups of Lie type in characteristic 2 arising from 0.2.3 could be included in the characteristic 2 type branch. But if we instead change that part of the proof into a separate conclusion, the Dichotomy Theorem could be formulated as a trichotomy (in the generic situation when $m_2(G) \geq 3$):

THEOREM 1.3.10 (Trichotomy Theorem for GW type). *Assume* $G = O^2(G)$, $O_{2'}(G) = 1$, *and* $m_2(G) \geq 3$. *Then one of the following holds:*

(I) G is of characteristic 2-type of 2-rank ≥ 3 (and so is not of GW type); or
(II) For some involution $z \in G$, $L_{2'}(C_G(z)) \neq 1$ (i.e., G is of component type); or
(III) G is of 2-uniqueness type (i.e., G has a strongly embedded subgroup).

Now a similar trichotomy emerged in the Gorenstein-Lyons analysis of the Even Case, as we shall begin to discuss at 2.2.1. But there the analogous third branch was not already handled in advance—instead it was subsequently treated by Aschbacher's Uniqueness Case Theorem 8.0.2.

We conclude the section with a more technical discussion of some other points. The reader primarily interested in a broad overview may wish to just skip ahead to the next section.

REMARK 1.3.11 (Further remarks on the proof of the Dichotomy Theorem).

(1) For a group X, the elementary observation B.1.9 shows $L_{2'}(X) = 1$ if and only if X is 2-*constrained*; that is, a Sylow 2-subgroup T of $O_{2',2}(X)$ satisfies $C_X(T) \leq O_{2',2}(X)$. Hence in the literature, the condition that G is of *non-component type* (that is, not of of component type) is often stated in the form that every involution centralizer is 2-constrained. In particular, this is the form used in the hypothesis of Theorem 2.96 in [**Gor83**], discussed below.

(2) Our "elementary" proof of the Dichotomy Theorem given in Section B.5 is essentially a slight simplification of that of Theorem 2.96 in Section 2.5 of Gorenstein's outline [**Gor83**]. We briefly discuss that simplification.

We note that in addition to 0.2.3 and 0.3.15, Gorenstein's proof also quotes Aschbacher's result on groups G with $\Lambda_2(G)$ disconnected—which we are presenting as a "later" (i.e. more advanced) result at 1.5.10. So a primary purpose of Section B.5 is to demonstrate that Gorenstein's use of 1.5.10 can in fact be avoided. In summary, our proof culminates in a strongly embedded subgroup, so that it then suffices instead to apply the "earlier" result Bender's Strongly Embedded Theorem 1.3.5.

In addition, our proof does not require the full strength of the hypothesis that G be fusion-simple in Gorenstein's 2.96; that is, the condition that $Z^*(G) = 1$ can be replaced by the condition that $O_{2'}(G) = 1$. ◊

1.4. The 2-Rank 2 Theorem

We now return to our discussion of groups of GW type, and specifically to the determination of the remaining groups of 2-rank at most 2—in order to complete the small case (1) of Definition 0.3.9. The material in this section corresponds to Sections 1.3–1.5 in Gorenstein's more detailed treatment in [**Gor83**].

Recall that after the Odd Order Theorem 1.2.1, *we may assume G has 2-rank at least* 1.

We first review the situation where G has 2-rank exactly 1. In that case it is easy to show (see A.1.7) that a Sylow 2-subgroup of G is either cyclic or generalized quaternion.

We start with the case where G has a cyclic Sylow 2-subgroup. Here an elementary argument (see Remark B.2.8) known already to Frobenius and Burnside shows:

LEMMA 1.4.1. *If G has cyclic Sylow 2-subgroups then G is 2-nilpotent (i.e., has a normal complement to a Sylow 2-subgroup). In particular the only such simple group is cyclic of order 2.*

Notice that the group of order 2, when added to those of odd prime order from the Odd Order Theorem 1.2.1, completes the list of abelian simple groups in conclusion (1) of the Main Theorem 1.0.1. So *from here on, by "simple" we will often implicitly mean* nonabelian *simple.*

We turn to the remaining case where G has a generalized quaternion Sylow 2-subgroup. In [**BS59**], Brauer and Suzuki show that no simple[11] group G has such a Sylow subgroup. One way to state their result is as follows:

THEOREM 1.4.2 (The Brauer-Suzuki Theorem). *If G is a finite simple group of 2-rank 1 then G is cyclic of order 2.*

The proof in [**BS59**] uses the theory of group characters; and in particular when G has a quaternion Sylow 2-subgroup of order exactly 8, it uses 2-modular character theory. Later, Glauberman in [**Gla74**] provided a proof for that subcase using only ordinary character theory.

We can now turn to simple groups G of 2-rank exactly 2. Here the analysis is much more elaborate; it is contained in the papers indicated in the remainder of this section.

A result of Alperin [**ABG73b**, see 7.1] shows that a Sylow 2-subgroup of G must be dihedral, semi-dihedral,[12] wreathed, or isomorphic to that of $U_3(4)$. We consider each of these cases in turn.

The case of a dihedral Sylow 2-subgroup was treated by Gorenstein and Walter in [**GW65a, GW65b, GW65c**]; they built on an earlier theorem of Brauer, Suzuki, and Wall [**BSW58**].

THEOREM 1.4.3 (The Dihedral Theorem). *Let G be a finite simple group with a dihedral Sylow 2-subgroup. Then either $G \cong A_7$, or $G \cong L_2(q)$ for an odd prime power $q \geq 5$.*

An alternate treatment of this result was given by Bender and Glauberman in [**Ben81, BG81**].

REMARK 1.4.4. As a corollary we can now describe the possibilities for a quasisimple group L with a Sylow 2-subgroup S of 2-rank 1: For S is not cyclic by 1.4.1, so that S must be a generalized quaternion group in view of A.1.7; and then by 1.4.2, we see that $Z(L)$ must be of even order. It follows that $L/Z(L)$ has dihedral Sylow

[11]in contrast to a more general *quasisimple* group with a nontrivial center, such as $SL_2(q)$ for odd q

[12]called *quasi-dihedral* in [**ABG73b**]. See Appendix Section A.1 for references on these classes of 2-groups.

2-subgroups and so appears in 1.4.3; hence (using e.g. A.3.5 on Schur multipliers) we find $L/O_{2'}(L)$ is either the double cover $2A_7$ of A_7, or $SL_2(q)$ for odd $q \geq 5$. $\diamondsuit$

The cases of a semi-dihedral or wreathed Sylow 2-subgroup were treated by Alperin, Brauer, and Gorenstein in [**ABG70, ABG73b, ABG73a**]. The corresponding simple groups appear in conclusion (3) below. Again we have stated the result in a form convenient for our purposes.

THEOREM 1.4.5 (The Semi-Dihedral/Wreathed Theorem). *Let G be a finite simple group of 2-rank 2. Then one of the following holds:*

(1) G has a dihedral Sylow 2-subgroup (and so appears in 1.4.3);
(2) A Sylow 2-subgroup of G is isomorphic to that of $U_3(4)$; or
(3) $G \cong M_{11}$, or $G \cong L_3(q)$ or $U_3(q)$ for odd q.

This result depends on some prior characterization theorems by Brauer, Suzuki, and O'Nan; a complete bibliography is provided in [**Gor83**, Sec. 1.4].

Finally Lyons in [**Lyo72**] treated the case remaining in conclusion (2) of 1.4.5—showing that $U_3(4)$ is the unique finite simple group having that Sylow 2-subgroup. Thus Lyons' result, along with the two previous results, completed the classification of groups of 2-rank exactly 2. Combining these theorems with the Brauer-Suzuki Theorem 1.4.2 we obtain:

THEOREM 1.4.6 (The 2-Rank 2 Theorem). *Let G be a finite simple group having 2-rank 1 or 2. Then either G is cyclic of order 2; or G is isomorphic to $L_2(q)$, $L_3(q)$, or $U_3(q)$, q odd; or $G \cong A_7$, M_{11}, or $U_3(4)$.*

A unified proof of the 2-Rank 2 Theorem is contained in the proof of the more general result called Theorem $\mathcal{C}_2$ in [**GLS05**].

In view of Theorems 1.2.1 and 1.4.6 covering groups of 2-rank at most 2, we have treated the small case (1) in the Definition 0.3.9 of small groups of *GW* type—so that we may proceed to the analysis of the generic groups in case (2). Thus *we may assume for the remainder of the chapter that our simple group G satisfies $m_2(G) \geq 3$, and is of component type.*[13]

In the language of Strategy 1.1.1, this says that we have completed Step 3. However before continuing to Step 4, where we obtain components, in the next section we will first complete our discussion of Step 1 by formulating the Preuniqueness Case and describing its reduction to the Uniqueness Case already handled by Bender.

1.5. (Level 1) The Sectional 2-Rank 4 Theorem and the 2-Generated Core Theorem

Thus before continuing with our discussion of the Generic Odd Subcase, consisting of the groups of component type, as in Section 1.3 we again widen our focus beyond the groups of *GW* type, in order to consider some further important level 1 results which are used frequently in both the Odd and Even Cases of the CFSG.

[13]As the chapter proceeds, we will often return to the example of the matrix groups $L_n(q)$ for q odd in Example 0.3.2. To guarantee that the 2-rank is at least 3, we take $n \geq 4$. For example, suitable conjugates of the involution t exhibited there generate a diagonal subgroup of 2-rank at least 3, as long as 4 divides $q-1$.

So again in this section, we will *not* be assuming that G is of GW type; but this time we will continue to assume that $m_2(G) \geq 3$.

As we follow the historical evolution of these results, we will encounter a number of early candidates for notions of "small" groups. One such condition is that our group have sectional 2-rank at most 4; and many of the results discussed in this section are closely related to that hypothesis. However the hypothesis that G is of 2-rank at most 2 is sufficient at level 0 for the Small Odd Subcase—primarily because the Signalizer Functor Method is available for generic analysis whenever the 2-rank of G is at least 3. So we will begin our discussion within a wider context, motivated by the relationship of signalizer functors with connected components of the commuting graph $\Lambda_2(G)$ on the Klein 4-subgroups of G—a theme we had already introduced in 0.3.17 in Section 0.3.

Connectivity of $\Lambda_2(G)$ and the Signalizer Functor Method. We provide further specific background on these topics in Appendix Section B.4 (with more general background on signalizer functors in Section B.3). For more detail, see e.g. Section 22 (as well as the latter part of Section 10) in [**GLS96**]—and especially Section 2.18 of [**GLS99**]. Here as usual we attempt only an expository overview in the special case of simple groups G of 2-rank at least 3.

In Definition 0.3.14 we saw that an A-signalizer functor θ is defined at involutions a of the elementary abelian 2-subgroup A of G. If $m_2(A) \geq 3$, completeness in the Signalizer Functor Theorem 0.3.15 shows that for every subgroup B of A,

$$\alpha(B) := \langle \theta(b) : b \in B^{\#} \rangle \text{ is a subgroup of } G \text{ of } \textit{odd} \text{ order.}$$

We refer informally to the function α, defined on noncyclic elementary 2-subgroups of G, as the "signalizer functor" determined by θ.[14]

It is now desirable (if possible) to show that α is constant not just on subgroups of A, but in fact on all 4-subgroups of G. If θ is chosen to be G-*equivariant* (Definition 0.3.16), as it usually is, then using the defining properties of θ, one has:

PROPERTIES 1.5.1.
(1) $\alpha(B) = \alpha(C)$ whenever $[B, C] = 1$ for another such 4-subgroup C; and
(2) $\alpha(B^g) = \alpha(B)^g$ for all $g \in G$. ◇

In particular when the commuting graph $\Lambda_2(G)$ is connected (cf. the discussion in 0.3.17), the independence of $\alpha(B)$ on B follows from (1).

Note that the equivariance condition (2) implies that $N_G(B) \leq N_G(\alpha(B))$. Furthermore since $\alpha(B) = \alpha(A)$ for elementary overgroups A of B by completeness, we see that $N_G(\alpha(B))$ also contains $N_G(A)$ for all such A. Thus $N_G(\alpha(B))$ can be regarded as a uniqueness subgroup in the language of Example 1.3.1.

Next since $m_2(G) \geq 3$,[15] we may choose a 4-group U normal in some fixed Sylow 2-subgroup S of G (A.1.8); and fix Λ_0 as the connected component of U in

[14] The distinction between θ and α is sometimes described in the language of "getting started" versus "moving around". (For example in the Even Case, see [**GL83**, p. 523].)

[15] Notice also that any 4-subgroup B which lies in an elementary abelian 2-subgroup A of rank at least 3 satisfies $m_2(C_G(B)) \geq 3$. In particular, this implies that the component of B contains vertices other than B itself (that is, such a B is not an *isolated* vertex in $\Lambda_2(G)$). Indeed since we are assuming that G is simple and $m_2(G) \geq 3$, no 4-subgroup B of G is isolated (e.g. [**GLS99**, 2.18.3.b]).

the graph $\Lambda_2(G)$. Then condition (1) implies that α is constant on Λ_0. It follows that $N_G(\Lambda_0) \leq N_G(\alpha(U))$. Furthermore by B.4.9.2,

$$\Lambda_2(G) \text{ is connected} \quad \text{iff} \quad N_G(\Lambda_0) = G \ ;$$

and in that case, the triviality of the functor α follows from the simplicity of G. (Cf. the role of $O_{p'}(G) = 1$ in applications of 0.3.17.)

So it remains to analyze the case when $N_G(\Lambda_0) < G$, which is equivalent to $\Lambda_2(G)$ disconnected. Thus, one main aim in this section will be to discuss *the classification of simple groups G with $m_2(G) \geq 3$ in which $\Lambda_2(G)$ is disconnected.*

The relation with proper 2-generated cores and the Preuniqueness Case. We had already introduced a relationship between $\Lambda_2(G)$ and the 2-generated 2-core $\Gamma_{2,S}(G)$ (of Definition 1.3.6)—at least implicitly, in our use of B.4.11 during the proof of 0.3.22. In fact disconnectedness of $\Lambda_2(G)$ leads naturally to consideration of yet a further variant of the 2-generated core:

DEFINITION 1.5.2 (The weak 2-generated core). Assume that $P \in Syl_p(X)$ with $m_p(P) \geq 3$. We define a subgroup of the group $\Gamma_{2,P}(X)$ in Definition 1.3.6, determined by those Q which satisfy a further restriction:

$$\Gamma^0_{2,P}(X) := \langle N_X(Q) : Q \leq P, m_p(Q) \geq 2, m_p(QC_P(Q)) \geq 3 \rangle .$$

We call $\Gamma^0_{2,P}(X)$ the *weak* 2-generated p-core of X. ◊

In particular, in our present situation with $p = 2$, $S \in Syl_2(G)$, G a simple group G of 2-rank at least 3, and Λ_0 the component of U chosen as above, it follows from 22.4.i in [**GLS96**] (and the remark for $p = 2$ after that statement) that:

$$\Gamma^0_{2,S}(G) \leq N_G(\Lambda_0) \leq N_G(\alpha(U)). \tag{1.5.3}$$

Thus $\Lambda_2(G)$ disconnected implies that $\Gamma^0_{2,S}(G) < G$. Furthermore when $\alpha(U)$ is nontrivial, $\Gamma^0_{2,S}(G)$ is an example of a uniqueness subgroup in the sense of Step 1 of Strategy 1.1.1.

So in the spirit of the Trichotomy Theorem 5.1.2 (or more specifically the Pretrichotomy Theorem 5.1.1) for the Even Case, we introduce:

DEFINITION 1.5.4 (The Preuniqueness Case for GW type). In our discussion of the Odd Case, we will refer to the situation $\Gamma^0_{2,S}(G) < G$ as the *Preuniqueness Case.*

And (in parallel with a result 8.0.1 in the Even Case) another major goal in this section will then be to essentially reduce the Preuniqueness Case to the Uniqueness Case of Theorem 1.3.5. Indeed the relevant result is:

THEOREM 1.5.5 (The Preuniqueness-implies-Uniqueness Theorem for groups of GW type). *Let G be a finite simple group, with $S \in Syl_2(G)$ of 2-rank at least 3. If G appears in the Preuniqueness Case for groups of GW type ($\Gamma^0_{2,S}(G) < G$), then either G satisfies the Uniqueness Case ($\Gamma_{1,S}(G) < G$—so that G is described in 1.3.5), or $G \cong J_1$.*

This theorem was achieved in two large steps:

Step A: If $\Gamma^0_{2,S}(G) < G$, then $\Gamma_{2,S}(G) < G$

Step B: If $\Gamma_{2,S}(G) < G$, then either $\Gamma_{1,S}(G) < G$ or $G \cong J_1$.

We shall be discussing these two steps in the remaining subsections of this section.

We can rephrase the problems in the language of commuting graphs; indeed Steps A and B can be obtained as consequences of the following somewhat more general program:

Step (a): Classify groups G in which $\Lambda_2(S)$ is disconnected.
Step (b): Classify groups G with $\Lambda_2(S)$ connected, but $\Lambda_2(G)$ disconnected.

For we note that Step A is already done in the subcase where $\Lambda_2(S)$ is connected: since then by B.4.16.b we have $\Gamma^0_{2,S}(G) = \Gamma_{2,S}(G)$. Thus if we have achieved the classification in (a), then, by treating the groups on the list in (a), we can complete the remaining subcase of Step A, and also treat one subcase of Step B. Then to complete Step B, we need only treat the remaining case where $\Lambda_2(S)$ is connected, where we may assume in addition that $\Lambda_2(G)$ is disconnected: for by 22.4.ii in [**GLS96**] (and the equality in the Remark after its proof), we have $N_G(\Lambda_0) = \Gamma_{2,S}(G)$. Thus determining the list of groups appearing in (b) amounts to the remaining verification for Step B.

We will discuss the treatment of each of these Steps in turn.

$\Lambda_2(S)$ disconnected and groups of sectional 2-rank at most 4. We observe first that condition (a) leads to yet another early measure of smallness for a Sylow 2-subgroup. For an easy argument [**Gor82**, Prop 1.37] shows that any 2-group S with a normal abelian subgroup of rank at least 3 must have $\Lambda_2(S)$ connected. The contrapositive of this statement says:

If $\Lambda_2(S)$ is disconnected then S has no normal abelian subgroup of rank at least 3.

The following result of MacWilliams [**Mac70**] in turn translates the condition into a more convenient alternative form:

THEOREM 1.5.6 (MacWilliams' Theorem). *Let S be a 2-group with no normal abelian subgroup of rank at least 3. Then every subgroup of S can be generated by at most four elements.*

The condition in the conclusion of 1.5.6 has the advantage of being inherited by quotients, and more generally by sections. A group G in which all 2-subgroups are generated by at most four elements is said to be of *sectional* 2-rank at most 4 (Definition A.1.6). As a consequence of MacWilliams' Theorem, a treatment of such groups will also treat Step (a).

Of course a group of sectional 2-rank at most 4 also has ordinary 2-rank at most 4; and such groups are not all covered by the 2-Rank 2 Theorem 1.4.6. Gorenstein and Harada classified the simple groups of sectional 2-rank at most 4 in [**GH74**]. We state their result in a form excluding the nonabelian simple groups which already appear in 1.4.6:

THEOREM 1.5.7 (The Sectional 2-Rank 4 Theorem). *Let G be a finite simple group of 2-rank at least 3 and sectional 2-rank at most 4. Then G is isomorphic to one of the following:*

(1) ${}^2G_2(3^{2n+1})$, $G_2(q)$, ${}^3D_4(q)$, $PSp_4(q)$, *q odd;*
$L_4^{\epsilon}(q)$,[16] *q odd,* $q \not\equiv \epsilon \pmod 8$; $L_5^{\epsilon}(q)$, *q odd,* $q \equiv -\epsilon \pmod 4$; *or*

(2) $L_2(8)$, $L_2(16)$, $L_3(4)$, *or* $Sz(8)$; *or*

(3) A_n, $8 \leq n \leq 11$; *or*

(4) M_{12}, M_{22}, M_{23}, J_1, J_2, J_3, McL, *or* Ly.

The proof in the lengthy monograph [**GH74**] of Gorenstein-Harada also makes use of several of their earlier papers, together with a large body of prior characterization theorems by Janko, Fong, W. Wong, D. Mason, and others. A complete tabulation of the prerequisite results is given by Gorenstein in Chapter 2 of [**Gor83**].

In fact Gorenstein's discussion of the proof of the Sectional 2-Rank 4 Theorem, beginning at Section 2.5 of [**Gor83**], presents an alternative approach to the original proof given in [**GH74**] in the subcase where all involution centralizers are 2-constrained, i.e. the noncomponent type situation as in Remark 1.3.11.1.

REMARK 1.5.8 (Remarks on the proof, continued). We can make further use here of a technical observation from Remark 1.3.11. Gorenstein's approach to Sectional 2-Rank 4 in Section 2.5 of [**Gor83**] uses his Theorem 2.96. Recall that his proof of the latter result makes use of Aschbacher's 2-Generated Core Theorem 1.5.10—whose proof in turn uses the Sectional 2-Rank 4 Theorem. But in fact there is no actual problem in Gorenstein's approach, since we observed at 1.3.11 that Aschbacher's result is not really required to establish Theorem 2.96. $\diamondsuit$

A more recent approach, establishing a result very similar to the Sectional 2-Rank 4 Theorem in the complementary subcase where some involution centralizer G is not 2-constrained, is provided in [**GLS05**, Thms $\mathcal{C}_2$, $\mathcal{C}_3$; see p. ix]—assuming the K-group hypothesis 0.5.4 that all simple sections of proper subgroups are known simple groups.

The Sectional 2-Rank 4 Theorem was used at many places in the CFSG. For example, in view of our discussion of Theorem 1.5.6, we can now as a corollary obtain the classification in Step (a). For under our hypothesis that $m_2(G) \geq 3$, one can simply check the list of groups in 1.5.7 for those in which $\Lambda_2(S)$ is disconnected, in order to obtain:

THEOREM 1.5.9 (Nonconnectedness Theorem). *Let G be a finite simple group containing a Sylow 2-subgroup S of 2-rank at least 3, such that $\Lambda_2(S)$ is disconnected. Then S is isomorphic to a Sylow 2-subgroup of J_2.*

Indeed using an earlier result of Gorenstein and Harada [**GH69**, Thm A] *(building on work of Janko* [**Jan69**]*) it follows that $G \cong J_2$ or J_3. And one checks (see the discussion at B.4.12) that $\Lambda_2(G)$ is in fact connected in both these cases.*

Furthermore for $G \cong J_2$ or J_3 one finds that $\Gamma^0_{2,S}(G) = \Gamma_{2,S}(G) = G$. Thus no exceptional groups arise in the Preuniqueness-implies-Uniqueness Theorem 1.5.5 when $\Gamma^0_{2,S}(G) \neq \Gamma_{2,S}(G)$. (Hence Steps A and B discussed above are vacuous here.)

---

[16] See after Remark A.2.1 for this notation combining linear and unitary groups.

It is the Nonconnectedness Theorem,[17] rather than the Sectional 2-Rank 4 Theorem, which in pp. 5–8 of Gorenstein's outline in [**Gor83**] completes the Small Odd Subcase of the CFSG. This is one reason why it was a significant simplification when Harada in [**Har81**] was able to give a much shorter proof of the Nonconnectedness Theorem, which was *independent* of the Sectional 2-Rank 4 Theorem although it assumed part of the K-group hypothesis 0.5.4.

But it should also be emphasized that the Sectional 2-Rank 4 Theorem continued to be applied in many other places in the CFSG: for example, it provides a characterization theorem for most simple groups of Lie type of BN-rank 2 defined over finite fields of odd order. As such, it is invoked in many later papers, notably Aschbacher's paper [**Asc77a**] on the Classical Involution Theorem 1.7.5.

$\Lambda_2(G)$ **disconnected and the 2-Preuniqueness Case.** Having completed Step (a) in which $\Lambda_2(S)$ is disconnected, we now turn to Step (b) where $\Lambda_2(S)$ is connected, but $\Lambda_2(G)$ is disconnected. Indeed since the Nonconnectedness Theorem 1.5.9 shows that when $\Lambda_2(G)$ is disconnected, G does not satisfy the hypothesis of Step (a), we see that Step (b) will in fact complete the classification of simple groups with $\Lambda_2(G)$ disconnected and $m_2(G) \geq 3$.

Also recall our discussion of the proof of the Preuniqueness-implies-Uniqueness Theorem 1.5.5, just before the previous subsection: Step (a) has completed Step A, and part of Step B; further Step (b) is equivalent to the remainder of Step B, where $\Gamma_{2,S}(G) < G$. The determination of groups with this latter condition is a result of Aschbacher in [**Asc74**, Thm 1].

As usual we specialize Aschbacher's result to the case where G is simple of 2-rank at least 3. The theorem determines those groups G where $\Gamma_{2,S}(G) < G$, so it supplies a solution to Step B. Notice only one group appears which does not have a strongly embedded subgroup.

THEOREM 1.5.10 (Proper 2-Generated Core Theorem). *Let G be a finite simple group with a Sylow 2-subgroup S of 2-rank at least 3, such that $\Gamma_{2,S}(G) < G$. Then either $\Gamma_{1,S}(G) < G$ (so that G is described in 1.3.5), or $G \cong J_1$.*

In particular if $\Lambda_2(G)$ is disconnected for some simple group G of 2-rank at least 3, then either $\Lambda_1(G)$ is disconnected or $G \cong J_1$.

As we have indicated, this result completes the proof of the Preuniqueness-implies-Uniqueness Theorem 1.5.5 for groups of *GW* type (and 2-rank at least 3). Of course, the Uniqueness Theorem 1.3.5 then completes the identification of the groups arising in the Preuniqueness Case. Thus only J_1 arises if we explicitly exclude groups of characteristic 2 type:

THEOREM 1.5.11 (The 2-Preuniqueness Theorem (for the Odd Case)). *Let G be a finite simple group of 2-rank at least 3, which is not of characteristic 2 type; and assume further that $\Gamma^0_{2,S}(G) < G$ for $S \in Syl_2(G)$. Then $G \cong J_1$.*

---

[17] While Gorenstein uses this name for the result in his outline at [**Gor83**, p. 6] in his earlier introductory volume, he had used a very similar name for a different result [**Gor82**, Thm. 4.33, index]—this latter result does not treat groups G with a disconnected Sylow 2-subgroup S, but instead those G in which $\Lambda_2(G)$ is disconnected. We record the latter result as Theorem 1.5.10. It appears as the first step ("A") of four steps in the CFSG described at pp. 54–55 of the introductory volume [**Gor82**]—and so could be regarded as the Small Odd Subcase in that treatment.

Moreover if A is a Sylow 2-subgroup of J_1, then there is no nontrivial A-invariant subgroup of J_1 of odd order. Hence there is no nontrivial signalizer functor in this case. Thus we have the following corollary:

COROLLARY 1.5.12. *Let G be a finite simple group of 2-rank at least 3, which is not of characteristic 2 type. Suppose that α is a G-equivariant signalizer functor defined on every 4-subgroup V of G such that $m_2(C_G(V)) \geq 3$. Then $\alpha = 1$.*

We are now in a position to give an overview of our progress up to this point in the chapter:

The work through Section 1.4 completed Step 2 in Strategy 1.1.1 for groups of *GW* type. Similarly the work in this section (notably the triviality of signalizers in Corollary 1.5.12) completes the treatment of Step 1 in the case of groups of *GW* type.

Moreover, Corollary 1.5.12 can be regarded as the essential idea[18] behind the proof of the Dichotomy Theorem 0.3.10—that is, to establish that either G is of characteristic 2 type, or G contains an involution t with $L_{2'}(C_G(t)) \neq 1$. Thus, when we return to groups of *GW* type in the next section, we may also assume that Step 3 of Strategy 1.1.1 is completed—and we are now ready to move on to Step 4 (namely proceeding from 2-components to components).

1.6. The B-Conjecture and the Standard Component Theorem

We now return to the consideration of a simple group G of *GW* type. Recall that Theorems 1.2.1 and 1.4.6 treat those groups G with $m_2(G) \leq 2$, so we are assuming that $m_2(G) \geq 3$. Thus we are in the generic case (2) of Definition 0.3.9; hence *for the remainder of the chapter, we may assume that G is of component type—namely $L_{2'}(C_G(z)) \neq 1$ for some involution z of G.*

This condition says that $C_G(z)$ has a 2-component—whose quotient modulo $O_{2'}(C_G(z))$ is quasisimple. But recall from our discussion at (0.4.2) that in the known simple groups, each 2-component in an involution centralizer is in fact quasisimple, and so is an *ordinary* component.

This distinction corresponds to proceeding from Step 3 to Step 4 in Strategy 1.1.1. Here one might hope to obtain an actual component "directly"—via the triviality of a suitable signalizer functor (e.g. as in Corollary 1.5.12). And indeed something in this direction can be established, when the 2-component L is large.[19] Thus much of the difficulty in this stage of the Odd Case arises when all 2-components L are small. But unfortunately, treating that case inextricably involves dealing also with larger 2-components; this is one way of summarizing the complications that will arise in our discussion in Sections 1.6–1.8 (e.g. in Theorem 1.8.7).

So in this section, we begin with a bit more background on components in local subgroups; for additional detail, see e.g. Sections 3.1–3.3 of Gorenstein [**Gor83**].

We can approach the problem in more generality: For a finite group H, Thompson defined[20] the subgroup $B(H)$ to be the product of all 2-components of H which

[18]Though again we recall, as in Remark 1.3.11.2, that our proof of the Dichotomy Theorem in Section 0.3 was designed to obtain this result about triviality of signalizers *without* having to quote 1.5.5; that is, using only the 2-Uniqueness Theorem 1.3.5, without the full 2-Preuniqueness Theorem 1.5.11.

[19]See for example the cases of Theorem 3 on pp. 84–85 and 85–88 in [**GLS02**].

[20]The B stands for the "bad" part of the 2-layer.

are *not* quasisimple. Thus $L_{2'}(H) = B(H)E(H)$ with $B(H) \cap E(H) \leq Z(E(H))$. Using this language, we can re-phrase the desired property with a more general conjecture:

CONJECTURE 1.6.1 (The B-Conjecture). *Let G be a finite group, and let H be any 2-local subgroup of G. Then $B(H) \leq B(G)$.*

In particular if $O_{2'}(G) = 1$, then $L_{2'}(C_G(z)) = E(C_G(z))$ for all involutions z of G. Hence a simple group G of GW type and of 2-rank at least 3 has an involution centralizer with a component.

Despite efforts by Thompson, no direct proof of this conjecture has been found; instead it was eventually established more indirectly, as a corollary of the Unbalanced Group Theorem 1.7.1. So we will refer to the conjecture thereafter as the B-Theorem 1.8.11.

In the analysis of groups of GW type, the B-Theorem is the principal way in which the "global" property $O_{2'}(G) = 1$ of the simple group G is exploited to obtain "local" information—that is, to describe 2-locals. By comparison, the Standard Component Theorem (which we will discuss next) exploits the global fact that $G = E(G)$ is simple, rather than just semisimple (i.e. the commuting product of more than one simple group); and that result delivers the striking local consequence of standard form, discussed below.

Long before the final proof of the B-Conjecture via the Unbalanced Group Theorem, Gorenstein and Walter had realized that not only should G possess some involution centralizer with a component, but in fact that such a component chosen "maximal" in a suitable sense[21] should usually satisfy many more properties, which in turn could be exploited to characterize G. This analysis was later refined by Aschbacher with the concept of *standard form*. (In our viewpoint of Strategy 1.1.1, this affords Step 5—which is really the central point of the whole process.) Before formulating Aschbacher's definitions, we will illustrate with our running example:

EXAMPLE 1.6.2 (Properties of some maximal components). We return to our earlier Example 0.3.2, with $G \cong L_n(q)$ for n and q odd, and $(n, q) \neq (3, 3)$. To obtain a component L which is as large as possible, we take $k = 1$. Recall then t is the central involution in $L \cong SL_{n-1}(q)$.

Next L is quasisimple and normal in $C_G(t)$, and indeed $C_G(t) = N_G(L)$. Furthermore from 0.3.2, we have $L = L' = C_G(t)'$. One consequence of the maximality of L is that $C_G(t)$ contains no other conjugate of L: For suppose that $L^g \leq C_G(t)$. Since also $L^g = (L^g)'$, we see that $L^g \leq C_G(t)' = L$. But then $L^g = L$. That is, we see L is *weakly closed* (cf. B.2.3) in $C_G(t)$ with respect to G. Thus $C_G(t) = N_G(L)$ can contain no distinct G-conjugate of L; and in particular L centralizes none of its G-conjugates.

If t' is an involution of $C_G(L)$, then t' must act as a scalar map on both eigenspaces for t, in view of the irreducibility of L. Since n is odd and $\det(t') = 1$, it follows that $t' = t$; thus t is the unique involution of $C_G(L)$. Next suppose that two conjugates $H := C_G(L)$ and $H^g = C_G(L^g)$ have an intersection of even order. That intersection must contain t, which must also then be the unique involution t^g of H^g; whence $g \in C_G(t) = N_G(L)$. Thus whenever $g \in G - N_G(L)$, $H \cap H^g$ must instead be of odd order.

---

[21] which we will not describe precisely here; see e.g. [**Gor83**, Defn 3.23].

Suppose next that instead n is even with $n \geq 6$, $q \equiv 3 \pmod 4$, and $k = 2$. Here referring back to Example 0.3.2, we can take $L \cong SL_{n-2}(q)$, and we find that $C_G(t)' = KL = LK$, with $K \cong SL_2(q)$. Again, since $n - 2 > 2$, there is no room for a distinct conjugate L^g inside $C_G(t)$. Moreover, under these hypotheses, we may again argue that t is the unique involution of $H := C_G(L)$, whence $H \cap H^g$ has odd order whenever $g \in G - N_G(L)$.

We also mention that we get similar conclusions when $G \cong A_n$ ($n \geq 9$), with $L \cong A_{n-4}$ and $H := C_G(L) \cong A_4$. In contrast to the cases above with $L_n(q)$, now t' could be an involution different from t in this A_4. Indeed, in this situation we have $C_G(t) < N_G(L) = HC_G(t)$; but we still get that $H \cap H^g$ has odd order for all $g \in G - N_G(L)$. ♢

Aschbacher was led to abstract the above properties in defining the following important concept:

DEFINITION 1.6.3 (standard component). Let L be a component of $C_G(z)$ for some involution z of G, and set $H := C_G(L)$. Then we will say that L is *standard* in G (or in *standard form*), if the following properties also hold:

(1) L commutes with none of its G-conjugates;

(2) $H \cap H^g$ has odd order for all $g \in G - N_G(L)$. (Hence $N_G(L) = N_G(H)$.)

In particular H is *tightly embedded* in G; that is, H is of even order but $H \cap H^g$ has odd order for all $g \in G - N_G(H)$.[22] ♢

Aschbacher's fundamental result [**Asc75a**, Thm I] based on this concept is frequently stated in the form we give below; its conclusion (1) shows that if the B-Conjecture can be established, then the classification of groups of *GW* type can essentially be reduced to solving *standard form problems*—namely given some particular quasisimple group L standard in G, identify the possibilities for G. And in fact that was the course of the subsequent history. The formulation below also uses some further development by Gilman in [**Gil76**, Thm 1].[23]

THEOREM 1.6.4 (initial form of Standard Component Theorem). *Let G be a finite simple group of GW type, of 2-rank at least 3. Assume further that the B-Conjecture 1.6.1 is valid in G. Then one of the following holds:*

(1) Some involution centralizer has a component L which is standard in G; or

(2) Some involution centralizer $C_G(z)$ has a component J which contains z and has generalized quaternion Sylow 2-subgroups.

In Example 1.6.2 if $n = 4$ and $q > 3$, then $G \cong L_4(q)$ supplies an example of the exceptional conclusion (2) where "J" could be either of the $C_G(z)$-conjugate components L and K isomorphic to $SL_2(q)$, neither of which is standard in G.

Later in 1.8.12 we will give a final form of the Standard Component Theorem—a form which was available once the B-Theorem was established; that form will also use part (1) of the Classical Involution Theorem 1.7.5, which we discuss in the next section, to identify the groups G arising in conclusion (2).

[22]This condition can be regarded as a generalization of strong embedding in Definition 1.3.4.

[23]A more detailed derivation of this formulation from Aschbacher's original statement appears in Section 3.4 of [**Gor83**]; see especially the Propositions 3.33 and 3.34 there, and the discussion of maximality for components.

REMARK 1.6.5. Aschbacher also established strong results describing the tightly embedded subgroup $C_G(L)$ in the standard form setup. Indeed for L a known simple group, one can in general deduce that $C_G(L)$ has 2-rank at most 2. Hence the structure of an involution centralizer $C_G(z)$ containing L is tightly controlled. ◇

1.7. The Unbalanced Group Theorem, the $2A_n$-Theorem, and the Classical Involution Theorem

However, we must back up. The B-Theorem, and hence also the Standard Component Theorem, in fact depends on the Unbalanced Group Theorem. We state that result below, and sketch its proof over the course of the next two sections. (The deduction of the B-Theorem as a corollary appears e.g. as Prop 3.26 in Gorenstein [**Gor83**].)

The Unbalanced Group Theorem, and the approach via unbalancing triples. The inspiration for the Unbalanced Group Theorem originally arose from work of Gorenstein and Walter, aimed at the control of normal subgroups of odd order ("cores")[24] in involution centralizers in simple groups—and more generally in their automorphism groups.

We saw in earlier examples that such nontrivial cores exist in many simple groups of GW type: In $G \cong L_n(q)$ in Example 1.6.2, there are typically linear maps of odd order of determinant 1 which act as scalars on both eigenspaces for the involution t. These will afford elements of $Z\big(C_G(t)\big)$ of odd order, and hence elements of $O_{2'}\big(C_G(t)\big)$. Also in $G = A_7$, a subgroup $A_3 \cong \mathbb{Z}_3$ will be normal in $C_G(t)$—and more generally for any "long" involution't in A_n, when $n \equiv 3 \pmod 4$. The existence of these examples is reflected in the conclusions of the Theorem:

THEOREM 1.7.1 (The Unbalanced Group Theorem). *Let G be a finite group with $F^*(G)$ simple, having an involution z such that $O_{2'}\big(C_G(z)\big) \neq 1$. Then one of the following conclusions holds:*

(1) $F^*(G)$ is of Lie type over a field of odd order (and most such groups do arise here); or

(2) $F^*(G)$ is an alternating group of degree $n \geq 7$, n odd; or

(3) $F^*(G) \cong L_3(4)$ or He.

Further $F^*(G) < G$ in (3), and in (2) when $n \equiv 1 \pmod 4$.

In our discussion of the proof, we will let (U) be an abbreviation for this result.

Elementary signalizer functor analysis establishes that a minimal counterexample G to (U) contains a certain local configuration; the definition is somewhat technical—cf. [**Gor83**, p. 191]:

DEFINITION 1.7.2 (unbalancing triple). A triple (x, y, L) in a finite group G is called an *unbalancing triple* if x and y are commuting involutions in G, and L is a 2-component of $C_G(x)$ such that

$$L = [L, y] = [\ L\ ,\ O_{2'}\big(C_G(y)\big) \cap C_G(x)\ ].$$

---

[24] The use of the term "unbalanced" in the result refers to the fact that nontrivial cores in involution centralizers are the obstruction to the condition of *balance* (e.g. (S2) in the definition 0.3.14 of signalizer functors) for the functor $O_{2'}\big(C_G(-)\big)$. For related terminology, see also Appendix section B.3 and its references—e.g. Section 4.4 of Gorenstein [**Gor82**], especially Theorem 4.54 and Definition 4.55.

◊

In this situation, set $H := L\langle y\rangle\big(\, O_{2'}\big(C_G(y)\big) \cap C_G(x)\, \big)$ and $\overline{H} := H/O_{2',2}(H)$. Then $F^*(\overline{H}) = \overline{L}$ is simple, and $\overline{y}$ is an involution in $\overline{H}$ with $O_{2'}\big(C_{\overline{H}}(\overline{y})\big) \neq 1$—so that $\overline{H}$ satisfies the hypotheses of (U). Furthermore $H \leq C_G(x)$, so that $H < G$ since $F^*(G)$ is simple. Thus, if G is a counterexample of minimal order, then $\overline{L}$ is of one of the simple groups listed in the conclusions of the Theorem.

So to prove (U), it is natural to treat the possible unbalancing triples (x, y, L), as $\overline{L}$ varies over the simple groups in the conclusion of (U).

Next we will indicate two significant early results by Solomon and Aschbacher: These results are proved under hypotheses different from that of the Unbalanced Group Theorem, but they are in the same spirit, in that they treat groups possessing an involution centralizer with a 2-component of known type. Further we will see that these results enabled Thompson to make an important reduction to the list of unbalancing triples which needed to be treated in the remainder of the proof of (U).

The $2A_n$ Theorem. So we now digress from our discussion of (U), to consider characterizations of simple groups G in terms of involution centralizers possessing 2-components L of certain specified types.

The cases with $L/O_{2',2}(L)$ an alternating group were treated for the most part by Solomon, in a sequence of papers begun prior to the formulation of the Unbalanced Group Theorem as a conjecture.

Historically the first situation Solomon treated was the subcase where $L/O_{2'}(L)$ is the double cover $2A_n$ of the alternating group. Using ideas of Goldschmidt, he established in [**Sol75**]:

THEOREM 1.7.3 (The $2A_n$ Theorem). *Let G be a finite simple group with an involution z such that $C_G(z)$ has a 2-component L containing z with $L/O_{2'}(L) \cong 2A_n$ for some $n \geq 8$. Then $G \cong McL$ or Ly. (In particular this treats the standard form problem for $L/O_{2'}(L) \cong 2A_n$, $n \geq 8$.)*

Solomon begins by constructing a signalizer functor on elementary 2-subgroups of $O_{2',2}\big(\, L_{2'}\big(C_G(z)\big)\, \big)$. The fusion of such subgroups can be controlled, allowing an invocation of Goldschmidt's 2-Fusion Theorem [**Gol74**]. Then further 2-fusion analysis leads to a case handled in earlier work [**GH71b**] of Gorenstein and Harada, which determines the groups in the conclusion.

The Classical Involution Theorem. Next we consider groups G containing an involution centralizer with a 2-component L such that $L/O_{2',2}(L)$ is of Lie type over a field of odd order. The characterization of a large subfamily of such groups by Aschbacher is an important step in the proof of (U).

We begin with a little more history. Thompson's efforts on the B-Conjecture helped to shape research on groups of GW type during the mid-1970s. In particular, he focused attention on groups G containing a generalized quaternion subgroup Q which is a TI-subgroup—that is (cf. Definition A.1.2), we get $Q \cap Q^g = 1$ for all $g \in G - N_G(Q)$.

Aschbacher was in turn led to consider a subgroup J of $C_G(z)$, where z is the unique involution in a generalized quaternion Sylow 2-subgroup of J, under some additional conditions on J and/or the embedding of J in G. For example one such condition demands that J be tightly embedded in G; that is, $J \cap J^g$ is of odd order for $g \in G - N_G(J)$.

Typically $J/O_{2'}(J)$ is of the form $SL_2(q)$ for some odd prime q, but it is also possible that we have $J = O_{2'}(J)Q$ or that $J/O_{2'}(J) \cong 2A_7$ (as suggested by Remark 1.4.4). For example we saw in 0.3.2 that for $G \cong L_n(q)$ with n even, we could take t and $K \cong SL_2(q)$ in the roles of "z" and "J". (Similarly the condition there that $k = 2$ suffices, even when n is odd.) Indeed in most groups G of Lie type over a field of odd order, one can choose z so that $C_G(z)$ has a component J containing z and isomorphic to $SL_2(q)$ for a suitable odd q. Also, such a J and Q arose in conclusion (2) of Aschbacher's Standard Component Theorem 1.6.4.

These remarks motivate the following definition:[25]

DEFINITION 1.7.4 (classical involution). We say that an involution z of a finite group G is a *classical involution* if $C_G(z)$ has a subnormal subgroup J, with $z \in J$ and $J/O_{2'}(J) \cong 2A_7$ or $SL_2(q)$ for some odd q. ◇

Aschbacher handled this situation and more in a monumental work [**Asc77a, Asc77b**].[26] Our discussion will follow that of Gorenstein in [**Gor83**, Sec 3.6]; in particular our statement below incorporates two corollaries to Aschbacher's [**Asc77a**, Thm I]—which are given as Theorems 3.67 and 3.65 in Gorenstein.

THEOREM 1.7.5 (The Classical Involution Theorem). *Assume that G is a finite group with $F^*(G)$ simple, containing an involution z such that z lies in a subnormal subgroup J of $C_G(z)$, where J has a generalized quaternion Sylow 2-subgroup.*

(1) If z is a classical involution (i.e., if $J/O_{2'}(J) \cong 2A_7$ or $SL_2(q)$ for odd q), then $F^*(G)$ is either a group of Lie type in odd characteristic, or M_{11}. (In particular, this further hypothesis holds whenever the subnormal subgroup J of 2-rank 1 has $J/O_{2'}(J)$ quasisimple, using Remark 1.4.4.)

(2) If J is tightly embedded[27] in G, then $F^*(G)$ is either a group of Lie type in odd characteristic, or M_{11}, or M_{12}. (In particular, this further hypothesis holds whenever $J = C_G(L)$ with L standard in G, directly from Definition 1.6.3.)

The hypothesis that J has a generalized quaternion Sylow 2-subgroup is satisfied when $J/O_{2'}(J)$ is quasisimple and isomorphic to $2A_7$ or $SL_2(q)$ with $q \geq 5$. But it also also satisfied in some solvable groups such as $SL_2(3)$—and this group does in fact arise as a possible J in many of the groups of Lie type over the field of order 3 in the conclusions. Moreover, when $G \cong M_{11}$ or M_{12}, $C_G(z)$ is solvable; and we may take $J \cong SL_2(3)$ in M_{11}, and $J \cong Q_8$ in M_{12}. These observations yield the following easy but important corollary of the Theorem:

COROLLARY 1.7.6. *Let G be a finite simple group containing an involution z, such that $C_G(z)$ has a 2-component L and $C_G\big(L/O_{2'}(L)\big)$ has generalized quaternion Sylow 2-subgroups. Then G is of Lie type in odd characteristic.*

PROOF. By the definition A.1.18 of a 2-component, $L/O_{2'}(L)$ is quasisimple and subnormal in $C_G(z)$. Then G is not M_{11} or M_{12}, by our observation above

[25] Our definition here is more specific than the fairly informal usage of the term "classical involution" given by Gorenstein on p. 172 of [**Gor83**]; this specificity will be useful in stating conclusion (1) of Aschbacher's theorem 1.7.5, which is usually referred to as the Classical Involution Theorem.

[26] The short note [**Asc80a**] just indicates the modifications in the proof needed to adjust for the omission of $\Omega_8^+(2)$ in the list of conclusions of Theorem I in the original statement in [**Asc77a**].

[27] We mention that (as in [**Gor83**, Sec 3.6]) it suffices to assume only that J is tightly embedded in G with generalized quaternion Sylow 2-subgroups: since then for each involution z of J, $C_J(z)$ is normal in $C_G(z)$.

that $C_G(z)$ is solvable in those groups. Notice also since $L \le C_G(z)$ that we get $z \in C_G(L) \le C_G\big(L/O_{2'}(L)\big)$.

Suppose first that L is not normal in $C_G(z)$. Then some $c \in C_G(z)$ determines a distinct conjugate $L^c \neq L$, so that L^c is also a 2-component of $C_G(z)$; hence (e.g. A.1.20) $L^c\langle z\rangle$ centralizes $L/O_{2'}(L)$. Therefore as $C_G(L/O_{2'}(L))$ has generalized quaternion Sylow groups, it follows that $z \in L^c$ and L^c has 2-rank 1. Then as L and L^c are conjugate in $C_G(z)$, L is of 2-rank 1 with $z \in L$. This verifies the hypothesis for case (1) of Theorem 1.7.5, with L in the role of "J"—and then that result gives the desired conclusion, since we excluded M_{11} at the outset.

So we may assume that L is normal in $C_G(z)$. Then $J := C_G\big(L/O_{2'}(L)\big) \cap C_G(z)$ is also normal in $C_G(z)$ and $z \in J$. In particular $z \in Z(Q)$ for some Sylow 2-subgroup Q of $C_G\big(L/O_{2'}(L)\big)$. Hence'$Q \le J$, so that the generalized quaternion group Q is also Sylow in J. Indeed as $J \le C_G(z)$, the unique involution z of Q is also the unique involution of J. We claim that J is tightly embedded in G: For given $g \in G$ with $J \cap J^g$ of even order, it follows that z must also be the unique involution z^g in J^g—so that $g \in C_G(z)$. So since L is normal in $C_G(z)$, we see that g normalizes J, and hence J is tightly embedded in G. This in turn completes the verification of the hypothesis for case (2) of Theorem 1.7.5; and then that result completes the proof, since we excluded M_{11} and M_{12} at the outset. □

Here are some of the ideas involved in Aschbacher's original proof of [**Asc77a**, Thm I]. We assume the hypothesis and notation in our statement of 1.7.5 during this discussion.

Thompson had noticed that a set of commuting maximal cyclic subgroups, of Sylow 2-subgroups of a suitable set $\mathcal{J}$ of conjugates of J generate an abelian group T—which may be thought of as the 2-part of a maximal torus in a group of Lie type over a field of odd order. The subgroups $Aut_K(T)$, $K \in \mathcal{J}$, then generate a subgroup $\mu(G)$ of $Aut_G(T)$, which plays the role of a Weyl group. Aschbacher shows that $\mu(G)$ is a 3-transposition group,[28] and hence is known by Fischer's 3-transposition theorem A.6.3.

It is useful to consider the commuting graph Δ on the G-conjugates of J. Naively, the generic case of the proof of [**Asc77a**, Thm I] is as follows: Suppose that Δ is connected, and also that G contains the commuting product of three conjugates of J with distinct central involutions. Then an analogue of the signalizer functor deployed by Goldschmidt and Solomon can be used to prove that J is quasisimple. As we are in an inductive context, we can identify the subgroup:

$$K_z := < J^g \le C_G(z) \ : \ [J, J^g] = 1 >,$$

which usually is a large group of Lie type defined over a finite field of odd order. Now that we know K_z and the Weyl group $\mu(G)$, we can use the Curtis-Tits Theorem A.5.1, or a suitable analogue, to identify G.

There are substantial difficulties in treating the "small" cases when J is solvable, and when Δ is disconnected.

To identify the unitary groups, Aschbacher used results of Phan—see Remark A.5.2. (But since Phan's original proof in [**Pha77a, Pha77b**] may be flawed, we instead rely here on a revised proof given later by Bennett and Shpectorov in [**BS04**].)

[28]We review the theory of 3-transpositions in Appendix A around Definition A.6.2.

Finally, as mentioned earlier, Aschbacher's argument quotes the Sectional 2-Rank 4 Theorem 1.5.7, to handle certain small cases.

Thompson's reduction. We now return to our discussion of the proof of (U). Using the $2A_n$ Theorem 1.7.3 and the Classical Involution Theorem 1.7.5, Thompson provided a short and elegant reduction of the Unbalanced Group Theorem 1.7.1 to the following set of unbalancing 2-components L:

THEOREM 1.7.7 (Thompson's Reduction Theorem). *Let G be a minimal counterexample to the Unbalanced Group Theorem. Let (x, y, L) be any unbalancing triple in G. Then one of the following conclusions holds:*

(1) $L/O_{2'}(L) \cong A_n$, n odd, $n \geq 9$; or

(2) $L/O_{2'}(L) \cong A_7$ or $L_2(q)$, q odd, $q \geq 5$; or

(3) $L/O_{2'}(L) \cong He$ or $L/Z^*(L) \cong L_3(4)$.

The fundamental idea behind the reduction is that if some involution centralizer has a 2-component which possesses a classical involution, then G should also have a classical involution. This approach was implicit in some work of J. Walter at about that time. Thompson's reduction depended on certain properties of groups of Lie type, which were established by Burgoyne, and on an earlier result of Solomon [**Sol74**] on groups with involution centralizers of type $Spin_7(q)$. The Reduction Theorem was published by Burgoyne [**Bur77**], since Thompson's "Notes on the B-Conjecture" in fact remained unpublished. The reduction argument is also described on pp. 192–199 of Gorenstein [**Gor83**].

1.8. Finishing the Unbalanced Group Theorem and the B-Theorem

To complete the proof of (U), after Thompson's reduction 1.7.7, it will often be helpful to assume the K-group Hypothesis 0.5.4, provided by induction in the proof of the CFSG. *Indeed this simplifying assumption applies to much of our discussion from here on in the chapter.*[29] For convenience we re-state the hypothesis here; and we will abbreviate it by (K) in the discussion that follows:

K-Group Hypothesis 0.5.4 Assume that all simple sections of proper subgroups of G are known simple groups, i.e., are on the list of conclusions 0.1.1 of the Classification of the Finite Simple Groups.

Historically, this hypothesis was in fact avoided in the original proof of (U). However in that process, the interdependence among the papers became rather complicated. We can substantially reduce these problems by instead assuming (K); in particular this will alleviate some of the difficulties indicated on pp. 193–194 and in Section 4.3 in Gorenstein [**Gor83**], describing the original proof which did not assume (K).

One particular simplification provided by (K) is that a minimal counterexample G to (U) is simple (as opposed to the more general condition that $F^*(G)$ be simple in the hypothesis of (U)). This will allow us to restrict various intermediate results to the case G simple, where the statements are considerably less complicated.

[29] And Hypothesis 0.5.4 is also used throughout most of the classification of groups of characteristic 2 type in our later chapters.

Alternating 2-components and the reduction of Gilman and Solomon. Case (1) of Theorem 1.7.7, where $L/O_{2'}(L)$ is an alternating group of degree at least 9, was handled by Solomon, who provided a characterization of larger alternating groups in terms of alternating 2-components of involution centralizers in the papers [**Sol76b, Sol77, Sol78a**]. Solomon also relied on work of Aschbacher [**Asc08**] (which had remained unpublished until recently). Here is Solomon's result—stated, as mentioned above, in the case where G is simple.

THEOREM 1.8.1 (Alternating Theorem). *Assume (K), with G a finite simple group. Suppose that x is an involution in G and L is a 2-component of $C_G(x)$ with $L/O_{2'}(L) \cong A_n$ for some $n \geq 8$. Suppose further that either*

(a) G is a minimal counterexample to (U);[30] *or*

(b) G is not a counterexample to (U), and L is quasisimple and standard in G.

Then $G \cong A_{n+4}$, and in particular (a) does not hold.

In conjunction with the $2A_n$ Theorem 1.7.3, this completes the treatment of the standard form problem for $L/Z(L) \cong A_n$ when $n \geq 8$.

In brief, the proof of (a) proceeds by a reduction to the standard form problems for a quasisimple component of alternating type in (b).

In more detail: In case (a), if the 2-component L is chosen with $L/O_{2'}(L) \cong A_n$ with n maximal, then L shares many of the properties of a standard component. For example, Solomon in [**Sol76a**] proved that L is not permutable with any of its distinct G-conjugates. These properties are exploited, in conjunction with signalizer functor methods, to reduce to the case where L is a quasisimple component standard in G. Then the solution of that standard form problem determines G as A_{n+4}, completing the proof of (b). It also completes the proof for (a)—since when G is an alternating group, we have $O_{2'}(C_G(z)) \neq 1$ only when $n \equiv 3 \pmod 4$, and this case appears in conclusion (2) of (U).

REMARK 1.8.2. This way of treating certain 2-components in the context of (U), via a reduction to a corresponding standard form problem, comes up again in various results later in this section. It illustrates a modification of the approach in our Strategy 1.1.1: Namely we have already indicated that the completion of Step 4 requires the proof of (U), but now we see that the proof of (U) requires that we continue through Steps 5 and 6 for certain 2-components L.

Once (U), and hence Step 4, are completed, it remains to complete Steps 5 and 6, but only for those components L not already covered in the proof of (U). $\diamondsuit$

In general, there is a notion of maximality for 2-components of involution centralizers of G (cf. our remarks before 1.6.3 for ordinary components), based on a certain partial ordering $<<$. In particular, L is maximal with respect to $<<$ if $|L/O_{2'}(L)|$ is maximal. As noted above, maximal 2-components under this partial ordering share some of the desirable properties of standard components.

The Alternating Theorem 1.8.1 reduces (U) to cases (2) and (3) of 1.7.7, i.e., to unbalancing triples (x, y, L) with $L/O_{2'}(L)$ isomorphic to A_7, He, $L_2(q)$, or to a covering group of $L_3(4)$. So we seem to be near the finish line. But unfortunately, an unbalancing 2-component L from this list might be less than a maximal 2-component K, which is not a member of an unbalancing triple (cf. [**Gor83**, middle of p. 193]). Indeed, for certain groups of GW type which have multiple identities

[30] Notice by 1.7.7 that we can in fact take $n \geq 9$ odd in this branch of the hypothesis.

including that of a group of Lie type in characteristic 2,[31] as well as $L_3(4)$ which is not of GW type, K might well be a large group of Lie type in characteristic 2.

To circumvent this problem, Gilman and Solomon in [**GS79a**] proved that if L comes from case (2) or (3) of 1.7.7, and is chosen maximal in a different sense, related to unbalancing triples (see [**Gor83**, Defn 3.29]), then L is *also* maximal with respect to $<<$. More detail on the result below can be found in Section 4.6 of [**Gor83**]; we mention also that the K-group hypothesis simplifies the analysis—since (cf. [**Gor83**, bottom of p. 193]) it can be used inductively to identify the possibilities for maximal 2-components.

THEOREM 1.8.3 (Gilman-Solomon Reduction Theorem). *Assume (K), with G a minimal counterexample to (U). Then there is an involution $x \in G$ and a 2-component L of $C_G(x)$ such that L lies in a maximal unbalancing triple and L is also a maximal 2-component in G. Further one of the following holds:*
(1) $L/O_{2'}(L) \cong A_7$ or $L_2(q)$, q odd, $q \geq 5$; or
(2) $L/O_{2'}(L) \cong He$ or $L/Z^*(L) \cong L_3(4)$.

That is, to complete the proof of (U), we have now replaced the list in 1.7.7 with the list in 1.8.3—but now we may also assume that the 2-component L is maximal in *both* senses.

2-components of type $L_2(q)$ or A_7, and the Odd Lie Type Theorem. We next record several results which handled case (1) of 1.8.3.

First Foote in [**Foo78**] dealt with the case of $C_G\big(L/O_{2'}(L)\big)$ of 2-rank at least 2 (cf. Remark 1.6.5); he obtained the groups in conclusion (2) below:

THEOREM 1.8.4. *Assume (K), with G a finite simple group having a maximal 2-component L in $C_G(x)$ such that $L/O_{2'}(L) \cong A_7$ or $L_2(q)$, for odd $q \geq 5$. Then either*
(1) $C_G\big(L/O_{2'}(L)\big)$ has 2-rank 1; or
(2) $G \cong A_9$, A_{10}, A_{11}, or J_2 with L isomorphic to $L_2(5) \cong A_5$, $L_2(9) \cong A_6$, A_7, or $L_2(5)$), respectively.

Thus in case (1) of 1.8.3, it remains to treat the subcase $C := C_G\big(L/O_{2'}(L)\big)$ of 2-rank 1. In fact Corollary 1.7.6 of the Classical Involution Theorem handled the case where C has generalized quaternion Sylow 2-subgroups, so we can assume in view of A.1.7 that C has cyclic Sylow 2-subgroups.

Before giving the results used to treat this case, we first digress briefly with a few remarks about their proofs. Since C has cyclic Sylow 2-subgroups, L is the only 2-component of $C_G(x)$, and in particular is normal in $C_G(x)$; so $C_G(x)$ induces a group of automorphisms on $\overline{L} := L/O_{2'}(L)$. Also since $\overline{L}$ is simple in case (1) of 1.8.3, we have $x \notin L$; indeed since L is of 2-rank 2, the subgroup $\langle x \rangle \times L$ of $C_G(x)$ has 2-rank 3. The case where $C_G(x)$ has 2-rank exactly 3 can be treated by working to obtain the hypotheses that G contains a self-centralizing E_8-subgroup (elementary 2-subgroup of rank 3) in a result of Harada [**Har75**, Thm 2]—which reduced the problem to checking the list of groups in the Sectional 2-Rank 4 Theorem. This leaves the case where $C_G(x)$ has 2-rank at least 4, where one checks that some involution in $C_G(x)$ induces an outer automorphism on $\overline{L}$, which is either a field automorphism

[31] because of standard isomorphisms listed in Remark A.3.6, such as $L_2(4) \cong L_2(5)$, $L_3(2) \cong L_2(7)$, and $Sp_4(2)' \cong L_2(9)$.

of $L_2(q)$, or in $S_7 \setminus A_7$. A particularly difficult subcase occurs when $\langle x \rangle$ is Sylow in C, and $L/O_{2'}(L) \cong L_2(9)$; in that case $|L|_2 = 8$.

Now we list the theorems used to determine the groups which arise in case (1) of 1.8.3, where $C_G\big(L/O_{2'}(L)\big)$ has cyclic Sylow 2-subgroups. In fact the treatment of the entire subcase where $|L|_2 = 8$ was obtained from more general work of Harris and Solomon [**HS77, Har77**] and, independently, of Fritz [**Fri77a, Fri77b**]. Below we state the principal case of Fritz's result—in which L is allowed to live in the slightly larger list in the 2-Rank 2 Theorem 1.4.6:

THEOREM 1.8.5. *Let G be a finite simple group containing an involution t, such that $C_G(t)$ has a 2-component L where $\overline{L} := L/O_{2'}(L)$ is simple of 2-rank 2, but $\overline{L} \ncong U_3(4)$. Suppose that $C_G(\overline{L})$ has cyclic Sylow 2-subgroups, and either*

(a) $|L|_2 = 8$; or
(b) $|G|_2 \leq 2^{10}$.

Then one of the following holds:

(1) G is a group of Lie type in odd characteristic; or
(2) $L \cong L_2(9) \cong A_6$ and $G \cong HS$.

REMARK 1.8.6 (Standard alternating components, continued). Let's see how the theorems just discussed, starting with Foote's result 1.8.4, essentially complete the treatment of standard subgroups L with $L/Z(L) \cong A_n$ and $n = 5, 6, 7$: First, the subcases in which $L/O_{2'}\big(Z(L)\big) \cong 2A_n$ are covered (since then $z \in Z(L)$) by the Classical Involution Theorem 1.7.5. Second, appealing to the list of Schur multipliers (e.g. A.3.5), we are reduced to the cases where $Z(L)$ has order 1 or 3. So for $n = 6, 7$ we have $|L|_2 = 8$, a case covered in case (a) of Theorem 1.8.5.

In fact the case $n = 5$, where $|L|_2 = 4$, is also covered by 1.8.7 below. (But also a short argument treating the subcase where $L \cong A_5$ and $C_G(L)$ has a cyclic Sylow 2-subgroup of order greater than 2, may be found in [**GLS05**, pp. 94–96]; while when $|C_G(L)|_2 = 2$, Harada's result [**Har75**, Thm 2] applies.) ◇

Since we have treated the subcase of 1.8.3 where $|L|_2 = 8$, we have dealt in particular with the case $L/O_{2'}(L) \cong A_7$ or $L_2(q)$ for *certain* q. The groups $L_2(q)$ with $|L|_2 = 4$ or $|L|_2 > 8$, were included in the following much more general result of Harris[32] in [**Har81b**], which allows $L/O_{2'}(L)$ to be *any* simple group of Lie type of odd characteristic.[33]

[32]This material was also treated independently by Walter; though [**Wal86**] did not appear until somewhat later.

[33]In fact it was a considerable surprise in the field when Harris' initial efforts to treat $L_2(q)$ turned out to require this more general characterization.

THEOREM 1.8.7 (Odd Lie Type Theorem). *Assume (K), with G a finite simple group. Suppose that x is an involution of G and L is a 2-component of $C_G(x)$, with $L/O_{2'}(L)$ a quasisimple group of Lie type defined over a field of odd order. Suppose further that either*

(a) G is a minimal counterexample to (U); or

(b) G is not a counterexample to (U) and L is quasisimple and standard in G.

Then G is one of the following:

(1) A simple group of Lie type defined over a field of odd order; or

(2) A_n, $n = 9$ or 10 with L isomorphic to $L_2(5) \cong A_5$ or to $L_2(9) \cong A_6$; or

(3) M_{12}, J_1, J_2 with $L \cong L_2(5)$, or HS with $L \cong L_2(9)$.

In particular (a) does not hold.

Furthermore this result treats the standard form problems for $L/Z(L)$ a simple group of Lie type in odd characteristic.

The proof relied on the solution of the standard form problems for some other fairly small components: The cases where $L/Z(L)$ is $L_3(3)$ or $U_3(3) \cong G_2(2)'$ were handled by Fritz in [**Fri77a, Fri77b**] when $|G|_2 \leq 2^{10}$, and in general by Harris in [**Har80a, Har81a**]. Most cases where $L/Z(L)$ is an orthogonal group over $\mathbb{F}_3$ of dimension 5 through 8 were handled in a unified way by Aschbacher in [**Asc82a**], although $L \cong 2\Omega_5(3) \cong 2U_4(2)$ was treated by Seitz in [**Sei81**].

Completion of the Unbalanced Group Theorem, and consequences. The result above finishes case (1) of 1.8.3; thus it remains to consider an unbalancing 2-component L in case (2) of 1.8.3, i.e., with $L/Z^*(L) \cong L_3(4)$ or He. Treating these cases again involved the treatment of the corresponding standard form problems. The solution of the standard form problem for He is included in the result 1.8.9 below of Griess-Solomon. The solution of the standard form problem for $L_3(4)$ involved work of various authors—including the relevant parts of Aschbacher-Seitz [**AS76b, AS81**] and Seitz [**Sei79a, Sei79b**]; and the Ph.D. thesis of Kai-Nah Cheng,[34] which was later published in two joint papers [**CH81, CH85**]. with Held.

THEOREM 1.8.8. *Assume that G is finite simple group with a standard component L such that $L/Z(L) \cong L_3(4)$. Then $G \cong Suz$, He, or $O'N$.*

Griess and Solomon in [**GS79b**], making use of the above result and signalizer functor methods, then completed the proof of (U) by eliminating the final case (2) of 1.8.3:

THEOREM 1.8.9. *Assume (K), with G a minimal counterexample to (U). Then G does not contain an unbalancing triple (x, y, L) with $L/Z^*(L) \cong L_3(4)$ or He.*

Furthermore no simple group G contains an involution centralizer with a standard component isomorphic to He.

Thus the Unbalanced Group Theorem 1.7.1 holds.

REMARK 1.8.10. In particular, *we may assume from here on that our simple group G of GW type satisfies $O_{2'}\big(C_G(z)\big) = 1$ for all involutions $z \in G$:* for the simple groups other than $L_3(4)$ in 1.7.1 appear as conclusions of the Main Theorem 1.0.1, while $L_3(4)$ is not of *GW* type. Some of the results that follow, notably 1.10.1, will include the hypothesis that all cores are trivial. ◇

[34] This thesis is mentioned at various points in the discussion of Gorenstein [**Gor83**]—but using Nah rather than Cheng for the surname.

In addition recall that, using [**Gor83**, Prop. 3.26], we get the B-Conjecture 1.6.1, as a corollary of (U). Thus we have proved:

THEOREM 1.8.11 (The B-Theorem). *Let G be a finite group and let H be any 2-local subgroup of G. Then $B(H) \leq B(G)$.*

In particular if $O_{2'}(G) = 1$, then $L_{2'}(C_G(z)) = E(C_G(z))$ for all involutions z of G. Hence a simple group G of GW type and of 2-rank at least 3 has an involution centralizer with a component.

As also noted earlier, one fundamental application of the B-Theorem is to remove the B-Conjecture as a hypothesis in the initial form 1.6.4 of the Standard Component Theorem. In addition part (1) of the Classical Involution Theorem 1.7.5 allows us to sharpen conclusion (2) of 1.6.4:

THEOREM 1.8.12 (The Standard Component Theorem (final form)). *Let G be a finite simple group of GW type of 2-rank at least 3. Then one of the following holds:*

(1) Some involution centralizer has a component L which is standard in G; or

(2) G has a classical involution and G is isomorphic to one of $PSp_4(q)$, $U_4(q)$ or $L_4(q)$, q odd.

Hence the classification of groups of GW type is reduced to the solution of standard form problems.

We have now arrived at the stage in Strategy 1.1.1 mentioned at the end of Remark 1.8.2: The B-Theorem 1.8.11 completes Step 4, and the Standard Component Theorem 1.8.12 finishes Step 5. Thus it remains now only to complete Step 6, by treating the standard form problems. Moreover many standard form problems were already treated in the course of proving 1.7.1. Thus the treatment of the remaining possible standard components is the theme of the next two sections.

1.9. The Odd Standard Component Theorem and the Aschbacher-Seitz reduction

As our discussion of the proof of (U) has emphasized, we have in fact already handled many of the standard form problems. Indeed, we have covered most of the cases where $L/Z(L)$ is of GW type—namely all but $U_3(4)$ and the sporadic groups. (The standard form problems for $U_3(4)$, and for the sporadics other than He will be included among the larger list of remaining cases considered later at 1.10.1).

So we summarize the results[35] obtained so far in the following theorem—where we use "Odd" in the name to indicate that the standard components covered are primarily those in our Odd Case, consisting of the groups of GW type.

[35]The individual cases are covered by 1.7.3, 1.8.1, 1.8.4, 1.8.5, 1.8.7, 1.8.8, and 1.8.9; with the Classical Involution Theorem 1.7.5.1 used as indicated in Remark 1.8.6, and in the form of Corollary 1.7.6 as indicated after 1.8.4.

THEOREM 1.9.1 (Odd Standard Component Theorem). *Assume (K), with G a finite simple group. Suppose that L is standard in G with $L/Z(L)$ isomorphic to one of:*

(a) A_n, $n \geq 5$, or
(b) a simple group of Lie type defined over a finite field of odd order, or
(c) $L_3(4)$ or He.

Then one of the following conclusions holds:

(1) $G \cong A_n$, $n \geq 9$; or
(2) G is a simple group of Lie type defined over a finite field of odd order; or
(3) $G \cong M_{12}$, J_1, J_2, HS, McL, Ly, Suz, He, or $O'N$.

REMARK 1.9.2. This result includes a number of cases (notably $L_3(4)$ in (c)) where $L/Z(L)$ is not of GW type. In particular, cases such as $L_3(4)$ (as well as He) need not be treated again later at 1.10.1, when we consider cases which are either sporadic or of Lie type in characteristic 2. Likewise, several groups with multiple identities as indicated in A.3.6, such as $U_4(2) \cong PSp_4(3)$ and $L_4(2) \cong A_8$, need not be treated again. ◊

So the classification of groups of GW type is reduced to standard form problems for $L/Z(L)$ not appearing in the list of the Odd Standard Component Theorem 1.9.1. We sketch the treatment of the remaining problems in this section and the next; for further detail, see Chapter 5 of Gorenstein [**Gor83**].

Aschbacher in [**Asc75a**] had established strong restrictions on the structure of $C_G(L)$ when L is standard in G (recall the restrictions on 2-rank mentioned in Remark 1.6.5). Using these results and his earlier work [**Asc08**] on standard components of alternating type, he and Seitz in [**AS76b, AS81**] established a further major reduction—namely to the case where $C_G(L)$ is of 2-rank 1; in fact they showed that when $C_G(L)$ is of 2-rank at least 2, the only "new" groups not already appearing in 1.9.1 are Ru and Co_1:

THEOREM 1.9.3 (Aschbacher-Seitz Reduction Theorem). *Let G be a finite simple group and suppose that L is standard in G with $C_G(L)$ of 2-rank at least 2. Then one of the following holds:*

(1) $G \cong A_n$, $n \geq 9$; or
(2) $G \cong J_2$, Suz, He, Ru, or Co_1.

Their proof relied on Aschbacher's general results on tightly embedded subgroups, combined with their detailed study of the centralizers of involutory automorphisms of finite groups of Lie type defined over fields of characteristic 2, in [**AS76a**].[36]

It also relied on the solution of certain standard form problems: the standard form problem for $L/Z(L) \cong L_3(4)$, leading to $G \cong Suz$ and He, in 1.8.8; the problem for A_n by Aschbacher [**Asc08**], which we also mentioned earlier; and those for $L/Z(L)$ of Lie type and characteristic 2 with BN-rank 1 (but not $L_2(4)$), which were treated by Griess, D. Mason, and Seitz in [**GMS78**]. They showed:

THEOREM 1.9.4. *Suppose G is a simple group with a standard component L such that $L/Z(L) \cong L_2(2^m)$, $Sz(2^{2m-1})$, or $U_3(2^m)$, $m \geq 2$, but $L/Z(L) \not\cong L_2(4)$. Then $L \cong Sz(8)$ and $G \cong Ru$.*

---

[36]The short note [**AS78**] just corrects some minor errors in the original [**AS76a**], notably in 14.2(iii) and 14.3(iii).

This will be a convenient place to summarize our various reductions so far.

Indeed our work may seem to be almost complete—in the sense that most of the groups in the conclusion of the Main Theorem 1.0.1 have already arisen: In conclusions (1)–(4) of the summary below, we list all the groups of *GW* type which have appeared in earlier results;[37] and this leaves only the seven sporadic groups appearing in the conclusion of later 1.10.1. Unfortunately, much more argument was still needed to finish—as we will see in the following section.

Conclusion (5) in the summary below arises from a brief examination of the Aschbacher-Seitz reduction 1.9.3 to $C_G(L)$ of 2-rank 1: For the Sylow 2-subgroups of $C_G(L)$ are then by A.1.7 either cyclic or generalized quaternion. But in the generalized quaternion case, we may apply the corollary 1.7.6 of the Classical Involution Theorem to identify G as being in conclusion (2) below. So we may assume that $C_G(L)$ has cyclic Sylow 2-subgroups, so that $C_G(L)$ is 2-nilpotent by 1.4.1. Indeed if L is a component of $C_G(x)$ for an involution x, then the subgroup $C := C_{C_G(x)}(L)$ of $C_G(L)$ is also 2-nilpotent. It follows, first, that L is the unique component of $C_G(x)$, and so is normal in $C_G(x)$; and, second, that the normal 2-complement $O_{2'}(C)$ lies in $O_{2'}\big(C_G(x)\big)$, since C is then also normal in $C_G(x)$. But we saw in Remark 1.8.10 that in view of the Unbalanced Group Theorem 1.7.1, we may assume that $O_{2'}\big(C_G(x)\big) = 1$, so that C is a cyclic 2-group. Of course $C_G(x)/C$ is isomorphic to a subgroup of $Aut(L)$.

Thus we obtain:

COROLLARY 1.9.5. *Let G be a finite simple group of GW type. Then one of the following conclusions holds:*

(1) $G \cong A_n$, $n \geq 5$; or

(2) G is a simple group of Lie type defined over a finite field of odd order; or

(3) G is $U_3(4)$; or

(4) $G \cong M_{11}$, M_{12}, J_1, J_2, HS, McL, Ly, Suz, He, Ru, $O'N$, or Co_1; or

(5) There exists an involution $x \in G$ and a component L of $C_G(x)$ such that $C_{C_G(x)}(L)$ is a cyclic 2-group, and $L/Z(L)$ is either a sporadic simple group, or a simple group of Lie type defined over a finite field of characteristic 2.

For most choices of L in (5) above, it was easy to show that $C = C_{C_G(x)}(L)$ is just $\langle x \rangle$, using a criterion of Finkelstein [**Fin77c**]; hence in those cases we see that $|Z(L)| = 1$ or 2 (according to whether $x \notin L$ or $x \in L$). So this left a fairly precise list of involution centralizer problems to tackle.

1.10. The Even Standard Component Theorem

During our discussion, we will informally describe the list of components remaining to be treated as constituting the Even Case—in contrast with those in 1.9.1, which are primarily of *GW*-type, and hence constitute the Odd Case.

In fact in (3) below we pare down the list of possibilities for $L/Z(L)$, indicated in case (5) of Corollary 1.9.5, a bit further, using (c) of 1.9.1, and Remark 1.9.2. Hence the classification 1.0.1 of finite simple groups of *GW* type will be completed by establishing the following result. (Again we recall Remark 1.8.10 that the Unbalanced

[37]For this list, in view of the Standard Component Theorem 1.8.12 it suffices to examine just the summary standard-form result 1.9.1 along with 1.9.3, as well as the earlier 2-Rank 2 Theorem 1.4.6.

Group Theorem 1.7.1 makes available the hypothesis below that $O_{2'}\big(C_G(z)\big) = 1$—and this in turn implies that $Z(L)$ is a 2-group.)

THEOREM 1.10.1 (Even Standard Component Theorem). *Let G be a finite simple group such that $O_{2'}\big(C_G(z)\big) = 1$ for all involutions z of G. Suppose that L is standard in G and satisfies the following conditions:*

(1) $L/Z(L)$ is either a sporadic group or a group of Lie type over a field of characteristic 2;

(2) $C_G(L)$ has cyclic Sylow 2-subgroups; and

(3) $L/Z(L)$ is not isomorphic to $L_2(4)$, $L_3(2)$, $L_4(2)$, $Sp_4(2)'$, $G_2(2)'$, $U_4(2)$, $L_3(4)$, or He.

Then G is isomorphic to Co_3, HN, Fi_{22}, Fi_{23}, Fi'_{24}, BM or M.

We complete our exposition by indicating a few of the intermediate results used in proving this theorem.

The principal reduction was achieved by Seitz in the papers [**Sei79a, Sei79b**]. We state the reduction below under Hypothesis (K), although it actually only relies on certain other standard form problems. In effect, Seitz considered the subcase of 1.10.1 where $Z(L) = 1$ and L is of Lie type; and he eliminated all cases of BN-rank rank greater than 2 with the exception of the four listed in conclusion (3). Note that the restriction in conclusion (2) to BN-rank exactly 2 is implicitly invoking the earlier Griess-Mason-Seitz Theorem 1.9.4, which treated the BN-rank 1 case. (In particular note that the centralizer of the rank-1 group $Sz(8)$ in Ru has 2-rank 2, so that Ru does not appear as a conclusion in the main theorem 1.10.1.)

THEOREM 1.10.2 (Seitz Reduction Theorem). *Assume (K), with G a finite simple group having a subgroup L which is standard in G. Assume that L is simple and is described in the hypothesis of 1.10.1. Assume further that $C_G(L)$ has cyclic Sylow 2-subgroups. Then one of the following holds:*

(1) L is a sporadic group; or

(2) L is a group of Lie type of BN-rank exactly 2, defined over a field of characteristic 2; or

(3) $L \cong Sp_6(2)$, $U_6(2)$, $\Omega_8^-(2)$, or $\Omega_8^+(2)$.

So for $L/Z(L)$ as described in 1.10.1, it remains to treat the small list of cases in 1.10.2 (which have $Z(L) = 1$) as well as cases in which $Z(L)$ is a nontrivial 2-group—so that the Schur multiplier of $L/Z(L)$ has a nontrivial 2-part, which also leads to a fairly short list of possibilities.

Furthermore the cases where $L/Z(L)$ is of BN-rank exactly 2,[38] whether $Z(L)$ is trivial or not, were mostly handled in a collection of papers by Seitz [**Sei77**], Gomi [**Gom78b, Gom78a**], I. Miyamoto [**Miy79, Miy80, Miy82**], and Yamada [**Yam79a, Yam79b, Yam79c, Yam85**].[39] The case where L is the Tits group ${}^2F_4(2)'$ was handled by Aschbacher in [**Asc82b**]. Their results can be summarized as eliminating the subcase where L is simple:

---

[38]We mention that Gorenstein at [**Gor83**, (vi), bottom p. 336] indicates a possibility of $L \cong F_4(2)$ in $G \cong BM$—but in fact, that component $F_4(2)$ turns out to be *not* standard in BM.

[39]The bibliography in Gorenstein [**Gor83**] indicated an expected journal for the preprint of Yamada; but the work was eventually published in a different journal, as [**Yam85**].

THEOREM 1.10.3. *Let G be a finite simple group having a standard component L such that $L/Z(L)$ is a simple group of Lie type of BN-rank 2 defined over a finite field of characteristic 2. Then $|Z(L)|$ is even.*

So the cases in 1.10.1 involving groups of Lie type are reduced to: the four simple cases in part (3) of 1.10.2; along with those cases where $L/Z(L)$ is of BN-rank at least 2 with a nontrivial 2-multiplier.

In fact the possible nontrivial 2-groups $Z(L)$ are all "exceptional" multipliers in the tables of [**GLS98**, Sec 6.1]. Examination of those tables shows that the Even Standard Component Theorem 1.10.1 has now been reduced to the following list of cases:

(1) $L/Z(L)$ is one of the 26 sporadic simple groups (indeed we may omit He—cf. 1.10.1.3); or

(2) $|Z(L)| = 2$ and $L/Z(L) \cong G_2(4)$, $F_4(2)$, or ${}^2E_6(2)$; or

(3) $L/Z(L) \cong Sp_6(2)$, $U_6(2)$, or $\Omega_8^+(2)$, with $Z(L)$ a 2-group (possibly trivial); or $L \cong \Omega_8^-(2)$.

Thus there are $25 + 3 + 4 = 32$ remaining cases for $L/Z(L)$, with some further subcases corresponding to the 2-subgroups of $Z(L)$. In the next few pages, we provide tables (with some updates of Sections 5.11 and 5.12 of Gorenstein [**Gor83**]),[40] indicating the papers in which these cases are handled.

The argument in each of the indicated papers proceeds from the structure of L to some final identification of G—this last step is typically achieved via an appeal to some existing recognition theorem for G, quoted from some other paper in the literature. We do *not* attempt to record here the recognition theorems involved, since the more detailed discussion in Gorenstein [**Gor83**] usually supplies that information.

[40]We mention that in Gorenstein [**Gor83**], the possibility of HN was inadvertently omitted from the list in Prop 5.155 on p. 470—but notice that it does then appear (in the alternative notation of F_5) in the relevant discussion on p. 471.

REMARK 1.10.4 (Solutions of standard sporadic component problems). Though the case of $L \cong He$ is not required in (1) above, for completeness of the list of sporadics we do mention it below (in parentheses).

std L	simple G?	where handled
M_{11}		covered by Sectional 2-Rank 4 Theorem 1.5.7
M_{12}	Co_3	Finkelstein-Solomon [**FS79a**]
$2M_{12}$		Finkelstein-Solomon [**FS79a**]; + Harada-Solomon [**HS08**]
M_{22}		Finkelstein [**Fin77a**]
$2M_{22}$		Harada, unpublished [**Har**]; Harada-Solomon [**HS08**]
$4M_{22}$		Griess, unpublished [**Gri**]; Harada-Solomon [**HS08**]
M_{23}		Finkelstein [**Fin76a**]
M_{24}		Egawa [**Ega81**]
J_1		Finkelstein [**Fin75**]
J_2		Finkelstein [**Fin76b**]
$2J_2$		Finkelstein-Solomon [**FS79a**]
J_3		Finkelstein [**Fin76b**]
J_4		Finkelstein [**Fin77b**]
HS		Solomon [**Sol77**]
$2HS$	HN	Solomon [**Sol77**]
McL		Solomon [**Sol75**]
Ly		Solomon [**Sol75**]
Sz		Solomon [**Sol78b**]
$2Suz$		Finkelstein-Solomon [**FS79a**]
(He)		(Griess-Solomon [**GS79b**])
Ru		Solomon [**Sol78b**]
$2Ru$		Finkelstein [**Fin77c**]
ON		Solomon [**Sol78b**]
Co_3		Finkelstein-Solomon [**FS79a**]
Co_2		Manferdelli, unpublished Ph.D. thesis [**Man79**]
Co_1		Solomon [**Sol78b**]
$2Co_1$		Finkelstein [**Fin77c**]
Fi_{22}		Solomon [**Sol78b**]
Fi_{23}		Solomon [**Sol78b**]
Fi'_{24}		Solomon [**Sol78b**]
$2Fi_{22}$	Fi_{23}, Fi'_{24}	Davis-Solomon [**DS81**]
HN		Solomon [**Sol78b**]
Th		Syskin [**Sys81**]
BM		Solomon [**Sol78b**]
$2BM$	M	Davis-Solomon [**DS81**]
M		Solomon [**Sol78b**]

REMARK 1.10.5 (Solution of standard small Even Case component problems). Finally we indicate the remaining components of Lie type in characteristic 2:[41]

standard L	in simple G?	where handled
$2G_2(4)$		Seitz [**Sei81**]
$2F_4(2)$		Seitz [**Sei81**]
$2\cdot {}^2E_6(2)$	BM	Stroth [**Str81**]
$Sp_6(2)$		Gomi [**Gom80**]
$2Sp_6(2)$	Co_3	Seitz [**Sei81**]
$U_6(2)$		Yamada [**Yam79d**]
$2U_6(2)$	Fi_{22}	Davis-Solomon [**DS81**]
$\Omega_8^+(2)$		Egawa, unpublished Ph.D. thesis [**Ega80**]
$2\Omega_8^+(2)$		Egawa-Yoshida [**EY82**]
$\Omega_8^-(2)$		Alward, unpublished Ph.D. thesis [**Alw79**]

Summary: Statements of the major subtheorems

The first two results cover case (1) of Definition 0.3.9 of Gorenstein-Walter type; namely they classify simple groups of 2-rank at most 2:

The Odd Order Theorem 1.2.1 Let G be a finite simple group of 2-rank 0. Then G is cyclic of odd prime order.

The 2-Rank 2 Theorem 1.4.6 Let G be a finite simple group having 2-rank 1 or 2. Then either G is cyclic of order 2; or G is isomorphic to $L_2(q)$, $L_3(q)$, or $U_3(q)$, q odd; or $G \cong A_7$, M_{11}, or $U_3(4)$.

Thus from now on we can assume that G has 2-rank at least 3, and appears in case (2) of Definition 0.3.9 of Gorenstein-Walter type; thus $L_{2'}(C_G(z)) \neq 1$ for some involution z of G—so that $C_G(z)$ has a 2-component, and G is of component type.

The hypotheses of the next two results do not demand that the groups considered be of GW type, but their conclusions involve groups of GW type, and the results were crucial in completing various cases in the classification of groups of GW type.

The Classical Involution Theorem 1.7.5 Assume that G is a finite group with $F^*(G)$ simple, containing an involution z such that z lies in a subnormal subgroup J of $C_G(z)$ having a generalized quaternion Sylow 2-subgroup.

(1) If z is a classical involution (i.e., if $J/O_{2'}(J) \cong 2A_7$ or $SL_2(q)$ for odd q), then $F^*(G)$ is either a group of Lie type in odd characteristic, or M_{11}. (In view of 1.4.4, this further hypothesis holds whenever the subnormal subgroup J of 2-rank 1 has $J/O_{2'}(J)$ quasisimple.)

(2) If J is tightly embedded in G, then $F^*(G)$ is either a group of Lie type in odd characteristic, or M_{11}, or M_{12}. (In particular, this further hypothesis holds whenever $J = C_G(L)$ with L standard in G.)

[41] We mention that Table 5.7 on p. 355 of Gorenstein [**Gor83**] inadvertently omits $G \cong Co_3$ in the case $L \cong 2Sp_6(2)$—but note that this case does appear later in the treatment at Prop 5.145 on p. 453.

The Unbalanced Group Theorem 1.7.1 Let G be a finite group with $F^*(G)$ simple, having an involution z such that $O_{2'}(C_G(z)) \neq 1$. Then one of the following conclusions holds:

(1) $F^*(G)$ is of Lie type over a field of odd order (and most such groups do arise here); or

(2) $F^*(G)$ is an alternating group of degree $n \geq 7$, n odd; or

(3) $F^*(G) \cong L_3(4)$ or He.

In particular, we may assume from now on that $O_{2'}(C_G(z)) = 1$ for involutions z of G.

The Unbalanced Group Theorem has the following corollary:

The B-Theorem 1.8.11 Let G be a finite group and let H be any 2-local subgroup of G. Then $B(H) \leq B(G)$.

In particular if $O_{2'}(G) = 1$, then $L_{2'}(C_G(z)) = E(C_G(z))$ for all involutions z of G. Hence a simple group G of GW type and of 2-rank at least 3 has an involution centralizer with a component.

The B-Theorem was in turn used by Aschbacher to reduce the classification of groups of GW type to standard form problems:

Standard Component Theorem 1.8.12 Let G be a finite simple group of Gorenstein-Walter type of 2-rank at least 3. Then one of the following holds:

(1) Some involution centralizer has a component L which is standard in G; or

(2) G has a classical involution and is isomorphic to one of $PSp_4(q)$, $U_4(q)$ or $L_4(q)$, q odd.

Thus we may henceforth assume that for some involution $z \in G$, $C_G(z)$ has a component L which is standard in G.

In fact, some of the results established during the proof of the Unbalanced Group Theorem had also treated standard form problems in which $L/Z(L)$ is of GW type (plus some others):

Odd Standard Component Theorem 1.9.1 Assume (K). Let G be a finite simple group. Suppose that L is standard in G with $L/Z(L)$ isomorphic to one of:

(a) A_n, $n \geq 5$, or

(b) a simple group of Lie type defined over a finite field of odd order, or

(c) $L_3(4)$ or He.

Then one of the following conclusions holds:

(1) $G \cong A_n$, $n \geq 9$; or

(2) G is a simple group of Lie type defined over a finite field of odd order; or

(3) $G \cong M_{12}$, J_1, J_2, HS, McL, Ly, Suz, He, or $O'N$.

Aschbacher and Seitz showed (for general L) that if $C_G(L)$ has 2-rank at least 2, then the only groups not arising in previous results on GW type are Ru and Co_1:

Aschbacher-Seitz Reduction Theorem 1.9.3 Let G be a finite simple group and suppose that L is standard in G with $C_G(L)$ of 2-rank at least 2. Then one of the following holds:

(1) $G \cong A_n$, $n \geq 9$; or

(2) $G \cong J_2$, Suz, He, Ru, or Co_1.

Thus it remains to consider the cases where $C_G(L)$ is of 2-rank 1—and $L/Z(L)$ is either sporadic or of Lie type in characteristic 2 (where a few of these cases are already covered by previous results). In view of A.1.7, we may apply Corollary 1.7.6 of the Classical Involution Theorem to further reduce to the case where $C_G(L)$ has cyclic Sylow 2-subgroups. We also add the hypothesis that involution centralizers have trivial cores, available in view of the Unbalanced Group Theorem 1.7.1. So the final step in the proof of the Main Theorem 1.0.1 is provided by the following result, due to Seitz and various other authors:

Even Standard Component Theorem 1.10.1 Let G be a finite simple group such that $O_{2'}(C_G(z)) = 1$ for all involutions z of G. Suppose that L is standard in G and satisfies the following conditions:

(1) $L/Z(L)$ is either a sporadic group; or a group of Lie type over a field of characteristic 2;

(2) $C_G(L)$ has cyclic Sylow 2-subgroups; and

(3) $L/Z(L)$ is not isomorphic to $L_2(4)$, $L_3(2)$, $L_4(2)$, $Sp_4(2)'$, $G_2(2)'$, $U_4(2)$, $L_3(4)$, or He.

Then G is isomorphic to Co_3, HN, Fi_{22}, Fi_{23}, Fi'_{24}, BM or M.

Finally we mention two important results at level 1—which were not about groups of GW type per se, but which were used at various places in the classification of groups of GW type (and indeed elsewhere in the CFSG).

The first is a major result of Gorenstein and Harada:

The Sectional 2-Rank 4 Theorem 1.5.7 Let G be a finite simple group of 2-rank at least 3 and sectional 2-rank at most 4. Then G is isomorphic to one of the following:

(1) ${}^2G_2(3^{2n+1})$, $G_2(q)$, ${}^3D_4(q)$, $PSp_4(q)$, q odd; $L_4^\epsilon(q)$, q odd, $q \not\equiv \epsilon \pmod 8$; $L_5^\epsilon(q)$, q odd, $q \equiv -\epsilon \pmod 4$; or

(2) $L_2(8)$, $L_2(16)$, $L_3(4)$, or $Sz(8)$; or

(3) A_n, $8 \le n \le 11$; or

(4) M_{12}, M_{22}, M_{23}, J_1, J_2, J_3, McL, or Ly.

(Of course the larger list of all groups of sectional 2-rank at most 4 is obtained by adding those from the 2-Rank 2 Theorem 1.4.6.)

The second theorem combines the first with results of Bender and Aschbacher, describing when the commuting graph on four-subgroups can be disconnected:

The Proper 2-Generated Core Theorem 1.5.10 Let G be a finite simple group with a Sylow 2-subgroup S of 2-rank at least 3, such that $\Gamma_{2,S}(G) < G$. Then either $\Gamma_{1,S}(G) < G$ (so that G is described in 1.3.5), or $G \cong J_1$.

In particular if G is simple of 2-rank at least 3 and $\Lambda_2(G)$ is disconnected, then either $\Lambda_1(G)$ is disconnected or $G \cong J_1$.

CHAPTER 2

Overview: The classification of groups of characteristic 2 type

Recall that the Dichotomy Theorem 0.3.10 divides the proof of the CFSG into two major subcases, which we have been informally referring to as the Odd and Even Cases.

In the previous chapter we gave an overview of the Odd Case, where the groups of Gorenstein-Walter (GW) type are classified. This chapter provides a similar overview of the treatment of the groups of characteristic 2 type. However here, for the most part, we will restrict attention to the level 0 results whose union gives the Even Case classification; each of those major subdivisions of the proof will be discussed later in greater detail in a corresponding chapter, at which point we will also indicate some level 1 results.

So to begin: In view of Theorem 1.0.1, which classifies groups of GW type, and the Dichotomy Theorem 0.3.10, *from here on any simple group G can be assumed to be of characteristic* 2 *type of* 2*-rank at least* 3.

This is essentially the "Conclusion" that Gorenstein records at the end of Volume 1, see p. 475 of [**Gor83**]; that statement also includes the inductive K-group hypothesis 0.5.4, which will also be assumed in most of our discussion from here on. We repeat it here for convenience.

K-Group Hypothesis 0.5.4 Assume that all simple sections of proper subgroups of G are known simple groups, i.e., are on the list 0.1.1 of conclusions of the Classification of the Finite Simple Groups. (When this hypothesis holds, we say that G is *K-proper*.) ◊

The Main Theorem for groups of characteristic 2 type

We recall from Definition 0.3.5 that a finite group G is of *characteristic* 2 *type* if $F^*(N) = O_2(N)$ for each 2-local subgroup N of G. As we indicated just before that definition, this is a fundamental property of the groups of Lie type over fields of characteristic 2. Moreover these groups appear as the generic conclusion (1) in the following classification of groups of characteristic 2 type and 2-rank at least 3.

THEOREM 2.0.1 (**Main Theorem for groups of characteristic 2 type**). *Let G be a finite K-proper simple group of characteristic 2 type such that G has 2-rank at least 3. Then one of the following conclusions holds:*
(1) G is a simple group of Lie type,[1] defined over a field of characteristic 2; or
(2) G is isomorphic to $G_2(3)$ or $U_4(3)$; or
(3) G is one of the following seven sporadic simple groups:

$$M_{22}, M_{23}, M_{24}, J_3, J_4, Co_2, Th.$$

We record a few observations about the boundary between the groups of characteristic 2 type in Theorem 2.0.1, and the groups of GW type in Theorem 1.0.1. (See also the discussion of exceptional isomorphisms among such groups at Remark A.3.6.) As we are assuming that G has 2-rank at least 3, a number of smaller groups which are of characteristic 2 type and 2-rank 2 (for example M_{11}), appear instead on the list of groups of GW type. On the other hand, conclusions (2) and (3) of 1.0.1 excluded four groups, which now appear on the list of groups of characteristic 2 type: two of them are listed explicitly in conclusion (2), while A_8 and $PSp_4(3)$ appear implicitly in conclusion (1) as $L_4(2)$ and $U_4(2)$.

In the remainder of the chapter, we present an introductory overview of the proof of the classification in Theorem 2.0.1 of the groups of characteristic 2 type with 2-rank at least 3.

As in the previous chapter, we begin with an informal summary of the main stages of the proof. Again the summary reflects essentially the historical subdivision of the proof. We postpone to the later sections a discussion of how this organization corresponds to Strategy 1.1.1, which emphasizes parallels with the treatment of groups of GW type.

In contrast to the various unexpected twists which arose during the evolution of the proof of the classification of the groups of GW type, described in the previous chapter, the overall shape of the proof of the theorem classifying the groups of characteristic 2 type was comparatively well understood in advance. For example, the brief outline supplied below is essentially contained in Gorenstein's 16-Step Program announced in his (unpublished) University of Chicago lectures in 1972.

Here are the major stages in the classification of the groups of characteristic 2 type:

OUTLINE 2.0.2 (Summary of the characteristic 2 type classification).
I. (small case: $e(G) \leq 2$) The Quasithin Theorem 3.0.1
II. (generic case: $e(G) \geq 3$)
  A. The $e(G) = 3$ Theorem 4.0.1
  B. The Trichotomy Theorem 5.1.2 for $e(G) \geq 4$.
  The three branches of the trichotomy were handled by:
    i. The Standard Type Theorem 6.1.1
    ii. The $GF(2)$ Type Theorem 7.0.1
    iii. The Uniqueness Case Theorem 8.0.2 ◇

[1]The full list of groups of Lie type arising here can be obtained from Remark A.2.1. As the groups are defined over fields of characteristic 2, groups of type ${}^2G_2$ do not arise, and the groups of type PSp_{2n} and $P\Omega_{2n+1}$ are isomorphic.

Chapters 3–8 contain a more detailed discussion of each of these six results. In the remainder of this chapter, we will state the results and introduce related concepts and notation.

As our standard example of a group of characteristic 2 type, we take the special linear groups $SL_n(q)$ for $q = 2^m$ (cf. Example 0.3.3). This group is of 2-rank at least 3 whenever $m + n > 4$.

2.1. The Quasithin Theorem covering $e(G) \leq 2$

As we indicated after Definition 0.5.2, the quasithin groups (that is, groups with $e(G) \leq 2$) arose naturally as a small case in Thompson's early work on N-groups [**Tho68**]. Later they came to constitute the groups in the Small Even Subcase of the eventual CFSG.

We also recall that for a group G of Lie type in characteristic 2,[2] the value of $e(G)$ is a good approximation to the BN-rank of G. Thus the main "target" groups in the study of groups with $e(G) \leq 2$ are the groups of BN-rank at most 2. In particular, among our standard examples, $SL_2(2^m)$ and $SL_3(2^m)$ are the special linear groups of BN-rank at most 2.

Historical remarks. Some pioneering efforts in the 1960s and 1970s, notably by Janko and Thompson, were devoted to the very smallest case given by the thin groups—where $e(G) \leq 1$. The "generic" thin groups are the groups of Lie type of BN-rank 1 over fields of characteristic 2, i.e. the Bender groups discussed earlier in Section 1.3. The heart of the problem, however, was the recognition of certain rank 2 groups defined over $GF(2)$. The full classification of the simple groups with $e(G) \leq 1$ was achieved by Aschbacher in [**Asc78b**].

Using similar methods, Mason undertook the study of groups with $e(G) \leq 2$. He produced a typescript [**Mas**] of about 800 pages, addressing much of the problem; but unfortunately this work was never completed or published.

Although the other parts of the CFSG had appeared by the early 1980s, the case involving the quasithin groups remained unresolved for many years thereafter. Most group theorists believed, at least initially, that the problem had been essentially handled by Mason, and would in due course be published. In addition it was hoped that a more elementary treatment of the quasithin case might emerge from the amalgam methods[3] which were being developed at around that time.

But as the years passed, the gap in the proof involving the quasithin groups became increasingly problematic—especially for researchers in various areas who quoted (or wished to quote) the CFSG.

In mid-1995, Aschbacher and Smith resolved to close the gap; and in early 1996, they started work on a new treatment of the quasithin problem. The project required a concerted effort over a period of years, with the result finally appearing in late 2004 in the volumes [**AS04a, AS04b**]. The actual proof is contained in the second volume [**AS04b**], of about 700 pages. The length of this volume is primarily a consequence of the large number of cases that must be handled. But the treatment also reflected several other factors, such as the desire to provide more details in work designed to close a notorious gap in the proof of this important theorem (especially after the length of time elapsed since the more active days of the CFSG).

---

[2] Except over the smallest field $\mathbb{F}_2$, in which case the diagonal subgroup is trivial.

[3] Cf. our discussion in the latter part of Section 3.3.

In particular, in order to make the overall treatment reasonably self-contained, a considerable amount of preliminary theory was developed independently in the first volume [**AS04a**] of about 500 pages, rather than just quoting from the rather scattered literature.

The treatment of the quasithin problem. In part to accommodate the GLS notion of groups of even type (cf. the discussion of later Theorem 3.12.1), and in part as an experiment to see how far certain techniques could be pushed, Aschbacher and Smith chose to consider a class of groups larger than the class of groups of characteristic 2 type from the original CFSG. Thus they replaced the condition in Definition 0.3.5 on all 2-locals, with the same restriction on only those 2-locals containing a full Sylow 2-subgroup:

DEFINITION 2.1.1 (Even characteristic). A group G is of *even characteristic* if we have:
$F^*(N) = O_2(N)$ for each 2-local subgroup N of G satisfying $|G : N|_2 = 1$.
And G is a *QTKE-group* if it is quasithin, K-proper, and of even characteristic. $\diamondsuit$

We now state the main result of [**AS04b**]. Note that the expected examples with BN-rank at most 2 appear in conclusion (1)—which is called the "generic case" in that work.

Quasithin Theorem 3.0.1 The finite nonabelian simple QTKE-groups are:
(1) (Generic case) Groups of Lie type of characteristic 2 and BN-rank at most 2, but $U_5(q)$ only for $q = 4$.
(2) (Certain groups of rank 3 or 4) $L_4(2)$, $L_5(2)$, $Sp_6(2)$.
(3) (Small alternating groups) A_5, A_6, A_8, A_9.
(4) (Small groups of odd Lie type) $L_2(p)$, p a Mersenne or Fermat prime, $L_3(3)$, $U_3(3)$, $L_4(3)$, $U_4(3)$, $G_2(3)$.
(5) (Sporadic) M_{11}, M_{12}, M_{22}, M_{23}, M_{24}, J_2, J_3, J_4, HS, He, Ru.

To apply this result to complete the proof of the original CFSG, one simply eliminates, from the list of groups in the Quasithin Theorem, those groups which are not of characteristic 2 type and 2-rank at least 3. The groups in (1) and (2), as well as the group $A_8 \cong L_4(2)$ in (3), appear in conclusion (1) of 2.0.1. Two of the groups in case (4) of the Quasithin Theorem appear in case (2) of 2.0.1, and the groups in case (5) of characteristic 2 type form the sublist M_{22}, M_{23}, M_{24}, J_3, J_4 in conclusion (3) of Theorem 2.0.1.

The fact that the Quasithin Theorem is a result about groups of even characteristic, rather than about groups of characteristic 2 type, makes possible an application to the corresponding quasithin problem in the GLS program: Namely the Even Type (Quasithin) Theorem in [**AS04b**, Ch. 16] shows that if one weakens the hypothesis of even characteristic to the GLS condition of even type, then the only further simple group to arise is J_1.

A more detailed discussion of the proof of the Quasithin Theorem appears in Chapter 3.

2.2. The trichotomy approach to treating $e(G) \geq 3$

We move next to the generic case ($e(G) \geq 3$) of the classification of groups of characteristic 2 type. The approach in that case was roughly analogous to the treatment of the groups of GW type in Chapter 1; Strategy 1.1.1 supplies a point of view from which to describe that analogy.

Recall that in applying Strategy 1.1.1 in the classification of groups of GW type, we focused on the centralizer of an involution, and on a "standard form" for certain centralizers. The generic class of examples in that case consisted of the groups of Lie type over fields of odd characteristic. In the case of the groups of characteristic 2 type, generic examples are the groups of Lie type over fields of characteristic 2. Hence the appropriate analogue of an involution is an element of *odd* prime order p. Once again we wish to show that the centralizer of some such element is in "standard form".

The treatment of quasithin groups has dealt with the small groups in Step 2 in Strategy 1.1.1. Now under the assumption that $e(G) \geq 3$, it is possible to define appropriate signalizer functors on elementary abelian p-subgroups of 2-local subgroups of G.

Recall by way of comparison that in the Odd Case, the signalizer analysis led to a trichotomy—indeed we saw that Theorem 0.3.10 could instead have been stated in the form:

GW-type Trichotomy Theorem 1.3.10 Let G be a finite simple group of 2-rank at least 3. Then one of the following holds:

(I) G is of characteristic 2 type (of 2-rank at least 3); or

(II) G is of 2-component type; i.e. for some involution $z \in G$, $L_{2'}(C_G(z)) \neq 1$.

(III) G is of 2-uniqueness type (i.e., G has a strongly embedded subgroup).

This theorem is usually stated instead in the form of the Dichotomy Theorem 0.3.10, with conclusion (III) omitted—since the Bender-Suzuki 2-Uniqueness Theorem 1.3.5 had already been proved by the time the Dichotomy Theorem was formulated by Gorenstein, and showed that all groups of 2-uniqueness type are in fact of characteristic 2 type. That is, (III) is subsumed in (I).

By analogy, similar signalizer analysis[4] yields the following trichotomy theorem in the Even Case. It is weaker than the result in the Odd Case in the sense that case (III) below involves a *pre*uniqueness condition like that in Definition 1.5.4, rather than the full Uniqueness Case of Definition 2.7.2.

THEOREM 2.2.1 (Weak Trichotomy Theorem (for groups of Characteristic 2 type)). *Let G be a finite simple group of characteristic 2 type with $e(G) \geq 3$. Then there exists an odd prime p with $m_{2,p}(G) \geq 3$ such that one of the following holds:*

(I) G is of characteristic p type (of p-rank at least 3); or

(II) G is of p-component type, i.e. for some $t \in G$ of order p, $L_{p'}(C_G(t)) \neq 1$; or

(III) G is of p-preuniqueness type (i.e. for $P \in Syl_p(G)$, $\Gamma^0_{2,P}(G)$ lies in a 2-local subgroup of G.)

For the case $e(G) \geq 4$, a much stronger version appears as the Trichotomy Theorem 5.1.2 of Gorenstein-Lyons in [**GL83**], which we begin to discuss in Section 2.3. A strong trichotomy is also implicit in Aschbacher's treatment of the case $e(G) = 3$ (as we indicate in Section 2.4).

[4]We do not know if this Weak Trichotomy Theorem has been explicitly stated in the literature. It could be regarded as a special case of the stronger trichotomy theorems mentioned below, requiring only fairly elementary parts of their proofs.

The treatment of the uniqueness branch (III) presents some contrasts with the Odd Case: In fact there are *no* finite simple groups G of characteristic 2 type with $e(G) \geq 3$, such that G is of p-preuniqueness type for all odd primes p with $m_{2,p}(G) \geq 3$. However this result was only established (by work of Aschbacher, Gorenstein, and Lyons which we describe in Chapter 8) *after* the Gorenstein-Lyons Trichotomy Theorem. Hence it appears as an independent branch of the trichotomy, which was not subsumed into any other branch. The resulting difficulties, in the treatment of the Uniqueness Case in Chapter 8, arise essentially because there is no analogue for odd p (independent of the Classification itself) of the Bender-Suzuki Strongly Embedded Theorem 1.3.5, to list the groups with a strongly p-embedded subgroup.

REMARK 2.2.2 (The Klinger-Mason argument). On the other hand, branch (I) had in effect been eliminated before the Trichotomy Theorem by Klinger and G. Mason, who extended methods introduced by Thompson in his N-Group Paper: The Klinger-Mason Dichotomy (see B.9.1) shows that if G is a finite simple group which is both of characteristic 2 type and of characteristic p type, then $m_{2,p}(G) \leq 2$—and so $e(G) \leq 2$ if (I) holds for all odd p, reducing to the small case.

Moreover at an early stage in the proof of the Klinger-Mason Dichotomy, it emerges that if G is of characteristic $\{2, p\}$ type with $m_{2,p}(G) \geq 3$, then $p = 3$ and G is of $GF(2)$ type, in the sense of Definition 2.3.7. And the groups of $GF(2)$ type will make an appearance for other reasons later. ♢

We can summarize the role of the Weak Trichotomy Theorem 2.2.1 in Strategy 1.1.1 by saying that, together with Steps 1 and 2, the Weak Trichotomy Theorem completes Step 3. In more detail: Step 2 is completed by the Quasithin Theorem 3.0.1, so that $e(G) \geq 3$; and then the Klinger-Mason Theorem B.9.1 eliminates the possibility that Case (I) could hold for any odd p with $m_{2,p}(G) \geq 3$. Step 1 is completed by the Uniqueness Case Theorem 8.0.2—which eliminates the possibility that Case (III) holds for all p with $m_{2,p}(G) \geq 3$. Thus Case (II) holds for some p with $m_{2,p}(G) \geq 3$—so that Step 3 is achieved, namely the existence of a p-component in the centralizer of some p-element t.

We could express this progress through Step 3 in the form:

THEOREM 2.2.3 (p-Component Theorem). *Let G be a finite simple group of characteristic 2 type with $e(G) \geq 3$. Then there is an odd prime p with $m_{2,p}(G) \geq 3$, and an element $t \in G$ of order p, such that $L_{p'}(C_G(t)) \neq 1$.*

Then the implementation of Steps 4 and 5 of the Strategy leads to the much stronger Trichotomy Theorem 5.1.2 of Gorenstein-Lyons. And indeed in order to achieve the desired "standard form" in Step 5, the groups of $GF(2)$ type (mentioned in connection with the Klinger-Mason analysis in branch (I) above) re-emerged in the Gorenstein-Lyons analysis, as a new exceptional branch (I) to be handled.

Thus the treatment of the groups G with $e(G) \geq 3$ is subdivided into the following branches:

(I) (exceptional:) $GF(2)$ type
(II) (generic:) standard type
(III) (impossible:) the Uniqueness Case.

The main result establishing this subdivision was the Gorenstein-Lyons Trichotomy Theorem 5.1.2 covering $e(G) \geq 4$. The trichotomy in Aschbacher's treatment of $e(G) = 3$ was more implicit (though the role of the Uniqueness Case was explicit). We will comment further on that aspect later at several points.

In the remaining sections of the chapter, we will state the major subtheorems on groups of characteristic 2 type. We begin with the theorem establishing the basic trichotomy, and then move on to those treating the resulting cases.

2.3. The Trichotomy Theorem for $e(G) \geq 4$

The proof of the Trichotomy Theorem 5.1.2 is fairly lengthy—an AMS Memoir [**GL83**] of over 700 pages. However, as in the case of the first quasithin monograph [**AS04a**], a large initial segment (of over 400 pages) develops basic properties of the known simple groups, which were not readily available in the literature at the time. The ensuing 300-page proof then produces very detailed and useful information in the branches leading to standard form and the uniqueness case, culminating in precise definitions of those branches. Because of the amount of detail available, the definitions of the two cases are rather lengthy. In this section, to simplify the exposition, we will give only an approximation to the definitions; we will provide the precise definitions in later chapters.

Before arriving at the statement of the result at the end of the section, we will explore various concepts, typically introduced via examples. Toward that end, it is natural to first treat those branches that arise most often "in nature"; thus we first discuss the generic case, then the exceptional case, and finally the impossible case.

Recall that we may assume by induction that our simple group G of characteristic 2 type satisfies the K-group hypothesis 0.5.4. Furthermore, in view of the Quasithin Theorem 3.0.1, we also assume that p is an odd prime such that $m_{2,p}(G) \geq 3$.[5]

The generic case: the notion of standard type. Because $m_{2,p}(G) \geq 3$, we may also assume that we are given an elementary abelian p-subgroup B of rank at least 3, contained in a 2-local subgroup of G. To get some feeling for the generic situation appearing as conclusion (II) of the Trichotomy Theorem 5.1.2, we will continue our practice of illustrating various concepts using matrix groups as examples.

EXAMPLE 2.3.1 ($SL_4(2^m)$ as an example). Most of the relevant concepts are visible in G given by $SL_4(2^m)$. To guarantee that $e(G) \geq 3$, we will assume that $m \geq 2$. Set $q := 2^m$, and $F := \mathbb{F}_q$.

We take our elementary abelian p-group B of maximal rank $m_{2,p}(G)$ to be a group of diagonal matrices.[6] So we fix a prime p dividing $q - 1$. Then we take B to be the subgroup consisting of the elements of order p in the diagonal subgroup:

$$B := \text{ all } \begin{pmatrix} \alpha & & & \\ & \beta & & \\ & & \gamma & \\ & & & \frac{1}{\alpha\beta\gamma} \end{pmatrix}, \text{ where } \alpha, \beta, \gamma \text{ have order 1 or } p \text{ in } F^\times.$$

[5]Of course we can take $m_{2,p}(G) \geq 4$ in the situation considered by Gorenstein-Lyons; but much of our discussion here will apply also to Aschbacher's case in which $m_{2,p}(G) = 3$.

[6]The use of notation "B" here (as in Gorenstein-Lyons) should not be confused with our use elsewhere of B for a Borel subgroup. (Often the B here will lie in a Cartan subgroup of a Borel subgroup.)

Visibly B has rank 3, and it lies in a Borel subgroup of G, which is a 2-local. For an odd prime r, the r-rank of G is at most 3, since any abelian r-subgroup of G embeds in a diagonal subgroup of $SL_4(E)$ for some extension field E of F. Thus $e(G) = m_{2,p}(G) = 3$ in $G = SL_4(F)$.

We wish to examine components in centralizers of elements of B. Since our elements are of determinant 1 we replace the element "t" in Example 0.3.2, which for the minimal value $k = 1$ had a unique nontrivial eigenvalue λ, by a member of B having the form:

$$x := \left(\begin{array}{c|ccc} \alpha & & & \\ \hline & \beta & & \\ & & \beta & \\ & & & \beta \end{array}\right), \text{ where } \alpha\beta^3 = 1, \text{ but } \beta \neq 1.$$

Then $\alpha \neq \beta$, and $C_G(x)$ is isomorphic to $GL_3(F)$, since we can take any invertible matrix A in the lower-right block, with $\frac{1}{\det(A)}$ in the upper-left block. In particular $E\big(C_G(x)\big)$ is a component, which we denote by:

$$L := \left(\begin{array}{c|c} 1 & \\ \hline & SL_3(F) \end{array}\right).$$

Notice furthermore that $C_G(L)$ is cyclic. This suggests one notion of maximality for the component L (arising from our "minimal" choice of x), leading to the following definition of a standard component: ◊

DEFINITION 2.3.2 (Triples (B, x, L) of $\mathcal{S}^*(p)$). Let $\mathcal{B}_{max}(G;p)$ denote the set of elementary abelian p-subgroups B of G such that B lies in a 2-local subgroup of G, and $m_p(B) = m_{2,p}(G)$.

Let $\mathcal{S}^*(G;p)$ (or just $\mathcal{S}^*(p)$ when the choice of G is clear) denote the set of triples (B, x, L) satisfying:

(1) $B \in \mathcal{B}_{max}(G;p)$;
(2) $x \in B^\#$;
(3) L is a component of $C_G(x)$ of order divisible by p; and
(4) $C_G(L)$ has cyclic Sylow p-subgroups.

Under these conditions we will say that L is a *standard component* of $C_G(x)$. ◊

REMARK 2.3.3 (The language of odd standard form). Although we have not yet reached the final definition of standard type, we mention that properties (3) and (4) above are essentially the definition in [**GL83**, pp. 436–437] of the intermediate concept of "odd standard form". This simpler notion of a standard component is used elsewhere in the literature.[7] Note that it follows from properties (3) and (4) that L is the unique component of $C_G(x)$ of order divisible by p. ◊

Now the maximal components L, defined by property (4), should in fact be large enough to have significant intersection with "adjacent" components J in centralizers of other elements y of B.

We illustrate with the example of $G = SL_4(F)$ and the triple $(B, x, L) \in \mathcal{S}^*(p)$ exhibited in Example 2.3.1. Let $y \in B$ be the conjugate of x with the eigenvalue of multiplicity 1 in the final position. Then $J = E(C_G(y))$ is given by:

[7]Notably in Aschbacher's work [**Asc81b, Asc83a**] on the case $e(G) = 3$ (discussed in Chapter 4), and in the Finkelstein-Frohardt work treating certain standard-form problems (discussed in Section 6.2).

$$y := \left(\begin{array}{ccc|c} \beta & & & \\ & \beta & & \\ & & \beta & \\ \hline & & & \alpha \end{array}\right), \text{ with } J := \left(\begin{array}{c|c} SL_3(F) & \\ \hline & 1 \end{array}\right).$$

Of course the triple (B, y, J) is also a member of $\mathcal{S}^*(p)$ and $C_B(J) = \langle y \rangle$.

We can compare the components L and J from the viewpoint of the subgroup of B generated by x and y, namely:

$$D := \langle x, y \rangle = \left(\begin{array}{c|cc|c} \alpha & & & \\ \hline & \epsilon & & \\ & & \epsilon & \\ \hline & & & \delta \end{array}\right), \text{ where } \alpha\epsilon^2\delta = 1,\ \alpha^p = \delta^p = \epsilon^p = 1.$$

We see for example that $C_L(D) = C_L(y)$ has $E\big(C_L(D)\big)$ given by

$$K := \left(\begin{array}{c|c|c} 1 & & \\ \hline & SL_2(F) & \\ \hline & & 1 \end{array}\right).$$

Now K is quasisimple and so is a component of $C_L(D)$; we may regard K as a "subcomponent" of both L and J, which are "adjacent" because of the existence of K.

Here are some further properties: First as J is quasisimple, J is generated by conjugates of K. Second, $C_L(K)$ is cyclic.

As before, these properties of our example suggest the following definitions in a general group G.

DEFINITION 2.3.4 (Neighbors adjacent via standard subcomponents). Assume that $(B, x, L) \in \mathcal{S}^*(G; p)$.

A *standard subcomponent* of (B, x, L) is a pair (D, K) such that the following conditions hold:

(1) $x \in D \leq B$ with $|D| = p^2$;
(2) $K := L_{2'}\big(C_L(D)\big)$ is quasisimple of order divisible by p; and
(3) $D = C_B(K)$.

We say that the triple (B, y, J) is a *neighbor* of (B, x, L) if the following conditions hold:

(1) $y \in B - \langle x \rangle$ and J is a p-component of $C_G(y)$;
(2) $C_B\big(J/O_{p'}(J)\big) = \langle y \rangle$;
(3) If $D := \langle x, y \rangle$ and $K := L_{2'}\big(C_L(D)\big)$, then (D, K) is a standard subcomponent of (B, x, L);
(4) J is contained in the normal closure of K in $L_{p'}\big(C_G(y)\big)$. ◊

Note that condition (2) for a neighbor J guarantees when $K \leq J$ in (4) that in fact $K < J$.

Later we will incorporate more detailed information in our definition of *standard type* in Definition 5.1.4. For the moment, for expository purposes, we content ourselves with a notion which we will call *simplified* standard type, which captures the spirt of that definition. In the remainder of this chapter the reader should be able, without serious loss, to use this simplified version in place of the real definition of standard type.

DEFINITION 2.3.5 (simplified standard type). Let G be a group, p an odd prime, and take some $(B, x, L) \in \mathcal{S}^*(G; p)$. We say that G is of *simplified standard type* with respect to (B, x, L) if and only if the following hold:

(1) $L/Z(L) \in Chev(2)$;

(2) There exists a standard subcomponent (D, K) of (B, x, L); and

(3) For every standard subcomponent (D, K) of (B, x, L), there exists a neighbor (B, y, J) of (B, x, L) with respect to (D, K) with $J/Z(J) \in Chev(2)$. ◊

Note that we have already checked these conditions in the case of $G = SL_4(q)$ and our triple (B, x, L) of Example 2.3.1.

In particular we emphasize that it is usually possible to produce a triple satisfying condition (1) in the Even Case. Further this restriction is useful, since it limits the number of components L which must be treated—in contrast with the Odd Case, where all quasisimple components L had to be dealt with.

We remark that conditions (2) and (3) above serve roughly as a replacement for Glauberman's Z^*-Theorem B.2.1 (which for odd primes p has no direct analogue independent of the CFSG). First, here is how those conditions would arise if we were in the Odd Case: Assume L is a standard component of $C_G(x)$, with x an involution and $L \in Chev(p)$ (p odd), and $C_G(L)$ has 2-rank 1. Then the Z^*-Theorem guarantees that x is not weakly closed in $C_G(x)$, so there is a G-conjugate y of x with $y \in C_G(x)$ but $y \notin C_G(L)$. And then, setting $D := \langle x, y \rangle$, we may in general infer the existence of an Odd-Case standard subcomponent (D, K) of (B, x, L), and a neighbor (B, y, J) of (B, x, L) with J a G-conjugate of L, just as in (2) and (3). Second, in the Even Case, we may argue essentially in the opposite direction: the existence of a neighbor (B, y, J) of (B, x, L) as in (3) can in general be used to deduce that $\langle x \rangle$ is not normal in $N_G(B)$, so in particular, $\langle x \rangle$ is not weakly closed in $C_G(x)$ with respect to G.

We will begin the discussion of the classification of groups of standard type later in Section 2.5, followed by more discussion in Chapter 6. But for now we continue our introduction to the Trichotomy Theorem, by turning from the generic to the exceptional branch.

The exceptional case: the notion of groups of $GF(2)$ type. We can get some idea of how the groups of $GF(2)$ type arise in conclusion (I) of the Trichotomy Theorem, by re-examining the diagonal matrix x in Example 2.3.1. For notice that if the field $\mathbb{F}_q$ is just the prime field $\mathbb{F}_2$, then there are no field elements to serve as nontrivial eigenvalues for x. Thus our element x of odd order must be some non-diagonal matrix.

In fact $SL_4(2)$ is quasithin, and so does not serve as a generic example. So consider instead G given by $SL_6(2)$, which has $e(G) = m_{2,3}(G) = 3$. If we take an element x of order 3 which is trivial on a 4-dimensional $\mathbb{F}_2$-subspace of the 6-space on which G acts, then $C_G(x)$ has a standard component $L \cong SL_4(2)$ acting on that subspace. However, when we choose $y \in B$ as in the discussion leading up to 2.3.4, we find that the "subcomponent" K is isomorphic to $SL_2(2)$—which is solvable rather than quasisimple. Hence G fails to be of simplified standard type in the sense of Definition 2.3.5.

The sporadic group Th provides an even more dramatic illustration. If G is Th, then G is of characteristic 2 type with $e(G) = m_{2,3}(G) = 3$, and the only component in the centralizer of an element x of order 3 is $L \cong G_2(3)$—which is not in $Chev(2)$, as in part (1) of Definition 2.3.5 (although $G_2(3)$ is a group of characteristic 2 type).

We now consider an example of the exceptional subcase in the Gorenstein-Lyons Trichotomy Theorem—leading to the notion of groups of $GF(2)$ *type* in Definition 2.3.7 (where $GF(2)$ is just the more classical notation for the field $\mathbb{F}_2$ of order 2). In fact Thompson had already identified the class of groups of $GF(2)$ type as constituting a difficult extremal case in his early work on N-groups [**Tho68**].

We return to the group $G = SL_4(q)$ of Example 2.3.1. The root subgroups of $SL_n(q)$ are the groups of transvections with a fixed axis and center; each root subgroup is isomorphic to a copy of the additive group of the field $\mathbb{F}_q$. Thus when $q = 2$ the root subgroups are of order 2; whereas for $q > 2$, the root groups are noncyclic and admit the nontrivial action of a Cartan subgroup. Indeed the absence of suitable noncyclic 2-groups was the feature defining the extremal case in Thompson's analysis, and thereafter throughout work on groups of characteristic 2 type.

Recall from Example 0.3.3 that if we take $G = SL_4(\mathbb{F}_q)$, then we obtain t and $U = O_2\big(C_G(t)\big)$ having the form

$$t = \left(\begin{array}{c|cc|c} 1 & & & \\ \hline & 1 & & \\ & & 1 & \\ \hline 1 & & & 1 \end{array}\right) \quad \text{and} \quad U = \left(\begin{array}{c|cc|c} 1 & & & \\ \hline * & 1 & & \\ * & & 1 & \\ \hline * & * & * & 1 \end{array}\right). \tag{2.3.6}$$

One can check that U has many noncyclic abelian subgroups, which are normal in U precisely when they contain the root group of t. But the only abelian subgroup *characteristic* in U (that is, normalized also by all automorphisms of U) is the root group itself (which is $Z(U)$). But the root group is of order q, so when $q = 2$, U is an example of a 2-group of *symplectic type* (Definition A.1.11), in that it has no noncyclic abelian characteristic subgroup.

This is the critical property in the definition of Thompson's extremal situation (cf. [**GL83**, p. 3]):

DEFINITION 2.3.7 ($GF(2)$ type). A finite group G is of $GF(2)$ *type*[8] if for some involution t of G, we have $R := F^*\big(C_G(t)\big)$ is a 2-group of symplectic type, i.e., R has no noncyclic characteristic abelian subgroup. $\diamondsuit$

Since our group G is of characteristic 2 type, the condition that $F^*\big(C_G(t)\big)$ is a 2-group is of course automatic for us.

In fact by the time Gorenstein-Lyons began their analysis, all groups of $GF(2)$ type (whether of characteristic 2 type or not) had been determined by the combined efforts of a number of researchers; we will indicate that result in Section 2.6. (And that work was independent of work in both the Even and Odd Cases of the CFSG.) So this branch of the Trichotomy Theorem had already been handled.

We complete our introduction to the Trichotomy Theorem, by turning to its final branch.

[8]Also in the literature G is sometimes said to have a *large* 2-subgroup of symplectic-type.

The impossible case: the branch leading to the Uniqueness Case. Since no groups arise in this situation (which appears as conclusion (III) of the Trichotomy Theorem), we cannot introduce the concepts by means of examples.

Instead we call attention to analogies with some parts of the treatment of groups of GW type. Recall from the discussion of the preuniqueness condition in Definition 1.5.4 that, when signalizer functors are nontrivial, then G has a proper weak 2-generated core $\Gamma^0_{2,S}(G)$ over a Sylow 2-subgroup S of G.

Similarly in branch (III) of the Trichotomy Theorem, when signalizer functors are of even order, we obtain the following uniqueness subgroups:

DEFINITION 2.3.8 (The Preuniqueness Case (for G of characteristic 2 type)). We say that the *Preuniqueness Case* holds for G and a set σ of odd primes p, if for each $p \in \sigma$ and $P \in Syl_p(G)$, $\Gamma^0_{2,P}(G)$ is contained in a 2-local subgroup M of G.

Here is a heuristic argument that the Preuniqueness Case should be impossible: We might expect, as in the Strongly Embedded Theorem 1.3.5, that if M is a uniqueness subgroup of G then G is a rank 1 group of Lie type in characteristic p and M is a Borel subgroup of G, so that in particular M is a p-local satisfying $F^*(M) = O_p(M)$—rather than a 2-local.

In fact the Trichotomy Theorem actually delivers considerably more detailed information about the Preuniqueness Case. But we will not require those details, which are used in the definition of the full Uniqueness Case, until later at Definition 2.7.2—which we will see (from [**GL83**, Thm C]) is in fact a consequence of (and hence equivalent to) the Preuniqueness Case.

In contrast to the Odd Case, the Uniqueness Case in characteristic 2 type had not already been treated early in the process; but instead essentially functioned as a conclusion to which reductions were made. The uniqueness branch was only eliminated later by Aschbacher in [**Asc83b, Asc83c**], as we will discuss in Section 2.7.

A statement of the Trichotomy Theorem. With the above definitions in place, we are finally in a position to give a formal statement of the main result of Gorenstein and Lyons in [**GL83**].

The strongest version is the result called the Trichotomy Theorem on p. 10 of [**GL83**]—and its conclusion (III) is stated in terms of the full Uniqueness Case of Definition 2.7.2.

However, for expository purposes, we will only state the version called the Pretrichotomy Theorem on p. 12 of [**GL83**]—where conclusion (III) now has an easier statement in terms of the Preuniqueness Case discussed above. For this reason, *we will for the moment abuse language by using the term Trichotomy Theorem to refer to the statement 5.1.1 (which Gorenstein and Lyons call the Pretrichotomy Theorem).* (We give the statement of the actual Trichotomy Theorem, in terms of the Uniqueness Case rather than the Preuniqueness Case, at 5.1.2.)

As in the Quasithin Theorem 3.0.1, results from this point on will typically continue to assume the inductive Hypothesis 0.5.4 that G is K-proper; that is, that proper subgroups of G have known composition factors.

Below we state a slight variant of the Trichotomy Theorem 5.1.1, which we have tailored to the level of our exposition in this chapter—by using in conclusion (II) our notion of simplified standard type in Definition 2.3.5, rather the the full standard type of Definition 5.1.4. Again the reader can, without much loss, use this version rather than 5.1.1 in the remainder of this chapter.

THEOREM 2.3.9 ((Simplified) Trichotomy Theorem for $e(G) \geq 4$). *Assume that G is a K-proper simple group of characteristic 2 type, with $e(G) \geq 4$. Let $\sigma(G)$ denote the (nonempty) set of odd primes p such that $m_{2,p}(G) \geq 4$. Then one of the following holds:*

(I) G is of $GF(2)$ type;
(II) For some $p \in \sigma(G)$, G is of simplified standard type with respect to some member of $\mathcal{S}^*(p)$.
(III) For every $p \in \sigma(G)$ and $P \in Syl_p(G)$, $\Gamma^0_{2,P}(G)$ is contained in a 2-local subgroup of G, (that is, the Preuniqueness Case holds.)

A more detailed discussion of the proof of the Trichotomy Theorem for $e(G) \geq 4$ will be given in Chapter 5.

As we suggested earlier, conclusion (I) of $GF(2)$ type in fact arises only when the prime p is 3.

In the next section there will also be a trichotomy (though somewhat more implicit) in our discussion of the case $e(G) = 3$. After that the remaining sections of the chapter will indicate the results which dealt with the three branches resulting from the Trichotomy Theorem.

2.4. The $e(G) = 3$ Theorem (including trichotomy)

For expository purposes, we have chosen in this chapter to discuss $e(G) = 3$ after $e(G) \geq 4$, instead of before it—to take advantage of our discussion of the Trichotomy Theorem.

Aschbacher classified the simple groups G of characteristic 2 type which satisfy $e(G) = 3$ in two papers [**Asc81b, Asc83a**], and we will state his result below as the $e(G) = 3$ Theorem 4.0.1.

There is one further piece of notation to introduce: In contrast to the convention of Gorenstein and Lyons in the Trichotomy Theorem, Aschbacher uses $\sigma(G)$ (or just σ) to denote a certain non-empty (but possibly proper) subset of the collection of all odd primes p which achieve the maximal value $m_{2,p}(G) = 3$. (That larger collection Aschbacher now denotes by τ.)

In Aschbacher's analysis, there are important technical reasons to make a careful choice of p—cf. Remark 4.1.6. So Aschbacher in fact begins by defining certain relevant subsets of τ. For expository purposes, it will suffice to specify just one of those sets, namely the set σ of primes which will satisfy the Preuniqueness Case in conclusion (5).

DEFINITION 2.4.1. Assume that $e(G) = 3$, so that the set τ of primes p satisfying $m_{2,p}(G) = 3$ is nonempty. Define a nonempty subset σ of τ as follows:
If $\tau \not\subseteq \{3, 5\}$, then we set $\sigma := \tau \setminus \{3, 5\}$ (so that σ contains some $p \geq 7$).
Otherwise, if $\{5\} \subseteq \tau \subseteq \{3, 5\}$, then we set $\sigma := \{5\}$.
The only remaining case is $\tau = \{3\}$, where we set $\sigma := \tau = \{3\}$.
We may refer to this σ more fully as $\sigma(G)$. ◇

We are now in a position to state the Main Result from [**Asc81b**]:

The $e(G) = 3$ Theorem 4.0.1 Let G be a finite K-proper simple group of characteristic 2 type with $e(G) = 3$. Then one of the following conclusions holds:

(1) $G \cong L_4(2^n)$ or $Sp_6(2^n)$, $n > 1$; or
(2) $G \cong \Omega_8^-(2^n)$, $n \geq 1$; or
(3) $G \cong U_5(2^n)$, $n \neq 2$; or
(4) $G \cong L_6(2)$, $L_7(2)$, $U_6(2)$, $Sp_8(2)$, $\Omega_8^+(2)$, $F_4(2)$, Co_2, or Th; or
(5) For every $p \in \sigma(G)$ and $P \in Syl_p(G)$, $\Gamma^0_{2,P}(G)$ is contained in a 2-local subgroup of G. (That is, the Preuniqueness Case holds.)

A more detailed discussion of the proof of the $e(G) = 3$ Theorem will be given in Chapter 4; but we will include a few general remarks at this point.

The analysis when σ contains a prime other than 3 is fairly similar in flavor to the "generic" analysis in the Gorenstein-Lyons Trichotomy setup, and this case is handled in Aschbacher's first paper [**Asc81b**]. However the remaining case where $\sigma = \{3\}$ (namely where $\tau = \{3\}$) requires more elaborate methods, which are developed in the second paper [**Asc83a**]. This division is also relevant at points in our discussion below.

The trichotomy interpretation. The $e(G) = 3$ Theorem is essentially a reduction to the Preuniqueness Case: for beyond the groups explicitly identified in conclusions (1)–(4) of the theorem, the remaining conclusion (5) is also conclusion (III) of the Trichotomy Theorem—namely the Preuniqueness Case of Definition 2.3.8. In his proof, Aschbacher treats the branches that correspond to conclusions (I) and (II) in the Trichotomy Theorem using his own arguments as well as recognition theorems from the literature, and, in the end, he achieves a final identification of the groups in those branches in his conclusions (1)–(4).

This trichotomy interpretation is not really explicit in the papers [**Asc81b, Asc83a**]; but it is referred to in the Gorenstein-Lyons trichotomy work [**GL83**, p. 12] and in Aschbacher's Uniqueness Case work [**Asc83b**, p. 384]. So we provide a brief indication of the emergence of the branches.

Recall that Aschbacher's first paper [**Asc81b**] treats the case where the set σ of Definition 2.4.1 is not $\{3\}$, so that the prime $p \in \sigma$ under analysis is not 3. Then, as in the Trichotomy Theorem for $e(G) \geq 4$, the $GF(2)$ type branch does not arise since $p \neq 3$. Thus in [**Asc81b**], instead of a trichotomy, the proof leads to only a dichotomy between odd standard form and the Preuniqueness Case (emerging essentially at [**Asc81b**, (9.13)]).

However the second paper [**Asc83a**] deals with the case $\sigma = \{3\}$, so that the prime p is now definitely 3. At (6.4) in that paper, the proof subdivides according to whether or not a certain set $\mathcal{U}(A)$, of noncyclic 2-subgroups, is empty. If it is empty, then G is shown to be of $GF(2)$ type and this leads via (3.10) of the paper to the five groups Co_2, Th, $U_5(2)$, $U_6(2)$, and $\Omega_8^+(2)$. Actually there are further groups of $GF(2)$ type in the conclusion of the Theorem—but they appear later in the odd standard form branch; for in the case where $\mathcal{U}(A)$ is empty, further analysis leads to a dichotomy between standard form and the uniqueness case (similar to that of the previous paragraph) which emerges essentially at (17.29) of the paper.

In the remainder of the chapter we turn to the treatment of the three branches in the Trichotomy Theorem for $e(G) \geq 4$; and in the Uniqueness Case branch, we also include $e(G) = 3$.

2.5. The Standard Type Theorem

We begin with a brief discussion of the treatment of groups of standard type in conclusion (II) in the Trichotomy Theorem 5.1.2. We limit ourselves to a statement and some brief remarks; a more detailed discussion appears in Chapter 6. Recall that in this chapter we use the notion of simplified standard type defined at 2.3.5; we will give the full definition of standard type later at 5.1.4.

In contrast to the large literature required in the case of the groups of GW type for the various standard form problems (which we summarized in Sections 1.8–1.10), the standard type problems needed for characteristic 2 type were treated in just a few papers.

Standard type for $e(G) \geq 4$. Indeed all standard type cases in the Gorenstein-Lyons Trichotomy Theorem 5.1.2 for $e(G) \geq 4$ were covered by the following result of Gilman and Griess in [**GG83**]:

The Standard Type Theorem 6.1.1 Let G be a K-proper finite simple group characteristic 2 type, with $e(G) \geq 4$. Suppose that G is of standard type with respect to some $(B, x, L) \in \mathcal{S}^*(G;p)$ for some odd prime p such that $m_{2,p}(G) \geq 4$. Then $G \in Chev(2)$.

The proof of Gilman and Griess will be discussed in more detail in Section 6.1.

Certain odd standard form problems for $e(G) = 3$. We mentioned earlier that in the $e(G) = 3$ Theorem, Aschbacher included his own treatment of many recognition problems, which he phrased in the language of odd standard form (recall Remark 2.3.3). However, Aschbacher also quoted work of Finkelstein-Frohardt in [**FF84, FF79, FF81b**] to cover a few odd standard form problems. Their work is at level 1 rather than level 0 in our viewpoint, and so will be described in Section 6.2 rather than here.

2.6. The $GF(2)$ Type Theorem

We consider next the treatment of groups of $GF(2)$ type in conclusion (I) in the Trichotomy Theorem 5.1.2. As we have already mentioned, this classification was already available before the Trichotomy Theorem, and was also quoted elsewhere in the Even Case of the CFSG. Also work on the groups of $GF(2)$ type did not make use of the K-group Hypothesis 0.5.4.

Our main discussion of the classification of groups of $GF(2)$ type appears in Chapter 7. Here we will provide a statement of the main theorem, together with a few brief remarks. Recall we defined the class of groups of $GF(2)$ type at 2.3.7.

Groups of $GF(2)$ *type*. A number of researchers worked on this problem, mainly in the 1970s; in particular, important reductions were made by Janko and his students (such as Dempwolff and Wong in [**DW77b, DW78**]), and by Aschbacher in [**Asc76a, Asc77**]. Subsequently fundamental work of Timmesfeld [**Tim78a**] reduced the problem to a number of specific configurations. These were then handled by several authors, notably Smith [**Smi80b, Smi80a, Smi79b**]. The paper [**Smi79b**] contains a summary of the final classification, and the literature leading up to it. The main result is:

The $GF(2)$ Type Theorem 7.0.1 Let G be a finite simple group of $GF(2)$ type. Then G is one of

(1) $L_n(2), n \geq 3$; $U_n(2), n \geq 4$; $\Omega^{\pm}_{2n}(2), n \geq 4$; ${}^3D_4(2)$, $E_6(2)$, ${}^2E_6(2)$, $E_7(2)$, $E_8(2)$;
(2) $L_2(q)$, $q = 2^m \pm 1, m \geq 3$; $L_3(3)$, $U_3(3)$, $L_4(3)$, $U_4(3)$, $G_2(3)$, $\Omega^+_8(3)$;
(3) A_6, A_8, A_9;
(4) One of the following 16 sporadic groups:
$M_{11}, M_{12}, M_{24}, J_2, J_3, J_4, Co_2, Co_1, HS, Suz, He, Fi'_{24}, HN, Th, BM, M$.

REMARK 2.6.1. As usual (cf. Remark A.3.6) there are a few isomorphisms among groups in this list: notably $L_2(7) \cong L_3(2)$, $A_6 \cong L_2(9)$, and $A_8 \cong L_4(2)$.

The result is of course more general than is actually required for our present purpose—namely the proof of the Main Theorem 2.0.1 for groups of characteristic 2 type. We summarize how to obtain from 7.0.1 the sublist of groups satisfying the hypotheses of 2.0.1, consisting of the groups of characteristic 2 type of 2-rank at least 3 with $e(G) \geq 3$: In conclusion (1) take all cases with $n \geq 6, 5, 4$ (for L_n, U_n, $\Omega^{\pm}_{2n}$ respectively), but exclude ${}^3D_4(2)$; exclude conclusions (2) and (3); and the only cases to arise in (4) are the sporadic groups Co_2 and Th. ◇

From conclusion (4), we see that the majority of the sporadic groups arise as groups of $GF(2)$ type. Hence the classification of the groups of $GF(2)$ type was regarded by a number of group theorists (say around 1977) as suggesting that there might not be any more sporadics left to discover—so that the classification of the finite simple groups might actually be within reach.

Groups of $GF(2^n)$ type. Aschbacher's second paper [**Asc83a**] on groups G with $e(G) = 3$ also makes use of a generalization of the class of groups of $GF(2)$ type that we call groups of $GF(2^n)$ type. We will discuss the classification of this larger class of groups in Section 7.5; but will not state the result here, since it is not among the level 0 results whose union gives the Even Case classification.

2.7. The Uniqueness Case Theorem

It remains to indicate how the Preuniqueness Case—giving conclusion (III) of the Trichotomy Theorem for $e(G) \geq 4$ in the form of 2.3.9, as well as conclusion (5) of the $e(G) = 3$ Theorem 4.0.1—led via work of Aschbacher, Gorenstein, and Lyons to the more detailed situation called the Uniqueness Case; which was in turn then eliminated by Aschbacher.

Earlier, we had observed that for the prime 2, uniqueness subgroups are often strongly embedded subgroups, as in Definition 1.3.4. Now changing the underlying prime to an odd prime p, and assuming the uniqueness subgroup is a 2-local subgroup, we are led to the notion of almost strong embedding, which includes certain technical constraints arising in the Gorenstein-Lyons analysis.[9]

DEFINITION 2.7.1 (Almost strongly p-embedded). Assume that G is of characteristic 2 type with $e(G) \geq 3$, and that M is a maximal 2-local subgroup

[9]We follow the format of the statement in Aschbacher [**Asc83b**, p. 385], which is slightly terser than the original statement of Gorenstein-Lyons [**GL83**, p. 8].

of G. Assume for some odd prime p that $m_p(M) \geq 3$, and for $P \in Syl_p(G)$ that $\Gamma_{2,P}(G) \leq M$.[10]

We say that M is *almost strongly p-embedded* in G if in addition one of the following three conditions is satisfied:[11]

(1) $\Gamma_{1,P}(G) \leq M$. (In which case M is in fact *strongly p-embedded* in G.)

(2) $p \geq 5$, M is solvable, and there is a subgroup P_0 of order p in P, which is weakly closed in P with respect to G, such that $L := E(C_G(P_0)) \cong L_2(p^m)$ for some $m \geq 2$, with $L \not\leq M$; furthermore $O_{p'}(M)C_G(L)$ is a Frobenius group with kernel $O_2(M)$ and complement $C_G(L)$; and finally for each subgroup X of order p in P other than P_0, $N_G(X) \leq M$.

(3) $p = 3$, M is solvable, $P \cong \mathbb{Z}_3$ wr $\mathbb{Z}_3$, and $\Gamma_{1,J(P)}(G) \leq M$; and if X is of order 3 in P but $X \not\leq J(P)$, then for a P-invariant Sylow 2-subgroup Q of $O_{3'}(M)$ we have $C_Q(X) \cong Q_8 * \mathbb{Z}_{2^m}$ for some m. ◇

The almost strongly p-embedded property is the crucial condition (1) in the definition below (see [**GL83**, p 9]) of the Uniqueness Case: it is applied for p in the set of primes $\sigma(G)$ used either by Gorenstein-Lyons in the Trichotomy Theorem 5.1.2 when $e(G) \geq 4$, or by Aschbacher in the $e(G) = 3$ Theorem 4.0.1:

Definition 2.7.2 (The Uniqueness Case). Assume that G is a K-proper simple group of characteristic 2 type with $e(G) \geq 3$. We say that G is in the *Uniqueness Case* if

(1) For each $p \in \sigma(G)$, G possesses an almost strongly p-embedded maximal 2-local M.

(2) For any 2-local subgroup H of G, $m_p(H \cap M) \geq 2$ implies $H \leq M$.

We mention that (2) goes beyond the condition $\Gamma_{2,P}(G) \leq M$ for $P \in Syl_p(G)$ which is part of the definition of almost strongly p-embedded. ◇

Now condition (2) was established, assuming the Preuniqueness Case 2.3.8, by work of Aschbacher, Gorenstein, and Lyons—in a result which we will discuss later as 8.1.3. (It appeared in Sections II.3.4 and II.3.6 of [**GL83**], and had earlier been stated as Theorem E in [**AGL81**].)

Thus it remained to derive condition (1) above from the Preuniqueness Case. (Indeed Aschbacher at [**Asc83b**, p 385] refers *only* to condition (1) as the Uniqueness Case, and hence quotes [**AGL81**, Thm E] in effect as an independent property of the Preuniqueness Case. Similarly, our later exposition will tend to emphasize the almost strongly p-embedded property as being the essence of the Uniqueness Case.)

And correspondingly, in Chapter II.3 of their Trichotomy Memoir [**GL83**], Gorenstein and Lyons made use of further work in the above paper [**AGL81**] with Aschbacher, to show in their Theorem C (which we discuss in the final subsection of Section 8.1) that when either $e(G) \geq 4$ or $e(G) = 3$, the Preuniqueness Case leads to an almost strongly p-embedded subgroup, namely condition (1) of the Uniqueness Case. Thus the Preuniqueness and Uniqueness Cases are in fact equivalent:

---

[10]Notice that this condition is already somewhat stronger than the Preuniqueness Case in Definition 2.3.8, which only required containment in a 2-local subgroup of the *weak* 2-generated core $\Gamma^0_{2,P}(G)$.

[11]Certain background definitions are used in these statements: For terminology related to a *Frobenius group*, see Definition A.1.3. For the *wreath* product of groups (and also the *central* product $A * B$), see the references after Definition A.1.6. In (3) the Thompson subgroup $J(P)$ is just the obvious normal subgroup $(\mathbb{Z}_3)^3$ of P; more generally see Definition B.6.4.

The Preuniqueness-implies-Uniqueness Theorem 8.0.1 Let G be a K-proper simple group of characteristic 2 type with $e(G) \geq 3$. If G satisfies the Preuniqueness Case for all $p \in \sigma(G)$, then G satisfies the Uniqueness Case for all $p \in \sigma(G)$.

Soon thereafter, Aschbacher in [**Asc83b, Asc83c**] demonstrated that this branch of the underlying trichotomy does not in fact lead to any further simple groups:

The Uniqueness Case Theorem 8.0.2 No K-proper finite simple group satisfies the hypotheses of the Uniqueness Case.

Our discussion of the Uniqueness Case Theorem appears in Chapter 8.

Conclusion: The proof of the Characteristic 2 Type Theorem

We can now give a proof of the Main Theorem for groups of characteristic 2 type, based on the level 0 theorems stated above. First we recall the statement of the result:

Main Theorem for groups of characteristic 2 type 2.0.1 Let G be a finite simple K-proper group of characteristic 2 type of 2-rank at least 3. Then one of the following conclusions holds:

(1) G is a finite simple group of Lie type defined over a field of characteristic 2; or
(2) G is isomorphic to $G_2(3)$ or $U_4(3)$; or
(3) G is one of the following seven sporadic simple groups:

$$M_{22}, M_{23}, M_{24}, J_3, J_4, Co_2, Th.$$

PROOF. If $e(G) \leq 2$, then by the Quasithin Theorem 3.0.1, G is isomorphic to one of the groups listed above.

Thus we may assume that $e(G) \geq 3$. Now by the Uniqueness Case Theorem 8.0.2 (using also 8.0.1 to reduce to the Preuniqueness Case), there exists a prime $p \in \sigma(G)$ such that G does not have a p-preuniqueness subgroup. That is, the Preuniqunes Case does not hold.

So now if $e(G) = 3$, then by Aschbacher's $e(G) = 3$ Theorem 4.0.1, G is isomorphic to one of the groups in his conclusions (1)–(4), and these groups are in the list above.

Hence we may assume that $e(G) \geq 4$. Then by the Gorenstein-Lyons Trichotomy Theorem 5.1.2, G satisfies conclusion (I) or (II) of that theorem; that is, either G is of $GF(2)$ type, or there exists a prime $p \in \sigma(G)$ such that G is of standard type with respect to some member of $\mathcal{S}^*(p)$.

In the former case, it follows from the $GF(2)$ Type Theorem 7.0.1 (using the hypotheses that G is of characteristic 2 type and $e(G) \geq 4$) that G is isomorphic to one of the groups listed above.

In the latter case, it follows from the Standard Type Theorem 6.1.1 that G is a finite simple group of Lie type defined over a field of characteristic 2, and so is listed in (1).

This completes the proof of the main theorem for groups of characteristic 2 type. □

Finally, we note that this result together with Theorem 1.0.1 on groups of *GW* type and the Dichotomy Theorem 0.3.10 yield the complete Classification Theorem:

CFSG Theorem 0.1.1 A finite simple group is either cyclic of prime order; or one of the following nonabelian groups:

(1) An alternating group $A_n, n \geq 5$.

(2) A simple group $G(q)$ of Lie type defined over a finite field $\mathbb{F}_q$ of prime power order $q = p^r$; i.e., one of:

$$L_n(q), U_n(q), Sp_{2n}(q), \Omega^+_{2n}(q), \Omega^-_{2n}(q), \Omega_{2n+1}(q),$$
$$G_2(q), F_4(q), E_6(q), E_7(q), E_8(q),$$
$${}^2B_2(2^{2m+1}), {}^2G_2(3^{2m+1}), {}^2F_4(2^{2m-1})', {}^3D_4(q), {}^2E_6(q)\ ,$$

for most values of the parameters q, $n \geq 2$, and $m \geq 1$.

(3) One of 26 sporadic groups, denoted by:

$$M_{11}, M_{12}, M_{22}, M_{23}, M_{24}, J_1, J_2, J_3, J_4,$$
$$Co_3, Co_2, Co_1, Fi_{22}, Fi_{23}, Fi'_{24},$$
$$HS, Mc, Ly, Suz, He, Ru, O'N, HN, Th, BM, M.$$

PROOF. Suppose not, and let G be a minimal counterexample. Then by induction on the order of G, G is K-proper. Moreover by the Dichotomy Theorem 0.3.10, either G is a K-proper finite simple group of *GW* type, or G is a K-proper finite simple group of characteristic 2 type. In the former case, G is in the list of the CFSG by Theorem 1.0.1, contrary to assumption. In the latter case, G is in the list of the CFSG by Theorem 2.0.1, again contrary to assumption. This final contradiction completes the proof. □

Outline of the classification of groups of characteristic 2 type

CHAPTER 3

$e(G) \leq 2$: The classification of quasithin groups

We recall from our Introduction, and the overview in Chapter 2, that the "small" groups in the classification of the groups of characteristic 2 type are the *quasithin* groups: that is the groups G with $e(G) \leq 2$ (recall Definition 0.5.2). We discuss such groups in this chapter. It should also be mentioned that the groups G with $e(G) = 3$ represented a borderline case, whose treatment by Aschbacher will be discussed in Chapter 4.

We also recall that a classification of the quasithin groups of characteristic 2 type was first undertaken by G. Mason in [**Mas**], after Aschbacher's earlier determination of the thin groups (the groups G with $e(G) \leq 1$) in [**Asc78b**]; but Mason's treatment was not complete, and was never published. A proof of a theorem classifying the simple quasithin groups was eventually published by Aschbacher and Smith in [**AS04b**]. For a number of reasons (see e.g. [**AS04b**, pp. 485, 496]), Aschbacher and Smith replaced the hypothesis of characteristic 2 type with the weaker hypothesis of *even characteristic* in Definition 2.1.1—so that the resulting classification of simple QTKE-groups, given as Theorem 3.0.1 below, is in fact somewhat more general than was needed for the "original" CFSG. Indeed, as we shall see in Section 3.12, the stronger result can be used to treat the quasithin problem in the GLS program, where the groups satisfy the related hypothesis of *even type*. The main result of [**AS04b**] is:

THEOREM 3.0.1 (Quasithin Theorem). *The finite nonabelian simple QTKE-groups consist of:*

(1) (Generic case) The simple groups of Lie type over fields of characteristic 2 *and with* BN*-rank at most* 2*, but* $U_5(q)$ *only for* $q = 4$.
(2) (Some groups of BN*-rank* 3 *or* 4*)* $L_4(2)$, $L_5(2)$, $Sp_6(2)$.
(3) (Some small alternating groups) A_5, A_6, A_8, A_9.
(4) (Some small groups of Lie type over fields of odd characteristic) $L_2(p)$*,* p *a Mersenne or Fermat prime,* $L_3(3)$, $U_3(3)$, $L_4(3)$, $U_4(3)$, $G_2(3)$.
(5) (Some sporadic groups) M_{11}, M_{12}, M_{22}, M_{23}, M_{24}, J_2, J_3, J_4, HS, He, Ru.

The set of types of BN-rank at most 2 in conclusion (1) consists of: L_2, L_3, U_3, U_4, U_5, Sp_4, G_2, ${}^2B_2$, ${}^2F_4$, and ${}^3D_4$.

In the context of the original CFSG, where the stronger hypothesis of characteristic 2 type is assumed, all the groups in conclusions (1) and (2) still arise, but the stronger hypothesis eliminates some groups of component type in the remaining cases: A_9 from conclusion (3); $L_4(3)$ from conclusion (4); and M_{12}, J_2, HS, He, and Ru from conclusion (5).

Furthermore some of the groups have 2-rank 2, and so also appeared in Theorem 1.0.1, which classified the groups of GW type: namely $L_2(4) \cong A_5$, $L_3(2) \cong L_2(7)$, $Sp_4(2)' \cong A_6$, $G_2(2)' \cong U_3(3)$, $U_3(4)$, $L_2(p)$, $L_3(3)$, and M_{11}.

An introduction to the proof of the Quasithin Theorem is given in the Introduction (Chapter 0) of [**AS04b**], to which the reader is also directed; see especially the outline in Section 0.3 of [**AS04b**]. Our exposition will largely follow that treatment, with suitable adjustments to accommodate the conventions of the present book.

For a full list of references used in the proof of the Quasithin Theorem, we refer the reader to the bibliography[1] given in [**AS04b**]. Hence our discussion in this chapter will refer to such papers only in passing, rather than listing them formally.

3.1. Introduction: The Thompson Strategy

We briefly summarize the discussion in Subsection 0.3.1 of [**AS04b**], which describes the approach used in the bulk of the proof of the Quasithin Theorem 3.0.1. (We mention that the Thompson strategy, in the form used here, largely evolved from its use by Aschbacher in his papers on thin groups and the Uniqueness Case Theorem 8.0.2; the reader may wish to compare the discussion here with that in Section 8.2, especially beginning around Remark 8.2.3.)

The approach goes back to Thompson's work [**Tho68**] on N-groups: essentially it involves exploiting a pair of maximal 2-local subgroups M and N over a fixed Sylow 2-subgroup T of G. Of course the maximality of these subgroups guarantees that the subgroup $G_0 := \langle M, N \rangle$ they generate is not contained in a 2-local subgroup of G. Indeed we typically expect that G_0 is all of G. Since G_0 is a completion of the amalgam[2] determined by M and N, the general approach is referred to in [**AS04b**] as the *Thompson amalgam strategy*, or just the Thompson strategy. In particular, determining the possible pairs M and N will provide a case division for the eventual classification.

First of course it is necessary to guarantee that *distinct* maximal 2-local subgroups M and N are actually available; that is, to treat the case where T is contained in a unique maximal 2-local M. This situation is handled in Chapter 2 of [**AS04b**], which we describe in Section 3.4. The generic examples in this case are the Bender groups, namely the groups of Lie type of BN-rank 1 over fields of characteristic 2.

After Chapter 2 of [**AS04b**], we may assume that there exist distinct maximal local overgroups M and N of T. The generic examples in this case are the groups G of Lie type of BN-rank 2 over fields of characteristic 2, where M and N are maximal parabolics.

As the Thompson strategy is implemented in [**AS04b**], the maximal 2-local M is the unique maximal 2-local subgroup of G which contains $\langle L, T \rangle$, for a suitable subnormal subgroup L of M; consequently L is called a *uniqueness subgroup*. In [**AS04b**], this uniqueness property is expressed using the notation:[3]

$$M := N_G(L^T) = !\mathcal{M}(\langle L, T \rangle). \tag{3.1.1}$$

This notion of a uniqueness subgroup is roughly in the spirit of 1.3.1, as M contains all 2-local subgroups containing both L and T.

[1] That bibliography follows the GLS convention (as does our present work) of distinguishing between papers actually used in the proof, and papers referenced only for expository purposes.

[2] For more on amalgams and completions, see our discussion around Remarks 3.3.14 and 3.3.15.

[3] The more classical notation for the property would be $\mathcal{M}(\langle L, T \rangle) = \{M\}$; cf. Remark 8.0.4.

Typically L has the following structure: $F^*(L) = O_2(L)$ and $L/O_2(L)$ is quasisimple. The proof is then partitioned into cases, essentially indexed by the possible quasisimple quotients $L/O_2(L)$, with a further refinement corresponding to possibilities for a certain normal elementary abelian subgroup V of M lying in $Z(O_2(M))$ and not centralized by L. (This is the "Fundamental Setup", which we describe further at Remark 3.5.2.) We set:

$$\overline{L} := L/C_L(V).$$

The second maximal 2-local N is replaced by a subgroup H chosen *minimal* with respect to $H \not\leq M$, i.e. lying in the following set:

DEFINITION 3.1.2. We define $\mathcal{H}_*(T, M)$ as the set of subgroups H which are minimal with respect to the properties $H \geq T$, $O_2(H) \neq 1$, and $H \not\leq M$. ♢

When $N_G(T) \leq M$, the subgroups H in $\mathcal{H}_*(T, M)$ are in fact *abstract minimal parabolics* over T:

DEFINITION 3.1.3. Let T be a p-subgroup. A group H is an *abstract minimal parabolic* over T if T is a Sylow p-subgroup of H with the properties that T is not normal in H, but T is contained in a unique maximal subgroup of H. ♢

In a typical example H of an abstract minimal parabolic with $p = 2$, we get $H/O_2(H) \cong L_2(2^n)$.

Further case-subdivision is determined by properties of H. The group defined by $U := \langle V^H \rangle$ also plays an important role. Often U is elementary abelian, and so defines an $\mathbb{F}_2H$-module.

The Thompson strategy proceeds by showing that each choice for the pair $\overline{L}$, V and the abstract minimal parabolic H leads either to a contradiction, or to an identification of $\langle M, H \rangle$ as one of the groups in the conclusion of the Theorem.

Most of the later sections in this chapter will be devoted to sketching the treatment in [**AS04b**] of the various choices for $\overline{L}$ and V in the proof of the Quasithin Theorem 3.0.1. In particular, those sections will correspond roughly with the subdivision of [**AS04b**] into "Parts".

Furthermore we adopt the following notational convention: *We will ordinarily reserve M, L, and H (as well as G and T) for the groups described above in the Thompson strategy; and we also normally reserve V and U for the $\mathbb{F}_2$-modules indicated above for those groups.*

REMARK 3.1.4 (Digression on preliminary results and techniques). Before proceeding to our discussion of how the Thompson strategy is implemented in [**AS04b**], we will first devote the next two sections to introducing[4] some background for that discussion—namely some theory used to describe the structure of local subgroups of quasithin groups, and general methods for analyzing groups of even characteristic, developed in the preliminary volume [**AS04a**], prepared by Aschbacher and Smith as a companion to the main proof in [**AS04b**].

In particular, our digression here will be in contrast to our convention in most other chapters—where we typically proceed as directly as possible to the main argument, while postponing most discussion of quoted results to the end of those chapters.

[4]In fact we also provide further detail on most of these methods in Appendix B.

One reason for deviating from that convention is that this is the first chapter where we outline proofs in more detail—which involves the use of an underlying mathematical foundation which we have not yet discussed. A second reason is the greater complexity of the quasithin work itself.

Indeed the substantial length (almost 500 pages) of [**AS04a**], which establishes much of that foundation, reflects an effort to provide a supplement making the treatment in [**AS04b**] reasonably self-contained. Thus [**AS04a**] contains not only specialized results designed explicitly for the quasithin problem, but also additional background material on more general techniques which are used repeatedly in the arguments in [**AS04b**]. That technical material is often stated (and even proved) in [**AS04a**], in preference to just giving a reference to the literature.

Much as for [**AS04b**], we also recommend that the reader consult the introductory chapter of [**AS04a**], which discusses the preparatory material in more detail. Our exposition below will focus only on a number of high points. ♢

3.2. Preliminaries: Structure theory for quasithin 2-locals (SQTK-groups)

In this section, we review certain concepts from Chapter A of [**AS04a**], which are fundamental for the analysis of the subgroups of a quasithin group G (especially 2-local subgroups).

Note first that, since our hypothesis of even characteristic is not inherited by subgroups (and sections) of G, the composition factors of such subgroups may be quasithin simple groups which do not appear among the QTKE-groups in the conclusion of the Quasithin Theorem 3.0.1. Hence A.2.2[5] of [**AS04a**] (nicknamed Theorem B there) gives the full list of known quasithin groups.

Next from Definition 0.5.2, the 2-local subgroups in a quasithin group satisfy the following hypothesis, which is somewhat stronger than the condition defining the quasithin groups:

DEFINITION 3.2.1 (strongly quasithin). A group X is *strongly quasithin* if we have $m_p(X) \leq 2$ for all odd primes p. Further X is an *SQTK-group* if in addition all simple sections of X are known. ♢

The list of such simple sections is given in A.2.3 of [**AS04a**] (nicknamed Theorem C); for the convenience of the reader, we include this important list:

REMARK 3.2.2 (The Strongly Quasithin Simple Groups). The simple SQTK-groups are:

(1) (generic case) the groups in $Chev(2)$ of BN-rank at most 2 (but excluding $U_4(q)$ and $U_5(q)$);
(2) $L_4(2)$ and $L_5(2)$ (but excluding $Sp_6(2)$);
(3) the small alternating groups A_5, A_6, A_7, and A_8 (but excluding A_9);
(4) $L_2(p^e)$, $L_3(p)$, and $U_3(p)$ (for p any odd prime and $e \leq 2$);
(5) the sporadic Mathieu groups, J_1, J_2, J_4, HS, He, and Ru (but excluding J_3).

[5]In our discussions of proofs we will often reference results in the original papers via the labels used in those papers—such as this one in [**AS04a**]. (We may also write "there" with such a label—in order to distinguish from results "here", i.e. in this book.) Our intention in giving such references is to provide a more detailed guide for those readers who are also consulting the sources. But of course readers who are primarily interested just in a general outline of the arguments can safely ignore most of these labels.

Of course most groups on this list also appear in the Quasithin Theorem 3.0.1; the groups appearing here, but not there, are not of even characteristic. ◇

These lists are generated using facts about the known simple groups found in references such as [**GLS98**]. They are applied constantly throughout the proof, in describing proper subgroups of G whose composition factors are known by induction.

A qualitative description of the structure of SQTK-groups appears in Section A.3 of [**AS04a**]; this description depends upon a generalization of the notion of a $2'$-component, in effect a "component modulo a normal solvable subgroup":

DEFINITION 3.2.3. We say Y is a $\mathcal{C}$-*component* of X (and write $Y \in \mathcal{C}(X)$) if Y is a nontrivial subgroup of X minimal subject to $Y = [Y, Y]$ and Y subnormal in X. ◇

Observe that the minimality of Y implies that $Y/O_\infty(Y)$ is a nonabelian simple group, where $O_\infty(Y)$ denotes the solvable radical of Y. The importance of the concept is emphasized in A.3.9.1, which shows that a nonsolvable SQTK-group X is largely determined by the product of its $\mathcal{C}$-components. (This result is roughly an analogue of (8.5.4) in our discussion of the Uniqueness Case.) Further results in Section A.3 show that the $\mathcal{C}$-components of X have properties analogous to those of ordinary components, such as:

Distinct $\mathcal{C}$-components of X commute modulo $O_2(X)$.

A particularly strong consequence of the SQTK-hypothesis is that:

Each $\mathcal{C}$-component Y has at most two distinct conjugates in X.

There are strong constraints on the structure of $\mathcal{C}$-components in an SQTK-group—most are an extension of a 2-group by a quasisimple group which is a central extension of a group on the list of simple SQTK-groups in Remark 3.2.2. This fact implies that there are very few possible proper inclusions among $\mathcal{C}$-components of 2-local subgroups of G.

REMARK 3.2.4. These facts are used throughout [**AS04b**]. In particular, they are crucial in the analysis in Section 1.2 of [**AS04b**]—to be discussed more fully in our Section 3.5. For example we can choose the group "X" in Definition 3.2.3, to be the maximal 2-local M in the Thompson strategy, and take "Y" to be a $\mathcal{C}$-component L of M which affords a "nonsolvable uniqueness group" for M in a sense to be described below. This point of view is also used to describe the minimal parabolic H in the strategy, in results such as E.2.2 in [**AS04a**]. ◇

Specifically, assume that some maximal 2-local N over T is nonsolvable. Then the theory of $\mathcal{C}$-components from Section A.3 of [**AS04a**] can be applied as follows:

Since N is nonsolvable, there is some $X \in \mathcal{C}(N)$. Then 1.2.1.4 in [**AS04b**] shows, since we have $F^*(N) = O_2(N)$, that $O_2(X) \neq 1$, so that X lies in the set $\mathcal{L}(N,T)$ in Definition 1.2.3 of [**AS04b**]:

DEFINITION 3.2.5. For N a finite group and T a 2-subgroup of N, $\mathcal{L}(N,T)$ is the set of subgroups X of N such that

(1) X is a $\mathcal{C}$-component of $\langle X, T\rangle$,
(2) $T \in Syl_2(\langle X, T\rangle)$, and
(3) $O_2(\langle X, T\rangle) \neq 1$. ◇

Hence X also lies in $\mathcal{L}(G,T)$. Then we can embed X in some L in the set $\mathcal{L}^*(G,T)$ of maximal members of $\mathcal{L}(G,T)$. The possible members of $\mathcal{L}(G,T)$ are described using the theory in Section A.3 of [**AS04a**] (e.g. A.3.6), with the members of $\mathcal{L}^*(G,T)$ determined by results on inclusions among such groups (such as A.3.12).

Now Theorem 1.2.7 there establishes various important properties of such a subgroup L which is a member of $\mathcal{L}^*(G,T)$: For example:

$$L \text{ is a } \mathcal{C}\text{-component of } N_G(L^T) =: M.$$

Recall that L has at most two conjugates in M. Of crucial importance is the uniqueness property (mentioned earlier at (3.1.1)):

(3.2.6) $$\text{For } L \in \mathcal{L}^*(G,T), \text{ we have } M := N_G(L^T) = !\mathcal{M}(\langle L,T\rangle).$$

For this reason L is called a *(nonsolvable) uniqueness subgroup*. The local subgroup M is the maximal 2-local used to implement the Thompson strategy. The uniqueness result (3.2.6) is a rough analogue of (8.5.5) in our discussion of the Uniqueness Case. The relevance of uniqueness results to the weak closure method is discussed for example in the proof of Theorem B.8.3.

REMARK 3.2.7 (The Possible Nonsolvable Uniqueness Groups L). The possibilities for $L/O_2(L)$, with $L \in \mathcal{L}^*(G,T)$, are given in 1.2.1.4 of [**AS04b**]; we summarize them as:

(a) a simple SQTK-group from the SQTK-list in Remark 3.2.2;

(b) a triple cover of one of the groups in (a), with some further restrictions;

(c) an extension of E_{p^2}, $p > 3$, by $SL_2(p)$;

(d) an extension of a nilpotent group by $SL_2(5)$, with some further restrictions.

The examples we encounter in practice will usually appear in cases (a) and (b)—as we will see in a moment. ♢

In the generic quasithin examples, where G is of Lie type and BN-rank 2 over a field of order q with $q > 2$, M is a maximal parabolic and $L = [M,M]$. However, the proof must also deal with shadows of more complicated groups, such as wreath products of groups of BN-rank 2, or $L_4(2^n)$ extended by a graph automorphism, where L is not normal in M.

An important subset of the set $\mathcal{L}^*(G,T)$ is the set denoted by $\mathcal{L}^*_f(G.T)$, which consists of those $L \in \mathcal{L}^*(G,T)$ such that $[\Omega_1(Z(O_2(L))), L] \neq 1$; i.e. $\Omega_1(Z(O_2(L)))$ affords a nontrivial L-module. Typically this is a faithful module for $L/O_2(L)$, hence the letter f for faithful.

3.3. More preliminaries: Some general techniques

About half of [**AS04a**], consisting of the material in Chapters B–F, is devoted to background on a number of general concepts which we introduce briefly in this section. These notions are in fact used elsewhere in the classification of groups of characteristic 2 type; indeed we have described most of them in our Appendix B—to which the reader can also refer for more detail.

Our aim in this section will be to try to give at least some of the flavor of each technique; in the later sections describing the main part of the proof, we will often just refer briefly to such applications by labels such as "pushing-up" or "weak closure".

Our conventions for discussing general techniques. Usually in the subsections to follow we will, first, describe each technique as it is applied in the general finite group. Then, second, we discuss applications to the study of quasithin groups. Finally, third, in a subsequent Remark, we will give a few specifics, indicating how the technique is applied during the proof in [**AS04b**]; in so doing, we will often make use of the notation M, H, etc.—established earlier in our discussion of the Thompson strategy in Section 3.1. Usually M will denote the maximal 2-local defined via the condition $M = !\mathcal{M}(\langle L, T\rangle)$ in (3.2.6), for a suitable nonsolvable uniqueness subgroup L; while H denotes a member of $\mathcal{H}_*(T, M)$ as in Definition 3.1.2, i.e., typically an abstract minimal parabolic subgroup of G over T not contained in M.

Determining modules with q or $\hat{q}$ small. Chapter B of [**AS04a**] contains various results related to failure of factorization (FF) and some generalizations; the reader is referred to our exposition in Section B.6 for background on this material.

Specifically, Chapter B of [**AS04a**] discusses pairs $(\overline{X}, U)$ where $\overline{X}$ is an SQTK-group with $F^*(\overline{X})$ quasisimple and U is a faithful $\mathbb{F}_2\overline{X}$-module for which the parameter $q(\overline{X}, U)$ or $\hat{q}(\overline{X}, U)$ is small. These parameters are defined in our Appendix at B.6.6, but for the convenience of the reader we also reproduce them here:

DEFINITION 3.3.1. Let U be a faithful $\mathbb{F}_2\overline{X}$-module. We define $q(\overline{X}, U)$ to be the minimum of the ratio:

$$\frac{m(U/C_U(A))}{m(A/C_A(U))}$$

as A varies over those nontrivial elementary abelian 2-subgroups of $\overline{X}$ which act quadratically on U, i.e., such that $[U, A, A] = 0$.

We define $\hat{q}(\overline{X}, U)$ to be the minimum value of the same ratio, where now we allow A to vary over those nontrivial elementary abelian 2-subgroups of $\overline{X}$ which act cubically on U, i.e. such that $[U, A, A, A] = 0$.

(Thus $\hat{q}(\overline{X}, U) \leq q(\overline{X}, U)$.) ◊

REMARK 3.3.2. If $q(\overline{X}, U) \leq 1$, we say that U is an *FF-module* for $\overline{X}$. Here FF stands for "failure of factorization". In our discussion of failure of factorization in the Appendix, following Definition B.6.4, G is a finite group and V is a normal elementary abelian 2-subgroup of G, so that V in fact affords a faithful $\mathbb{F}_2$-module for $\overline{G} = G/C_G(V)$, and $\overline{G}, V$ plays the the role of "$\overline{X}, U$" above. Further T is a Sylow 2-subgroup of G and $J(T)$ is the Thompson subgroup defined there. Then the failure of factorization modules (FF-modules) are those where the Thompson factorization:

$$G = C_G(V)N_G\big(J(T)\big),$$

given at (B.6.5), fails to hold. As suggested above, this implies that $q(\overline{G}, V) \leq 1$ (cf. Remark B.6.6).

Chapter D of [**AS04a**] supplies the motivation for determining in Chapter B the larger class of modules satisfying $\hat{q}(\overline{G}, V) \leq 2$; namely it proves and explores the consequences of Theorem D.1.5 in [**AS04a**], which is a version of the Meierfrankenfeld-Stellmacher qrc-Lemma:[6]

[6]We mention that "R_2" in the conclusion of [**AS04a**, D.1.3] should actually be "$C_{R_2}(U)$", as in the proof there.

THEOREM 3.3.3 (The *qrc*-Lemma). *Let G_1 and G_2 be finite subgroups of a group G, having a common Sylow 2-subgroup T, and such that $F^*(G_i) = O_2(G_i)$ for $i = 1, 2$. Set $R := O_2(G_1)$, and suppose that V is a normal subgroup of G_1 contained in $\Omega_1(Z(R))$ with $R = C_T(V)$. Assume furthermore that $O_2(Aut_{G_1}(V)) = 1$ and that G_2 is an abstract minimal parabolic. Finally, assume that no nontrivial subgroup of R is normal in $\langle G_1, G_2 \rangle$. Then one of the following conditions holds:*

(1) $V \not\leq O_2(G_2)$—so that $\hat{q}(G_1, V) \leq 2$ by E.2.13.2.

(2) $q(G_1, V) \leq 1$ (so that V is an FF-module for G_1).

(3) The dual of V is an FF-module for G_1—so that $q(G_1, V) \leq 2$ by B.5.13.

(4) $q(G_1, V) \leq 2$ (with further information on $U := \langle V^{G_2} \rangle$).

(5) A case with $R \cap O_2(G_2)$ normal in G_2 (which will not arise in practice).

REMARK 3.3.4. Later we learn that in a QTKE-group G, each of the conclusions of the *qrc*-Lemma implies that $\hat{q}(G_1/C_{G_1}(V), V) \leq 2$, and sometimes stronger restrictions hold.

REMARK 3.3.5. The principal applications of the *qrc*-Lemma occur in the Fundamental Setup (i.e. 3.5.2), where $M = !\mathcal{M}(\langle L, T \rangle)$, V is a normal elementary abelian 2-subgroup of M, $R = C_T(V)$, the role of "G_1" is played by $N_M(R)$, and the role of "G_2" is played by $H \in \mathcal{H}_*(T, M)$. The fact that $\langle L, T \rangle$ is a uniqueness subgroup supplies the hypothesis in the *qrc*-Lemma that no nontrivial subgroup of R is normal in $\langle G_1, G_2 \rangle$. ♢

The basic general results on failure of factorization and related modules are reviewed in Sections B.1 and B.2 of [**AS04a**]. More specific results are then deduced in the quasithin case. First Theorem B.4.2 in [**AS04a**] lists the possible pairs $(\overline{X}, U)$, when $\overline{X}$ is an SQTK-group with $F^*(\overline{X})$ quasisimple, and U is a faithful irreducible FF-module (that is, $q(\overline{X}, U) \leq 1$). The generic groups $F^*(\overline{X})$ which arise are:

$$L_2(2^n), SL_3(2^n), Sp_4(2^n), G_2(2^n),$$

with U the natural module. Also arising are certain small examples involving:

$$A_5, A_6, A_7, 3A_6, L_4(2), L_5(2).$$

Since some of the original work determining the FF-modules for all nearly simple groups (e.g. by Cooperstein–Mason and McClurg) was never published, the material here quotes the more recent and comprehensive treatment by Guralnick and Malle [**GM02, GM04**]. The deduction from their work appears in Chapter K of [**AS04a**].

Then Theorem B.4.5 in [**AS04a**] extends that list to modules with $\hat{q}(\overline{X}, U) \leq 2$, in the language of Definition 3.3.1. The new groups which arise are:

$$U_3(2^n), Sz(2^n), M_{12}, M_{22}, M_{23}, M_{24}, 3M_{22}, J_2.$$

Moreover, certain additional modules arise for the groups $L_2(2^n)$, $L_3(2^n)$, $Sp_4(2^n)$, $G_2(2)'$, and A_7.

Again the result is deduced in Chapter K of [**AS04a**] from the list determined by Guralnick and Malle [**GM02, GM04**]. Section B.5 of [**AS04a**] extends the above results, in particular removing the condition that $\overline{X}$ is irreducible on U; see Theorems B.5.1 and B.5.6 in [**AS04a**].

Other results in Chapter B analyze particular groups $\overline{X}$ and modules U in some detail. The case where $\overline{X}$ is not of Lie type and characteristic 2 involves estimates on additional parameters associated to the representation, discussed in Section A.4

of our Appendix A; Chapter H of [**AS04a**] continues this theme, building on the original computations of Aschbacher in [**Asc82**].

Pushing-up theorems for SQTK-groups. The theory of pushing-up is discussed in Chapter C of [**AS04a**]. For general background on this area, we refer the reader to the exposition in Section B.7 of Appendix B. Here we simply summarize the basic setup:

We consider a subgroup X of G having $F^*(X) = O_2(X)$ and Sylow 2-subgroup T. *Pushing-up* refers to the process of replacing X with $N_G(C)$, where C is a nontrivial characteristic subgroup of T which is normal in X. Pushing-up necessarily fails if $C(X,T) \leq Y < X$ for $T \in Syl_2(X)$; recall from B.7.1 that $C(X,T)$ is the subgroup of X generated by the normalizers in X of all nontrivial characteristic subgroups of T.

Section C.1 of [**AS04a**] gives a fairly self-contained development, for quasithin groups, of the basics of pushing-up; though some ideas and results are quoted from [**Asc81a**]. That treatment culminates in C.1.29, which is a version of the Local $C(G,T)$-Theorem, stating that X is the product of $C(X,T)$ with certain "obstructions" called χ_0-blocks. These χ_0-blocks are a subset of the χ-blocks in our Definition B.7.2.

DEFINITION 3.3.6. A χ_0*-block* of X is a subnormal subgroup L of X such that
(1) $F^*(L) = O_2(L) = \Omega_1(Z(O_2(L)))$,
(2) $L/O_2(L) \cong SL_2(2^n)$ for some $n \geq 1$, and
(3) For $U := [O_2(L), L]$, $U/C_U(L)$ is the natural module for $SL_2(2^n)$, or for A_5 (if $n = 2$).
Indeed, $U = O_2(L)$, unless perhaps $n = 2$ and $L/U \cong SL_2(5)$. ◊

Then C.1.29 gives the Local $C(G,T)$ Theorem B.7.3 in the special case of SQTK-groups:

THEOREM 3.3.7 (The SQTK Local $C(G,T)$-Theorem). *Assume X is an SQTK-group such that $F^*(X) = O_2(X)$. For $T \in Syl_2(X)$ assume $C(X,T) =: Y < X$.*
Then $X = YL_1 \dots L_r$, with $1 \leq r \leq 2$, and each L_i is a χ_0-block of X for which $(L_i \cap M)/O_2(L_i)$ is a Borel subgroup of $L_i/O_2(L_i)$, and $L_i = [L_i, J(T)]$.

Chapter C also considers more general situations where the Sylow 2-subgroup T of X is replaced by a suitable *proper* subgroup R of T for which $C(X,R) < X$. In such situations the obstructions can again be enumerated. They are more general than χ_0-blocks, but of the same flavor. One example is given by Theorem C.1.32, which quotes pushing-up theory for groups X with $X/O_2(X)$ of Lie type and BN-rank 2, due to Meierfrankenfeld–Stellmacher in [**MS93**]. In this result R is the preimage in X of the unipotent radical of a maximal parabolic of $X/O_2(X)$.

REMARK 3.3.8. The proof in [**AS04b**] of the Main Theorem uses pushing-up in many ways; here we will focus on one important class of applications which show certain 2-local subgroups X are contained in the maximal 2-local M of the Thompson strategy. First if $1 \neq R \trianglelefteq T$ is invariant under a uniqueness subgroup L as in Section 1.2, the uniqueness property (3.2.6) shows that $C(G,R) \leq M$. So if $R \leq X$ but $X \not\leq M$, then $C(X,R) \leq X \cap C(G,R) \leq X \cap M < X$, so that $X \cap M$ can play the role of "M_X" in Hypothesis C.2.8. Then if we can show that the obstructions in C.4.8 do not exist, we can conclude that $X \leq M$. Chapter 4 of [**AS04b**] uses such 2-local subgroups to show other subgroups Y are contained in M; for example in Theorems 4.4.3 and 4.4.14 there, "Y" is the normalizer of a suitable subgroup of odd order.

Weak Closure. Section E.3 of [**AS04a**] reviews weak closure. For background on the subject, we refer the reader to Section B.8 in Appendix B. The terminology arises in the following way. Let Y be a subgroup of a group X. Recall that a subgroup W of Y is said to be *weakly closed* in Y with respect to X if W is the unique member of its X-conjugacy class contained in Y. For example if Y is a Sylow p-subgroup of the finite group X, then the Thompson subgroup $J(Y)$ is weakly closed in Y with respect to X. Very roughly, given a family W_k of weakly closed subgroups of a Sylow p-subgroup T of X, and setting $C_k := C_T(W_k)$, weak closure arguments are designed to produce factorizations involving two values of k, of the form

$$X = X_1X_2 := C_X(C_i)N_X(W_j). \tag{3.3.9}$$

We consider such factorizations at B.8.6 in Appendix B; they generalize the Thompson factorization (B.6.5): For assume that $J(T) \leq C_T(V)$, for a suitable module V for X and take $W_j := J(T)$. We have $J(T) = J\big(C_T(V)\big)$, and then a Frattini argument yields:

$$G = C_G(V)N_G\big(C_T(V)\big) = C_G(V)N_G\big(J(T)\big),$$

so that the second factor has the form indicated in (3.3.9). Suppose also $Z(T) \leq V$, so that we can replace the first factor by $C_G\big(Z(T)\big)$; then taking $W_i := T$ we get $C_i = Z(T)$, and now the first factor is also as in (3.3.9).

Sufficient conditions for such factorizations are often given in terms of "$n(X)$" and various other parameters described in B.8.2 and B.8.4. Definition E.1.6 gives the precise definition of $n(X)$. Roughly speaking, $n(X) = n$ when $X/O_2(X)$ is of Lie type over $\mathbb{F}_{2^n}$, and $n(X) = 1$ otherwise.

In particular, when X is an abstract minimal parabolic, many arguments hinge on the value of $n(X)$; note that when $n(X) > 1$, we can take advantage of the existence of a nontrivial Cartan subgroup.

Weak closure factorizations. Section E.3 of [**AS04a**] begins with a review of the basics of the weak closure method, and quotes a number of results from [**Asc81e**]. For example, E.3.18 and E.3.19 in [**AS04a**] correspond to our B.8.5 and B.8.6, while E.6.3 in [**AS04a**] is a version of our B.8.3. These factorizations had been used extensively in [**Asc83b, Asc83c**] in the proof of the Uniqueness Case Theorem, which of course preceded work on the quasithin groups; the reader may wish to refer for comparison to the applications of weak closure discussed in Chapter 8—for example, around (8.2.5), (8.4.3), and 8.5.23. In more "classical" applications of weak closure, "X" is our second 2-local subgroup H: so if we get the factorization $H = H_1H_2$, and we also have uniqueness theorems (in the sense of Remark 8.0.4) to force $H_i \leq M$, then we have obtained a contradiction to our choice of $H \not\leq M$.

Some further developments of the method. Subsection E.3.3 of [**AS04a**] is then largely devoted to a discussion of how weak closure is applied in [**AS04b**]. But in addition, this subsection extends the classical factorization theory used in the Uniqueness Case—roughly, the factorizations discussed in the previous paragraph are assumed to fail as $H \not\leq M$, and this failure is used to establish inequalities among the weak closure parameters. So we now give a brief indication of the flavor of those extensions.

We recall the Thompson strategy of Section 3.1, in the context of our discussion of weak closure in Section B.8 of Appendix B: we are studying an elementary 2-subgroup V normal in the maximal 2-local subgroup M of G containing a Sylow 2-subgroup T of G. We also consider some $H \not\leq M$ and set $\overline{M} := M/C_M(V)$. Recall from Definition B.8.1 the following family $\{W_i(T,V)\}$ of weakly closed subgroups of T. First we define:

$$\Gamma_i(T,V) := \{A \leq T : A \leq V^g \text{ and } m(V^g/A) = i \text{ for some } g \in G\};$$
$$W_i(T,V) := \langle \Gamma_i(T,V) \rangle \text{ and } C_i(T,V) := C_T\big(W_i(T,V)\big).$$

We next introduce another parameter w. As we saw at (B.6.3), weak closure factorizations are often predicated on showing that the j-th weak closure $W_j(T,V)$ of V centralizes some module U. We now restrict to the special case $V = U$; and set:

DEFINITION 3.3.10. $w = w(G,V) := \min\{j : W_j(T,V) \not\leq C_G(V)\}$.

Thus the indices j for which we can typically expect a factorization of the form given in 3.3.9 satisfy $j < w$, so we are primarily interested in the situation where $w > 0$.

When $w > 0$ (and also when V is not an FF-module), E.3.29 in [**AS04a**] establishes a relationship among certain parameters called the Fundamental Weak Closure Inequality (FWCI):

(3.3.11) If $w > 0$, then $m_2(\overline{M}) + w > r$,

where we recall from B.8.2 that $r = r(G,V)$ is defined as:

$$r(G,V) := min\{m(V/U) : U \leq V \text{ and } C_G(U) \not\leq N_G(V)\}.$$

Since often we can estimate the parameters r and $m_2(\overline{M})$, we obtain a lower bound on w, useful in establishing the exact value of w. (And the FWCI frequently turns out to be an equality in examples in the Quasithin Theorem, as well as in shadows which satisfy most of its hypotheses.) The parameter w is related (and often even equal) to the parameter $n(H)$ for our subgroup H in the Thompson strategy.

In certain circumstances Proposition B.8.6 supplies a family of factorizations. Here is an example: Take $i = 0$, and suppose that $k = n(H) < s(G,V)$ (where s is defined in Remark B.8.2). Then by Proposition B.8.6:

$$H = C_H(C_k)N_H(W_0).$$

Suppose that $k < w$. Then V centralizes W_k, so that $V \leq C_T(W_k) = C_k$; hence:

$$C_H(C_k) \leq C_G(V) \leq N_G(V) = M.$$

As $w > 0$, $W_0 \leq C_T(V)$, and so W_0 is weakly closed in $C_T(V)$; and then it follows that $N_G\big(C_T(V)\big) \leq N_G(W_0)$. But we have $M = !\mathcal{M}(N_G\big(C_T(V)\big))$ as a consequence of (3.2.6), since usually:

$$C_T(V) = O_2(\langle L,T \rangle) \trianglelefteq \langle L,T \rangle.$$

We conclude that $N_G(W_0) \leq M$. Thus both factors of H are contained in M, contrary to $H \not\leq M$. This leads us to expect instead that $w \leq k = n(H)$.

The uniqueness condition $M = !\mathcal{M}(N_G\big(C_T(V)\big))$ is part of Hypothesis E.3.36 in [**AS04a**], and a parameter $n'(\overline{M})$ analogous to $n(H)$ is defined in Definition E.3.37 of [**AS04a**]. As an example of the inequalities obtained, we record part (1) of E.3.39 in [**AS04a**]:

(3.3.12) Assume E.3.36, with $n'(\overline{M}) < r$. Then $w \leq n(H) \leq n'(\overline{M})$.

We can now give a rough indication of how these inequalities are often used in practice:

REMARK 3.3.13. In the proof in [**AS04b**], weak closure is the primary tool used in Chapters 7 and 8; and it is frequently useful elsewhere.

The usual first step, in weak closure involving a subgroup V, is to establish some "control" of the centralizers of subgroups of V; cf. our discussion in Remark B.8.2. For the maximal 2-local M of the Fundamental Setup, with module V for the uniqueness subgroup L, we need to show, for suitable choices of $v \in V$, that we get $C_G(v) \leq N_G(V) = M$. We often establish such results using uniqueness theorems, which in turn may involve pushing-up arguments, as in Remark 3.3.8. This process leads to lower bounds on the value of r.

These bounds can lead in turn to a lower bound on w, supplied by the FWCI inequality (3.3.11). Furthermore we can often also apply E.3.39 of [**AS04a**] (compare also (3.3.12)) to obtain a lower bound on the parameter $n(H)$, for $H \in \mathcal{H}_*(T, M)$ in the Thompson strategy.

These bounds will sometimes lead to a contradiction; and at other times, to a unique solution for the values of the various parameters, which can be used to identify G as one of the groups in the conclusion of the Quasithin Theorem. $\diamondsuit$

Further preliminaries. The remainder of Chapter E in [**AS04a**] develops some estimates for a number of the weak closure parameters; these estimates depend in part on results mentioned in our Section A.4, notably those from [**Asc82**]; the values in Chapter H of [**AS04a**] are also relevant.

Amalgams. Chapter F of [**AS04a**] discusses the amalgam method. This theory developed, for the most part, after the apparent completion of the CFSG around 1981; thus the work on quasithin groups in [**AS04a, AS04b**] is the only part of the work on the classification of groups of characteristic 2 type to really exploit the amalgam method.

Basic amalgams: weak BN-pairs of rank 2. Section F.1 of [**AS04a**] quotes some results from the "Green Book" of Delgado, Goldschmidt, and Stellmacher [**DGS85**]. The amalgams in the Green Book are the weak BN-pairs of rank 2. We will describe these pairs more formally later at Definition 3.6.3; but for our present purposes, a weak BN pair is roughly defined as a quotient of a free product (with amalgamation) of two subgroups G_1 and G_2, which resemble minimal parabolics in a group of Lie type of BN rank 2—amalgamated over their intersection, which resembles a Borel subgroup. So in particular such a pair arises in the groups of Lie type of BN rank 2, which are the generic conclusions in the Quasithin Theorem 3.0.1. The corresponding weak BN-pair amalgams α are listed in F.1.12.

REMARK 3.3.14 (Amalgams and completions). In the more general language of the amalgam method, the *amalgam* records only the inclusion of $G_1 \cap G_2$ in G_1 and G_2; while a group X generated by G_1 and G_2 (subject of course to the relations arising from the common subgroup $G_1 \cap G_2$) is called a *completion* of that amalgam. The most general completion is given by the full free product with amalgamation; while a particular completion arising inside some G will be a quotient satisfying some further relations, and so typically cannot be identified without specifying suitable further information. $\diamondsuit$

Next Sections F.2–F.5 develop some techniques for identifying a particular group X of Lie type of BN rank 2 with input given by its weak BN-pair α, along with information such as an involution centralizer in X. Here the final recognition of X depends on the Tits-Weiss work [**TW02**] on Moufang polygons (or on the earlier Fong-Seitz work [**FS73**] on split BN-pairs of rank 2). Section F.6 is devoted to the important special case of the *Goldschmidt amalgams* (appearing in [**Gol80**]) where the subgroups G_1 and G_2 satisfy $G_i/O_2(G_i) \cong S_3$; Theorem F.6.18 gives a list of SQTK-groups X that can be generated by such subgroups.

REMARK 3.3.15. In applications in [**AS04b**], the subgroups "G_1" and "G_2" are often essentially the groups M and H of the Thompson strategy when those groups resemble minimal parabolics. The results above are then used for the final identification of the *completion* "X" of the amalgam: that is, the group generated by the subgroups G_1 and G_2, generated by that amalgam. In the Generic Case of the Quasithin Theorem in our discussion in Section 3.6, X is usually a group of Lie type of BN rank 2. In other cases, X may be a proper subgroup. ◇

Further amalgam methodology. The arguments in [**DGS85**] were subsequently extended to the more general *amalgam method*, investigating the graph Γ given by "incident" cosets of intersecting subgroups G_1 and G_2 which now are no longer required to resemble minimal parabolics.

Sections F.7–F.9 of [**AS04a**] develop a version of the amalgam method for SQTK-groups. The method analyzes paths H_0, H_1, ..., H_n in the above graph. Furthermore the assumptions in Hypothesis F.7.6 include a certain module V which is a subgroup of G_1, and the analysis focuses on a corresponding parameter:

DEFINITION 3.3.16 (The graph parameter b). Set:

$$b = b(\Gamma, V) := \min\{n : V \not\leq H_{n+1}\},$$

for some path H_0, H_1, ..., H_{n+1} in the graph. ◇

In groups of Lie type of BN-rank 2, this parameter b is related to the composition length of a unipotent radical in a minimal parabolic; while in the abstract, the size of b supplies information about the structure of the groups G_i.

Section F.8 develops some methods for the important special case where $b \geq 3$ is odd; in particular the analysis involves a subgroup V_1 playing the role of a 1-dimensional subspace (over some field of characteristic 2) for the action of G_1 on V. Section F.9 then pursues a parallel development in the case where such a 1-subspace V_1 is of order exactly 2.

REMARK 3.3.17. In applications in [**AS04b**], the roles of "G_1, G_2" are often played by the 2-locals M (or even just $\langle L, T\rangle$) and H in the Thompson strategy of Section 3.1, with V a suitable subgroup of our usual module for L (which we have also been calling V). The case with V_1 of order 2 in Section F.9 is particularly useful in Chapters 12–14 of [**AS04b**], when $\overline{L}$ is a group of Lie type over $\mathbb{F}_2$; it leads to some restrictions on H and the group $U = \langle V^H\rangle$. Furthermore the subcase where U is abelian usually corresponds to the situation $b \geq 3$ odd, where results proved under Hypothesis F.9.8 restrict the H-module $\tilde{U}$—for example, F.9.16 says that $q(Aut_H(\tilde{U}), \tilde{U})) \leq 2$. ◇

3.4. The degenerate case: A Sylow T in a unique maximal 2-local

Having established some preliminary background in the previous two sections, we are now ready to return to the main part of the proof in [**AS04b**] of the Quasithin Theorem.

As indicated in Section 3.1, the Thompson strategy is based on the existence of distinct maximal 2-locals over T; so it is not available until the "small" case in Theorem 2.1.1 of [**AS04b**] is treated:

THEOREM 3.4.1. *Assume that G is a simple QTKE-group such that a Sylow 2-subgroup T of G is contained in a unique maximal 2-local subgroup M.*
Then $G \cong L_2(2^m)$, $U_3(2^m)$, $Sz(2^m)$, $L_2(p)$ for $p > 7$ a Fermat or Mersenne prime, $L_3(3)$, or M_{11}.

REMARK 3.4.2. To see why $p > 7$, we recall that $L_2(5) \cong L_2(4)$, and so it is included in the family $L_2(2^m)$; while $L_2(7) \cong L_3(2)$ has BN-rank 2, and so fails the uniqueness hypothesis. In fact $L_3(2)$ will arise later from the Thompson strategy, at the very end of the analysis—see the discussion in our Section 3.11, or Section 14.7 of [**AS04b**]. $\diamondsuit$

REMARK 3.4.3. We mention (cf. Remark 8.2.3 in the Uniqueness Case) that if G were of characteristic 2 type, we could use the Global $C(G,T)$ Theorem B.7.8 to quickly deal with the case where T is in a unique maximal 2-local, since that hypothesis implies $C(G,T) < G$. Instead G is only of even characteristic, so it is necessary to provide an independent (and rather lengthy) proof of the analogous Theorem 3.4.1. Nonetheless the proof uses ideas suggested by the proof of the Global $C(G,T)$ Theorem in the GLS project, as we see in a moment. $\diamondsuit$

The proof of Theorem 3.4.1 is outlined at [**AS04b**, p. 517], and carried out there in Chapter 2. We now discuss certain parts of that proof.

Identifying the Bender groups. First Proposition 2.2.2 and Lemma 2.2.5 of [**AS04b**] isolate a configuration leading to the generic case of Theorem 3.4.1, in which G is a Bender group—$L_2(2^m)$, $U_3(2^m)$, or $Sz(2^m)$.

Of course this conclusion would follow from Bender's Theorem 1.3.5 if M were strongly embedded in G: that is, if $N_G(D) \leq M$ for all nontrivial subgroups D of T. But, following an approach of GLS at [**GLS94**, p. 97], we assume only that $N_G(D) \leq M$ for D in the set $\mathcal{S}$ of nontrivial subgroups D of T such that $F^*(N_G(D)) = O_2(N_G(D))$. (Notice $\mathcal{S}$ might not consist of all the nontrivial subgroups of T, since we are not assuming that G is of characteristic 2 type.) Lemma 2.2.5 of [**AS04b**] shows that $\mathcal{S}$ contains the Alperin-Goldschmidt conjugation family (Definition B.2.4), while Theorem 2.2.2 uses the Alperin-Goldschmidt Fusion Theorem B.2.6 to produce a strongly closed abelian 2-subgroup (cf. B.2.3). At that point it would be possible to finish quickly, by identifying G as a Bender group using Goldschmidt's theorem in [**Gol74**], which classifies the groups with a strongly closed abelian 2-subgroup. But instead, a fairly easy direct argument now completes the proof; the argument largely parallels the GLS proof of Goldschmidt's Theorem, given as Theorem SA in Section 24 of Chapter 2 in [**GLS99**]. (That result in turn can be regarded as a special case of the Shult Fusion Theorem, cf. [**AS04a**, I.8.3].)

One advantage of the GLS approach using the conjugation family is that later in the analysis, certain difficult configurations involving involutory automorphisms of Bender groups, and wreath products of Bender groups, are avoided.

The pairs (U, H_U) and the maximal set Γ_0. The proof now focuses on certain pairs (U, H_U),[7] again suggested by ideas in the GLS project—this time in their proof of the Global $C(G,T)$ Theorem B.7.8; that treatment is outlined in Section 2.10 of [**GLS94**].

For the remainder of the section, we assume that G is not a Bender group. In particular by Theorem 2.2.2 discussed above, there is some $D < T$ satisfying $F^*(N_G(D)) = O_2(N_G(D))$, but $N_G(D) \not\leq M$. We may choose D maximal with these properties, and let S denote a Sylow 2-subgroup of $N_M(D)$.

The argument on p. 522 of [**AS04b**] shows that $(S, N_G(D))$ is a pair (U, H_U) with $O_2(H_U) \neq 1$ and $H_U \not\leq M$, together with certain technical conditions on U given in Notation 2.3.1 of [**AS04b**]—notably $U \in Syl_2(H_U)$ and $F^*(H_U) = O_2(H_U)$. (This last condition crucially uses at Lemma 2.3.3 the choice of $D \in S$ from Theorem 2.2.2.) So Notation 2.3.5 defines Γ as the set of subgroups H of G for which $O_2(H) \neq 1$, and $H \geq H_U$ for some pair (U, H_U) as above; while Γ_0 is the subset of members of Γ satisfying certain maximality conditions. The remainder of Section 2.3 establishes properties of members of Γ_0. Then the proof proceeds by investigating the possibilities for $H \in \Gamma_0$.

Identifying the remaining examples. Assume first that $F^*(H) = O_2(H)$ for some member H of Γ_0, Then Theorem 2.4.1 of [**AS04b**] shows that $G \cong L_2(p)$ for $p > 7$ a Fermat or Mersenne prime, $L_3(3)$, or M_{11}. (These are the groups in the conclusion of Theorem 3.4.1 which remain, after the identification of the Bender groups in Proposition 2.2.2 of [**AS04b**].)

The argument begins by invoking Lemma 2.3.8.4: this essentially makes use of the uniqueness hypothesis on M, to apply C.1.29 of [**AS04a**] (the Local $C(G,T)$ Theorem B.7.3 in the SQTK subcase 3.3.7) to H, and describes H in terms of certain χ_0-blocks L. (Definition 3.3.6).

Considerable argument is then required in the case of L an $L_2(2^m)$-block in Theorem 2.4.7, in order to eliminate certain "shadows"[8] which satisfy the hypotheses of Theorem 3.4.1 except for simplicity; for example, when $G = G_0\langle\gamma\rangle$, with $G_0 \cong L_3(2^m)$ and γ is an involutory graph or unitary automorphism. This subcase is eventually eliminated in the proof of 2.4.24, via Thompson Transfer B.2.9; but only after detailed reconstruction of much local structure. The arguments essentially begin by applying some more specialized pushing-up results such as C.6.1 and C.5.6 from [**AS04a**].

The case of L a normal A_3-block leads at 2.4.26 (again using C.6.1) to $H \cong S_4$, so that the Sylow 2-subgroup T is dihedral or semidihedral; and then the preliminary result [**AS04a**, I.4.3] (which essentially quotes the 2-Rank 2 Theorem 1.4.6) identifies G as $L_2(p)$, $L_3(3)$, or M_{11}.

At this point, H contains either a single A_5-block, or the product of two A_3-blocks. These can occur in shadows given by non-simple extensions of $L_4(3)$, and it

[7]In this subsection, we essentially follow the notation in [**AS04b**]; this involves a temporary deviation from our usual notational convention, where H and U were reserved for certain subgroups in the Thompson strategy in Section 3.1.

[8]For the informal terminology of *shadows*, see [**AS04b**, p. 484].

requires several more pages of analysis (again using C.5.6 and C.6.1) to eliminate these shadows.

The final contradiction. Thus in Section 2.5 of [**AS04b**], all the members H of Γ_0 now satisfy the condition $F^*(H) \neq O_2(H)$; and it remains to establish a contradiction. Here $E(H)$ must by 2.5.3 in fact contain a component K described in 2.3.9.7. The various possibilities for K are then successively ruled out—but the final eliminations are complicated by shadows discussed in 2.5.14 and 2.5.17: namely of further non-simple extensions of $L_4(3)$, and of wreath products of groups $L_2(p)$.

3.5. The Main Case Division (Possibilities for a suitable group $\overline{L}$ and module V)

By Theorem 3.4.1, we may assume henceforth (i.e. starting with Chapter 3 of [**AS04b**]) that:

For $T \in Syl_2(G)$, M is not *the only maximal* 2*-local containing T* .

So we are now in a position to implement the Thompson strategy described in Section 3.1. In particular, once we have chosen M, we can always find members H of the set $\mathcal{H}_*(T, M)$ defined in 3.1.2. As noted before, H is usually an abstract minimal parabolic (see Remark 3.5.3).

We emphasize that *many later arguments will obtain a contradiction by forcing H to be contained in M*—contrary to the condition $H \not\leq M$ in the definition of $\mathcal{H}_*(T, M)$.

Chapter 3 of [**AS04b**] establishes the primary case division in the proof—which corresponds essentially to possible choices of M, or more precisely, to choices of suitable subnormal subgroups of M.

Describing M via a uniqueness group L. Instead of trying to list all possibilities for M, it will be more practical to produce a list of possibilities for a "uniqueness subgroup" L in M. Since this part of the analysis is carried on within a 2-local overgroup M of T, it is independent of the results in Chapter 2 of [**AS04b**] which deliver the second 2-local H for the Thompson strategy; so this material is in fact developed earlier, in Chapter 1 of [**AS04b**].

Whenever some maximal 2-local over T is nonsolvable, we may proceed in the manner described in our discussion in Section 3.2; namely we use the theory of $\mathcal{C}$-components and the structure of SQTK-groups to get a maximal 2-local overgroup M of T satisfying (3.2.6)—recall this says:

$$L \in \mathcal{L}^*(G, T) \text{ with } M := N_G(L^T) = !\mathcal{M}(\langle L, T \rangle). \tag{3.5.1}$$

As noted before, L is called a *(nonsolvable) uniqueness subgroup*. Moreover in general $L/O_2(L)$ is quasisimple—as in cases (a) and (b) of Remark 3.2.7. Hence the bulk of the proof focuses on this situation—in the Fundamental Setup, which we will describe shortly.

So the non-quasisimple cases (c) and (d) of Remark 3.2.7 are essentially just a nuisance lurking in the background until late in the game, when they are eliminated in 13.1.7 of [**AS04b**]. But the nuisance is comparatively mild, since each contains a solvable uniqueness subgroup. Thus one can focus instead on a subgroup inside the solvable group $O_{2,F}(L)$ (the preimage of $F\big(L/O_2(L)\big)$), leading to the definition after 1.3.3 of [**AS04b**] of a set $\Xi^*(G, T)$ of *solvable uniqueness groups*. By 1.3.7 in [**AS04b**], we get for $X \in \Xi^*(G, T)$ an analogue $M = !\mathcal{M}(XT)$ of the uniqueness property (3.5.1) for L. This allows us to shift attention to members X of $\Xi^*(G,T)$. There are only two possibilities for $X/O_2(X)$—namely for some odd prime p, $X/O_2(X)$ is either elementary of order p^2, or extraspecial of order p^3. (This situation does not arise in actual examples, and does not significantly complicate the proof; so we will de-emphasize it in our exposition—and the reader can probably just ignore it.)

Of course, it is also eventually necessary to deal with the case where $\mathcal{L}^*(G,T)$ is empty, or equivalently, when all 2-local overgroups of T are solvable. This case is treated at the very end of the proof, in Section 15.4 of [**AS04b**]. The argument utilizes yet another set $\xi^*(G,T)$ of solvable uniqueness subgroups, which is somewhat similar to $\Xi^*(G,T)$—except that its members X instead have $X/O_2(X)$ of prime order.

The Fundamental Setup (the main cases for $L/O_2(L)$ and a suitable module V). We now return to the situation in Chapter 3 of [**AS04b**], where, in addition to a maximal 2-local M (chosen if possible to contain a uniqueness group L or X as above), we are also given a subgroup $H \in \mathcal{H}_*(T,M)$. We summarize the development in Section 3.2 of [**AS04b**]; cf. also the exposition in Subsection 0.3.3 of the Introduction to [**AS04b**].

Restrictions on the "initial" module W for a uniqueness group L. As M is a maximal 2-local and $H \not\leq M$, no nontrivial subgroup of T is centralized by both M and H. Thus we can hope to choose our pair so that M does not centralize $\Omega_1(Z(T))$. In that event there exists a certain $\langle L,T\rangle$-invariant elementary abelian subgroup W of T, which we regard as an $\mathbb{F}_2\langle L,T\rangle$-module. This choice corresponds roughly to choosing[9] L in the "faithful" subset $\mathcal{L}_f^*(G,T)$ of Section 1.2 of [**AS04b**]. We next explore L and W (we will return to H somewhat later).

A crucial consequence of this choice of L is that 3.2.2.9 of [**AS04b**] imposes strong restrictions on W: For details, see our Remark 3.3.5; in summary, the uniqueness property $M = !\mathcal{M}(\langle L,T\rangle)$ in (3.5.1) supplies the hypothesis of the *qrc*-lemma D.1.5 of [**AS04a**] (see our Lemma 3.3.3)—which shows that the parameter $\hat{q} = \hat{q}(Aut_{\langle L,T\rangle}(W),W)$ (in the language of Definition 3.3.1) is at most 2; and if $\hat{q} = 2$, then often also $q = q(Aut_{\langle L,T\rangle}(W),W) \leq 2$. Consequently the pair L,W appears on the list of such modules in B.4.5 in [**AS04a**].

In the discussion above, L was a nonsolvable uniqueness group; but a similar setup arises when we work instead with a suitable solvable uniqueness group X in $\Xi_f^*(G,T)$. The treatment of solvable uniqueness subgroups is usually fairly easy.

A corresponding module V for M (and possible cases). We will continue to concentrate our exposition on the case where $L/O_2(L)$ is quasisimple, as in conclusions (a) and (b) of Remark 3.2.7. With this addition, the previous discussion leads to:

REMARK 3.5.2 (The Fundamental Setup (FSU)). Most of the cases treated in the proof of the Quasithin Theorem appear in the *Fundamental Setup* defined in Definition 3.2.1 of [**AS04b**], which we now summarize somewhat roughly:

Assume there is some nonsolvable uniqueness group L for M such that $L/O_2(L)$ is quasisimple. Recall then we have $M = N_G(\langle L^T\rangle)$.

---

[9]In an example G of Lie type, the choice corresponds to choosing M to be a maximal parabolic other than the normalizer of a long-root group.

Assume in addition that $L \in \mathcal{L}_f^*(G,T)$ and choose W as above, so roughly W is 2-reduced (cf. Remark 8.3.7) and L is irreducible on $W/C_W(L)$.

The remaining results in Section 3.2 of [**AS04b**] analyze possibilities under the hypotheses of the FSU. In particular we shift attention from W to $V := \langle W^M \rangle$. Regard V as an $\mathbb{F}_2$-module for M, and adopt the bar convention for images in the quotient:

$$\overline{M} := M/C_M(V) .$$

The restrictions on W discussed above, such as the bound $\hat{q} \leq 2$ obtained from the *qrc*-lemma, are used to describe the possibilities for the $\mathbb{F}_2\overline{M}$-module V for each possible choice of the known quasisimple group $\overline{L}$.

The resulting list of modules is essentially that given in B.4.5 of [**AS04a**]: roughly it consists of a few small modules for each choice of $\overline{L}$, such as the natural module—cf. the list indicated informally after our Remark 3.3.5. Since not all the quasisimple groups in Remark 3.2.7(a)(b) have such modules, the possibilities for $\overline{L}$ are reduced to those on a certain sublist of the simple SQTK-groups in Remark 3.2.2.

The results are summarized in Theorems 3.2.5 and 3.2.6 of [**AS04b**]. Most of the proof of the Quasithin Theorem will be devoted to the situation of 3.2.5 of [**AS04b**], with 3.2.8 and 3.2.9 supplying a more explicit list of modules in the subcases where V is, and is not, an FF-module (that is, where $q \leq 1$, or $q > 1$). ♢

The results described above provide a solution of what at [**AS04b**, p. 492] is called the First Main Problem: a priori, M might be any strongly quasithin group; but the existence of the nonsolvable uniqueness subgroup L, and the restrictions on $\overline{L}$ and V, show that M more closely resembles a maximal 2-local in the groups in the conclusion of the Quasithin Theorem 3.0.1. So the bulk of the remaining work will consist of an analysis of these possibilities for $\overline{L}$ and V.

Other preliminary technical results. Theorem 3.3.1 of [**AS04b**] shows, with some effort, that for M chosen with respect to a uniqueness group L as described in the FSU (see our Remark 3.5.2), we get $N_G(T) \leq M$. (This can be compared with Theorem 8.1.5 for the Uniqueness Case.)

REMARK 3.5.3 (H as abstract minimal parabolic). A crucial technical consequence of Theorem 3.3.1 of [**AS04b**] is Corollary 3.3.2 of [**AS04b**], which shows that each member of $\mathcal{H}_*(T,M)$ is an abstract minimal parabolic (our Definition 3.1.3).

So in the Thompson strategy, the case division will correspond essentially to the Cartesian product of the list of possibilities for M, determined by the possibilities for $\overline{L}$ and V in the FSU (our 3.5.2), with the list of possibilities for the abstract minimal parabolic H in E.2.1 and E.2.2 of [**AS04a**]. The size of this list may help explain why the treatment of the "small" quasithin case should require two lengthy volumes. ♢

Chapter 4 of [**AS04b**] contains a collection of preliminary results, applying pushing-up in ways that build on the discussion in Remark 3.3.8: Recall that Sections C.2–C.4 of [**AS04a**] established some "local" pushing-up results; that is, they describe certain subgroups K in an SQTK-group X with $O_2(X) \neq 1$, that are obstructions to $C(X,R) = X$ for suitable 2-subgroups R of X. In Chapter 4 of [**AS04b**], those results are applied "globally": now the SQTK-group X is a 2-local in some QTKE-group Y, which we also wish to describe. Typically the aim is to force Y into the 2-local M of the Thompson strategy. This approach is often used in the proof of uniqueness theorems.

Indeed when M arises from a pair $\overline{L}$, V in the FSU, 4.2.4.3 in [**AS04b**] shows that the pushing-up hypothesis C.2.8 of [**AS04a**] is satisfied for a suitable 2-subgroup R which is normalized by L and T; in particular then $C(G, R) \leq M$, so that if $X \not\leq M$ then $C(X, R) \leq X \cap M < X$. Theorem 4.2.13 in [**AS04b**] then provides a refinement of the earlier list in C.4.8 of [**AS04a**] of possible obstructions K. The obstructions in 4.2.13 can be ruled out in various circumstances—for example when Y covers $L/O_2(L)$. Then forcing Y into M shows that M is not just the unique maximal 2-local above $\langle L, T\rangle$ (as in (3.5.1)), but also $M = !\mathcal{M}(J)$ for any J covering $\langle L, T\rangle$ modulo $O_2(\langle L, T\rangle)$. Several variations on this theme are established in Sections 4.3 and 4.4: for example, $M = !\mathcal{M}(L)$ (i.e., T is not required) when $\overline{L} \cong L_2(2^n)$ in Theorem 4.3.2; and M is the unique maximal 2-local subgroup above certain odd locals in Theorems 4.4.3 and 4.4.14.

Overview of the case analysis for the proof. Most of the remaining chapters in the proof in [**AS04b**] consist of analyses of the possibilities for $\overline{L}$ and V in the Fundamental Setup 3.5.2; we begin with an outline of their treatment:

First, the infinite families of examples (the "generic" case) are the groups G of Lie type over fields of characteristic 2 with BN-rank 2; aside from the groups defined over $\mathbb{F}_2$ these arise when:

(a) $\overline{L} \cong L_2(2^n)$ for $n > 1$, and V is the natural or orthogonal module.

This situation is treated in Chapters 5 and 6 of [**AS04b**], and is discussed here in Section 3.6. Next, most pairs $\overline{L}$ and V which do *not* lead to an example are eliminated in the treatment of:

(b) V is not an FF-module for $\overline{L}$,

appearing in Chapters 7–9 of [**AS04b**], and discussed here in Section 3.7. The effect of these results is to reduce to a small (indeed after Chapter 11 of [**AS04b**], a finite) number of configurations, in:

(c) V is an FF-module for $\overline{L}$.

However, the analysis of the various subcases of (c) is rather delicate and lengthy—it occupies Chapters 10–14 of [**AS04b**] (about 300 pages), and in our discussion it is subdivided into Sections 3.8–3.10, according to the possibilities for $\overline{L}$.

Finally Chapter 15 of [**AS04b**] deals with those cases where the set $\mathcal{L}_f^*(G, T)$ of faithful nonsolvable uniqueness groups L is empty, discussed in our Section 3.11. This case is of course outside the FSU 3.5.2.

3.6. The Generic Case—where $\overline{L} = L_2(2^n)$ with $n > 1$

Chapters 5 and 6 of [**AS04b**] contain most of the analysis for the case where we have $\overline{L} \cong L_2(2^n)$ in the Fundamental Setup 3.5.2. One reason to treat this case first is to immediately obtain "almost all" of the groups arising in the Quasithin Theorem 3.0.1: for Chapter 5 of [**AS04b**] produces most of the groups of Lie type of rank 2, which give the generic conclusion (1) of the Theorem. So we first review this case from the point of view of the FSU:

EXAMPLE 3.6.1. Suppose G is a group of Lie type of BN-rank 2, defined over the field $\mathbb{F}_{2^n}$. When $n = 1$, some of the parabolic subgroups are solvable, so we assume $n > 1$, in order to guarantee the existence of nonsolvable uniqueness groups in $\mathcal{L}^*(G, T)$. Then choosing L to lie in the faithful subcollection $\mathcal{L}_f^*(G, T)$ amounts to taking $L := [M, M]$, for M a maximal parabolic over T which does *not* normalize a long-root group.[10] One checks, in each case, that $\overline{L}$ is $L_2(2^n)$—or $L_2(2^{2n})$ in the unitary groups; and furthermore that the module V constructed in the FSU is either a natural module—or an orthogonal module for $L_2(2^{2n}) \cong \Omega_4^-(2^n)$ in the unitary groups. Finally our $H \in \mathcal{H}_*(T, M)$ is the second maximal parabolic over T; in particular $H/O_2(H)$ is a (possibly twisted) group of Lie type of rank 1, which is defined over $\mathbb{F}_{2^{kn}}$ with $kn > 1$.

In particular we have $n(H) = kn > 1$, in the language of Definition E.1.6 in [**AS04a**], which we discussed after (3.3.9): this parameter $n(H)$ is roughly $\log_2(q)$, when a component of $H/O_2(H)$ is a group of Lie type over a field of order $q > 2$, while $n(H) = 1$ otherwise. At various points in the analysis there will be case subdivisions on the value of $n(H)$; in particular when $n(H) > 1$, one can exploit the existence of nontrivial Cartan subgroups of $H \cap M$. ◊

Chapter 5 of [**AS04b**] begins a more general analysis of the case $\overline{L} \cong L_2(2^n)$ in the Fundamental Setup. Hypothesis 5.0.1 (called the Generic Case) also assumes that L is normal in M.

Some preliminary analysis is carried out in Section 5.1. Notably Lemma 5.1.3 reduces to five possibilities for the module V in the FSU 3.5.2. Further information is developed, especially in the case where $n(H) > 1$; as we just saw in Example 3.6.1, this condition sets up the study of "Cartan subgroups": The initial focus is on a nontrivial Hall $2'$-subgroup B of $H \cap M$, which by 5.1.5 must normalize a Hall $2'$-subgroup D of $N_L(T \cap L)$; their product BD in M will in due course be seen to be a Cartan subgroup in the group G of BN-rank 2—generated by the amalgam defined essentially by L and $O^2(H)$.

The final part of Section 5.1 of [**AS04b**] is devoted to a troublesome case where we have $H/O_2(H) \cong L_3(4)$; in particular there is already at 5.1.16 a preview of the more systematic use later in Section 5.2 of results on weak BN-pairs of rank 2 from Section F.1 of [**AS04a**] (cf. our Remark 3.3.15). Similarly 5.1.17 includes an early use of some techniques common in many later chapters, namely: the material from Section G.2 on almost-special groups, and the amalgam method of Section F.8—to obtain restrictions on H and on the module U (cf. our Remark 3.3.17).

The generic examples. The groups of Lie type of BN-rank 2 over fields of even order $q > 2$, together with one sporadic group, now arise in Section 5.2:

REMARK 3.6.2 (The groups from the Generic Case). Theorem 5.2.3 in [**AS04b**] shows that:

If $\overline{L} \cong L_2(2^n)$ with $n(H) > 1$ and $L \trianglelefteq M$, then one of the following holds:
(1) $n = 2$, and V is the sum of at most two A_5 modules
(with further information);
(2) $G \cong M_{23}$;
(3) G is of Lie type and BN-rank 2 over $\mathbb{F}_{2^m}$ for $m > 1$
(but $U_5(2^m)$ only for $m = 2$).

In fact in (3), we have $m = n$ except in the unitary groups, where $m = \frac{n}{2}$ for even $n > 2$. ◊

[10]When G has type $Sp_4(2^n)$, the notion of "long" root is a notational choice; so that *either* maximal parabolic over T could be chosen for M.

At the start of the proof, several awkward configurations lead to conclusion (1); thereafter it is assumed that conclusion (1) fails.

All remaining choices for L and H are shown to lead to a *weak BN-pair α of rank* 2, with BD as a Cartan subgroup, in the following sense:

DEFINITION 3.6.3 (Weak BN pair of rank 2). A weak BN pair of rank 2 in characteristic 2 is a subgroup amalgam $\alpha = (P_1, B, P_2)$ (having a completion given by $X = \langle P_1, P_2 \rangle$ which is realized in G) with the properties that $B = P_1 \cap P_2$, $\ker_B(X) = 1$, and for $i = 1, 2$ there exists a normal subgroup P_i^* of P_i such that the following hold:

(1) $F^*(P_i) = O_2(P_i) \leq P_i^*$ and $P_i = P_i^* B$; and

(2) $P_i^* \cap B$ is a Sylow 2-normalizer in P_i^*, and $P_i^*/O_2(P_i^*)$ is isomorphic to $SL_2(2^{n_i})$, $SU_3(2^{n_i})$, $U_3(2^{n_i})$, $Sz(2^{n_i})$, $n_i \geq 1$, or to a dihedral group of order 10.

(We had given a rough approximation of the definition earlier at Remark 3.3.15; of course more discussion appears in Section F.1 of [**AS04a**].)

Here the roles of "P_1, P_2" are played by L and $K := O^2(H)$, respectively. Then Proposition 5.2.6 lists the seven possibilities for α—corresponding to the seven groups of Lie type and BN-rank 2 in the Quasithin Theorem, and appearing in conclusion (3) in Remark 3.6.2 above. The determination of α uses F.1.12 in [**AS04a**], based on Delgado-Goldschmidt-Stellmacher [**DGS85**]. In the case of the unitary amalgams, the condition $n(H) > 1$ in fact forces $n \geq 4$ even; so when $n = 2$, the subcase $\overline{L} \cong L_2(4)$, which occurs in $G \cong U_4(2)$, still needs to be treated later (indeed not until 13.5.12 of [**AS04b**]). In particular, as in Example 3.6.1, it follows from the list in 5.2.6 that V is either a covering of the natural module for $\overline{L}$, or the orthogonal module for $\overline{L}$. Theorem 5.2.9 re-states the list of 5.2.6, in a form convenient for identifying G from α.

Since the amalgam α contains L, but not necessarily T (as it is constructed from $O^2(H)$ rather than H), the result 5.2.5, which gives $O_2(G_0) = 1$ for the completion G_0 of α, does not simply appeal to the uniqueness of M as an overgroup of $\langle L, T \rangle$ in (3.5.1). Instead it requires an appeal to the uniqueness of M as an overgroup of L, as guaranteed for $\overline{L} \cong L_2(2^n)$ by 4.3.2 of [**AS04b**], whose proof uses the pushing-up analysis in Chapter 4; indeed 4.3.2 is also used at several other points in Chapters 5 and 6.

In fact the branch of the argument corresponding to $G \cong M_{23}$, which properly contains the group $L_3(4)$ of rank 2 with its amalgam α, is distinguished in Theorem 5.2.9 by the condition that $O^2(H)$ is not maximal in $\mathcal{L}(G, T)$; and the rest of the section is then devoted to the proof of Theorem 5.2.10, which identifies G in this case. First the subgroup G_0 generated by L and $O^2(H)$ is identified as $L_3(4)$, from the amalgam α and some further local information—quoting the corollary F.4.26 of Theorem F.4.8 in [**AS04a**] (which in turn quotes Tits-Weiss [**TW02**]). Finally G is identified as M_{23} using the notion of a uniqueness system given in Section 37 of Aschbacher [**Asc94**].

For all other branches, Section 5.3 contains the final identification of G as a group of Lie type and BN-rank 2. Note that the universal completion of the amalgam α is an infinite group—having as a homomorphic image a finite group $\overline{G}$ of Lie type in characteristic 2. Under additional hypotheses, such as $C_G(z) \leq P_1$ for some involution $z \in Z\big(O^{2'}(P_1)\big)$, Theorem F.4.31 permits the identification $G \cong \overline{G}$; as in the previous paragraph, the identification quotes [**TW02**].

Reducing the remaining cases to $\overline{L} \cong \Omega_4^-(2)$. Now when $n(H) > 1$, Remark 3.6.2 reduces the $\overline{L} \cong L_2(2^n)$ problem to conclusion (1) of 3.6.2, where $n = 2$ and V is a sum of A_5-modules. Chapter 6 of [**AS04b**] completes this reduction, proving it in the remaining case where $n(H) = 1$:

REMARK 3.6.4. The analysis culminates at Theorem 6.2.19 of [**AS04b**], which shows:

When $L \trianglelefteq M$, $\overline{L} \cong L_2(2^n)$, and $n(H) = 1$, we have one of:

(1) $n = 2$, with V the sum of at most two A_5-modules
(with further information).

(2) $G \cong M_{22}$.

After this reduction, in case (1) when $\overline{L} \cong L_2(4)$, $\overline{L}$ can be regarded instead as $\Omega_4^-(2)$, and as such, it is treated along with other groups $\overline{L}$ defined over $\mathbb{F}_2$, starting at Chapter 12 of [**AS04b**]. ♢

Hypothesis 6.1.1 of [**AS04b**] assumes $n(H) = 1$, but V is *not* a sum of A_5-modules. In particular this restricts the possibilities for V in 5.1.3. Section 6.1 culminates in Theorem 6.1.27, which reduces to the case $n = 2$ with V a natural $L_2(4)$-module. The proof is fairly intricate; in particular there are early uses of techniques common in the later chapters: such as the amalgam method from Section F.7 of [**AS04a**], starting at 6.1.18 (cf. our Remark 3.3.17); as well as some weak closure arguments from Section E.3, based on control of centralizers of involutions in V, e.g. in 6.1.13.1 (cf. Remark 3.3.13).

Section 6.2 completes the proof: Again the analysis involves weak closure and the amalgam method, as well as results about almost-special groups from Section G.2 of [**AS04a**]. The configuration leading to M_{22} arises in Proposition 6.2.13; the identification of G from local subgroups is accomplished via a result of Ku [**Ku97**]. Elimination of the remaining configurations uses G.11.2 from [**AS04a**], which gives information about large-extraspecial subgroups in quasithin groups. (Results like G.11.2 are also used later at various points; so the treatment of the Quasithin Theorem does not need to quote the $GF(2)$ Type Theorem 7.0.1.)

3.7. Reducing to V an FF-module for $\overline{L}$

Chapters 7–9 of [**AS04b**] now turn to the elimination of "most" of the remaining possibilities for the module V in the FSU 3.5.2. Namely, it is assumed here that V is *not* a failure-of-factorization module for $\overline{L}$; thus in the language of Definition 3.3.1, the module parameters satisfy $\hat{q}(\overline{M}, V) \leq 2$ but $q(\overline{M}, V) > 1$—and these conditions determine a corresponding subset of pairs $(\overline{L}, V)$ from the list of 3.2.9 of [**AS04b**], discussed in our Remark 3.5.2. It turns out that J_4 is the only example emerging from this long list of cases.

The elimination of these non-FF modules V is comparatively easy, if somewhat technical—it primarily applies the weak closure methods of Section E.3 of [**AS04a**]; cf. our Remark 3.3.13.

Setup and preliminaries. The main result Theorem 7.0.1 of these chapters is actually a little weaker than we just suggested, since a few non-FF modules appear in conclusions (3) and (4):

REMARK 3.7.1. Theorem 7.0.1 in [**AS04b**] shows that the FSU 3.5.2 leads to one of:

(1) V is an FF-module for $Aut_{\langle L,T\rangle}(V)$.
(2) V is the cocode module for $M/V \cong M_{24}$, and $G \cong J_4$.
(3) V is the orthogonal module for $\overline{L} \cong \Omega_4^-(2^{n/2})$
(4) L is not normal in M, V is one of a number of exceptional non-FF modules, and $\overline{L} \cong L_2(2^n)$, $Sz(2^n)$, or $L_3(2)$.

The cases in (3) and (4) are treated later in Chapter 10 of [**AS04b**]; while the FF-modules in case (1) will occupy Chapters 12–15 (though some have already been treated in the Generic Case, discussed above in our Section 3.6). ◊

Hypothesis 7.0.2 of [**AS04b**] assumes the FSU 3.5.2, with V a non-FF module which is not one of the exceptions in (3) and (4) above. The goal now is to eliminate all cases except the one leading to $G \cong J_4$ in (2).

The pairs $(\overline{L}, V)$ to be treated are listed in Proposition 7.1.1. They include the natural modules for $U_3(2^n)$ and $Sz(2^n)$, tensor products and direct sums of two natural modules for small linear groups, as well as code and cocode modules for Mathieu groups. Proposition 7.2.1 provides some rough bounds on various weak closure parameters (cf. our discussion of Remark B.8.2). Next the short Sections 7.3 and 7.4 establish some inequalities among those parameters, using weak closure arguments, beginning to implement the strategy we discussed at Remark 3.3.13.

Modules not arising in any shadow. The sublist of modules V from 7.1.1 which do not arise in any shadow is then quickly treated in Section 7.5 of [**AS04b**]. The primary restriction on the parameters for this class of modules is the Fundamental Weak Closure Inequality (FWCI) at E.3.29 in [**AS04a**] (which we discuss as (3.3.11)—and which is re-stated for use in Section 7.5 of [**AS04b**] as 7.5.1). This inequality either already conflicts with earlier relations among the parameters, such as 7.4.1, or else leads to a contradiction after a bit more effort.

Modules arising in shadows and in J_4. The modules V now remaining from the list in 7.1.1 correspond to shadows or to the group J_4—so that there are groups which *do* contain a 2-local assembled using $\overline{L}$ and V. Hence there can be no purely numerical contradiction from the information in the FWCI inequality (3.3.11). Instead typically that inequality can be shown to be an equality, which then gives a unique solution for the values of the parameters, which we can exploit to obtain a contradiction, or identify J_4.

Chapter 8 of [**AS04b**] analyzes these modules V. The main cases remaining in 7.1.1 have $\overline{L}$ a Mathieu group M_{22}, M_{23}, or M_{24}; with V a module arising either from the Golay code, or from the "cocode" (also called the Todd module). There are also a few cases for $\overline{L} \cong SL_3(2^n)$.

Section 8.1 provides a fairly simple elimination of most cases of a cocode module. In these cases, 2-locals constructed from such a V and $\overline{L}$ arise in shadows from the Fischer groups Fi_{22}, Fi_{23}, and Fi'_{24}; the M_{24} case also arises in J_4. For certain involutions $x \in V$, it is possible to force $C_G(x) \leq M$. As in Remark 3.3.13, this now leads at 8.1.2 (via E.6.27 in [**AS04a**]) to improvements over the bounds in Chapter 7 on the weak closure parameters. These new bounds are used in Theorem 8.1.3 to eliminate the configurations corresponding to the Fischer groups; and to reduce the cocode module analysis to the case of $\overline{L} \cong M_{24}$—where in addition the FWCI (3.3.11) is an equality, with a unique solution for the parameters given by the values actually occurring in J_4.

Section 8.2 continues the analysis of the remaining cocode case, but also considers the code modules for the Mathieu groups—as well as most cases related to $SL_3(2^n)$. The code modules for M_{22} and M_{24} arise in the shadows Co_2 and Co_1, while certain cases involving $SL_3(2)$ appear in the shadows $U_6(2)$ and $U_7(2)$. In order to obtain a contradiction eliminating these cases, we construct a 2-local subgroup which is not strongly quasithin. Proposition 8.2.5 determines the structure of a subgroup I of H generated by V and a suitable conjugate V^h; this includes information about $A := V^h \cap O_2(I)$, and $W := C_V(A)$, which has rank 3 or 4. The 2-local $N := N_G(W)$ is then constructed. There is some use of 4.4.3 of [**AS04b**] to control odd locals. Sometimes N is not strongly quasithin, eliminating for example the code module for M_{24} and the shadow of $U_7(2)$. So at Lemma 8.2.10, there are just three remaining cases, including the case leading to J_4. Then the centralizer $C := C_G(z)$ of a 2-central involution z is constructed; this fails to be strongly quasithin in two of the cases, reducing to the configuration for J_4. Finally G is identified as J_4 from the centralizer C, using work of Aschbacher and Segev in [**AS91**].

Eliminating a few more special modules. The final section of Chapter 8 of [**AS04a**] eliminates one last possibility for V related to $L_3(2)$; the proof uses weak closure arguments that are a bit more subtle than those used in Chapters 7 and 8.

Then Chapter 9 eliminates the case where L is not normal in M and we have $\langle \overline{L}^M \rangle \cong \Omega_4^+(2^n)$ acting on its natural orthogonal module V; the final argument again uses weak closure arguments—but only after considerable additional preparation. For example, arguments involving E.6.27 of [**AS04a**] are replaced by suitable alternatives; and the control of odd locals supplied by 4.4.14 of [**AS04b**] is applied when $n(H) > 1$, to reduce at 9.2.7 to $n = 2$, with $n(H) \leq 2$. To eliminate the case where $n(H) = 2$, a certain overgroup K_0 of $O^2(H)$ is constructed, which cannot be contained in M since $H \in \mathcal{H}_*(T, M)$. Then preliminary results from [**AS04a**] are applied—notably F.1.12 on weak BN-pairs of rank 2, and C.1.32 on Meierfrankenfeld-Stellmacher rank-2 pushing-up—to determine the structure of K_0, obtaining a contradiction by forcing it into M. Thus now $n(H) = 1$ with $n = 2$; and here one can again obtain control of the centralizers of elements of V, and hence restrictions on weak closure parameters, leading fairly quickly to a contradiction using the standard weak closure factorization E.3.19 of [**AS04a**]. (We state a version as Proposition B.8.6.)

This completes the proof of the main result Theorem 7.0.1 of Chapters 7–9 of [**AS04b**], which we recorded in Remark 3.7.1.

REMARK 3.7.2 (Subsequent analysis of $\overline{L}$ in characteristic 2). The reduction to the modules V appearing in (1), (3), and (4) of Remark 3.7.1 now also provides useful information about the structure of $\overline{L}$: First, in conclusion (1), V is an FF-module, so it is well known that $\overline{L}$ cannot be sporadic; indeed B.4.2 of [**AS04a**] shows that $\overline{L}$ must either be of Lie type over a field of characteristic 2, or a covering of a small alternating group, such as $\hat{A}_6$ or A_7—which we can roughly regard (and treat) as being of Lie type and characteristic 2. Furthermore the exceptional conclusions (3) and (4) in 3.7.1 involve groups $\overline{L}$ of characteristic 2. Thus in the remaining cases of the FSU 3.5.2 that we need to analyze, we have also essentially reduced to the case where $\overline{L}$ is of Lie type over a field of characteristic 2. ◇

3.8. Cases with $\overline{L}$ over $\mathbf{F}_{2^n}$ for $n > 1$

By Remark 3.7.2, we have reduced to the case where each group $\overline{L}$ in the FSU 3.5.2 is a group of Lie type over a field of characteristic 2. The goal of Chapters 10–11 of [**AS04b**] is roughly to show that no further examples arise when $\overline{L}$ is defined over a field $\mathbb{F}_{2^n}$ for $n > 1$, providing a reduction to the case, in Chapters 12–15, where $\overline{L}$ is defined over $\mathbb{F}_2$.

We recall several finer points from our discussion of the case division, and of the treatment of various cases in the first nine chapters of [**AS04b**]: First, recall from Remarks 3.6.2 and 3.6.4 that when $\overline{L} \cong L_2(2^n)$ and $\overline{L}$ is normal in $\overline{M}$, then either $\overline{L} \cong L_2(4)$, which we *choose* to consider later, in the guise of the group $\Omega_4^-(2)$ defined over $\mathbb{F}_2$; or G is a group appearing on the list of examples of the Main Theorem. Second, from Remark 3.7.1, either V is an FF-module for $Aut_{\langle L,T\rangle}(V)$, or G is J_4, or one of the exceptional cases in conclusions (3) and (4) of the Remark holds. Thus we may assume that either V is an FF-module for $Aut_{\langle L,T\rangle}(V)$, and $\overline{L}$ is not $L_2(2^n)$ with $n > 2$; or L is not normal in M and one of the exceptional cases in conclusion (4) of Remark 3.7.1 holds. Third, by Remark 3.7.2, when V is an FF-module, either $\overline{L}$ is of Lie type over a field of characteristic 2, or $\overline{L}$ is $\hat{A}_6$ or A_7 (with FF-modules V listed in B.4.2 of [**AS04a**]) which we regard as "honorary" groups defined over $\mathbb{F}_2$. All the cases over $\mathbb{F}_2$ will be considered later, beginning in Chapter 12 of [**AS04b**] (and discussed in Section 3.9).

In summary, to complete our treatment of the FSU 3.5.2, we must consider three types of problems. First, we must deal with those cases where $L \trianglelefteq M$ and $\overline{L}$ is a group over $\mathbb{F}_2$, in an extended sense. Second, we must deal with those cases where $L \trianglelefteq M$ and V is an FF-module for $\overline{M}$, described in B.4.2 of [**AS04a**], with $\overline{L}$ a group of Lie type over a field of even order $q > 2$, of BN-rank 2, and possessing an FF-module—and hence isomorphic to $SL_3(2^n)$, $Sp_4(2^n)$, or $G_2(2^n)$ with $n > 1$. These cases will be eliminated in Theorem 11.0.1. Third, we must deal with the cases where L is not normal in M; these include the cases in conclusion (4) of Remark 3.7.1, but also some cases where V is an FF-module. All these configurations will be eliminated in Theorem 10.0.1.

The solution of the second and third problems makes use of techniques similar to some of those used in Chapters 5–9 of [**AS04b**]. The groups $\overline{L}$ over $\mathbb{F}_{2^n}$ with $n > 1$ have a nontrivial Cartan group, so some of the methods used in the treatment of the Generic Case for $L_2(2^n)$, discussed in Section 3.6, are applicable. Additional use is made of the preliminary pushing-up results for controlling odd locals from Section 4.4 of [**AS04b**]. The methods of weak closure used in Chapters 7–9, and discussed in Section 3.7, are still reasonably effective—partly because Chapters 10–11 are eliminating configurations which are not too close to examples in the Quasithin Theorem 3.0.1.

Reducing to $\overline{L}$ normal in $\overline{M}$. The main result of Chapter 10 is Theorem 10.0.1, showing that in the FSU 3.5.2, the subgroup L must be normal in M. The non-normal cases to be eliminated are listed in 10.1.1: in each, L^M has two members L_1 and L_2; and they include the exceptional cases in conclusion (4) of Remark 3.7.1, where V is not an FF-module for $\overline{M}$. In each case, $V = V_1 \oplus V_2$, where $V_i = [V, L_i] = C_V(L_{3-i})$. The proof is not too lengthy, perhaps because the only shadows are certain wreath products, which are neither simple nor quasithin. The analysis focuses on controlling the centralizers (in the sense of our Remark 3.3.13) of elements of the $\overline{M}$-module V—particularly elements in the summands V_1 and V_2. If one keeps in mind the wreath product shadows, this is the obvious place to look for a contradiction.

Preliminary analysis reduces at 10.2.3 to $n(H) \leq 2$; the study of a Hall $2'$-subgroup of $H \cap M$ in particular invokes 4.4.3 and 4.4.14 of [**AS04b**] to control certain odd locals. Bounds on weak closure parameters are obtained at 10.2.9, from control of centralizers of involutions in V_1 and V_2, using for example the preliminary pushing-up result C.1.24 of [**AS04a**]. Corollary 10.2.13 reduces the proof to a single case involving $\overline{L} \cong L_3(2)$. Then 10.3.3 shows that H by contrast has two $\mathcal{C}$-components with quotient $L_2(4)$; but a subsequent observation shows that H must also define a member of $\mathcal{L}_f^*(G, T)$, so that when we replace M by a maximal overgroup of H, we contradict the reduction to the case $\overline{L} \cong L_3(2)$.

Eliminating $\overline{L}$ of rank 2. Chapter 11 of [**AS04b**] is devoted to the case where $\overline{L}$ is $L_3(2^n)$, $Sp_4(2^n)$, or $G_2(2^n)$, with $n > 1$ and V an FF-module for $\overline{M}$. The analysis in this chapter is somewhat more complicated: For example, weak closure arguments are usually less effective on FF-modules. Furthermore there are shadows in which the rank-2 groups L are 2-components in‘2-locals; these shadows are higher-rank groups of Lie type which are not quasithin—namely for $q := 2^n$, the rank-3 groups $L_4(q)$, $Sp_6(q)$, and $\Omega_8^-(q)$, as well as an extension of $\Omega_8^+(q)$ by an outer automorphism. Thus the object is to obtain a contradiction by working in locals which, in the shadows, are not quasithin.

The weak closure analysis starting in Section 11.1 is somewhat analogous to that of Chapter 10, in focusing on control of centralizers of elements in certain subspaces V_i of V, for $i = 1, 2$—which are invariant under T rather than M, and have $\mathbb{F}_q$-dimensions i. The proof of 11.1.2 uses results from [**AS04a**] such as F.1.9 on weak BN-pairs of rank 2, and G.2.5 on almost-special groups; while 11.1.5 uses C.1.32 on rank-2 pushing-up as well as [**GLS96**, 14.2] on “Goldschmidt–O'Nan pairs”. Section 11.2 obtains some bounds on the weak closure parameters, such as 11.2.1 giving $n(H) \leq n$ in most cases; in particular this eliminates the shadow of $\Omega_8^-(q)$. Proposition 11.3.2 shows when $C_G(V_1) \leq M$ that $\overline{L}$ is $Sp_4(q)$ rather than $SL_3(q)$; this eliminates the shadow of $L_4(q)$, and one configuration in the shadow of $Sp_6(q)$—essentially because the strongly quasithin restriction forces an unnatural embedding in a group of rank 2 of elements that should really generate a Cartan group of rank 3. Next Lemma 11.4.4 reduces the proof to the case $n(H) = 1$; and this eliminates the remaining shadows, in which $n(H) = n > 1$. In fact the underlying reason for the elimination is again essentially that an involution centralizer in the shadow is not strongly quasithin. Finally the restriction $n(H) = 1$ leads to a contradiction, mainly via weak closure arguments (though the final case in 11.5.8 also uses the amalgam method from Section F.7 of [**AS04a**] to obtain restrictions on H and U—cf. Remark 3.3.17).

3.9. Cases with $\overline{L}$ over $\mathbf{F}_2$ (but not $L_3(2)$)

In light of the reductions discussed in Section 3.8, to complete the treatment of the FSU 3.5.2, we need only consider the case where $L \trianglelefteq M$ and V is an FF-module for $\overline{M}$, with $\overline{L}$ a group over $\mathbb{F}_2$. Roughly speaking, Chapters 12 and 13 of [**AS04b**] treat the larger and smaller possibilities for $\overline{L}$—though the smallest case where $\overline{L} \cong L_3(2)$ is postponed to Chapter 14.

From this point on, we can expect to encounter examples appearing in the conclusion of the Quasithin Theorem 3.0.1. We will also encounter quite a few shadows which are not quasithin, but still complicate the analysis. Thus the arguments will often be quite lengthy, so our discussion will be more sketchy.

In addition, the techniques used will change: For the groups $\overline{L}$ over $\mathbb{F}_2$ have only a trivial Cartan subgroup, so arguments based on the existence of nontrivial Cartan subgroups (such as in our discussion in Sections 3.6 and 3.8) are unavailable. Also the examples and shadows will render the weak closure arguments, prominent in the discussion in Section 3.7, successively less effective as we go on. To counterbalance this, increasingly frequent use will be made of the amalgam method, in ways indicated starting at F.7 of [**AS04a**] (cf. Remark 3.3.13).

We mention that for the three "parts" of the FSU which we have already discussed (namely Chapters 5–6, 7–9, and 10–11 in [**AS04b**]), the arguments are essentially independent; so that the order of their treatment could be changed. But this freedom no longer exists after Hypothesis 12.2.3—which assumes that G is not one of the examples already identified. Thus in particular, no $\overline{L}$ satisfies any of those earlier conditions; and so *all* $\overline{L}$ are defined over $\mathbb{F}_2$. Indeed as time goes on, we adopt hypotheses which continually shrink the set of possibilities for $\overline{L}$.

Preliminary analysis. Section 12.1 first eliminates an exceptional situation, in which $\overline{L}$ is $L_4(2)$ or $L_5(2)$, and V is the direct sum of a natural module V_1 and its dual V_2. Here it is possible (as in Chapter 10 of [**AS04b**] discussed in Section 3.8) to control centralizers of elements in the summands V_i, to facilitate weak closure arguments which lead to a contradiction.

After this reduction, $V/C_V(\overline{L})$ is an irreducible FF-module for $\overline{M}$ in all remaining cases of the FSU 3.5.2. From the discussion in Section 12.1 of [**AS04b**], we get a subset of the possibilities listed in 3.2.8 of [**AS04b**]; in particular we have $n = 1$ in conclusions (1)–(4) of that lemma, using the elimination of the cases with $n > 1$ in Chapter 11 of [**AS04b**], discussed in Section 3.8. This progress is summarized in Theorem 12.2.2 of [**AS04b**], where cases (1) and (2) list the examples identified earlier in Chapters 5, 6, and 8 of [**AS04b**], while case (3) is a more compact and convenient re-statement of the sublist of cases from 3.2.8 of [**AS04b**] which have not yet been treated. We will express this informally via:

REMARK 3.9.1 (The cases remaining). Hypothesis 12.2.3 in [**AS04b**] assumes that cases (1) and (2) of 12.2.2 do not occur; so that all remaining pairs $(\overline{L}, V)$ appear in case (3) of 12.2.2.

This hypothesis (and various stronger versions appearing later) will hold throughout Chapters 12–14, which complete the treatment of the FSU.

And so 12.2.2.3 provides the master list, which will be gradually reduced as various cases on the list are treated during those chapters. The list consists of the groups $\overline{L} \cong L_n(2)$ for $n = 3, 4, 5$, A_n for $5 \leq n \leq 8$, the triple cover $\hat{A}_6$, and $G_2(2)'$; with the possibilities for the modules V specified for each group. ♢

REMARK 3.9.2 (Treating the case where V is TI). After some preliminary results based on the hypothesis in Remark 3.9.1, Theorem 12.2.13 then treats the situation where V is a TI-set in G. In cases (2) and (3) below, we get examples appearing in the statement of the Quasithin Theorem 3.0.1; so the effect of the result is to reduce to the case in (1) below where V is not a TI-set in G:

(1) $C_G(v) \not\leq M$ for some $v \in V^{\#}$, so V is not a TI-set in G.
(2) L is an $L_3(2)$-block, and $G \cong L_4(2)$ or A_9.
(3) L is an $L_4(2)$-block, and $G \cong L_5(2)$ or M_{24}. ◇

The proof assumes that (1) fails, and it follows at 12.2.16 that V is a TI-set in G. Some elementary weak closure is applicable, and at 12.2.19 it is shown that H is solvable. This allows a reduction at 12.2.20 to $\overline{L} \cong L_n(2)$ for $n = 3, 4, 5$, with V the natural module; furthermore for a subgroup I of H generated by two conjugates of V, the group $W := O_2(I)$ induces the full group of transvections on V with axis given by the hyperplane $W \cap V$. Then 12.2.24 shows that L is an $L_n(2)$-block (in the sense of Definition B.7.2); while 12.2.25 eliminates the case $n = 5$. Finally Proposition 12.2.27 uses this information to identify G—when $n = 4$, by Theorem 41.6 in [**Asc94**]; and when $n = 3$, by I.4.6 in [**AS04a**] (which in turn quotes e.g. [**Asc03a**]).

Thus, from now on, we may assume that conclusion (1) of Remark 3.9.2 holds. Therefore, for example, if for some $v \in C_V(T)^{\#}$ we have $C_G(v) \not\leq M$, then we can choose H to lie in $C_G(v)$. We will often write G_1 for $C_G(v)$ in this situation.

Treating larger groups $\overline{L}$ over $\mathbb{F}_2$. The rest of Chapter 12 is devoted to treating the larger possibilities for $\overline{L}$ in the list of Remark 3.9.1, namely $\overline{L} \cong L_4(2)$, $L_5(2)$, A_7, and certain situations where $\overline{L} \cong \hat{A}_6$.

Treating various special modules. Section 12.3 eliminates the case $\overline{L} \cong A_7$; the argument is complicated by shadows of M_{23}, McL, and $\Omega_7(3)$. Here 12.3.3 reduces the problem to the case where V is of dimension 6 rather than 4; that is, A_7 arises as a subgroup of A_8, which is in turn regarded as $\Omega_6^+(2)$ rather than $L_4(2)$. The arguments use pushing-up results from [**AS04a**], such as C.2.5 and C.2.7. Then 12.3.6 and 12.3.8 control centralizers of certain involutions in V, as in Remark 3.3.13, making possible weak closure analysis. Finally $O^2(H)$ is identified and forced into $C_M(v)$ for suitable $v \in V$, for the usual contradiction to $H \not\leq M$ as $H \in \mathcal{H}_*(T, M)$.

Section 12.5 then eliminates the subcase for $\overline{L} \cong L_5(2)$ where V is a 10-dimensional exterior square of a natural module; the analysis is complicated (though only slightly) by shadows of $Sp_{10}(2)$, $\Omega_{10}^+(2)$, $\Omega_{12}^-(2)$, and the non-simple group $O_{12}^+(2)$. The argument proceeds roughly as in Section 12.3 described above: Normalizers of certain subspaces V_i of V are forced into M at 12.5.4, facilitating weak closure analysis. Then $O^2(H)$ is forced into $C_M(v)$ for some $v \in V$, contrary to $H \in \mathcal{H}_*(T, M)$. Some of the analysis uses G.2.2 of [**AS04a**] on almost-special groups.

Section 12.6 eliminates the analogous subcase for $\overline{L} \cong L_4(2)$, where V is a 6-dimensional exterior square of a natural module. Again there are shadows: of $\Omega_8^+(2)$, non-simple $O_{10}^+(2)$, $Sp_8(2)$, and $P\Omega_8^+(3)$. Also such an L is a $2'$-component in non-maximal subgroups, in larger orthogonal and symplectic groups, and in J_4 and Fi'_{24}. Again it is shown that $C_G(v) \leq M$ for certain $v \in V$ at 12.6.2—but now at the cost of considerable effort, in particular using C.1.29 of [**AS04a**] (the SQTK

Local $C(G,T)$-Theorem 3.3.7). Again this sets the stage for weak closure analysis.

At this point we interject a brief digression:

REMARK 3.9.3 (A standard case division). Here the argument divides into as many as three subcases, depending first on whether $V \leq O_2(H)$ or $V \not\leq O_2(H)$; and then (mainly when $V \leq O_2(H)$), on whether $U := \langle V^H \rangle$ is nonabelian or abelian; this division will recur frequently in the remainder of Chapters 12–15. The treatment involves amalgam methods discussed in Remark 3.3.17, particularly when U is abelian.

The case where U is a nonabelian 2-group is often treated using the theory of "almost-special" subgroups in Section G.2 of [**AS04a**]: Groups satisfying Hypothesis G.2.1 can be investigated via methods resembling those used in the treatment of large extraspecial groups discussed in our Chapter 7; and indeed this almost-special analysis is designed to make the quasithin work independent of any need to quote the results in that Chapter. Centralizers with large extraspecial subgroups in quasithin groups are essentially determined in Sections G.10 and G.11. $\diamondsuit$

We return to the analysis in Section 12.6 of [**AS04b**] with V of dimension 6 for $\overline{L} \cong L_4(2)$. The case of U nonabelian (or equivalently in this particular situation, where $V \not\leq O_2(H)$) is eliminated at 12.6.26 (and the shadows are also eliminated at this point): Hypothesis F.9.1 of [**AS04a**] holds, making possible an application of F.9.5, leading to a single possibility for $O^2(H)$, which in turn forces $|U|$ to exceed a bound determined by L. Finally in the case of U abelian, we see at 12.6.30 that Hypothesis F.9.8 of [**AS04a**] is satisfied with $b \geq 3$ odd; here F.9.16 shows that $q(Aut_H(\tilde{U}), \tilde{U}) \leq 2$, for a suitable section $\tilde{U}$ of the H-module U—which along with F.9.18 leads to a single possibility for H and U, contradicting other restrictions from F.9.16.

Section 12.7 treats the case $\overline{L} \cong \hat{A}_6$, assuming V is a faithful irreducible 6-dimensional $\mathbb{F}_2$-module. Here Theorem 12.7.1 shows:[11]

If $\overline{L} \cong \hat{A}_6$ and $m(V) = 6$, then G is M_{24} or He.

Lemma 12.7.5 then subdivides the problem into two cases: namely $C_G(t) \leq M$ and $C_G(t) \not\leq M$, for t a non-2-central involution in V; the proof uses pushing-up results such as C.4.8 and C.3.1 of [**AS04a**]. Theorem 12.7.7 then shows that the branch $C_G(t) \not\leq M$ leads to $G \cong He$: the centralizer $C_G(z)$ of a 2-central involution z is constructed (using F.6.11 of [**AS04a**] on Goldschmidt amalgams), and this centralizer is used to identify G as He by Theorem 44.4 in [**Asc94**]. Thus we may assume that $C_G(t) \leq M$. Next Theorem 12.7.14 shows that if $V \not\leq O_2(C_G(z))$ then $G \cong M_{24}$: further weak closure arguments, along with G.2.2 of [**AS04a**] on almost-special groups, pin down the structure of M and $C_G(z)$, and then G can be identified as M_{24}—again via Theorem 44.4 in [**Asc94**]. The remaining subcase $V \leq O_2(C_G(z))$ is eliminated using some weak closure arguments.

Treating natural modules for linear groups. Finally Sections 12.8 and 12.9 eliminate the remaining cases for $\overline{L} \cong L_4(2)$ or $L_5(2)$, in which V is the natural module. In summary, the analysis culminates in Theorem 12.9.1, which establishes a contradiction under the assumption that G is not one of the groups listed in Hypothesis 12.8.1. In particular here V is not a TI-set in G by our Remark 3.9.2.1. The

[11] The identification of M_{24} in the case $\overline{L} \cong \hat{A}_6$ is independent of an earlier appearance of M_{24} in Theorem 12.2.13 (indicated in Remark 3.9.2) when $\overline{L} \cong L_4(2)$.

argument is also complicated by shadows of Co_3 and F_3, as well as larger groups in which L is a $2'$-component of a local which is not maximal.

In fact the preliminary work in Section 12.8 also applies to the cases in Chapters 14 and 15, where $\overline{L}$ is allowed to be $L_3(2)$ or $L_2(2)$. It uses a notational convention that will also arise in later arguments:

(3.9.4) Set $G_1 := C_G(z)$, for z a 2-central involution in V, and $U := \langle V^{G_1} \rangle$.

Since L is transitive on $V^\#$, we have $G_1 \not\leq M$ by Remark 3.9.2.1, so we can take H to be a subgroup of G_1. Lemma 12.8.4 shows that Hypothesis G.2.1 of [**AS04a**] (on almost-special groups) is satisfied, so, as in Remark 3.9.3, the argument divides into two cases, depending on whether U is abelian or nonabelian. When U is nonabelian, Lemma 12.8.13 shows that a certain H-section of U is on a sublist of the list of G.11.2, setting the stage for a proof that G_1 is an almost-extraspecial centralizer. On the other hand when U is abelian, Lemma 12.8.6 shows that Hypothesis F.9.8 of [**AS04a**] is satisfied, with $b \geq 3$ odd (cf. Remark 3.3.17). This makes possible applications of F.9.16 and F.9.18, which give strong restrictions on the G_1-module U. These restrictions are used in Section 12.9, and also in Chapter 15, of [**AS04b**].

Section 12.9 now restricts attention to the cases $\overline{L} \cong L_4(2)$ or $L_5(2)$; and focuses on the normalizers $N_G(V_i)$ of T-invariant subspaces V_i of dimension i in V. First Lemma 12.9.4 rules out the possibility that U is nonabelian. Then amalgam arguments lead to a short list of possibilities for $O^2(H)$. Next 12.9.7 shows that for $i > 1$, $N_G(V_i) \leq M$, setting the stage for weak closure arguments. A final contradiction arises from studying the H-module $\langle V_2^H \rangle$.

Treating smaller groups $\overline{L}$ over $\mathbb{F}_2$. Chapter 13 of [**AS04b**] then turns to the smaller choices for $\overline{L}$ in Remark 3.9.1, given by the groups A_5, A_6 (including the case where the center of $L/O_2(L) \cong \hat{A}_6$ is in the kernel of the representation of L on V), and $U_3(3)$. These are indeed groups over $\mathbb{F}_2$ because of the isomorphisms $A_5 \cong \Omega_4^-(2)$, $A_6 \cong Sp_4(2)'$, and $U_3(3) \cong G_2(2)'$. Furthermore since $A_5 \cong L_2(4)$, A_5 also represents the only subcase of the case $\overline{L} \cong L_2(2^n)$ remaining after Remarks 3.6.2 and 3.6.4. The treatment of these cases essentially reduces the list in Remark 3.9.1 to $L_3(2)$, which will be treated later in Chapter 14 of [**AS04b**].

The arguments in Chapter 13 now become still more lengthy and detailed.

In the usual spirit, Hypothesis 13.1.1 now excludes the groups G which appeared as conclusions in earlier chapters. Thus for each $L' \in \mathcal{L}_f^*(G,T)$, $\overline{L}'$ is either in one of the above three cases, or $\overline{L}'$ is $L_3(2)$.

REMARK 3.9.5 (A convention for treating A_5 last). However, by convention we will only choose $\overline{L} \cong A_5$ when no other choice is possible—that is, when for each $L' \in \mathcal{L}_f^*(G,T)$, $\overline{L}' \cong A_5$. Thus formally A_5 is the last group to be treated, rather than $L_3(2)$. On the other hand, it is more convenient to treat A_5 in Chapter 13, because of numerous similarities with the treatment of A_6. But keep in mind that when $\overline{L}$ is A_5, we assume the stronger hypothesis. ♢

Preliminaries. The rest of Section 13.1 is mainly devoted to Theorem 13.1.7, reducing to the case that every $L \in \mathcal{L}_f^*(G,T)$ must in fact have $L/O_2(L)$ quasisimple—thus finally eliminating the annoying possibilities (c) and (d) from Remark 3.2.7.

The analysis involves some weak closure at 13.1.14. To control centralizers of elements of V at 13.1.12.3, there are appeals to pushing-up results such as C.2.4 and C.2.7 from [**AS04a**].

Section 13.2 establishes some preliminary results for A_5 and A_6. There is detailed study of the FF-module V, and use of C.1.37 of [**AS04a**] on rank-2 pushing-up.

An analysis of all cases for $\overline{L}$ starts in Section 13.3. Restrictions are obtained on the structure of arbitrary members K of $\mathcal{L}_f(G,T)$. Again the focus is on control of normalizers G_i of T-invariant subspaces V_i of rank i in V. Pushing-up results such as C.6.3 of [**AS04a**] are used. As before, the argument concentrates on:

$$U := \langle V^{G_1} \rangle.$$

As in (3.9.4), usually L is transitive on $V^{\#}$, so that $G_1 \not\leq M$ by Remark 3.9.2.1.

Initial reductions. Theorem 13.3.16 in [**AS04b**] eliminates the case $\overline{L} \cong G_2(2)'$. Information on the centralizers of various subspaces of V is obtained in 13.3.12.2, using the pushing-up result C.2.7 of [**AS04a**], and these results are extended in 13.3.17. The case where U is nonabelian is handled at 13.3.18, using F.9.7 under Hypothesis F.9.1 of [**AS04a**]. The case U abelian is then eliminated fairly quickly: Results involving weak closure, such as 13.3.19, force H into a centralizer lying in M, as usual contrary to $H \in \mathcal{H}_*(T,M)$.

Thus in the remainder of Chapter 13, $\overline{L}$ is isomorphic to A_5 or A_6.

Next Theorem 13.4.1 treats a configuration leading to one of the remaining examples, showing:

If $\overline{L} \cong A_6$ with $C_V(L) \neq 1$, then $G \cong Sp_6(2)$.

The example arises in the branch of the analysis where G_1 is solvable. Here we have $M/O_2(M) \cong Sp_4(2)$, with V an extension of the natural module, so M already resembles one maximal parabolic of $Sp_6(2)$. Next from 13.3.5.3 we may take $G_1 \not\leq M$. Then Lemma 13.4.8.3 constructs $H \in \mathcal{H}_*(T,M)$ with $H \leq G_1$ and $H/O_2(H) \cong S_3 \times S_3$—resembling the maximal parabolic of $Sp_6(2)$ determined by a middle node of its Dynkin diagram. To obtain a subgroup resembling the third maximal parabolic of $Sp_6(2)$, we take abstract minimal parabolics H_1 and H_2 (in the language of Definition 3.1.3) from M and H, which avoid the intersection $M \cap H$. The H_i in turn define a Goldschmidt amalgam, which can be identified using F.6.5 of [**AS04a**]. Then 13.4.7.1.ii shows for $H_0 := \langle H_1, H_2 \rangle$ that we have $H_0/O_2(H_0) \cong L_3(2)$, so that H_0 indeed resembles the remaining maximal parabolic of $Sp_6(2)$. Then by I.5.1.2 of [**AS04a**], the groups M, H, and H_0 generate a subgroup $E \cong Sp_6(2)$ of G. The proof that $E = G$ uses an extension, due to Aschbacher, of the theorem classifying groups with a strongly embedded subgroup; the result is called Corollary ZD on page 22 of [**GLS99**].

The remainder of Section 13.4 eliminates the case that G_1 is nonsolvable: A variant of the approach in the case G_1 solvable is now applied, to H along with an $H_0 \in \mathcal{H}_*(T,M)$ constructed as in 13.4.7; the pushing-up result C.1.13 of [**AS04a**] is used along the way. But this time H_0 is not quite compatible with M: in fact F.7.9 from the amalgam method of [**AS04a**] is applied with M and H_0 playing the roles of "G_1, G_2"; but weak closure arguments now lead to a contradiction.

In the remainder of the Chapter, we have $C_V(L) = 1$, so L is transitive on $V^{\#}$—and $G_1 \not\leq M$.

Refining the case division on U: the subcase W nonabelian. To complete the treatment of A_5 and A_6, we use the case division on $U = \langle V^{G_1} \rangle$ discussed in Remark 3.9.3. However, this time it will require a section or more to handle each case.

Indeed we introduce further subcases, defined in terms of the normal closure:

$$W := \langle V_3^{G_1} \rangle$$

of the T-invariant subgroup V_3 of rank 3 in V (V_3 is the space orthogonal to V_1 with respect to the symplectic or orthogonal form on V). We apply the amalgam method of Section F.9 of [**AS04a**] to V_3 in the role of "V_+" in Hypothesis F.9.1. We sometimes apply results in Section G.2 on almost-special groups, with V_3 in the role of "V".

Section 13.5 treats the case where W is nonabelian. All remaining examples with $\bar{L} \cong A_5$ or A_6 arise here; the main result is Theorem 13.5.12, which shows:

Assume $G \not\cong Sp_6(2)$, with $W = \langle V_3^{G_1} \rangle$ nonabelian.
If $\bar{L} \cong A_5 (\cong L_2(4))$, then $G \cong U_4(2)$ or $L_4(3)$.
If $\bar{L} \cong A_6$, then $G \cong U_4(3)$.

In the branch of the argument leading to the examples, it is shown at 13.5.15 that L is an A_5-block or A_6-block, and that $G_1/O_2(G_1)$ is a subgroup of $S_3 \times S_3$. Then when $\bar{L} \cong A_6$, G can be identified as $U_4(3)$ using M and G_1, by 45.11 in [**Asc94**]. When $\bar{L} \cong A_5$, two possibilities for the amalgam on M and H are determined by F.1.12 of [**AS04a**]. Then F.4.31 in [**AS04a**] (which quotes Tits-Weiss [**TW02**]) uses those amalgams to identify G as $U_4(2)$ or $L_4(3)$. The remainder of Section 13.5 eliminates the other branch; results such as F.9.5 and G.2.4 are used to determine the possibilities for $O^2(H)$, which are successively ruled out.

So in the remaining sections, we assume that W is abelian; and we work toward a contradiction.

Eliminating the cases when W is abelian. Section 13.6 now completes the treatment of the case where $\bar{L} \cong A_5$. For Theorem 13.6.7 in [**AS04b**] shows that $C_G(v) \le M$, for each non-2-central involution $v \in V$. This makes possible weak closure arguments which pin down a few possibilities for $H/O_2(H)$. Now amalgam methods establish that b is odd and $b \ge 3$, which in turn yields that $q\big(Aut_H(\tilde{W}), \tilde{W}\big) \le 2$ for a suitable H-section $\tilde{W}$ of W, whereas the possible groups $H/O_2(H)$ do not have such a module (by B.4.5 in [**AS04a**]).

Section 13.7 turns to the case of $\bar{L} \cong A_6$. Some preliminary restrictions on the possibilities for H and W are established using results such as F.9.2 and F.9.7. Then Theorem 13.7.8 eliminates the subcase (from our more usual case division in Remark 3.9.3) where U is nonabelian: the possibilities for $O^2(H)$ are successively ruled out, using the fact, established in 13.7.12, that certain H-sections of W are FF-modules for H.

Section 13.8 completes the treatment of $\bar{L} \cong A_6$, by eliminating the remaining subcase, where U is abelian: Again it is proved that Hypothesis F.9.8 is satisfied with V_3 in the role of "V_+", and $b \ge 3$ is odd. Then 13.8.8, F.9.13 and F.9.16 are used to show that sections of the H-module W are among those listed in B.4.2 and B.4.5 of [**AS04a**]. Eventually F.9.18 is used to produce a list of possibilities for $O^2(H)$ at 13.8.17. These are successively ruled out, using F.9.13, F.9.18, and a detailed analysis of the modules listed in B.4.2 and B.4.5.

Chapter 13 appendix: a solvable case. Section 13.9 pre-emptively treats a case where a solvable analogue of $\overline{L}$ is afforded by $O_4^+(2)$; the result is used in Chapter 15, after the treatment of the FSU 3.5.2 has been completed—when solvable uniqueness groups come into play.

Thus Theorem 13.9.1 eliminates the situation where $M \cong S_4 \wr \mathbb{Z}_2$—which is complicated by the shadow of the group A_{10}. The arguments are comparatively elementary and combinatorial; in particular there is an appeal to Thompson Transfer B.2.9, in eliminating a non-shadow case.

3.10. Cases with $\overline{L} = L_3(2)$, and analogues for $L_2(2)$

Chapter 14 of [**AS04b**] is primarily devoted to completing the analysis of the FSU 3.5.2, by treating the final remaining case where $\overline{L} \cong L_3(2)$. Recall from Remark 3.9.5 that at this point, for each $L' \in \mathcal{L}_f^*(G,T)$, $\overline{L}' \cong A_5$ or $L_3(2)$. But we only choose $L \in \mathcal{L}_f^*(G,T)$ with $\overline{L} \cong A_5$, if for each $L' \in \mathcal{L}_f^*(G,T)$, we have $\overline{L}' \cong A_5$—a situation treated in Chapter 13 of [**AS04b**]. Thus in this chapter, when $\mathcal{L}_f^*(G,T)$ is not empty, we assume that there is some $L' \in \mathcal{L}_f^*(G,T)$ with $\overline{L}' \cong L_3(2)$, so indeed $\overline{L} \cong L_3(2)$. Three new examples arise in this chapter with $\overline{L} \cong L_3(2)$; and there are also many shadows, so the analysis is correspondingly elaborate.

Chapter 14 also considers "solvable 2′-components" L, such that $\overline{L} \cong L_2(2)$: that is, L is not some 2′-component in the FSU, but instead a suitable solvable uniqueness group. The analysis of these solvable uniqueness subgroups will be completed in Chapter 15. Nonetheless six of the remaining examples will arise during Section 14.2; this leaves only two final groups, which make their appearance in Section 15.4.

The arguments continue to be quite lengthy and detailed. In particular, the amalgam method of Sections F.8–F.9 of [**AS04a**] remains prominent—with the usual bifurcation of Remark 3.9.3 into the cases $U := \langle V^H \rangle$ nonabelian or abelian.

Preliminary analysis for $L_2(2)$. Section 14.1 of [**AS04b**] sets the stage for an investigation of the $L_2(2)$ case, where there exists some $M \in \mathcal{M}(T)$ and some four-group $V \trianglelefteq M$, for which we have $\overline{M} = M/C_M(V) = GL(V) \cong L_2(2)$. This case will arise later in Chapter 15, where the set $\mathcal{L}_f^*(G,T)$ of nonsolvable "faithful" uniqueness groups is empty. (Note that this is the case where we encounter the few groups of Lie type of BN-rank 2 which did not appear in the Generic Case (i.e. conclusion (3) of Remark 3.6.2)—namely those defined over $\mathbb{F}_2$, where parabolics are solvable.) To produce such uniqueness subgroups (such as in (3.5.1) and its solvable version at 1.3.7 of [**AS04b**]), we consider 2-local overgroups of T maximal with respect to a partial ordering $\lesssim$, introduced in A.5.2 of [**AS04a**].

Much of the treatment of $\mathcal{L}_f^*(G,T)$ empty proceeds under Hypothesis 14.1.5. That Hypothesis assumes $\mathcal{L}_f^*(G,T)$ is empty, $M_c = !\mathcal{M}(C_G(Z))$ for $Z := \Omega_1(Z(T))$, and $|\mathcal{M}(T)| > 1$. There is no loss of generality in this last assumption by Theorem 2.1.1 of [**AS04b**] (Theorem 3.4.1 in our discussion). By 14.1.12 in [**AS04b**], there is $M \in \mathcal{M}(T) \setminus \{M_c\}$ which is maximal with respect to $\lesssim$. We will apply the Thompson strategy to M and an appropriate $H \in \mathcal{H}_*(T,M)$, which can be chosen in M_c. (Thus M_c is an analogue of the "G_1" in (3.9.4).)

In fact Section 14.2 (and much of the rest of Chapter 14) even assumes Hypothesis 14.2.1, which is a bit stronger than Hypothesis 14.1.5 (and can often be expected to hold, by 14.1.18): it assumes in particular that M is the unique maximal member of $\mathcal{M}(T)$ under $\lesssim$, that $m(V) = 2$ and $M/C_M(V) = GL(V) \cong L_2(2)$, and that $M_c := C_G(Z)$ is a maximal 2-local. Having established this context, we set $R_c := O_2(M \cap M_c)$ and $L := O^2(\langle R_c^M \rangle)$. From 14.2.2 in [**AS04b**], we now have $\mathcal{M}(T) = \{M, M_c\}$ and $\mathcal{M}(LT) = \{M\}$, with $O_2(L) = C_L(V)$ and the linear group $\overline{LR_c} \cong L_2(2)$, so that L is analogous to the members of $\mathcal{L}_f^*(G,T)$. As before, we study some $H \in \mathcal{H}(M,T)$, and the group $U := \langle V^H \rangle \leq O_2(H)$. When U is nonabelian, it is an almost-special 2-group, while if U is abelian, the amalgam method applies with $b \geq 3$ odd.

The remainder of Section 14.2 treats the subcases in which the six examples arise. First Theorem 14.2.7 shows (under Hypothesis 14.2.1):

If $n(H) > 1$ for some H, then $G \cong {}^3D_4(2)$, J_2, or J_3.

The assumption that $n(H) > 1$ makes possible an approach like that in the Generic Case discussed in Section 3.6: A nontrivial Hall $2'$-subgroup of $O^2(H) \cap M$ plays the role of a Cartan subgroup in a weak BN-pair α formed from M and H. Using F.1.9 of [**AS04a**], α is determined at 14.2.16 as the amalgam of type ${}^3D_4(2)$ or J_2. There are three finite simple groups with these amalgams; they are identified using I.4.7 and F.4.31 of [**AS04a**], which in turn quote [**Asc94**, Sec. 47] and [**Fro83**], as well as [**TW02**] mentioned earlier. Next Theorem 14.2.20 treats the case where $|H : T|$ is prime:

If $H/O_2(H) \cong D_{10}$ or $Sz(2)$, then $H/O_2(H) \cong Sz(2)$, and $G \cong {}^2F_4(2)'$.
If $H/O_2(H) \cong L_2(2)$, then $G \cong M_{12}$ or $G_2(2)'$.

In contrast to Theorem 14.2.7, here a Cartan subgroup of H is trivial. But again the amalgam α, obtained from M and H, is determined at 14.2.21 to be of type ${}^2F_4(2)'$, M_{12}, or $G_2(2)'$. Then G is identified from α using I.4.4 and I.4.6 in [**AS04a**] (which quote e.g. [**Asc02a**] and [**Asc03a**]) and F.4.31 (which quotes [**TW02**]).

This reduces the treatment of groups satisfying Hypothesis 14.2.1 to the case where $n(H) = 1$ and $|H : T|$ is not prime (which is eventually completed at 14.6.25).

The case U nonabelian. Section 14.3 now treats the last case in the FSU, where $\overline{L} \cong L_3(2)$, in parallel with the $L_2(2)$ case. Recall these cases also had parallel treatment of some preliminary results in Section 12.8 of [**AS04b**]. For each $1 \leq i \leq m(V)$, we let V_i be the T-invariant subspace of V of rank i, and we set $G_i := N_G(V_i)$. Thus $G_1 = C_G(Z)$. After a few preliminaries, it is assumed that U is nonabelian, and the treatment of this case continues through Section 14.4. When U is nonabelian we choose H to be G_1. Preliminary results include 14.3.13, which gives an initial list of possibilities for H, via an appeal to G.11.2 of [**AS04a**].

Then the elimination of the $L_2(2)$ case when U is nonabelian is completed in Theorem 14.3.16. First 14.3.15 reduces the list of possibilities for H in 14.3.13 to a single possibility. This determines a configuration corresponding to the shadow of an extension of $U_4(3)$: Namely by 14.3.20 in [**AS04b**], certain overgroups of T in H and M generate the normalizer N of a certain elementary 2-group A of rank 4, with $N_G(A)/A \cong A_6$, and $|T : N_T(A)| = 2$ by 14.3.18. Then a contradiction is obtained after 14.3.22, using Thompson Transfer B.2.9.

Thus we may assume that $\overline{L} \cong L_3(2)$.

A reduction to the case where U is extraspecial is then afforded by Theorem 14.3.26, which establishes (under Hypothesis 14.3.1, which assumes in particular that G is not $Sp_6(2)$) that:

For $\overline{L} \cong L_3(2)$ and U nonabelian but not extraspecial, we have $G \cong HS$.

The proof shows that the involution centralizer G_1 and the normalizer M of V have the structures of the corresponding 2-locals in HS. Then these local subgroups are used to identify G as HS via I.4.8 in [**AS04a**], which in turn quotes [**Asc03b**].

So now U is extraspecial, and the list from 14.3.13 is reduced at 14.4.2 to a smaller list of possibilities of width at least 2. The only case leading to an example appears in Theorem 14.4.3:

When $\overline{L} \cong L_3(2)$ and U is extraspecial of width 2, we have $G \cong G_2(3)$.

Again G_1 and M are shown to have the structure of 2-locals in $G_2(3)$, so that G can be identified as $G_2(3)$ via I.4.5 in [**AS04a**], which quotes [**Asc02b**] (which in turn uses the classification in [**DGS85**] of weak BN-pairs).

The rest of Section 14.4 completes the treatment of the case U nonabelian, by successively eliminating the remaining cases of width greater than 2 in 14.4.2; certain shadows such as HN and $\Omega_8^-(2)$ are ruled out by constructing a 2-local subgroup which is not strongly quasithin.

The case U abelian. In Section 14.5, and the remainder of Chapter 14, it is assumed that U is abelian. Only one example arises, namely Ru when $\overline{L} \cong L_3(2)$. Still the elimination of all other possibilities is lengthy. Set $\tilde{U} := U/Z$.

Section 14.5 contains various preliminary results (using sections F.8 and F.9 of [**AS04a**]) which obtain important restrictions on the action of H on the module U, such as $q\big(Aut_H(\tilde{U}), \tilde{U}\big) \leq 2$.

Finishing $L_2(2)$. Section 14.6 completes the elimination of $L_2(2)$ when U is abelian.

The preliminaries first restrict the centralizers of certain involutions u of U, and then construct overgroups I of $C_T(u)$ not contained in M or H. Lemma 14.6.10 makes a particular choice of such an I, generated by suitable subgroups I_1 and I_2 of M and H; and in conclusion (5) shows that if $C_T(u) \in Syl_2(I)$ and $I \leq C_G(u)$, then $m(\langle V^{I_2}\rangle) = 3$. This lemma will lead to a contradiction which eliminates both major subcases arising later in the section. The proof of 14.6.10 uses 14.6.6, which describes I via the theory of Goldschmidt amalgams in Section F.6 of [**AS04a**].

Next Theorem 14.6.11 eliminates the case where $O_{2'}\big(H/O_2(H)\big) \neq 1$, using the condition that $q\big(Aut_H(\tilde{U}), \tilde{U}\big) \leq 2$ in 14.5.18, and the list of such modules in D.2.17 of [**AS04a**], to reduce to a single possibility—in which the rank of $\langle V^{I_2}\rangle$ is 4, rather than 3 as established in 14.6.10.5.

Thus $O_{2'}\big(H/O_2(H)\big) = 1$. Then 14.6.22 and 14.6.24 essentially determine the possibilities for $H/O_2(H)$ and its action on $\tilde{U}$, again relying on the restrictions obtained from 14.5.18. Again the the rank of $\langle V^{I_2}\rangle$ is 4, contradicting 14.6.10.5. (Thus 14.6.25 completes the analysis of Hypothesis 14.2.1.)

Finishing $L_3(2)$. Finally Section 14.7 finishes the treatment of $\overline{L} \cong L_3(2)$ in the FSU, where U is abelian, culminating in Theorem 14.7.75 which shows:

When $\overline{L} \cong L_3(2)$ and U is abelian, we have $G \cong Ru$.

Again it is a lengthy process to eliminate all but this one possibility.

Theorem 14.7.14 eliminates the only case with H solvable, which had arisen already at 14.7.9.2. The analysis uses F.9.13 from the amalgam method of [**AS04a**].

Eventually H and M are essentially uniquely determined, forcing H into a 2-local subgroup lying in M, contradicting $H \in \mathcal{H}_*(T, M)$ as usual.

Next Theorem 14.7.52 shows that $O^2(H/O_2(H))$ is $G_2(2)'$ or A_5—by successively eliminating most cases from an initial list of possibilities in F.9.18 of [**AS04a**]. Again 14.5.18 is used to restrict the list of possibilities for the $\mathbb{F}_2H$-module $\tilde{U}$. Then Theorem 14.7.53 shows $O^2(H/O_2(H))$ is not $G_2(2)'$, via a detailed analysis of the natural module $\tilde{U}$ for $G_2(2)'$, obtaining a contradiction to the generalized-hexagon geometry in $\tilde{U}$ based on restrictions from results such as F.9.13.

We have reduced to the case $O^2(H/O_2(H)) \cong A_5$. It is shown at 14.7.71 that H has five chief factors on $O_2(H)$, with a total dimension of 11, including $\tilde{U}$ of dimension 4; this is the structure found in an involution centralizer in Ru. Then 14.7.72 shows that $O_2(M)$ is of order 2^{11}, and has a structure resembling that of the normalizer of an E_8-subgroup in Ru. Finally this information on M and H provides the hypothesis for the identification of G as Ru, established as Theorem J.1.1 in [**AS04a**].

REMARK 3.10.1 (Summary of the analysis of the FSU). Section 14.8 of [**AS04b**] summarizes the results from Chapters 13 and 14, by listing the groups satisfying Hypothesis 13.3.1 of [**AS04b**], and showing how the major theorems in those sections treat all cases arising under that hypothesis. Then Theorem 14.8.2 (nicknamed Theorem D) lists all the QTKE-groups in which $\mathcal{L}_f^*(G,T)$ is nonempty, and shows how Theorem D follows from Theorem 14.8.1. This completes the treatment of the Fundamental Setup 3.5.2. The groups G are:

(1) A rank-2 group over $\mathbb{F}_{2^m}$ for $m > 1$ (but $U_5(2^m)$ only for $m = 2$).
(2) $L_4(2) \cong A_8$, $L_5(2)$, $U_4(2)$, $Sp_6(2)$, A_9, $L_4(3)$, $U_4(3)$, $G_2(3)$.
(3) M_{22}, M_{23}, M_{24}, J_4, HS, He, Ru.

The discussion in this section also briefly reviews the analysis in Chapters 5–14; in particular (recall Remark 3.9.5) it indicates how the treatment of A_5, in Chapter 13, in effect occurs after the treatment of $L_3(2)$ in Chapter 14.

We also recall that the $L_2(2)$ case in Chapter 14 begins the treatment of the situation where the set $\mathcal{L}_f^*(G,T)$ of nonsolvable uniqueness groups is empty. Indeed most of the examples in that case already arose in Section 14.2 of [**AS04b**]:

$$G_2(2)' \cong U_3(3),\ {}^2F_4(2)',\ {}^3D_4(2),\ M_{12},\ J_2,\ J_3\ .$$

Thus the only examples still to be encountered (in Section 15.4 of [**AS04b**]) are the two smallest groups of BN-rank 2 over $\mathbb{F}_2$, namely $L_3(2)$ and $Sp_4(2)' \cong A_6$. $\diamondsuit$

3.11. The final case where $\mathcal{L}_f(G,T)$ is empty

Chapter 15 of [**AS04b**] completes the proof of the Quasithin Theorem 3.0.1, by treating the remaining case where the set $\mathcal{L}_f^*(G,T)$ of faithful nonsolvable uniqueness groups is empty.

This case was already begun in Chapter 14, and indeed the first three sections of Chapter 15 continue to assume Hypothesis 14.1.5. In particular, $M_c = !\mathcal{M}(C_G(Z))$, for $Z := \Omega_1(Z(T))$, and $|\mathcal{M}(T)| \ge 2$, so in the Thompson strategy we can choose our $M \in \mathcal{M}(T) - \{M_c\}$, $V = \langle Z^M \rangle$, and $H \in \mathcal{H}_*(T,M)$ to lie in M_c.

In fact the subcase where $m(V) = 2$ and $M/C_M(V) \cong L_2(2)$ was completely analyzed in Chapter 14. The six examples that arose in that case in Section 14.2 are recalled in conclusion (1) of Theorem 15.1.3. (And no further groups satisfy Hypothesis 14.1.5.)

Reduction to the $O_4^+(2)$ configuration. The purpose of Theorem 15.1.3 in [**AS04b**] is to show either $m(V) = 2$ or V is the 4-dimensional orthogonal module for $M/C_M(V) \cong O_4^+(2)$. The proof of this theorem appears in Sections 15.1 and 15.2.

Some of the arguments are analogous to the treatment in the FSU 3.5.2: For example, we had mentioned earlier that (1) of 14.1.12 shows that we may choose an $M \in \mathcal{M}(T) \setminus \{M_c\}$ to be maximal in the partial order $\lesssim$ on maximal 2-locals introduced at A.5.2 of [**AS04a**]. Set $\overline{M} = M/C_M(V)$. By (2) of 14.1.2, $N_M(C_M(V))$ is a uniqueness subgroup. Furthermore 15.1.1 shows that $\hat{q}(\overline{M}, V) \leq 2$, as in the FSU. (The proof invokes the qrc-lemma of [**AS04a**], but in the form of Theorem 3.1.6 of [**AS04b**]—rather than D.1.5 of [**AS04a**] which we have stated as Theorem 3.3.3.) This supplies an initial list of possibilities for $\overline{M}$ and its action on V. That list is re-stated at 15.1.7. It then remains to eliminate all of the cases on that list other than the case $m(V) = 4$ and $\overline{M} \cong O_4^+(2)$. (We mention also that the proof of 15.1.7 shows that the stronger variant Hypothesis 14.2.1 of Hypothesis 14.1.5 in fact holds, when needed to apply the work of Chapter 14.)

To begin, 15.1.11 shows for $H \in \mathcal{H}_*(T,M)$ that $U := \langle V^H \rangle$ is elementary abelian—so that here there will be no analogue of the "U nonabelian" cases in the standard case division appearing in Chapters 12–14 (discussed in Remark 3.9.3). Furthermore 15.1.12 produces some strong restrictions on the action of H on the module U. (To some extent these will be used in the way the restrictions of 14.5.18 were exploited in the case of U abelian in Chapter 14.)

The remainder of Section 15.1 eliminates three of the possibilities for $\overline{M}$ from the list of 15.1.7. Analysis focuses on the centralizer $G_1 \not\leq M$ of a certain subgroup Z_1 of V. Possibilities for a $\mathcal{C}$-component I of G_1 are determined, using pushing-up results from [**AS04a**] such as C.1.26, C.1.34, and C.2.7; eventually reducing to a list of four cases at 15.1.27. Then a contradiction is obtained by considering the action of I on a certain subgroup U_I of $\Omega_1(Z(O_2(I)))$; see 15.1.28 and 15.1.29.

Section 15.2 eliminates the four remaining possibilities for $\overline{M}$ from 15.1.7; this sublist is given at 15.2.1. In 15.2.8 we see that $H/O_2(H) \cong O_4^+(2)$; this leads at 15.2.12 to a list of possibilities for the normal closure K_c of $O^2(H)$ in M_c. The possibilities for K_c in turn lead to Theorem 15.2.15, which shows that we have $M_c = !\mathcal{M}(C_{M_c}(Z_S))$, for a certain subgroup Z_S of V—the proof eliminates possible obstructions from the lists in the pushing-up results C.2.4 and C.2.7.3 in [**AS04a**], and in 1.1.5.3 on non-local subgroups of G. This uniqueness result 15.2.15 is used repeatedly, along with restrictions such as those on M in 15.2.14, to establish further results on weak closure in 15.2.26. Now the amalgam method from Section F.7 of [**AS04a**] is applied, using also some arguments mimicking those in Section F.8; and the restrictions on the H-module U in 15.1.12 are exploited. Then a contradiction is obtained using the bounds in 15.2.26.

Eliminating the $O_4^+(2)$ configuration. Section 15.3 is devoted to the final case remaining under Hypothesis 14.1.5, where $\overline{M} \cong O_4^+(2)$. Here V has 2-dimensional summands V_1 and V_2 under $O^2(\overline{M}) \cong E_9$, and the analysis uses the 2-subgroup $S := N_T(V_1)$. Another important subgroup is $Y := O^2(\langle R_c^M \rangle)$ where $R_c := O_2(M \cap M_c)$. The case $Y \cong A_4 \times A_4$ is ruled out in 15.3.9, using Theorem 13.9.1 which dealt with the special case $M \cong S_4 \wr \mathbb{Z}_2$. An important technical result 15.3.45 now shows that M is the unique maximal 2-local above first YS, and then even above $C_Y(V_1S)$. The result and its proof are somewhat similar to 15.2.15: possible obstructions are eliminated, from initial lists provided by C.2.7.3 and B.5.6 of [**AS04a**]. Again the uniqueness theorems make possible results on weak closure. But now, unlike the situation in 15.1.11, we do get the usual case division of Remark 3.9.3, according to whether $U := \langle V^{M_c} \rangle$ is abelian or nonabelian. The case U nonabelian is again treated via almost-special methods; notably G.9.3 from [**AS04a**] in 15.3.58, to reduce H to situations resembling the shadows of extensions of $L_4(2)$, $L_5(2)$, and He—which are finally eliminated via Thompson Transfer B.2.9. In the case where U is abelian,[12] the earlier results on weak closure are now strengthened, leading at 15.3.67 to the construction of a certain $\{2,3\}$-subgroup $X \not\leq M$. Now subgroups of X and Y are used to define a Goldschmidt amalgam in 15.3.68, generating a group G_0 which can be identified using F.6.18 of [**AS04a**]. But this G_0 must have a nonsolvable subgroup lying in M, whereas M is solvable.

Obtaining the last two examples. The first three sections of Chapter 15 of [**AS04b**] have now treated Hypothesis 14.1.5. So Section 15.4 treats the remaining case, where Hypothesis 14.1.5 does not hold; hence in particular $C_G(Z)$ is contained in more than one maximal 2-local.

Section 15.4 begins by showing that then $\mathcal{L}(G,T)$ is empty—that is, all 2-local overgroups of T are solvable. In particular each $H \in \mathcal{H}_*(G,T)$ in the Thompson strategy will also be solvable. We sketch the argument in the section, which leads to Theorem 15.4.24—identifying G as either $L_3(2)$ or $Sp_4(2)' \cong A_6$ (though the groups appear in the proof already at 15.4.7).

In section 15.4 of [**AS04b**], a set $\xi^*(G,T)$ of "thin" solvable uniqueness subgroups X is defined, by the properties that $X = O^2(X)$ is T-invariant and $|XT : T|$ is prime. Again there are analogies with the FSU 3.5.2: The definition of $\xi^*(G,T)$ includes the *assumption* that we have $M = !\mathcal{M}(XT)$ (cf. (3.5.1) and 14.1.2.2 of [**AS04b**]); this property will be established in suitable situations e.g. in 15.4.3 and 15.4.12.

At 15.4.7, one branch of the argument leads either to $M \cong S_4 \wr \mathbb{Z}_2$—which was eliminated earlier in 13.9.1; or to $M \cong S_4$. In the latter case, it follows that $C_G(Z) = T \cong D_8$, and:

If $C_G(Z) = T \cong D_8$, then $G \cong L_3(2)$ or A_6.

For we can apply I.4.1.2 in [**AS04a**] to identify G.

In the remaining branches, Theorem 15.4.8 shows that $C_G(Z) = T$. This leads to useful properties of all maximal 2-locals; in particular, 15.4.12 shows each $M \in \mathcal{M}(T)$ contains a normal uniqueness group from $\xi^*(G,T)$ or $\Xi^*(G,T)$. Such uniqueness groups X and Y are now chosen from distinct maximal 2-locals, and used to define suitable subgroups G_1 and G_2, which are shown to form a Goldschmidt amalgam after 15.4.20. Hence the group they generate can be described using F.6.18 of [**AS04a**]. This in turn forces $O_2(G_1)$ and $O_2(G_2)$ to be contained in an abelian subgroup, whereas by the original construction these groups do not commute. This contradiction eliminates the remaining branches in Section 15.4.

---

[12] As in other arguments where U is abelian, F.9.16 is applied at 15.3.64 to get $q(Aut_H(\tilde{U}), \tilde{U}) \leq 2$; and this in turn is used to get $r(G,V) = 2$ in 15.3.65.2.

Theorem 15.4.24 then summarizes the work of Section 15.4, as leading only to the groups $L_3(2)$ and A_6. Finally Theorem 15.5.25 (nicknamed Theorem E) summarizes all the work on the case where $\mathcal{L}_f^*(G,T)$ is empty in Chapters 14 and 15; recall that the six groups which arose earlier were $G_2(2)'$, ${}^2F_4(2)'$, ${}^3D_4(2)$, M_{12}, J_2, and J_3.

We can now complete the proof of the Quasithin Theorem 3.0.1. By Theorem 3.4.1, we may assume $|\mathcal{M}(T)| > 1$. Then Theorem D in [**AS04b**], summarized in Remark 3.10.1, shows we may assume that $\mathcal{L}_f(G,T)$ is empty. Finally Theorem 15.4.25 in [**AS04b**] shows that the Quasithin Theorem holds when $\mathcal{L}_f(G,T)$ is empty, completing the proof of the Quasithin Theorem.

3.12. Bonus: The Even Type (Quasithin) Theorem for use in the GLS program

While [**AS04b**] was designed to classify quasithin groups of even characteristic, Chapter 16 also provides a treatment of the quasithin problem for use in the GLS project: namely Theorem 16.5.14 extends the Quasithin Theorem 3.0.1, by showing that only one further quasithin group satisfies the more general GLS hypothesis of *even type* (which extends characteristic 2 type somewhat, by allowing certain components in involution centralizers):

THEOREM 3.12.1 (Even Type (Quasithin) Theorem). *Assume that G is a K-proper quasithin simple group, which is of even type, but not of even characteristic. Then $G \cong J_1$.*

The proof uses a small subset of the theory of standard components; we discussed that theory in the latter half of our Chapter 1. In particular it quotes Lemma 3.4 of [**Asc75a**], which can also be deduced from Theorem ZD on page 21 of [**GLS99**].

Obtaining a standard component L. Section 16.1 of [**AS04b**] first reviews the definition of groups of even type; including the list of quasisimple groups allowed to be components of involution centralizers. The remainder of the section then establishes detailed properties of involution centralizers in the automorphism groups of those components, as well as possible embeddings among such components.

Since the hypothesis also specifies that G is *not* of even characteristic, Section 16.2 begins with the observation that there must then be some 2-central involution z whose centralizer denoted by $G_z := C_G(z)$ has a componen L. The remainder of the section develops properties of such pairs (z, L); for example, Theorem 16.2.4 shows that L is not just subnormal, but even normal in G_z. The proofs primarily involve properties of the known simple groups; but 16.2.10 quotes 3.4 of [**Asc75a**] mentioned above.

Then Section 16.3 shows that L is in standard form; recall from Definition 1.6.3 this means that L commutes with no conjugate, and that $C := C_G(L)$ is tightly embedded in $N_G(L) = N_G(C)$. The proof proceeds via showing that L is "terminal" in the sense of [**GLS99**]: in particular, for each involution t in C, L is also a component of G_t. For otherwise L is properly contained in the normal closure K of L in $E(G_t)$; and then an argument eliminates each of the possibilities for the embedding $L < K$ determined in Section 16.1. We mention that 16.3.13 uses results on balance stated as [**AS04a**, I.3.1], which in turn quotes [**GLS96**, 5.22]. (We discuss balance conditions in Section B.3.)

Determining L and identifying G. At this point it would be possible to complete the proof by quoting results from the literature on standard form in our Chapter 1. But to keep the result compatible with the GLS program, which does not assume those results, a more direct and elementary argument is now made in [**AS04b**] to complete the proof.

Section 16.4 studies the set Δ of conjugates C' of C such that $N_{C'}(C) = N_{C'}(L)$ is of even order; the Strongly Embedded Theorem 1.3.5 is quoted to guarantee that Δ is nonempty. Some subsequent arguments use elementary results such as I.7.2 and I.7.7 in [**AS04a**] on tight embedding. (The original proofs had appeared in [**Asc75a**] and [**Asc76b**].) After 16.4.9 it is possible to restrict to a nonempty subset Δ_0 in which involutions of a Sylow 2-subgroup R of $N_{C'}(L)$ induce only inner automorphisms of L. But also R is faithful on L by 16.4.2.1; so the size and structure of R are limited by the 2-structure of L. Finally 16.4.11 shows that Δ_0 defines a symmetric relation between C' and C, and that R is conjugate to $T_C := C_T(L)$. These are important conditions in the remaining analysis.

Section 16.5 makes a choice of C' and $R = T_C^g$. Proposition 16.5.1 now shows that the case of $L/Z(L)$ a Bender group leads to $G \cong J_1$. The proof applies I.7.2 of [**AS04a**] to $\Omega_1(T_C)\Omega_1(T \cap L)$ in the role of "V", constructing a "Goldschmidt–O'Nan pair" which can be described via [**GLS96**, 14.2]; and from that list only the case $L \cong L_2(4)$ is possible, with V of rank 3 and $\Omega_1(T_C)$ of order 2. Now as R is faithful and inner on L, it is elementary; hence so is its conjugate T_C, and we conclude that $T_C = \Omega_1(T_C)$ is of order 2. Thus $G_z = L \times T_C$, and this centralizer determines G as J_1 using I.4.9 in [**AS04a**], which quotes Janko [**Jan66**].

The remainder of Section 16.5 eliminates the possibilities for L remaining from the list in Section 16.1. These components are wider, so the requirement that $N_G(L)$ should be strongly quasithin imposes strong restrictions on R. Indeed in the cases in 16.5.3 and 16.5.4, T can be described explicitly, with the fusion of R and T_C then forcing a contradiction. In most of the remaining cases, Notation 16.5.5 constructs for $u \in \Omega_1(R)$ a certain nontrivial subgroup X of $C_L(u)$; this subgroup is exploited in a fairly uniform way to eliminate the remaining cases, again via restrictions from Section 16.2 on the size and structure of the group of automorphisms of L. The final contradiction via 16.5.10.2 (as well as certain earlier ones) is to the explicit structure of the set of 2-central involutions in the various groups L.

This contradiction completes the proof of the Even Type (Quasithin) Theorem 3.12.1.

Further comments on quoted and preliminary results. We discussed results and techniques from Chapters A–F in [**AS04a**] in our Sections 3.2 and 3.3. We conclude the chapter with a brief discussion of various specialized topics from the remaining Chapters in [**AS04a**].

Chapters G and H are devoted to facts about certain representations of SQTK-groups, including weak closure parameters; often these facts are extracted from the existing literature, as mentioned earlier. Section G.2 contains material on "almost-special" groups: this is a suitable version of large-extraspecial subgroup theory for quasithin group—developed independently here, so that the treatment of the Quasithin Theorem does not need to quote the $GF(2)$ Type Theorem 7.0.1.

Chapter I contains, for the convenience of the reader, the statements of various results from the literature which are quoted during the proof in [**AS04b**]; in our discussion of that proof, we have often explicitly indicated such references.

Chapter J proves as J.1.1 a characterization of the Rudvalis group Ru, based on hypotheses which arise in Section 14.7 of [**AS04b**]; this result was needed because characterizations in the existing literature used less convenient hypotheses.

CHAPTER 4

$e(G) = 3$: The classification of rank 3 groups

The principal theorem classifying the finite simple groups of characteristic 2 type with $e(G) = 3$ was proved by Aschbacher in two papers [**Asc81b, Asc83a**].

We will state that result below; but the reader may also wish to review our discussion of the result in Section 2.4. In particular, we recall the set $\tau = \tau(G)$ of all odd primes p with $m_{2,p}(G) = 3$; and the nonempty subset σ of large primes from Definition 2.4.1:

If $\tau \not\subseteq \{3,5\}$, set $\sigma := \tau \setminus \{3,5\}$
If $\{5\} \subseteq \tau \subseteq \{3,5\}$, set $\sigma := \{5\}$.
Finally if $\tau = \{3\}$, then we set $\sigma := \tau = \{3\}$.

The Main Result in [**Asc81b**] proves that a group G of characteristic 2 type with $e(G) = 3$ is either known, or appears in the Preuniqueness Case given in Definition 2.3.8.

THEOREM 4.0.1 ($e(G) = 3$ Theorem). *Let G be a finite K-proper simple group of characteristic 2 type with $e(G) = 3$. Then one of the following conclusions holds:*

(1) $G \cong L_4(2^n)$ or $Sp_6(2^n)$, for some $n > 1$; or
(2) $G \cong \Omega_8^-(2^n)$, for some $n \geq 1$; or
(3) $G \cong U_5(2^n)$, for some $n \neq 2$; or
(4) $G \cong L_6(2)$, $L_7(2)$, $U_6(2)$, $Sp_8(2)$, $\Omega_8^+(2)$, $F_4(2)$, Co_2, or Th; or
(5) for each $p \in \sigma(G)$ and $P \in Syl_p(G)$, $\Gamma_{2,P}^0(G)$ is contained in a 2-local subgroup of G. (That is, the Preuniqueness Case holds.)

Thus the classification of the groups with $e(G) = 3$ was in fact not completed until Aschbacher's proof of the Uniqueness Case Theorem 8.0.2, and the proof by Aschbacher, Gorenstein and Lyons of the Preuniqueness-implies-Uniqueness Theorem 8.0.1.

In the latter part of Section 2.4, we expressed Aschbacher's work on groups G with $e(G) = 3$ in the form of a trichotomy, whose three branches correspond to those in the Trichotomy Theorem 5.1.2 on groups G with $e(G) \geq 4$. But that viewpoint is at best only implicit in Aschbacher's exposition in [**Asc81b, Asc83a**], so our discussion in this chapter will instead be closer to the development in those papers.

In particular, we emphasize that the trichotomy in our Section 2.4 is *not* the "three part division" referred to in Aschbacher's introductory exposition at [**Asc81b**, p. 359]—instead that three part division consists of the work of Gorenstein and Lyons on groups G with $e(G) \geq 4$, and the two parts of Aschbacher's analysis of groups G with $e(G) = 3$, corresponding to the cases when $\sigma(G)$ is, or is not, just $\{3\}$ (in [**Asc83a, Asc81b**], respectively).

4.1. The case where $\sigma(G)$ contains a prime $p \geq 5$

Aschbacher's first paper [**Asc81b**] handles the subcase where $\sigma(G) \neq \{3\}$: that is, where $\sigma(G)$ contains a prime $p \geq 5$. The corresponding classification appears in Theorem 1 of that paper:

THEOREM 4.1.1 ($e(G) = 3$ subtheorem I). *Let G be a finite K-proper simple group of characteristic 2 type, with $\sigma(G)$ containing some prime $p > 3$. Then one of the following holds:*

(1) $G \cong L_4(2^n)$ or $Sp_6(2^n)$, for some $n \geq 3$; or $\Omega_8^-(2^n)$, for some $n \geq 2$; or $U_5(2^n)$, for some $n \geq 4$; or

(2) for each $p \in \sigma(G)$ and $P \in Syl_p(G)$, $\Gamma^0_{2,P}(G)$ is contained in a 2-local subgroup of G.

Recall that conclusion (2) was later eliminated by Aschbacher's work on the Uniqueness Case, and the work of Aschbacher, Gorenstein, and Lyons on the Preuniqueness Case.

The bulk of the proof of Theorem 4.1.1 involves proving that if G does not appear in the Preuniqueness Case, then G is one of the groups in conclusion (1); hence we shall assume throughout the remainder of this section, sometimes without explicit comment, that the Preuniqueness conclusion (2) does not hold. Then this assumption is used to establish the triviality of appropriate signalizer functors. This is in the spirit of the Gorenstein-Walter Alternative, appearing at Remark 1.1.2 in our discussion of Step 3 in Strategy 1.1.1.

The Signalizer Analysis. We begin by comparing this analysis to that in the Gorenstein-Lyons work on the Trichotomy Theorem 5.1.2, to be discussed in Chapter 5. In that work, signalizer functor theory is applied to establish the first major dichotomy (our Theorem 5.2.2, which is Theorem B in [**GL83**, Ch II.2])—which shows that when G is not in the Preuniqueness Case, then the relevant signalizer functor Δ is trivial, which in due course leads in general to a centralizer of standard type. (A third branch, consisting of groups of $GF(2)$ type, splits off from the standard type branch at a later point in the analysis.)

Aschbacher is able to obtain a similar dichotomy for groups G with $e(G) = 3$ when σ contains a prime $p > 3$: His first major reduction is the result [**Asc81b**, 9.13], which we state below as Proposition 4.1.4; it establishes the triviality of a certain signalizer functor α, a condition which will later produce centralizers in standard form like those in the groups in conclusion (1). We now expand on this brief summary.

As Aschbacher observes at [**Asc81b**, p. 359], when $e(G) = 3$ he is not able to use the moving-around functor "$\theta^{(2)}$" used by Gorenstein and Lyons (appearing in Theorem 5.2.7) when $e(G) \geq 4$. Instead he begins with a getting-started functor which uses the group generated by the following p-signalizers for p an odd prime:

DEFINITION 4.1.2. For B an elementary abelian p-subgroup of G of rank 2, we define

$$\alpha(B) = \alpha_G(B) := \langle\, [\, O_2(C_G(b))\ ,\ B\,] : b \in B^{\#}\,\rangle\ \Big(\bigcap_{b\in B^{\#}} O_2(C_G(b))\Big).$$

In addition, for any X such that $B \leq X \leq G$, we define $\alpha_X(B)$ by the same expression, replacing G by X throughout. ♢

One must start (as we will see shortly) by showing that the function α is obtained from a G-equivariant A-signalizer functor γ for some $B \leq A \cong E_{p^3}$ when we set $\alpha(B) := \langle \gamma(b) : b \in B^{\#} \rangle$, and hence has the properties in 1.5.1; then the functor must be analyzed, ideally leading to a proof of the alternative: either the functor is trivial or the Preuniqueness Case holds. We need the following definition:

DEFINITION 4.1.3. For $X \leq G$, let $\theta_p(X)$ denote the largest *solvable* normal p'-subgroup of X. (We write $\theta(X)$ when the choice of p is clear.)

Aschbacher's main result on signalizers [**Asc81b**, 9.13] establishes various properties, including the triviality of α:

PROPOSITION 4.1.4. *Assume that the hypotheses of Theorem 4.1.1 hold, but that its conclusion (2) fails. Then there exists $p \in \sigma(G)$ such that for every elementary abelian p-subgroup A of G of rank 3, for each hyperplane B of A, and each $c \in A^{\#}$, the following conclusions hold:*

(a) $\alpha(B) = 1$;

(b) A centralizes a Sylow 2-subgroup of $\theta(C_G(c))$; and

(c) $\bigcap_{b \in B^{\#}} \theta(C_G(b))$ has odd order.

The properties (b) and (c) will be important later in establishing the existence of a centralizer in standard form.

To prove this result, Aschbacher first defines (see [**Asc81b**, p. 364]), for each B, a signalizer functor γ_B on A such that $\alpha(B)$ is the completion $\langle \gamma_B(a) : a \in A^{\#} \rangle$ of γ_B. Invoking completeness in the Signalizer Functor Theorem 0.3.15, he establishes as [**Asc81b**, 3.10] the following result:

PROPOSITION 4.1.5. *Let G be a finite group, p an odd prime, and A an elementary abelian p-subgroup of G of rank 3. Suppose that for every hyperplane B of A and every element $c \in A^{\#}$, we have $\alpha_{C_G(c)}(B) \leq O_2(C_G(c))$.*

Then $\alpha(B)$ is a 2-group, and is independent of the choice of hyperplane B of A.

Proposition 4.1.5 reduces the proof of Proposition 4.1.4 to the verification (at [**Asc81b**, 9.10]) of the "local balance" condition:

$$\alpha_{C_G(c)}(B) \leq O_2(C_G(c))$$

for all hyperplanes B of A and all $c \in A^{\#}$. This is similar to the $\frac{3}{2}$-*balance* condition of Gorenstein and Lyons (see e.g. [**GLS96**, Defn. 20.2]) localized at the prime 2— we will return to this theme in a moment. (We also discuss balance conditions in Section B.3.) There are two steps in verifying this condition. Set $X := \alpha_{C_G(c)}(B)$ and $\overline{C} := C_G(c)/\theta(C_G(c))$. Note that $X = O^{2'}(X)$ by definition of α.

Step I: Show $\overline{X} = 1$. This step is carried out using 9.2 and 9.9 of [**Asc81b**]. The proof of 9.2 reduces, via 8.3 of [**Asc81b**], to consideration of a component $\overline{L}$ of $\overline{C}$ such that $[\overline{L}, \overline{B}] = \overline{L} = [\overline{L}, \overline{X}]$. The possibilities for $\overline{L}$ are essentially enumerated in Theorems 2 and 3 in section 5 of [**Asc81b**]. Then an analysis of the automorphism groups of these groups in section 6 of [**Asc81b**], culminating in 6.11, shows $\overline{X}$ is of odd order, and hence $\overline{X} = 1$. Note since $\theta(L)$ is a p'-group that the preimage L is a p-component of $C_G(c)$.

We pause to indicate one feature of Aschbacher's functor α. Suppose first that we were instead working with a more classical signalizer functor, defined using some variation on $O_{p'}(C_G(-))$; then for each b, c in A with $\overline{C} := C_G(c)/\theta(C_G(c))$, we would want $\overline{A}$ to centralize the image $\overline{X}$ in $\overline{C}$ of $O_{p'}(C_G(b)) \cap C_G(c)$. This is in the background in establishing balance for the signalizer functors in the Gorenstein-Lyons Trichotomy work—cf. our discussion following Definition 5.2.5. In the present setup, when $\overline{L}$ is our standard example of a group of Lie type in characteristic 2, the only obstruction to the local $\frac{3}{2}$-balance condition for $p > 3$ occurs when $\overline{X}$ lies in a torus $\overline{T}$ centralized by $\bar{b}$, and some b_1 in A acts as a field automorphism on $\overline{L}$ and on $\overline{T}$. In general b_1 might not centralize the subgroup $\overline{X}$ of $\overline{T}$, but in this case, b_1 does centralize a Sylow 2-subgroup of $\overline{T}$, so in particular it centralizes $O_2(C_{\overline{L}}(\bar{b}))$. This gives a rough indication of why the usual obstruction is avoided by Aschbacher's use of $O_2(C_{\overline{L}}(-))$ instead of $O_{p'}(C_{\overline{L}}(-))$ in the construction of α in Definition 4.1.2.

Step II: Show $X \leq O_2(C_G(c))$. Having proved in Step I that $X \leq \theta(C_G(c))$, it is now necessary to deal with solvable subgroups of the form YXB, where Y is an XB-invariant q-group for some odd prime $q \neq p$, and $[X, B]$ acts non-trivially on Y. This step also uses 9.2 and 9.9 of [**Asc81b**]; but also appeals to Theorem 9.3 of [**Asc81b**]. In the proof of Theorem 9.3, the leverage is provided by the fact that $e(G) = 3$, using various preliminary results from Section 3 of the paper.

Both Steps above have parallels in [**GL83**]: for Step I, the existence proof for a 2-balanced E_{p^4}-subgroup—cf. the discussion of "getting started" after 5.2.6 below; for Step II, the use of the functor γ on B to control $\Gamma_{2,B}(G)$ in "moving around"—cf. 5.2.10 and the sentence following it.

REMARK 4.1.6. Indeed Lemma 9.4, the first result in the proof of Theorem 9.3, uses 3.6 of [**Asc81b**], which supplies an easy contradiction when $p > 7$. But for $p \in \{5, 7\}$, some delicate analysis is needed; and it is for precisely this reason that 5 and 7 play a special role in the definition of the set $\sigma(G)$ of primes.

The Component Analysis. Let us briefly review Strategy 1.1.1. The Pre-uniqueness Case represents Step 1 in that strategy for the groups of characteristic 2 type, while the Quasithin Theorem 3.0.1 represents Step 2. The signalizer functor analysis, culminating in Proposition 4.1.4, goes a long way toward completing Step 3 in the case considered in this section. Conclusions (a)–(c) of the Proposition are the foundation for the analysis, in Steps 3–5 of the Strategy, of p-components in the centralizers of p-elements. Then a centralizer in standard form will lead in Step 6 to the construction of a large Lie type subgroup I of G, and finally to the idenitification $G = I$, much like the identification of a standard-type group G by Gilman and Griess in [**GG83**], as we discuss in the first subsection of Section 6.1.

Odd standard form. Using the facts that $p > 3$ and G is of characteristic 2 type, a short argument [**Asc81b**, 10.5] building on Proposition 4.1.4 completes Step 3 of Strategy 1.1.1:

LEMMA 4.1.7. *For every elementary abelian p-subgroup A of G with $|A| = p^3$, and each $a \in A^\#$, there is $a_1 \in A^\#$ and a p-component L of $C_G(a_1)$ with $L = [L, a]$.*

Fixing A as in Lemma 4.1.7, Aschbacher makes the following definitions.

DEFINITION 4.1.8. For $a \in A^\#$, $\mathcal{L}(a)$ is the set of p-components of $C_G(a)$, and

$$\mathcal{L}(A) := \bigcup_{a \in A^\#} \mathcal{L}(a).$$

(At [**Asc81b**, 12.1] this definition is suitably extended to an elementary abelian p-subgroup E containing A.)

He defines an ordering on $\mathcal{L}(A)$ (a variant of one of the standard orderings, such as the one mentioned after Remark 1.8.2):

$$L << K \qquad \text{iff} \qquad L = \theta(L)(L \cap K),$$

and considers the following subsets of $\mathcal{L}(A)$:

DEFINITION 4.1.9.
(1) $\mathcal{L}^*(A)$ is the set of "maximal" members of $\mathcal{L}(A)$ under $<<$;
(2) $\mathcal{L}_0(A)$ is the set of $L \in \mathcal{L}(A)$ such that L contains a 2-subgroup S with $1 \neq S = [S, A]$;
(3) $\mathcal{L}_1(A)$ is the set of $L \in \mathcal{L}(A)$ such that A induces inner automorphisms on $L/\theta(L)$; and
(4) $\mathcal{L}_i^*(A) = \mathcal{L}_i(A) \cap \mathcal{L}^*(A)$.

Using the facts that G is of characteristic 2 type and $e(G) = 3$, he establishes the proposition below. (See 10.4, 10.6, 10.7, and 11.2 in [**Asc81b**].) Note that (2) corresponds to Step 4 of Strategy 1.1.1; while (1) and (3) correspond to Step 5, by providing a component in standard form with properties similar to 2.3.2.4 and 2.3.5.1.

PROPOSITION 4.1.10. *The following conditions hold:*
(1) if $L \in \mathcal{L}^*(A)$, *then* $|C_A(L/\theta(L))| = p$;
(2) if $L \in \mathcal{L}_0(A)$, *then* L *is quasisimple;*
(3) if $L \in \mathcal{L}_0^*(A)$, *then* $L/Z(L) \in Chev(2)$; *and*
(4) $\mathcal{L}_0^*(A) \cap \mathcal{L}_1(A) \neq \emptyset$.

Identifying a large Lie-type subgroup "I". We remark that Lemma 4.1.7 actually establishes that $\mathcal{L}_0^*(A) \neq \emptyset$. The first three assertions[1] of Proposition 4.1.10 are fairly easy. The last entails more work, analyzing to a contradiction the case when $L \in \mathcal{L}_0^*(A)$ with $m_p(L) = 1$, and with some $a \in A^\#$ inducing a field automorphism on L.

Once Proposition 4.1.10 is established, it remains to complete Step 6 of Strategy 1.1.1 by identifying G. We have a component $L \in \mathcal{L}_0^*(A) \cap \mathcal{L}_1(A)$ in odd standard form, as in Remark 2.3.3, so we are in a situation roughly comparable to that of Gilman and Griess in [**GG83**]. Like Gilman and Griess, Aschbacher embeds A in a maximal elementary abelian p-subgroup E of $C_G(A)$, and studies $W^* = N_G(E)/C_G(E)$, which will be the Weyl group[2] when G is identified as a group of Lie type. His analysis thereafter is also similar. He reduces at [**Asc81b**, 12.6] essentially to the cases where either W^* acts irreducibly on E, or $O_p(W^*) \neq 1$.

The latter case would arise if G were not simple and some $c \in E$ induced a nontrivial field automorphism on $F^*(G)$ of Lie type. In this case, Aschbacher in [**Asc81b**, Theorem 12.9] builds a Sylow p-subgroup of G and obtains a contradiction, via transfer, to the simplicity of G. (Cf. our discussion in Section 6.1 of Prop. 4.1 in Gilman and Griess [**GG83**].)

[1]For the parallels in [**GL83**], cf. the results we state as 5.3.5—5.3.10.

[2]Cf. our discussion of the proof of the Standard Type Theorem 6.1.1, and especially of the automizer $A_G(B^*)$ playing the role of the Weyl group there.

In the remaining case, using McLaughlin's Theorem A.6.1 on groups generated by transvections during the proof of [**Asc81b**, 12.7], he arrives at the following situation in 13.1 of [**Asc81b**]:

PROPOSITION 4.1.11. *The following conditions hold:*
(1) W^* *acts irreducibly on* E*;*
(2) W^* *contains no transvections; and*
(3) $\mathcal{L}_0^*(A) \subseteq \mathcal{L}_1(A)$.

A detailed analysis ensues. In particular, in the "Weyl group" W^*, the set D^* of reflections consists of the involutions induced in members of $\mathcal{L}_0^*(E)$. If $A < E$, Aschbacher shows in [**Asc81b**, 13.7] that D^* is a set of 3-transpositions in $X^* := \langle D^* \rangle$. Then Fischer's Theorem A.6.3 is invoked in 13.8 (in the special case considered in 4.4) to identify X^* as S_n for some $n \in \{5, 6, 7\}$. And in any case, he eventually succeeds in identifying, by means of the Curtis-Tits Theorem A.5.1 (stated as 7.3) and Phan's Theorem (stated as 7.4, see our Remark A.5.2), the following subgroup I of G—as summarized at [**Asc81b**, 15.2]:

THEOREM 4.1.12. *Set* $I := \langle L : L \in \mathcal{L}_0^*(E) \rangle$.
Then $I \cong L_4(q)$, $Sp_6(q)$, $\Omega_8^-(q)$, *or* $U_5(q)$, *for some* $q = 2^n$.

The final step: Identifying G as I. Having determined in Theorem 4.1.12 the possible isomorphism types for the subgroup I, it remains to show that $I = G$. Of course as G is simple, if $I < G$ then also $N := N_G(I) < G$. Aschbacher establishes as [**Asc81b**, 15.15] the following pushing-up result. It is phrased in terms of a variant of the group $C(G,T)$ in Definition B.7.1—which when proper in G gives the main hypothesis of the Global $C(G,T)$-Theorem B.7.8.

PROPOSITION 4.1.13. *Let* $T \in Syl_2(N)$ *and* $S = C_T\big(\Omega_1(Z\big(J(T)\big))\big)$. *Then*

$$C^*(G,T) \leq N,$$

where $C^*(G,T) := \langle C_G(\Omega_1\big(Z(T)\big)), N_G(S_0) : 1 \neq S_0 \text{ char } S \rangle$.

Once Proposition 4.1.13 is established, Aschbacher quotes the Global $C(G,T)$-Theorem for a contradiction when $N < G$, completing the proof of Theorem 4.1.1.

To prove Proposition 4.1.13, Aschbacher first controls centralizers of p-elements. In [**Asc81b**, 15.8], he establishes (in the language of Definition 1.3.6) that $\Gamma_{2,A}(G)$ lies in N. Then using the Seitz Generation Theorem A.3.4—which, as $p > 3$, yields that $K = \Gamma_{2,A}(K)$ for any A-invariant section $K \in Chev(2)$, apart from only a couple of exceptional configurations—he shows $\mathcal{L}(A) \subseteq N$ in 15.9. Next he establishes in 15.10 that $\Gamma_{1,E}(G) \leq N$. Then he switches to 2-locals, and obtains in 15.11 that every 2-local overgroup of A (and, with one exception, of any hyperplane of A) is contained in N. Finally a short argument at 15.15 completes the proof that $C^*(G,T) \leq N$, and hence also the proof of Theorem 4.1.1. In other words, we now know that Theorem 4.0.1 holds when $\sigma(G)$ contains a prime $p > 3$.

Comments on quoted results. We now mention a number of the background results used in [**Asc81b**], particularly those that have not as yet arisen explicitly in our exposition in the section. Many of these results are listed by Aschbacher in Sections 3–8 of [**Asc81b**]. And this list is, for the most part, a subset of the longer list of references quoted in the second paper [**Asc83a**], which we will discuss at the end of the next section.

Certain papers are mentioned in the introduction to the paper, just for expository purposes, to motivate the classification of groups G with $e(G) = 3$, and place them in the context of the CFSG. These include the Gorenstein-Lyons Trichotomy [**GL83**] for groups G with $e(G) \geq 4$, and the Uniqueness Case in Aschbacher's [**Asc83b, Asc83c**] (using [**AGL81**] to proceed from the Preuniqueness Case). There is also mention of the Finkelstein-Frohardt papers [**FF84, FF79, FF81b**]—although these references are not actually used until the second paper [**Asc83a**].

There are some references to standard results on coprime action (cf. our Section B.1). For example, the Thompson Dihedral Lemma B.1.7 is 3.4, and the proof of 10.9 uses the Bender-Thompson Signalizer Lemma B.1.10.

The results 3.12–3.14 quote properties of known groups—in particular, on subgroups of linear groups of dimensions 2, 3, and 4. There are references to Dickson's Theorem for $SL_2(q)$, to Bloom [**Blo67**] for $PSL_3(q)$, and to Mitchell [**Mit14**] for certain subgroups of $Sp_4(q)$.

We mentioned earlier that 13.8 invokes a special case 4.4 of Fischer's Theorem A.6.3 classifying 3-transposition groups. In fact Section 4 also quotes some specific properties of 3-transposition groups, which are proved in [**Fis71**]; an alternative reference could now be Aschbacher's book [**Asc97**] on Fischer's classification.

Some further references for properties of known groups include: at 5.10, use of the list of centralizers in Table 1 of the Aschbacher-Seitz work [**AS76b, AS81**] on standard components of known type; at 6.6, use of Aschbacher-Seitz [**AS76a**] on involutions in characteristic 2; and at 12.6, use of cohomology of known groups in Jones-Parshall [**JP76**].

We already mentioned the use of 7.3 and 7.4, the Curtis-Tits Theorem and Phan's Theorem, to identify G. Finally the Global $C(G,T)$-Theorem is used at 15.15, as indicated in our discussion of Proposition 4.1.13. Aschbacher lists the papers which established that result; we give that list later in our discussion of the $C(G,T)$ Theorem in B.7.3.

4.2. The case $\sigma(G) = \{3\}$

In his second paper [**Asc83a**], Aschbacher treats the case where $\sigma(G) = \{3\}$. Recall this means that $m_{2,3}(G) = 3$, but $m_{2,p}(G) \leq 2$ for any $p > 3$. His main theorem is the following result:

THEOREM 4.2.1 ($e(G) = 3$ subtheorem II). *Let G be a K-proper finite simple group of characteristic 2 type such that $e(G) = 3 > m_{2,p}(G)$ for all primes $p > 3$. Then one of the following conclusions holds:*

(1) $G \cong L_6(2)$, $L_7(2)$, $Sp_8(2)$, $\Omega_8^-(2)$, or $F_4(2)$; or
(2) $G \cong U_5(2)$, $U_6(2)$, $\Omega_8^+(2)$, Co_2, or Th; or
(3) $G \cong L_4(4)$, $Sp_6(4)$, or $U_5(8)$; or
(4) for each $P \in Syl_3(G)$, $\Gamma^0_{2,P}(G)$ is contained in a 2-local subgroup of G.

Recall that conclusion (4) was later eliminated by Aschbacher's work on the Uniqueness Case.

Although the analysis for Theorem 4.2.1 is often more delicate than that for Theorem 4.1.1, the general approach can still be described in terms of the same broad subdivisions used in the previous section—namely Signalizer Analysis and Component Analysis, culminating in the identification of G as I.

So the exposition in this section will parallel that in the previous section. In beginning with the signalizer analysis, we are deviating somewhat from Aschbacher's order: namely we temporarily postpone our discussion of Section 6 in [**Asc83a**]—which consists largely of preliminary lemmas used in the Component Analysis. In particular, we are deferring consideration of a sub-branch of the component analysis which did not arise in the previous section: namely the branch leading to the groups of $GF(2)$ type, analogous to branch (I) in the Gorenstein-Lyons Trichotomy Theorem 5.1.2 on groups G with $e(G) \geq 4$.

The Signalizer Analysis. As in the case of Theorem 4.1.1 in the previous section, the proof of Theorem 4.2.1 essentially assumes that G does not appear in the Preuniqueness Case, and then goes on to prove that G satisfies one of conclusions (1)–(3) of Theorem 4.2.1. Hence we shall assume throughout the remainder of this section, sometimes without explicit comment, that the Preuniqueness conclusion (4) does not hold.

In the spirit of the Gorenstein-Walter Alternative (see Remark 1.1.2, in our discussion of Step 3 of Strategy 1.1.1), the assumption that the Preuniqueness Case does not hold should lead to the triviality of appropriate signalizer functors and thence to the quasisimplicity of 3-components in the centralizers of elements of order 3.

For comparison, recall from the previous section that in Aschbacher's first paper [**Asc81b**, 9.13], such an analysis led to the uniqueness/standard form dichotomy in Proposition 4.1.4. In the second paper [**Asc83a**], a corresponding dichotomy emerges (with more effort) at Theorem 17.29, which shows that one of two sets, $\mathcal{S}$ and $\mathcal{T}$, must be nonempty. Then $\mathcal{S} \neq \emptyset$ leads to a centralizer in standard form, while $\mathcal{T} \neq \emptyset$ leads to the Preuniqueness Case (and hence eventually to a contradiction as described in Chapter 8). We begin a series of definitions, leading to the definition of $\mathcal{S}$ and $\mathcal{T}$.

Definitions related to the signalizer analysis. First we specify the relevant sets of elementary abelian 3-subgroups:

DEFINITION 4.2.2. Let $\mathcal{E}$ denote the set of elementary abelian 3-subgroups of G of rank 3 which normalize some nontrivial 2-subgroup of G, i.e., which are contained in a 2-local subgroup of G. This set is the analogue of the set $\mathcal{B}_{max}(G;p)$ (Definition 2.3.2) in the Gorenstein–Lyons work on trichotomy. Notice that $\mathcal{E}$ is nonempty since $m_{2,3}(G) = 3$.

Next, $\mathcal{F}$ denotes the set of maximal elementary abelian 3-subgroups of G which contain some member of $\mathcal{E}$. Of course, $\mathcal{F}$ is also nonempty. ◊

For $A \in \mathcal{E}$, the possible existence of A-invariant sections of G isomorphic to the semidirect product $E_{p^2} : SL(2,3)$, with $p > 3$ prime, is a fundamental obstruction to forcing A-invariant 2-subgroups of $O_{3'}(C_G(a))$ into $O_2(C_G(a))$, for $a \in A^\#$. This is why, in the signalizer analysis in the case $\sigma(G) = \{3\}$, Aschbacher must replace the 2-group $O_2(-)$ with the solvable $3'$-group $\theta(-)$ defined by (cf. 4.1.3):

DEFINITION 4.2.3. For a group X, $\theta(X)$ is the largest solvable normal $3'$-subgroup of X. ◊

A major price is paid later for this replacement, when he must deal with configurations involving field automorphisms in the Signalizer Analysis.

We can now define the crucial sets $\mathcal{S}$ and $\mathcal{T}$.

DEFINITION 4.2.4. $\mathcal{S}$ denotes the set of all $A \in \mathcal{F}$ such that, for each $a \in A^{\#}$, and each rank-2 subgroup B in A, the solvable A-invariant $3'$-subgroups:

$$[\theta(C_G(a)), A] \text{ and } \Delta_G(B) := \bigcap_{b\in B^{\#}} \theta(C_G(b))$$

are of *odd* order. ♢

We remark that if $A \in \mathcal{S}$, then fairly standard signalizer arguments, together with the fact that $m_{2,q}(G) \leq 2$ for all primes $q > 3$, will force the quasisimplicity of all 3-components of $C_G(a)$ for all $a \in A^{\#}$; this will lead later to a centralizer in standard form during the Component Analysis. However more complicated Signalizer Analysis will be carried out before that—in the case where $\mathcal{S}$ is empty.

Moreover, if $A \in \mathcal{F} \setminus \mathcal{E}$ then $m(A) \geq 4$, so that $A \in \mathcal{S}$ since $m_{2,3}(G) = 3$. Thus, in the difficult case where $\mathcal{S}$ is empty, we have $\mathcal{E} = \mathcal{F}$.

DEFINITION 4.2.5. $\mathcal{T}$ is the set of all $A \in \mathcal{E}$ with $\Gamma_{2,A}(G)$ lying in a 2-local subgroup of G. ♢

Showing that $\mathcal{T}$ is nonempty (assuming $\mathcal{S}$ is empty) is the first step in the direction of proving that the Preuniqueness Case holds.

The principal object of interest in the Signalizer Analysis is the following signalizer functor, analogous to the one studied in Theorem 4.1.1 (Cf. Definition 4.1.2, with $O_2(-)$ replaced by $\theta(-)$.)

DEFINITION 4.2.6. For B an elementary abelian p-subgroup of G of rank 2, we define

$$\alpha(B) = \alpha_G(B) := \langle\ [\ \theta(C_G(b))\ ,\ B\ ] : b \in B^{\#}\ \rangle\ \Delta_G(B). \qquad ♢$$

Notice that if $A \in \mathcal{E} \setminus \mathcal{S}$, then either $\Delta_G(B)$ has even order for some subgroup B of A of rank 2, or $[\theta(C_G(b)), A]$ has even order for some $b \in A^{\#}$. In either case, this implies that $\alpha(B)$ has even order for some subgroup B of A of rank 2. In particular this will hold in the part of the analysis where $\mathcal{S}$ is empty.

The fundamental dichotomy on $\mathcal{S}$ and $\mathcal{T}$. The initial phase of the Signalizer Analysis shows that if $\mathcal{S}$ is empty, then $\mathcal{T}$ must be nonempty. This corresponds roughly with the "getting started" phase of the proof of Theorem B in [**GL83**]—cf. our discussion of Theorem 5.2.6 in the next chapter. Aschbacher's result is Theorem 17.29 in [**Asc83a**], which we state in the form:

THEOREM 4.2.7. *Let G be a counterexample to Theorem 4.2.1. Then $\mathcal{S} \cup \mathcal{T}$ is nonempty.*

The proof of this Theorem occupies Sections 8–17 in [**Asc83a**]. The argument, especially in Sections 14–17, develops sophisticated variations on standard signalizer functor theory. It is an extension of the Gorenstein-Walter Alternative into extremely difficult territory, where the line between the two alternatives must be drawn very delicately, and powerful tools such as pushing-up techniques are required. The complexity and depth of these sections is the major reason why the case $\sigma = \{3\}$ requires a separate paper in the solution of the $e(G) = 3$ problem. So we will now devote several pages to a discussion of this part of the proof.

We assume throughout our discussion of the proof of Theorem 4.2.7 that $\mathcal{S} \cup \mathcal{T}$ is empty.[3] Since $\mathcal{S}$ is empty, we saw that $\alpha(B)$ has even order for some rank-2 subgroup B of some member $A \in \mathcal{E}$, which we fix; furthermore we saw that $\mathcal{E} = \mathcal{F}$, so that also $A \in \mathcal{F}$.

In the following discussion, we regard the proof as divided into four distinct stages.

(Stage 1: Determining obstructions to balance). We begin by summarizing the argument at the beginning of Section 13 of [**Asc83a**]:

We just saw that the signalizer function α is nontrivial. But α cannot in fact be the completion of a signalizer *functor*—for otherwise 6.2 in [**Asc83a**] shows that $A \in \mathcal{T}$, contrary to our assumption. So by standard signalizer functor analysis, this implies the existence of some local obstruction to balance (cf. the discussion of Proposition 4.1.5, and of Definition B.3.20); and the identification of such obstructions will be important later. In fact the obstructions take the following very specific shape, using the contrapositive of Proposition 4.21.2 in [**Asc83a**]:

PROPOSITION 4.2.8. *The set $\Omega(A)$ is nonempty, where*

$$\Omega(A) := \{F \leq A : |F| = 3 \textit{ with } \alpha_{C_G(F)/F}(A/F) \not\leq \theta\big(C_G(F)/F\big) \}.$$

A failure of balance in turn reflects an obstruction corresponding to a 3-component of $C_G(F)$, as recorded in Proposition 4.22. Examining the known quasisimple groups which are possible 3-components, we obtain the following list of obstructions in Theorem 13.1 of [**Asc83a**] (with the component shown to be normal in the centralizer later at 13.6):

PROPOSITION 4.2.9. *Let $A \in \mathcal{F}$ and let $\langle c\rangle = F \in \Omega(A)$. Set $H := N_G(\langle c\rangle)$. Let $L \in \mathcal{L}_A(c)$, the set of all 3-components of $C_G(c)$ such that $\alpha_{Aut_H(\overline{L})}(A/F) \neq 1$, where $\overline{L} := L/\theta(L)$.*

Then $\overline{L}/Z(\overline{L}) \cong A_7$, M_{12}, M_{22}, HJ, or $L_3^{\pi}(q)$ with $\pi \in \{+1, -1\}$ and $q \equiv \pi$ (mod 3). (Here $L_3^{+1}(q) = L_3(q)$ and $L_3^{-1}(q) = U_3(q)$.) Moreover $LA = L \times \langle c\rangle$, and A can be chosen so that for any c, L is uniquely determined and in particular is normal in H.

We emphasize that the proof of the proposition depends on several earlier reductions. Indeed, one major class of obstructions to α-balance consists of 3-components $L \in \mathcal{L}_A(c)$ with $\overline{L}$ a group of Lie type defined over a field of cardinality 2^{3m}, and with some element $a \in A$ inducing a nontrivial field or graph-field (i.e. of order 3) automorphism on $\overline{L}$. These cases are eliminated in Sections 8–12 of [**Asc83a**], primarily by a delicate transfer argument leading to the contradiction $O^3(G) < G$. However, there is an additional complication in the elimination of the case $\overline{L} \cong (S)U_3(2^{3n})$ in Proposition 8.6: Aschbacher studies the embedding of an E-invariant 2-subgroup S of L, where $E \in \mathcal{E}$ and $E \leq A$;[4] he establishes that S is a maximal normal elementary abelian 2-subgroup of a maximal 2-local subgroup M of G, and S is a TI-set in G. He is then able to invoke (as 3.10)

[3]But see Remark 4.2.19 regarding the more general technical hypothesis Aschbacher uses in this area of [**Asc83a**].

[4]Of course we have $E = A$ during our proof of Theorem 4.2.7, where we are assuming that $\mathcal{E} = \mathcal{F}$; but that is not assumed in Aschbacher's more general hypothesis for Section 8 of [**Asc83a**].

the $GF(2^n)$ Type Theorem 7.5.2 classifying groups with such TI-subgroups, to obtain a contradiction in this case.

Further reductions are obtained in Sections 12 and 13, preparing the way for the second stage of the argument, which begins in Section 14.

(Stage 2: Controlling certain 3-signalizers). Aschbacher's strategy now is to eventually force the contradiction $A \in \mathcal{T}$, by carefully analyzing the obstructions to the putative signalizer functor which should be associated to α. To do so, he defines a certain set of A-invariant $3'$-subgroups, and studies the proper subgroups of G generated by such signalizers. Thus the argument bears some resemblance to the *proof* of the Signalizer Functor Theorem.

We begin our study of such subgroups, following Aschbacher at the beginning of Sections 14 and 15 (as well as p. 112). To simplify the exposition, we will ignore a term "L_B" that Aschbacher includes in his definition of $\mathcal{Q}(G)$; it is required only in one special case involving a 3-component L_b with $L_b/O_2(L_b) \cong HJ$. We continue to fix:

$$A \in \mathcal{F},$$

as in Proposition 4.2.9.

DEFINITION 4.2.10. $\mathcal{Q}(G)$ is the set of all A-invariant $3'$-subgroups Q of G which satisfy:

$$C_Q(B) \leq O_{3'}(C_G(B))O_{3'}(C_G(A))$$

for each hyperplane B of A. For any proper A-invariant subgroup Y of G, we set:

$$\mathcal{Q}(Y) := \{Q \in \mathcal{Q}(G) : Q \leq Y\},$$

and set $\Delta(Y) := \langle Q : Q \in \mathcal{Q}(Y)\rangle$.

For π a set of primes, let $\mathcal{Q}(Y,\pi)$ be the set of π-subgroups in $\mathcal{Q}(Y)$, and let $\mathcal{Q}^*(Y,\pi)$ denote the maximal members of $\mathcal{Q}(Y,\pi)$ under inclusion.

For any subgroup M of G, we will write $\mathcal{Q}(Y) \subseteq M$ to mean that $Q \leq M$ for all $Q \in \mathcal{Q}(Y)$. ◊

Note that $\mathcal{Q}(G)$ is clearly invariant under $N_G(Q)$.

If a signalizer functor associated to α existed, then the subgroup generated by members of $\mathcal{Q}$ would be a $3'$-group. However in our situation, there exist $\mathcal{Q}$-generated subgroups Y, which instead involve obstructing 3-components from the list of pairs (c, L) in Proposition 4.2.9.

Nevertheless, Aschbacher is able to deal with such obstructions; he begins by proving a weak analogue of the Thompson Transitivity Theorem ([**Gor82**, 4.14]), which is the first key ingredient in the proof of a Signalizer Functor Theorem. This analogue is contained in Aschbacher's Proposition 14.11.1:

THEOREM 4.2.11 (Aschbacher Transitivity Theorem). *$N_G(A)$ transitively permutes the members of the set $\mathcal{Q}^*(G,2)$ under conjugation.*

Much as quasi-equivalence of linear representations is weaker than equivalence, Aschbacher's Transitivity Theorem establishes that any two elements of $\mathcal{Q}^*(G,2)$ can be considered equivalent, *but only modulo an unspecified automorphism of the operator group A.* While useful, the theorem does not yet answer the question addressed in Stage 3 below: why do elements of $\mathcal{Q}(C_G(c),2)$, for different elements $c \in A$, generate a 2-group or even a $3'$-group?

In fact at 16.2, Aschbacher is able to improve 4.2.11 above, establishing $C_G(A)$-transitivity at least for pairs of elements of $\mathcal{Q}^*(G,2)$ connected by a chain where adjacent elements satisfy a nontrivial intersection condition. It is convenient to give the statement here; the definition of $\mathcal{Y}_0^*$ will be given just below:

PROPOSITION 4.2.12. *Assume* $Y_1, Y_2 \in \mathcal{Y}_0^*$. *If* $\mathcal{Q}\big(N_G(Y_1) \cap N_G(Y_2), 2\big)$ *contains a nontrivial member, then* Y_1 *and* Y_2 *are conjugate under* $O_{3'}\big(C_G(A)\big)$.

(Stage 3: Uniqueness of a certain maximal signalizer Y). In the proof of the Signalizer Functor Theorem 0.3.15, the stronger transitivity under $C_G(A)$ eventually leads to the completeness of the functor; cf. the discussions of [**Gor82**, pp. 185,198] and [**GLS96**, pp. 130–135]. In particular, the "closure" part of the completeness argument shows that members of an orbit on suitable signalizers from the Transitivity Theorem are permutable—and so their product also defines such a signalizer, which is therefore the unique maximal signalizer of this type. This outcome can be phrased in the following language:

> If θ is an A-signalizer functor, then the set $\mathcal{N}_\theta^*(A)$ of maximal A-invariant p'-subgroups Q of G, such that $C_Q(a) = Q \cap \theta\big(C_G(a)\big)$ for all $a \in A^\#$, contains a unique member.

Aschbacher then works to show that an another set $\mathcal{Y}_0^*$ of maximal signalizers contains a unique member Y; he approaches this set in Section 15 by restricting attention to certain subgroups Y of G such that $Y = \Delta(Y)$, in the notation of Definition 4.2.10:

DEFINITION 4.2.13. Let $\mathcal{Y}$ denote the set of all A-invariant subgroups Y of G. Let $\mathcal{Y}_2$ denote those $Y \in \mathcal{Y}$ with $O_2(Y) \neq 1$, let:

$$\mathcal{Y}_4 := \{Y \in \mathcal{Y} : Y = O^{3'}(Y) = \Delta(Y) \text{ and } Y/O_2(Y) \text{ is quasisimple }\},$$

and let $\mathcal{Y}_0 := \mathcal{Y}_4 \cap \mathcal{Y}_2$. Finally, let $\mathcal{Y}_0^*$ denote the set of maximal members of $\mathcal{Y}_0$ under inclusion. $\diamondsuit$

Aschbacher eventually establishes at 16.3.3 that if $\mathcal{Y}_0 \neq \emptyset$, then $\mathcal{Y}_0^*$ has a unique member. Then in Theorem 16.9.1, he shows that $\mathcal{Y}_0$ is nonempty. Now we state an omnibus result—combining these assertions with others which will emerge during our discussion of the proof over the next several pages. Recall that (c, L) denotes some obstructing pair in Proposition 4.2.9.

THEOREM 4.2.14. $\mathcal{Y}_0^*$ *contains a unique member* Y, *and* $N_G(\langle c \rangle) \leq N_G(Y)$. *Moreover, either* $Y = O_2(Y)L$ *and* $\Omega(A) = \{\langle c \rangle\}$; *or* $L \leq Y$ *with* $\overline{L} \cong A_7$, *and* $Y/O_2(Y) \cong A_{10}$.

The first principal step in the proof of Theorem 4.2.14 is an analysis in Section 15 of the structure and embedding of members of $\mathcal{Y}_0^*$. A key result 15.3.1 asserts:

PROPOSITION 4.2.15. *Assum* $Y \in \mathcal{Y}_0^*$. *Then* $N_G(\ O^{3'}\big(\Delta(N_G(Y)\big)\ )$ *is the unique maximal 2-local overgroup of* YA *in* G.

It should be noted that such uniqueness theorems in the sense of Remark 8.0.4 (and more of these arise later in Sections 15 and 16) make use of the theory of pushing-up—in particular, they use the preliminary results 3.15 and 3.16 in [**Asc83a**], which quote the literature on pushing-up—as we will indicate later, in comments at the end of this section.

The uniqueness of the 2-local in 4.2.15 is then used repeatedly in Sections 15–17; in particular in setting up the following key result (Theorem 15.14), which pins down the possible structures for a $Y \in \mathcal{Y}_0^*$ such that $L \cap Y$ contains a subgroup $Y_0 \in \mathcal{Y}_4$:

THEOREM 4.2.16. *Assume $Y \in \mathcal{Y}_0^*$, and suppose that $L_0 := L \cap Y$ contains a subgroup $Y_0 \in \mathcal{Y}_4$. Then one of the following holds:*

(1) $Y = O_2(Y)L_0$, $L = O_2(L)L_0$, and $\Omega(A) = \{\langle c \rangle\}$; or

(2) $Y/O_2(Y) \cong A_7$ or A_{10}, $\overline{L} \cong A_7$, and $N_G(A)/C_G(A) \cong S_4$ or $\mathbb{Z}_2 \times S_4$ acts transitively on $\Omega(A)$, with $|\Omega(A)| = 3$; or

(3) $Y/O_2(Y) \cong A_7$, $\overline{L} \cong U_3(5)$, $Y = O_2(Y)L_0$, and $\Omega(A) = \{\langle c \rangle\}$.

The next principal step (Aschbacher's 16.3 and 16.4) establishes most of Theorem 4.2.14—including the fact that $\mathcal{Y}_0^*$ contains a unique member.

THEOREM 4.2.17. *Assume $Y \in \mathcal{Y}_0^*$. Then the following conclusions hold:*

(1) $N_G(A) \le N_G(Y)$ and $N_G(\langle c \rangle) \le N_G(Y)$;

(2) $\mathcal{Y}_0^* = \{Y\}$;

(3) $L \le Y$;

(4) Either $Y/O_2(Y) \cong A_{10}$ or $\Omega(A) = \{\langle c \rangle\}$.

We devote the next few paragraphs to a discussion of the proof of Theorem 4.2.17. It repeatedly uses the transitivity results 4.2.11 and 4.2.12.

To begin the proof of 4.2.17, Aschbacher appeals to 15.1 in [**Asc83a**] in order to use the list from 14.8 (pruned by 15.23) to see that there is always *some* choice for the pair (c, L) such that $L \cap Y$ is a member of $\mathcal{Y}_4$, so that the Structure Theorem 4.2.16 can be applied. Aschbacher shows that case (3) of 4.2.16 cannot hold, while in the remaining cases, $L = (L \cap Y)O_2(L)$. As $L \cap Y \in \mathcal{Y}_4$, it follows that L (or $LO_2(Y)$ when $O_2(L) = 1$) is a member of $\mathcal{Y}_4 \cap \mathcal{Y}_2 = \mathcal{Y}_0$; and then $L \le Y_a$ for some $Y_a \in \mathcal{Y}_0^*$. Thus $Y_a \cap Y$ contains $L \cap Y$, which is generated by 2-groups satisfying the hypothesis of 4.2.12; and so Y is conjugate under $O_{3'}(C_G(A))$ to Y_a. But $C_G(A)$ normalizes L by 4.2.9, so we conclude that $L \le Y$. Now in case (1) of 4.2.16, we have $\Omega(A) = \{\langle c \rangle\}$, and the choice (c, L) is uniquely determined by 4.2.9; so here conclusion (3) of 4.2.17 holds. But obtaining conclusion (3) for *all* possible choices of (c, L) requires some additional analysis of the situation in case (2) of 4.2.16, i.e. where $\overline{L} \cong A_7$. Recall that conclusion (1) or (2) of 4.2.16 holds, so we have also established conclusion (4) of 4.2.17, as well as the final statement of 4.2.14.

The next step is to prove that $W := N_G(A) \le N_G(Y)$, as required for conclusion (1). We set $X := O_{3'}(C_G(A))$. By (3), L is contained in each member of $\mathcal{Y}_0^*$, so it follows from 4.2.12 that

$$X \text{ is transitive on } \mathcal{Y}_0^* . \tag{4.2.18}$$

Then a Frattini argument gives $W = XN_W(Y)$, so it will now suffice to establish that $X \le N_G(Y)$. Suppose R is an X-invariant member of $\mathcal{Q}(G, 2)$. By 15.3.2 in [**Asc83a**], $N_G(Y)$ contains a member of $\mathcal{Q}^*(G, 2)$, so by Aschbacher's Transitivity Theorem, we may take $R \le N_G(Y)$. Then by careful analysis of the various choices for L, Aschbacher shows that, for a suitable choice of R, $P := O_2(\langle R, L \rangle) \ne 1$. Then applying 15.5 in [**Asc83a**], with LP, Y in the roles of "Y, K", he can conclude that $X \le N_G(Y)$—completing the proof that $W \le N_G(Y)$.

This inclusion now leads to the remaining conclusions of 4.2.17: First since we saw that $X \le N_G(Y)$, (4.2.18) supplies the crucial uniqueness of Y in (2). Next LA is normal in $H := N_G(\langle c \rangle)$ by 4.2.9, so we get $H = LN_H(A) = L(H \cap W)$ by a Frattini argument; hence as $W \le N_G(Y)$ and $L \le Y$ from (3), we get $H \le N_G(Y)$, completing the proof of (1)—and indeed of 4.2.14, aside from the existence of Y.

Thus the crucial missing piece in the proof of Theorem 4.2.14 is to show that $\mathcal{Y}_0 \neq \emptyset$. This is the content of Aschbacher's Theorem 16.9.1. The contrary assertion, that $\mathcal{Y}_0$ is empty, is equivalent to the assertion that every $Y \in \mathcal{Y}$ with $Y = \Delta(Y)$ and $O_2(Y) \neq 1$ is a $3'$-group. Roughly speaking, this reduces to the case where Y is solvable. At 16.8 Aschbacher applies the Glauberman Triple Factorization [**Gla77**, Thm B] to obtain strong restrictions on such a subgroup Y, which eventually lead to a contradiction.

(Stage 4: Forcing $\Gamma_{2,A}(G)$ into $N_G(Y)$ to finish). Having finally established the existence and uniqueness of $Y \in \mathcal{Y}_0^*$ in Theorem 4.2.14, Aschbacher now sets $M := N_G(Y)$. We have $N_G(A) \leq M$ by 4.2.17.1. Also $Y = O^{3'}(\Delta(N_G(Y)))$, so by Proposition 4.2.15, M is the unique maximal 2-local subgroup of G containing YA. The goal of Section 17 is to prove that:

$$\Gamma_{2,A}(G) \leq M,$$

so that $A \in \mathcal{T}$, contradicting the assumption that $\mathcal{S} \cup \mathcal{T}$ is empty, and completing the proof of Theorem 4.2.7. Of course $\Gamma_{2,A}(G)$ is generated by $N_G(A)$ together with the normalizers $N_G(B)$ of the hyperplanes B of A; so as $N_G(A) \leq M$, we may assume $N_G(B) \not\leq M$ for some hyperplane B of A.

Lemma 17.2 of [**Asc83a**] shows that some element of $\mathcal{Q}(G)$ is not contained in M. The proof goes as follows: For $b \in B^{\#}$, set:

$$X_b := O_{3'}(C_G(b))O^{3'}(\Delta(C_G(b))).$$

Aschbacher shows that either $X_b = O_{3'}(C_G(b))$, or we have $\langle b \rangle \in \Omega(A)$ where also $X_b = O_{3'}(C_G(b))O^{3'}(C_G(b)^{(\infty)})$. Therefore $N_G(B)$ permutes $\{X_b : b \in B^{\#}\}$, so $A \leq N_G(B) \leq N_G(X_B)$ where $X_B := \langle X_b : b \in B^{\#} \rangle$. Furthermore 17.1 in [**Asc83a**] shows that $Y \leq X_B$. If $\mathcal{Q}(G) \subseteq M$, then Y is normal in M, and also $Y \leq XB$, so $1 < O_2(Y) \leq O_2(X_B)$. Thus $N_G(O_2(X_B))$ is a 2-local containing $N_G(B)$ and YA. But $M = !\mathcal{M}(YA)$, so $N_G(B) \leq M$, contrary to the choice of B.

So the remainder of Section 17 is devoted to the obtaining a contradiction by showing instead that $\mathcal{Q}(G) \subseteq M$. This completes the initial phase of the Signalizer Analysis, establishing Theorem 4.2.7 that $\mathcal{S} \cup \mathcal{T}$ is nonempty.

REMARK 4.2.19. For those reading the proof in [**Asc83a**] in detail, we mention that in order to apply various lemmas from Sections 14–17 later in the paper, many of those lemmas are proved under the more general hypothesis (in the language of Section 13) that "Hypothesis S or T" holds: For in proving Theorem 4.2.7, one may assume that $\mathcal{T}$ is empty, which implies Hypothesis T—and that in turn implies Hypothesis R of Section 15. The final contradiction proving Theorem 4.2.7 in fact shows at 17.28 that Hypothesis R is never satisfied. But several lemmas will also be used in Section 19, where in fact Hypothesis S holds.

The case $\mathcal{S} = \emptyset$: From $\mathcal{T}$ to the Preuniqueness Case. Theorem 4.2.7 shows that $\mathcal{S} \cup \mathcal{T}$ is nonempty. Later, in the Component Analysis, we will consider the case $\mathcal{S}$ nonempty. But first, we consider the case $\mathcal{S}$ empty—so that there is some $A \in \mathcal{T}$, which will lead to the Preuniqueness Case.

This is handled in Sections 18–20 of [**Asc83a**]—which correspond to the "moving around" phase of the Signalizer Analysis, in the language of Gorenstein and Lyons in Theorem B of their Trichotomy work [**GL83**] (see our discussion starting after Theorem 5.2.2, e.g. at 5.2.10): that is, controlling the normalizers of elementary abelian 3-subgroups B of rank 2 which are connected via paths in $\Lambda_2(G)$ to A. The principal goal of these three sections is the proof of Theorem 20.11 in [**Asc83a**]:

THEOREM 4.2.20. *Let G be a counterexample to Theorem 4.2.1, with $\mathcal{S}$ empty. Let $P \in Syl_3(G)$. Then $\Gamma^0_{2,P}(G)$ is contained in a 2-local subgroup of G. (That is, the Preuniqueness Case holds for $\sigma = \{3\}$; so G is* not *in fact a counterexample to Theorem 4.2.1.)*

As in the portion of [**GL83**] mentioned above, the argument is by contradiction, the point of attack being a pair of subgroups:

$A \cong A_1 \cong E_{27}$ such that $|A \cap A_1| = 9$ and $\Gamma_{2,A}(G) \leq M$, but $\Gamma_{2,A_1}(G) \not\leq M$.

The thrust of the analysis is to relate $O_2(M)$, 3-signalizers in M, and 3-components of M to their counterparts $O_2(C_G(a))$, etc., in centralizers of elements $a \in A \cup A_1$ of order 3. The goal is to prove $\Gamma_{2,A_1}(G) \leq M$, contrary to the choice of A_1.

In order to emphasize some of the major steps in the proof in Sections 18–20 of [**Asc83a**], our discussion below will deviate somewhat from Aschbacher's order of presentation.

Those sections assume explicitly that $\mathcal{S}$ is empty. Then Theorem 4.2.7 is invoked implicitly, to guarantee the existence[5] of some $A \in \mathcal{T}$; recall from Definition 4.2.5 that $A \cong E_{27}$ and $\Gamma_{2,A}(G)$ is contained in some maximal 2-local subgroup of G. We will make a "choice" of such a pair (A, M). Moreover we denote by $\mathcal{M}$ the set of maximal 2-local subgroups of G, and for $X \subseteq G$ we set $\mathcal{M}(X) := \{M \in \mathcal{M} : X \subseteq M\}$.

(Section 20: Reducing to a pushing-up hypothesis). We begin with Aschbacher's final argument in the proof of Theorem 4.2.20, which is carried out in Section 20 of [**Asc83a**]. The aim is Theorem 20.10, which verifies Hypothesis 19.1, the working hypothesis for section 19. Indeed Hypothesis 20.1.1 assumes[6] that for some choice of (A, M), there is an overgroup H of A, with $O_2(H) \neq 1$, in a certain set $\mathcal{H}^*$, defined with respect to (A, M) in Section 18, consisting of obstructions to Hypothesis 19.1. The argument in Section 20 uses detailed properties of members of $\mathcal{H}^*$, which are developed in the latter part of Section 18 of [**Asc83a**]; eventually 20.9 supplies a contradiction to an earlier result 20.4.

This contradiction shows that Hypothesis 20.1.1 cannot hold. Thus for each choice of (A, M), and for each H in $\mathcal{H}^*$, we have $O_2(H) = 1$. As a consequence Theorem 20.10 shows that Hypothesis 19.1 in [**Asc83a**] holds:

PROPOSITION 4.2.21. *For each choice of (A, M) and overgroup X of A in G with $O_2(X) \neq 1$, we have $O_2(X \cap M) = O_2(X)$.*

PROOF. Since the deduction is implicit in [**Asc83a**], we give some details supplied by Aschbacher: Assume that for some choice of (A, M) and some overgroup X of A with $O_2(X) \neq 1$, that instead $O_2(X \cap M) \neq O_2(X)$.

---

[5] In particular note that the statement "Assume A ..." at the start of Section 18 and thereafter in [**Asc83a**] is not really an assumption, but just establishes notation corresponding to some *choice* of such an A.

[6] In the alternative case Hypothesis 20.1.2, Section 20 proves a technical result (Theorem 20.12) used in [**AGL81**] which reduces the Preuniqueness Case to the Uniqueness Case (discussed in our Section 8.1). Indeed the arguments in Section 20 quote a number of common preliminary results, which are established separately in the two cases, in Sections 18 and 19 respectively.

These hypotheses imply that $X < G$ and $O_{3'}(X \cap M) \neq O_{3'}(X)$, so we have the hypotheses of 18.1 in [**Asc83a**]. (We mention that the proof of 18.1 uses part (1) of the important Theorem 4.44 in [**Asc83a**], with the pair X, $X \cap M$ in the roles of "G, M".) Note also by Coprime Action 0.3.11.2 that $O_2(X) = \Gamma_{2,A}(O_2(X))$, so that $O_2(X) \leq X \cap M$ and hence $1 < O_2(X) < O_2(X \cap M)$.

Now 18.1.1 of [**Asc83a**] provides a member H_1 of the set $\mathcal{H}$ of Section 18 (whose maximal members comprise $\mathcal{H}^*$), satisfying $O_2(X \cap M) \leq H_1 \leq X$; so we conclude from the previous paragraph that $1 < O_2(X) \leq O_2(H_1) =: Q_1$.

Next embed $H_1 \leq H \in \mathcal{H}^*$ and set $L_1 := H_1^\infty$ and $L := H^\infty$. Then we get $L = L_1 O_2(L)$ from 18.1.2; so since $[Q_1, L_1] \leq O_2(L_1)$, we have $[Q_1, L] \leq O_2(L)$. Now the definition of $\mathcal{H}$ includes the property $H = LAO_2(H \cap M)$; so we conclude that $1 < Q_1 \leq O_2(H)$. But this is a contradiction as $O_2(H) = 1$ because of the failure of 20.1.1. □

(Section 19: Pushing-up to obtain the Preuniqueness Case). In section 19, Aschbacher uses the conclusions of Proposition 4.2.21 to deduce that G is in the Preuniqueness Case. This conclusion comes out at Theorem 19.20.

One elementary but strong consequence of 4.2.21 is the following proposition (which appears as 19.2 in [**Asc83a**]); it has various applications, but is especially useful later in the "moving around" analysis involving the graph $\Lambda_2(G)$.

PROPOSITION 4.2.22. *Assume that A_1 and A_2 are elementary abelian 3-subgroups of rank 3 in G with $\Gamma_{2,A_i}(G) \leq M_i \in \mathcal{M}$, $i = 1, 2$, and that $A_1 \leq M_2$.*

Then $M_1 = M_2$. (In particular then $M_1 = !\mathcal{M}(\Gamma_{2,A_1}(G))$.)

The final statement follows by taking $A_1 = A_2$.

Furthermore, Proposition 4.2.21 enables Aschbacher to use pushing-up results from [**Asc81d**]. (The theory of pushing-up is discussed in our Appendix Section B.7; see also the comments on quoted results at the end of this section.)

Notice that Proposition 4.2.22 suggests that we might wish to first show that when $A \cong E_{27}$, is contained in some 2-local, then in fact $\Gamma_{2,A}(G)$ is also contained in a 2-local: i.e. $\mathcal{E} \subseteq \mathcal{T}$. Indeed Theorem 19.12 of [**Asc83a**] establishes this.

With this result in hand, Theorem 19.20 proves that the Preuniqueness Case holds as a consequence of 4.2.22. Again since the deduction is implicit in [**Asc83a**], we provide some details. We begin with our usual choice of a pair (A, M); our argument uses the graph $\Lambda_2(G)$, so we consider some elementary abelian 3-subgroup A_1 of rank 3, intersecting A in B of rank 2. Then $A_1 \leq C_G(B) \leq \Gamma_{2,A}(G) \leq M$, so $A_1 \in \mathcal{E}$. Then Theorem 19.12 shows that $A_1 \in \mathcal{T}$, so that:

$$\Gamma_{2,A_1}(G) \leq M_1 \text{ for some } M_1 \in \mathcal{M}.$$

It now follows from Proposition 4.2.22 that $M = M_1$. Of course by induction on the distance of A_1 from A in the graph $\Lambda_2(G)$, this also holds for any A_1 of rank 3 in P which is "2-connected" to A via a path in $\Lambda_2(G)$. Hence as $m_3(P) = 3$, it follows (B.4.8) that all non-isolated rank-2 subgroups of P lie in a single connected component of $\Lambda_2(P)$ which contains A. That is:

$$\Gamma^0_{2,P}(G) \leq M,$$

so the Preuniqueness Case holds, completing the proof of Theorem 4.2.20.

The Component Analysis. Since Theorem 4.2.20 has treated the case $\mathcal{S}$ empty, the final three sections (Sections 21–23) of [**Asc83a**] assume that:

$$\mathcal{S} \text{ is not empty.}$$

Under this assumption, it is shown that there is a centralizer in odd standard form, and that centralizer is used to identify G.

As $\mathcal{S}$ is nonempty, there exists a maximal elementary abelian 3-subgroup A of G containing a rank-3 subgroup $E \in \mathcal{E}$, with the property that both $[\theta(C_G(a)), A]$ and $\Delta_G(B)$ are of odd order, for all $a \in A^\#$ and all subgroups B of A of rank 2. We choose $A \in \mathcal{F} \setminus \mathcal{E}$ if possible; so when $A \in \mathcal{E}$, we are in the subcase $\mathcal{F} = \mathcal{E}$. We fix A and E; and then set $W := N_G(A)$ and $W^* := W/C_G(A)$—which (as the previous section) will be the Weyl group of G when G is identified as a group of Lie type.

The $GF(2)$-type branch. At the start of Section 21, Aschbacher quotes 6.4.1 in [**Asc83a**], which treats the possibility that the set $\mathcal{U}(E)$ of E-invariant 4-subgroups of G is empty.[7] In this case, for any maximal 2-local subgroup M of G containing E, all characteristic abelian subgroups of $O_2(M)$ are cyclic, so Philip Hall's theorem shows at Theorem 5.1.2 in [**Asc83a**] that $O_2(M)$ is of symplectic type. Hence G is of $GF(2)$-type, and he can simply apply the classification of such groups. (Aschbacher at 3.10 quotes the relevant literature; we discuss the result as the $GF(2)$ Type Theorem 7.0.1.) Under our hypotheses, one obtains the groups Co_2, Th, $U_5(2)$, $U_6(2)$, and $\Omega_8^+(2)$; these are the groups in conclusion (2) of Theorem 4.2.1. (Some of the groups in the other conclusions of 4.2.1 are also of $GF(2)$ type—but have $\mathcal{U}(E)$ nonempty, and so will be identified later by other means.)

The non-$GF(2)$-type branch: initial 3-component analysis. Thus, for the remainder of the proof, Aschbacher may fix some $U \in \mathcal{U}(E)$. In particular the existence of U is used in 6.12 in [**Asc83a**], which is the next lemma quoted in Section 21. This is an argument of Klinger-Mason type, in the sense of Remark 2.2.2: roughly, since G is not of $GF(2)$-type, but G is of characteristic 2 type with $e(G) = 3$, G cannot also be of characteristic 3 type—and so the centralizer of some 3-element should have a 3-component. (In particular, this completes Step 3 of Strategy 1.1.1.) Lemma 6.12 in [**Asc83a**] establishes the existence of a pair $(c, L) \in \mathcal{L}_3^*(A, U)$: Here, $\mathcal{L}^*(A)$ denotes the set of 3-components of centralizers of elements $a \in A^\#$, which are maximal in a natural ordering on this set. Then $\mathcal{L}_3^*(A, U)$ is the set of all pairs (c, L) such that $c \in C_E(U)^\#$, and L is a 3-component of $C_G(c)$ such that $L = [L, U]$ and $L \in \mathcal{L}^*(A)$. (There are some further technical conditions, which we omit to simplify the exposition.) Note also in the groups $L_4(4)$ and $Sp_6(4)$ appearing in conclusion (3) of Theorem 4.2.1 which are defined over $\mathbf{F_4}$, that U is a root subgroup of L and $E \cap L$ is a Cartan subgroup of L.

After these preliminaries, Section 21 begins by enumerating at 21.1 the possible isomorphism types for $\overline{L} = L/\theta(L)$, with $(c, L) \in \mathcal{L}_3^*(A, U)$. Whereas in the earlier paper [**Asc81b**, 10.7], all such sections were in $Chev(2)$, this time some sporadic groups are also on the list. Nonetheless from this list it is fairly easy for Aschbacher at 21.3 to establish the following important fact:

THEOREM 4.2.23. *L is quasisimple.*

We sketch the proof: Since $A \in \mathcal{S}$, A centralizes a Sylow 2-subgroup of $\theta(L)$. Then as $L = [L, A]$, L also centralizes such a Sylow subgroup. Hence $\theta(L)$ is 2-nilpotent,

---

[7]There should be no confusion between this use of $\mathcal{U}$ and the usage $\mathcal{U}(M)$ in Chapter 8. Here the argument is an elementary abelian 3-group E; there it is a maximal 2-local subgroup M.

so $L/O_{2'}(L)$ is quasisimple. But also $m_{2,q}(G) \leq 2$ for all primes $q > 3$, so 4.4 in [**Asc83a**] shows that the sections $\overline{L}$ in 21.1 are too large to act nontrivially on a chief section of $O_{2'}(L)$. Hence L is in fact quasisimple.

In particular, Step 4 of Strategy 1.1.1 has been completed. Moreover, the odd standard form condition in Step 5 (e.g., the fact that $m_3(C_G(L)) = 1$) follow easily from the fact that $e(G) = 3$, specifically in 6.6.

Several possibilities for $\overline{L}$ in 21.1 are then eliminated, leading to the next main result 21.15:

THEOREM 4.2.24. *$W \not\leq H := N_G(\langle c \rangle)$; i.e. $\langle c \rangle$ is not weakly closed in A with respect to G.*

We sketch the main case of the proof: Lemmas 21.4 and 21.5 in [**Asc83a**] show for a generic pair (c, L) that we can replace our original U with $U_0 \in \mathcal{U}(E)$, such that $U_0 \leq L$ (so c centralizes U_0), and there is $a \in C_E(U_0) \setminus \langle c \rangle$ and $X \in \mathcal{L}(a)$ such that $(a, X) \in \mathcal{L}_3^*(A, U_0)$. In particular, X is a 3-component of $C_G(a)$ for which we have $X = [X, U_0]$ and $X \in \mathcal{L}^*(A)$. Now if the theorem fails, $X \cap W \leq H$. Inspecting the list of 21.1 for such sections, we conclude that $\overline{X} \cong L_3(2^n)$ and U_0 induces inner automorphisms on $\overline{X}$. But also $C_{\overline{X}}(c)$ has odd order, so as c centralizes U_0, U_0 acts trivially on $\overline{X}$—contrary to $X = [X, U_0]$.

At this point, Aschbacher can finish the treatment of the cases where $\overline{L}$ is isomorphic to $L_4(2)$, $L_5(2)$, or $Sp(6, 2)$: Namely Theorems 4.2.23 and 4.2.24 complete the verification of the hypotheses used by Finkelstein and Frohardt for these and other odd standard form problems in [**FF84, FF79, FF81b**]. So he can then quote their results at 21.16[8] to identify the groups $L_6(2)$, $L_7(2)$, $Sp_8(2)$, $\Omega_8^-(2)$, and $F_4(2)$; these are the groups in conclusion (1) of Theorem 4.2.1. We discuss the Finkelstein-Frohardt results as Theorem 6.2.1—in view of their similarity to the treatment of standard type by Gilman and Griess, which we describe in Section 6.1.

Identifying a large Lie-type subgroup "I". The remainder of Section 21 is devoted to the treatment of the cases determined by the sections $\overline{L}$ in 21.1 of [**Asc83a**] which remain possible after 21.16. Much as in the Component Analysis in Section 4.1, Aschbacher again establishes a version of odd standard form (corresponding to Step 6 of Strategy 1.1.1); eventually he identifies the three groups in conclusion (3) of Theorem 4.2.1.

As before (cf. 4.1.11), the process requires a careful analysis of the structure of the "Weyl group" W^* acting on A; this further refines the possibilities for L, and is summarized by Aschbacher at 21.31 and 22.1:

THEOREM 4.2.25. *One of the following conclusions holds:*
(1) $m_3(A) = 4$ and $L \cong U_4(2^n)$, $n > 1$, n odd; or
(2) $m_3(A) = 3$, with $O^2(W(C_3)) \leq W^* \leq W(C_3)$; and either
(i) $L \cong Sp_4(2^n)$, $U_4(2^n)$, $SL_3(2^n)$, or $U_5(4)$, n even; or
(ii) $L/\theta(L) \cong SL_3(4)$, and $C_L(a)^{(\infty)} \cong L_2(4)$ for each $a \in A \setminus \langle c \rangle$ with $C_L(a)$ nonsolvable.

Here again $W(C_3) \cong S_4 \times \mathbb{Z}_2$ is the Weyl group of type C_3.

[8] Aschbacher's appeal to these papers is in 21.2, but that statement is not actually used until 21.16.

Now Aschbacher considers the set D^* of all reflections in the Weyl group W^* of A, which are induced by elements of components $X \in \mathcal{L}_3^*(A, U)$—where now U may vary in $\mathcal{U}(E)$. For $t \in W$ with $t^* \in D^*$, set:[9]

$$K_0(t) := O^{3'}\big(E(C_G(C_A(t)))\big) \text{ and } I := \langle K_0(t) : t^* \in D^* \rangle,$$

except for one special case where the definition of I is modified slightly, which we omit for simplicity of exposition. The point is that t should centralize a hyperplane A_t of A, and $E\big(C_G(A_t)\big)$ should be, in a group G of Lie type in characteristic 2, a subgroup $L_2(2^n)$ generated by a pair of opposite root groups. Detailed analysis (at 22.6 and 22.9) reduces the cases in 4.2.25 to:

$m_3(A) = 4$, $L \cong U_4(2^n)$, $n > 1$ odd, and $\langle D^* \rangle \cong S_5$; or
$m_3(A) = 3$, $L \cong SL_3(4)$, and $\langle D^* \rangle \cong S_4 = W(A_3)$; or
$m_3(A) = 3$, $L \cong SL_3(4)$ or $Sp_4(4)$, and $W^* = \langle D^* \rangle \cong W(C_3)$.

When $m_3(A) = 4$, Phan's Theorem (see Remark A.5.2) is used for the identification of $I \cong U_5(8)$. And when $m_3(A) = 3$, I is identified (respectively in the two cases above) as isomorphic to $L_4(4)$ or $Sp_6(4)$, using the Curtis-Tits Theorem A.5.1.

These three subgroups I are the groups in conclusion (3) of Theorem 4.2.1.

The final step: Identifying G as I. This step is accomplished in Section 23 of [**Asc83a**]; it parallels Gilman and Griess's final step in determining the groups of standard type with $e(G) \geq 4$ (see the end of Section 6.1). As in the proof of Theorem 4.1.1, Aschbacher considers the subgroup $N := N_G(I)$, and assumes that $I < G$, so that $N < G$ by the simplicity of G.

In contrast to the procedure in the first paper [**Asc81b**], discussed in Section 4.1, this time there is no appeal to the Global $C(G, T)$-Theorem. Instead, Aschbacher shows at 23.3 that N is strongly 3-embedded in G. Next, for z a long-root involution of I, he shows at 23.4 that $C_G(z) \leq N$; for otherwise $C_N(z)$ is strongly 3-embedded in the $\mathcal{K}$-group $C_G(z)$, which leads to a contradiction to the list of such $\mathcal{K}$-groups in 24.1 of Gorenstein-Lyons [**GL83**]. Then it follows that z fixes a unique point in the permutation representation of G on the cosets of N. This gives a contradiction to Holt's Theorem B.7.16.

Hence $I = G$, completing the proof of Theorem 4.2.1, and hence also of the $e(G) = 3$ Theorem 4.0.1.

Further comments on quoted results. Certain papers from the literature are quoted primarily for expository purposes: The introduction to this second paper [**Asc83a**] of course refers back to the first paper [**Asc81b**] for much of its background and notation. The broader discussion of the $e(G) = 3$ result also refers, for context, to the Gorenstein-Lyons Trichotomy [**GL83**] for $e(G) \geq 4$; and to the Uniqueness Case in [**Asc83b, Asc83c**] (using [**AGL81**] to proceed from the Preuniqueness Case).

In addition, the first paper [**Asc81b**] is quoted to make use of earlier results and arguments, at various points in the course of the second paper [**Asc83a**].

When appealing to results from other sources in the literature, Aschbacher frequently includes an appropriate statement and reference in the preliminary Sections 3–5 and 7—so that later sections can reference these lemmas.

Several of these are standard facts about coprime action (cf. Section B.1). For example, the proof of 5.8 uses the Thompson Dihedral Lemma B.1.7; and 3.7 is the Bender-Thompson Signalizer Lemma B.1.10. But many of the references fall

[9]We mention that the "C_G" in this expression is inadvertently omitted in [**Asc83a**].

into one or more general classes of less elementary techniques, which we will discuss briefly below. (Most of these techniques are also discussed in Appendix Chapter B.)

At 3.18 there is a reference to an unspecified result of G. Higman on fixed-point-free action; this appears in [**Hig57**], and is given as (3) of [**Hup67**, V.8.8]; though Aschbacher mainly uses the special case (2) for the prime 3 given by the Burnside reference [**Bur55**, p. 90] there.

Properties of known groups. We mention a number of specific references, which go beyond our discussion in Section A.3.

There are a number of appeals to properties of groups of Lie type. The Aschbacher-Seitz work [**AS76a**] on involutions in characteristic 2 is used e.g. at 20.5. The tables from the Aschbacher-Seitz work [**AS76b, AS81**] on standard components of known type are used at 21.4 and 23.4. The Seitz Generation Theorem A.3.4 is used at a number of points, such as 4.23, 4.28, and 19.14. In addition, properties of $SL_3(q)$ are quoted at 3.11; Aschbacher refers to a treatment by Bloom [**Blo67**].

There are a number of references to properties of particular sporadic groups, notably at 4.14–4.17, and also at 4.43 in the proof of Theorem 4.34. These references include Conway [**Con71**] and Held [**Hel69**] on the Mathieu groups, Janko [**Jan76**] and Finkelstein-Rudvalis [**FR73, FR74**] on the Janko groups, Magliveras [**Mag71**] on HS, O'Nan [**O'N76b**] on $O'N$, and Parrott [**Par76**] on Ru.

Several results make use of the Schur multipliers of the groups of Lie type, referring to Griess [**Gri73**]; and of cohomology of groups of Lie type, referring to Jones-Parshall [**JP76**]

Recognition theorems. (Cf. Section A.5.) We already mentioned the results of Curtis-Tits and Phan, stated at 7.6 and 7.7 in [**Asc83a**], which are used by Aschbacher to identify most groups in his main result; as well as the particular cases handled by Finkelstein-Frohardt at 21.2.

The standard theorem A.6.1 of McLaughlin identifying groups generated by transvections is quoted at 21.7. Furthermore several results in Section 7 (referring back to Section 4 of the first paper [**Asc81b**]) make use of Fischer's theory of 3-transposition groups (cf. our discussion of A.6.3) to identify various sections (usually operating faithfully on low-rank elementary abelian p-groups).

Transfer. (Cf. Section B.2.) The local analysis sometimes makes use of general results on transfer. In particular 11.14 and 11.31 use a result of Yoshida in [**Yos78**].

Failure of factorization. A considerable amount of the local analysis makes use of the theory of failure of factorization—cf. Section B.6 in our Appendix B. In particular that theory involves some very detailed analysis of properties of specific 2-modular representations of known simple groups, as in our Sections A.3 and A.4. Among such results are 3.18–3.22; and these results are applied later, especially in Sections 18–20.

In particular 3.21 quotes standard results on failure of factorization from the unpublished preprint [**CM80**] of Cooperstein and Mason; their results have now been included in the work of Guralnick and Malle [**GM02, GM04**]. The results 3.18 and 3.22, as well as later results such as 15.5, 18.9, and 20.4, make reference to Aschbacher's detailed analysis of $GF(2)$-representations of some known groups in [**Asc82**]. (Cf. Section A.4.)

The Glauberman Triple Factorization [**Gla77**, Thm B] is used at 16.5 and 16.8.

Pushing-up. Our discussion of Sections 14–20 of [**Asc83a**] touched upon the more systematic use in this paper of pushing-up techniques. Some of Aschbacher's references go beyond those indicated in our discussion of pushing-up in Section B.7, so we indicate them here.

As 3.15 Aschbacher quotes Baumann's result [**Bau79**] on pushing-up $L_2(2^n)$.

To establish 3.16, Aschbacher appeals to the classification of groups of characteristic 2 type in which some maximal 2-local possesses a $\mathcal{Z}$-block. In addition to papers listed in our discussion of the Global $C(G,T)$-Theorem B.7.8 he quotes J. Hall [**Hal82**] and Harada [**Har80b**].

Signalizer functors. We have emphasized the fundamental role of signalizer functors in [**Asc83a**]. It is worth noting that, while [**Asc83a**] normally uses only the standard solvable case of the Signalizer Functor Theorem 0.3.15, the proof of 14.6 does make use of the nonsolvable extension by McBride (see Remark B.3.14).

The uses of the classification of groups of $GF(2^n)$ type. As 3.10 in [**Asc83a**], Aschbacher quotes the papers in the literature giving the classification of groups of $GF(2^n)$ type. We shall discuss this result as Theorem 7.5.2, indicating the relevant papers at that point.

Aschbacher uses 3.10 at various points in [**Asc83a**]; we now give a brief summary of those appeals.

Our discussion of the component analysis emphasized that the classification of groups of $GF(2)$ type (which we shall discuss as Theorem 7.0.1) is quoted "in the foreground"—namely in the proof of 6.4.1, to identify the groups in conclusion (2) of Theorem 4.2.1.

Three of the groups in conclusion (1) of Theorem 4.2.1, namely $L_6(2)$, $L_7(2)$, and $\Omega_8^-(2)$, are also of $GF(2)$ type—though with $\mathcal{U}(A)$ nonempty. So Aschbacher does not identify them "directly" via the classification 3.10—instead the identification is in terms of a centralizer in standard form, quoting at 21.16 (via 21.2) the Finkelstein-Frohardt Theorem 6.2.1. However, as we mentioned in our discussion of the proof of 6.2.1, Finkelstein and Frohardt use the $GF(2)$ type classification to identify those three groups.

Aschbacher's other applications of 3.10 are "in the background": namely they are made only to obtain a contradiction in various situations, not to identify examples. One use of the case $n = 1$ in the background appears in the proof of 6.11.

Several other background uses of 3.10 do require the case $n > 1$, that is, the full classification 7.5.2 of groups of $GF(2^n)$ type. Such applications occur right after 8.6.6, and in the proof of 17.7.

CHAPTER 5

$e(G) \geq 4$: The Pretrichotomy and Trichotomy Theorems

5.1. Statements and Definitions

In this chapter we give a precise statement, and an outline of the proof, of the Pretrichotomy Theorem of Gorenstein and Lyons [**GL83**]. In addition we state the stronger Trichotomy Theorem, which follows from the Pretrichotomy Theorem using the Preuniqueness-Implies-Uniqueness Theorem 8.0.1 of Aschbacher, Gorenstein, and Lyons (Theorem C in [**GL83**], which in turn draws on [**AGL81**]).

Statements of the results. Some of the definitions needed to interpret the statement of the following result already appear in our introduction to the Trichotomy Theorem in Section 2.3; we will supply the remaining definitions after we state the theorem:

THEOREM 5.1.1 (Pretrichotomy Theorem). *Assume that G is a simple group of characteristic 2 type, with $e(G) \geq 4$, in which all proper subgroups are $\mathcal{K}$-groups. Then one of the following holds:*

(1) G is of $GF(2)$-type; or

(2) G is of standard type with respect to some member of $\mathcal{S}^*(G;p)$ for some $p \in \sigma(G)$; or

(3) for every $p \in \sigma(G)$ and $P \in Syl_p(G)$, $\Gamma^0_{2,P}(G)$ is contained in a 2-local subgroup of G. (That is, the Preuniqueness Case holds.)

We now begin a more detailed discussion of the various terms used in the statement of this result; in particular, this will allow us to proceed from the notion of "simplified" standard type, which we used for expository purposes in our Section 2.3, to the more technical notion of *standard type* actually used in [**GL83**]. But we will start with some discussion of the Preuniqueness Case:

First, we recall from 2.3.9 the set of primes:

$$\sigma(G) := \{p : p \text{ is an odd prime, and } m_{2,p}(G) \geq 4\},$$

which is nonempty since $e(G) \geq 4$. We shall restrict our attention in this chapter to such primes p, and to the associated p-local subgroups of G.

In particular, conclusion (3) of the Pretrichotomy Theorem asserts the existence of a "weak uniqueness subgroup" $\Gamma^0_{2,P}(G)$ for every $p \in \sigma(G)$. Gorenstein and Lyons say, in this case, that G is of *preuniqueness type* with respect to $\sigma(G)$; that is, we have the Preuniqueness Case of Definition 2.3.8 for $\sigma(G)$. In that case, the Preuniqueness-Implies-Uniqueness Theorem 8.0.1 of Aschbacher, Gorenstein, and Lyons (Theorem C in [**GL83**], along with Theorem E in [**AGL81**]) shows that G must in fact be of uniqueness type with respect to $\sigma(G)$—that is, we have the Uniqueness Case of Definition 2.7.2. So along with this result, the Pretrichotomy Theorem yields the full Trichotomy Theorem:

THEOREM 5.1.2 (Trichotomy Theorem). *Assume that G is a simple group of characteristic 2-type, with $e(G) \geq 4$, in which all proper subgroups are $\mathcal{K}$-groups. Then one of the following holds:*

(I) G is of $GF(2)$-type; or

(II) G is of standard type with respect to some member of $\mathcal{S}^*(G;p)$ for some $p \in \sigma(G)$; or

(III) for every $p \in \sigma(G)$, G possesses an almost strongly p-embedded subgroup.

(That is, the Uniqueness Case holds.)

The classification of groups G satisfying conclusions (I), (II), or (III) of the Trichotomy Theorem will be discussed in Chapters 6, 7, and 8, respectively. Moreover in Section 8.1 we discuss Theorem C of [**GL83**]. So for the remainder of the present chapter, we focus on the Pretrichotomy Theorem.

Recall from Definition 2.3.2 that the set $\mathcal{B}_{max}(X;p)$ consists of all elementary abelian p-subgroups B of X such that B normalizes a nontrivial 2-subgroup of X and $m_p(B) = m_{2,p}(X)$. Note that the normalizing condition is equivalent to the statement that B is contained in a 2-local subgroup of X. We can now give the more precise form of conclusion (1) of the Pretrichotomy Theorem which actually arises in Theorem A of [**GL83**] (see 5.3.1): namely G is of $GF(2)$-type *with respect to a maximal 2-local M* of G if $3 \in \sigma(G)$ and $O_2(M)$ is a 2-group of symplectic type containing some B in $\mathcal{B}_{max}(G;3)$. Notice in that event that G is of $GF(2)$-type in the usual sense of Definition 2.3.7, since $M = C_G(t)$ for t the involution in $Z(O_2(M))$. In particular, the groups in conclusion (1) of the Pretrichotomy Theorem (and (I) of the Trichotomy Theorem) are determined using the $GF(2)$ Type Theorem 7.0.1.

Definitions related to standard triples and neighbors. The proof of the Pretrichotomy Theorem proceeds by assuming that G is neither of $GF(2)$-type nor of preuniqueness type, and then establishing that G is of standard type as in conclusion (2). The key conditions in the definition of "standard type" are the existence of standard triples and neighbors. We begin with a concept generalizing Definition 2.3.2. Let $\hat{\mathcal{S}}(G;p)$ denote the set of all triples (B,x,L) such that:

(1) $B \in \mathcal{B}_{max}(G;p)$ and $x \in B^\#$; and

(2) L is a p-component of $C_G(x)$ such that $C_B(L/O_{p'}(L)) = \langle x \rangle$.

Thus the set $\mathcal{S}^*(G;p)$ of Definition 2.3.2 is the subset of $\hat{\mathcal{S}}(G;p)$ consisting of those triples (B,x,L) where L is quasisimple and $m_p(C_G(L)) = 1$. We call the members of the set $\mathcal{S}^*(G;p)$ *standard triples* in G (for the prime p).

Next we recall concepts from Definition 2.3.4: First if $(B,x,L) \in \mathcal{S}^*(G;p)$, then a *standard subcomponent* of (B,x,L) is a pair (D,K) such that

(1) $x \in D \leq B$ with $m_p(D) = 2$;

(2) $K := L_{p'}(C_L(D))$ is a component of $C_L(D)$, and

(3) $C_B(K) = D$.

We now add the technical restriction (which we did not assume in Definition 2.3.4) that if $p = 3$, then if possible K is chosen not isomorphic to $U_4(2)$ or A_6; this condition avoids certain generational difficulties at some points in the proof.

We next recall the very important concept of a neighbor—again adding details omitted from Definition 2.3.4. Assume that we are given $(B,x,L) \in \mathcal{S}^*(G;p)$, and $(B, x^*, L^*) \in \hat{\mathcal{S}}(G;p)$. We call (B, x^*, L^*) a *neighbor* of (B, x, L) in G (with respect to (D, K)) provided the following conditions hold:

(1) (D, K) is a standard subcomponent of (B, x, L) and $D := \langle x, x^* \rangle$;

(2) $L^* \leq \langle K^{L_{p'}(C_G(x^*))} \rangle$; and

(3) x does not centralize $L^*/O_{p'}(L^*)$.

Since $(B, x, L) \in \mathcal{S}^*(G;p)$, L must be quasisimple. Definition 5.1.4 will require that for G to be of standard type, the p-component L^* in any neighbor (B, x^*, L^*) of (B, x, L) must also be quasisimple. Notice furthermore that condition (3) implies that $L^* \neq (L \cap L^*)O_{p'}(L^*)$. Thus both L and L^* are distinct proper *pumpups* (cf. [**GLS94**, 1.7.1]) of the standard subcomponent K. One can then identify G as $\langle L, L^* \rangle$ for a suitable choice[1] of L^*, as in the work of Gilman and Griess [**GG83**], discussed in Chapter 6.

Splitting primes; the definition of standard type. Gorenstein and Lyons show that, for some $p \in \sigma(G)$, there exist standard triples (B, x, L) which have neighbors (B, x^*, L^*) such that L^* is quasisimple (i.e. roughly speaking, G is of "simplified" standard type with respect to (B, x, L), as in Definition 2.3.5). Indeed they establish many more properties of a suitable standard triple, which they bundle into the definition of "standard type."

In particular, both L and L^* will be covering groups of simple groups of Lie type, defined over some field $\mathbb{F}_{2^n}$ of characteristic 2. Furthermore B can be chosen to intersect L and its neighbors in a torus of largest possible p-rank. But this torus might not be split—especially when L is a twisted group; in general, p will be a divisor of either $2^n - 1$ or $2^n + 1$. This leads to two further definitions:

DEFINITION 5.1.3 (Splitting and half-splitting primes). Let p be an odd prime. Let J be a finite simple group of Lie type defined over the field $\mathbf{F}_{2^n}$. Let $\hat{J}$ be any perfect central extension of J. We say that p is a *splitting prime* for $\hat{J}$, or p *splits* $\hat{J}$, if and only if one of the following holds:

(1) J is a Chevalley (i.e., untwisted) group, and p divides $2^n - 1$; or

(2) J is a twisted group of Lie type, but $J \not\cong {}^3D_4(2^n)$, and p divides $2^n + 1$; or

(3) $J = {}^2D_r(2^n)$ or ${}^3D_4(2^n)$, and p divides $2^n - 1$; or

(4) $n = 1$ and $p = 3$.

And p is a *half-splitting prime* for $\hat{J}$, or p *half-splits* $\hat{J}$, if and only if one of the following holds:

(1) p splits J; or

(2) $J = B_r(2^n)$, $D_r(2^n)$, $F_4(2^n)$, $E_7(2^n)$, or $E_8(2^n)$, and p divides $2^n + 1$; or

(3) $J = {}^2A_5(2^n)$ or $J = {}^2E_6(2^n)$, and p divides $2^n - 1$; or

(4) $J = A_5(4)$, $Z(\hat{J}) = 1$, and $p = 5$; or $J = A_8(2)$ and $p = 7$; or $J = E_6(2)$ and $p = 7$. ◊

The half-splitting primes include all primes p for which $Inndiag(J)$—the subgroup of $Aut(J)$ generated by all inner and diagonal automorphisms—has p-rank at least $\max(e(J) - 1, 3)$; cf. [**GL83**, I.16-2].

With this language in hand, we are finally in a position to give the definition of standard type, and hence complete our discussion of the terminology used in the statements of the Pretrichotomy and Trichotomy Theorems.

---

[1]If e.g. $G = F_4(q)$, some L^*'s give $\langle L, L^* \rangle = B_4(q)$.

Indeed a rough initial summary of conclusion (2) of the Pretrichotomy Theorem 5.1.1 would be that we can choose our triple (B, x, L) so that L is split by p, while all neighbors J of L are half-split by p, except possibly when x induces a field automorphism of J. Here is the actual statement:

DEFINITION 5.1.4 (standard type). Let G be a group, p an odd prime, and assume that $(B, x, L) \in \mathcal{S}^*(G; p)$. We say that G is of *standard type* (with respect to (B, x, L)) if and only if the following conditions hold:

(1) $L/Z(L)$ is a finite simple group of Lie type defined over a field of characteristic 2;

(2) p is a splitting prime for L;

(3) Every element of B induces an inner-diagonal automorphism of L;

(4) There exists a B-invariant 2-subgroup T of $C_G(x)$ with $[T, B] \neq 1$;

(5) For every neighbor (B, x^*, L^*) of (B, x, L) in G, B normalizes L^*, $L^*/Z(L^*)$ is a finite group of Lie type defined over a field of characteristic 2, and either p half splits L^*, or x induces a nontrivial field automorphism on $L^*/Z(L^*)$; and

(6) For every standard subcomponent (D, K) of (B, x, L), there exists a neighbor (B, x^*, L^*) of (B, x, L) with respect to (D, K); and $[K, O_{p'}(C_G(d))]$ is of odd order for all $d \in D^{\#}$. ♢

EXAMPLE 5.1.5. In fact we had already checked many of these conditions when $G = SL_4(q)$, for the triple (B, x, L) of Example 2.3.1. The 2-group in condition (4) can be taken to be the upper unitriangular matrices in $C_G(x) \cong GL_3(q)$. And if in (6), d is not a power of x or x^*, then $C_G(d)$ will be the commuting product of K with a diagonal subgroup containing D. ♢

Note that the existence of a neighbor (B, x^*, L^*) for every standard subcomponent (D, K) of (B, x, L) will serve in Chapter 6 as a substitute for the existence of a G-conjugate x^* of x in $C_G(x) \setminus \langle x \rangle$—which, when $p = 2$, is a consequence of Glauberman's Z^*-Theorem B.2.1.

5.2. The Signalizer Analysis

Part I of [**GL83**] is devoted primarily to the proof of many properties of the known finite simple groups and their covering groups. For the most part, we mention such results as they come up during the proof, though we will add some further comments on quoted results at the end this chapter.

Our exposition will therefore concentrate on Part II of [**GL83**], which contains the proof of the Pretrichotomy Theorem itself. From the viewpoint of our outline, the proof begins with signalizer analysis, which we discuss in this section, although that analysis actually appears later in [**GL83**], in their Chapter 2.[2] Then in a subsequent section, we will discuss the remainder of the proof of the Pretrichotomy Theorem—which is carried out in Chapter 1 of [**GL83**]. (Chapter 3 of [**GL83**] is devoted to the proof of the Preuniqueness-Implies-Uniqueness Theorem 8.0.1, which we discuss in Section 8.1.)

[2]Normally in this chapter, our references to the chapters in [**GL83**] will be to Part II of [**GL83**]; however, we will sometimes use numbering such as II.2 or I.3 when we need to distinguish chapters in the two Parts.

The main dichotomy (Theorem B). The approach used by Gorenstein and Lyons is best regarded as an extension or variation of the Gorenstein-Walter Alternative, which is in turn a dichotomy between the triviality of a suitable signalizer functor and the existence of uniqueness subgroups (see Remark 1.1.2, in our discussion of Step 3 of Strategy 1.1.1). We introduce some notation for the corresponding conditions in the current situation:

DEFINITION 5.2.1. Let G be a group and p a prime. Then $\mathcal{E}_2^p(G)$ is the set of elementary abelian p-subgroups of G of rank 2, and for any $D \in \mathcal{E}_2^p(G)$:

$$\Delta_G(D) := \bigcap_{d \in D^\#} O_{p'}(C_G(d)).$$

Now assume that G is of characteristic 2 type and p is odd. As D varies, the subgroups $\Delta_G(D)$ will be used in building the functor Θ_2 for the proof of Theorem 5.2.6. Then, instead of the "trivial functor" alternative of Gorenstein-Walter, the following weakened version is used. The "O" is for "odd":

$(\Delta O)_p$: For every $B \in \mathcal{B}_{max}(G;p)$ and every $D \in \mathcal{E}_2^p(B)$, $\Delta_G(D)$ is of odd order if $p > 3$, and $[\Delta_G(D), B]$ is of odd order if $p = 3$.

The extra condition for $p = 3$ will not arise in our discussion until the proof of Theorem 5.3.4.

The alternative corresponding to the existence of a uniqueness subgroup is:

$(\Gamma_2^0)_p$: $m_p(G) \geq 3$, and for every $P \in Syl_p(G)$, $\Gamma_{2,P}^0(G)$ lies in a 2-local subgroup of G. ◊

Proving that these alternatives establish a dichotomy for certain primes $p \in \sigma(G)$ is the main goal of Chapter 2 of [**GL83**]; in fact Theorem B in [**GL83**] establishes the dichotomy for *all* primes p for which $m_{2,p}(G)$ has a particular critical value:

THEOREM 5.2.2. *Assume G is a $\mathcal{K}$-proper simple group of characteristic 2 type with $e(G) \geq 4$. Suppose that for some integer $m \geq 4$, $(\Gamma_2^0)_q$ holds for every odd prime q such that $m_{2,q}(G) > m$.*

Then for all odd primes p with $m_{2,p}(G) = m$, either $(\Gamma_2^0)_p$ or $(\Delta O)_p$ holds.

To relate this to the Pretrichotomy Theorem 5.1.1, note that conclusion (3) of that result could be reformulated as:

$(\Gamma_2^0)_p$ holds for every odd prime p with $m_{2,p}(G) \geq 4$.

Hence if we assume that the preuniqueness conclusion (3) fails, then Theorem 5.2.2 will produce for us an odd prime p such that $(\Delta O)_p$ holds—and indeed with $m_{2,p}(G)$ as large as possible subject to this condition. And it is for this choice of p that Gorenstein and Lyons prove in Chapter 1 that G is of standard type with respect to some member of $\mathcal{S}^*(G;p)$, unless G is of $GF(2)$-type and $p = 3$.

In a nutshell, the idea for the proof of Theorem 5.2.2 is the following: Assume that the result fails for the odd prime p with $m_{2,p}(G) = m \geq 4$. Then the failure of $(\Delta O)_p$ will permit us to define a p-signalizer functor whose *completion* W (namely the subgroup generated by its values; cf. Remark B.3.8 and the *closure* in [**GLS96**, Defn 21.2]) is a p'-subgroup of G of even order. Moreover there will be some 2-subgroup $1 \neq W_0 \trianglelefteq W$, with $N_G(W_0)$ containing $\Gamma_{2,P}^0(G)$ for a $P \in Syl_p(G)$, contrary to the failure of $(\Gamma_2^0)_p$.

The proof is actually achieved in two steps, which the authors call "getting started" and "moving around," and which require different signalizer functors. As the names suggest, the first stage gains control of the normalizers of certain subgroups of a fixed $B \in \mathcal{B}_{max}(G;p)$; while the second stage extends this control to all elementary abelian p-subgroups of G of rank 2 which are "2-connected" to B, that is, connected via a path in the commuting graph $\Lambda_2(G)$ (recall Definition 1.3.8). It is in the second stage that W_0 is used.

REMARK 5.2.3. The definition of $(\Gamma_2^0)_p$ includes the anomalous-looking condition that G should have p-rank at least 3. After all, the condition $p \in \sigma(G)$ means that $m_{2,p}(G) \geq 4$. The definition is formulated this way so that it will also be appropriate in the discussion of Theorem C of [**GL83**] in Section 8.1. $\diamondsuit$

Getting started. We first summarize some of the definitions related to the signalizer functors which will be used. The getting-started functor will be the "2-balanced" functor Θ_2—to be defined shortly—on a suitable $B \in \mathcal{B}_{max}(G;p)$. (We discuss balance conditions in Section B.3.) For this to be effective, signalizer functor theory requires G to be 2-balanced with respect to B. Moreover, Gorenstein and Lyons are able to arrange for B to be a witness to the failure of $(\Delta O)_p$. The set of all such witnesses is denoted $\mathcal{B}^1_{max}(G;p)$:

DEFINITION 5.2.4. $\mathcal{B}^1_{max}(G;p)$ is the subset of $\mathcal{B}_{max}(G;p)$ consisting of those B such that $\Delta_G(D)$ is of even order for some $D \in \mathcal{E}_2(B)$. $\diamondsuit$

Once 2-balance is established for such a subgroup B, signalizer functor theory will immediately give $\Gamma_{3,B}(G) \leq N$, where N is the normalizer of the completion W of Θ_2; moreover W is a p'-group, and owing to the condition $B \in \mathcal{B}^1_{max}(G;p)$, W has even order. As a result N will contain significant portions of the centralizers of all elements $b \in B^\#$. Indeed the usual suspects obstructing the slickest use of signalizer functors—p-components of $C_G(b)$ that are not locally 1-balanced with respect to B—are rounded up into N, and by $L_{p'}$-balance (i.e. I.18.2 in [**GL83**]; we had used the language of "L-balance" at 0.5.5) into $L_{p'}(N)$. Thus, the structure of the *single* $\mathcal{K}$-group $L_{p'}(N)$ places strong restrictions on the *full set* of non-locally-1-balanced components of centralizers of all elements of $B^\#$, a conclusion that is critical for the "moving around" portion of the proof of Theorem 5.2.2. The group $L_{p'}(N)$ even turns out to contain a somewhat larger class of p-components of $C_G(b)$, $b \in B^\#$, as in the following definition:

DEFINITION 5.2.5. Assume $B \in \mathcal{B}_{max}(G;p)$ and $b \in B^\#$, and set $C_b := C_G(b)$.

Define a p-component L of $L_{p'}(C_b)$ to be *strongly locally* 1-*balanced* in C_b with respect to B if every $N_B(L)$-invariant p'-subgroup of $N_{C_b}(L)$ centralizes $L/O_{p'}(L)$.

Then define $\Lambda(b;B)$ to be the product of all p-components L of $L_{p'}(C_b)$ which are *not* strongly locally 1-balanced in C_b with respect to B. (If there are no such L, then set $\Lambda(b;B) := 1$.) Also set

$$\begin{aligned} \Lambda^*(b;B) &:= \Lambda(b;B)\, O_{p'}(C_b)\, O_{p'}\big(C_G(B)\big), \text{ and} \\ \Lambda^*(G,B) &:= \big\langle \Lambda^*(b,B) : 1 \neq b \in B \big\rangle. \end{aligned}$$

$\diamondsuit$

What will generally pull $\Lambda^*(G;B)$ into N is a $\mathcal{K}$-group property verified by Gorenstein and Lyons and summarized in Section I.27 of [**GL83**]: with very few exceptions, quasisimple $\mathcal{K}$-groups which are not strongly locally 1-balanced with respect to B are instead "well-generated", i.e. by centralizers of subgroups of B of rank at least 3. In this way, they eventually obtain the following getting-started theorem (Theorem 1 of [**GL83**], established in Sections 2–8 of Chapter 2):

THEOREM 5.2.6. *Assume the notation and hypotheses of Theorem 5.2.2, and also that* $(\Delta O)_p$ *fails. Then for some* $B \in \mathcal{B}^1_{max}(G;p)$, G *possesses a maximal* 2-*local subgroup* M *with the following properties:*

(1) $\Lambda^*(G;B) \leq M$; *and*

(2) $\Gamma_{3,B}(G) \leq M$.

In more detail, the proof proceeds as follows, starting with any $B \in \mathcal{B}^1_{max}(G;p)$. We may hope, since $m_p(B) \geq 4$, to use preliminary results such as I.21-3 of [**GL83**] (with 4 in the role of "$k+2$") to show that the getting-started functor defined on $B^\#$ by:

$$\Theta_2(b) := \langle \Delta_G(D) \cap C_G(b) : D \in \mathcal{E}_2(B) \rangle$$

(which is the 2-balanced functor by definition) is in fact a B-signalizer functor, whence Θ_2 is in particular (1)-balanced in the sense of (S2) of Definition 0.3.14. This is established at Proposition II.2.4.2 of [**GL83**], unless certain explicit configurations of centralizers occur. The application of I.21-3.1 of [**GL83**] in turn requires establishing 2-balance for G and B (e.g. [**GL83**, Proposition II.2.4.1.I]); and the failure of this condition can be analyzed locally, and hence in $\mathcal{K}$-groups. Such an analysis appears in results such as I.19.10 in [**GL83**]. Indeed the more general analysis in [**GLS02**, 3.3.1, 3.3.2] shows that the principal local obstructions are components of the following type in $C_G(b)$ for some $b \in B^\#$:

(a) $\overline{L} := L/O_{p'}(L) \cong A_{kp^2+r}$, $2 \leq r < p$, where $B/C_B(\overline{L})$ acts with k regular orbits of size p^2; or

(b) $\overline{L} \cong PSL^\epsilon_p(q)$, where B induces a group of order p^2 of inner-diagonal automorphisms (i.e., a subgroup of $PGL^\epsilon_p(q)$) on $\overline{L}$ whose preimage in $GL^\epsilon_p(q)$ is nonabelian.

The key point is that, in these examples, there are almost always elementary abelian p-subgroups in the picture, of far larger rank than that of the subgroup B in (a) or (b)—subgroups contradicting the fact that we chose $B \in \mathcal{B}_{max}(G;p)$. For example if $p = 3$, then $L \cong A_{11}$ cannot arise as an obstruction: for then $B/C_B(L)$ must by (a) be a group of rank 2, acting regularly on 9 points, and centralizing a nontrivial 2-group in $N_G(L)/C_G(L) \cong \Sigma_{11}$; whereas L contains elementary abelian 3-groups C of rank 3 centralizing the same 2-group, so $m_p\big(C_B(L)C\big) > m_p(B)$, contradicting the maximal rank of B. And this disparity of rank only increases in larger examples. The obstructions of type (b) above with $p = 3$ and $m_3(B) = 4$ are especially troublesome; a lengthy delicate analysis is required in their presence to find, somewhere in G, a different $B \in \mathcal{B}^1_{max}(G;3)$ for which 2-balance can be verified.

For such a B, signalizer functor theory immediately implies that Θ_2 is a balanced signalizer functor. In particular, the completion $W := \langle \Theta_2(b) : b \in B^\# \rangle$ is a p'-group by the Signalizer Functor Theorem 0.3.15.1, and, as noted above, W is also of even order since B was chosen to be in $\mathcal{B}^1_{max}(G;p)$. On the other hand, failure of strong local 1-balance from Definition 5.2.5—or even failure just of local 1-balance—is a problem everywhere, so there is no hope of defining a 1-balanced B-signalizer functor. (There are, however, fairly general circumstances in which a so-called $\frac{3}{2}$-balanced (or Aschbacher-Goldschmidt) functor can be defined; and Gorenstein and Lyons do indeed use this functor to handle some difficult cases.)

The drawback of the 2-balanced functor Θ_2, as compared with 1-balanced functors (had they been available!) is that for the normalizer $N := N_G(W)$ of the completion, one can directly obtain only conclusion (2) of Theorem 5.2.6, that $\Gamma_{3,B}(G) \leq N$, rather than the ideal stronger conclusion $\Gamma_{2,B}(G) \leq N$. Nevertheless, in view of the good generational properties of quasisimple groups which are not strongly locally 1-balanced (cf. the use of II.2.3.2 at [**GL83**, p 542], which in turn goes back via II.2.3.1 to I.27.1), it is possible to show that $\Lambda^*(G; B) \leq N$.

To complete the proof of Theorem 5.2.6, it then remains to establish that:

$N = N_G(W)$ normalizes some nontrivial 2-subgroup R of G;

for then the theorem is satisfied by any maximal 2-local subgroup M of G containing $N_G(R)$. This final step is accomplished at [**GL83**, Sec. 2.8], the proof resting mainly, but not entirely, on the facts that $\Gamma_{3,B}(G) \leq N$ and that G is of characteristic 2 type. The subgroup $X := O_{p'}(N)$ of even order is first analyzed via its intersections with 2-local subgroups of the form $N_G(T)$, where T is a Sylow 2-subgroup of some B-invariant subgroup of X of even order. The above facts yield that $O_2(N_G(T)) \leq \Gamma_{3,B}(G) \leq N$ and then that the socle of $N_N(T)$ is a 2-group. For instance, if $N_N(T)$ covers $L/O_{2'}(L)$ for some 2-component L of X, then $O_2(L) \neq 1$. This type of analysis enables Gorenstein and Lyons to show that either $O_2(N) \neq 1$—and hence the theorem holds—or else $F^*(X/O_{2'}(X))$ is either a 2-group or a product of components permuted transitively by B; moreover in these two cases, B acts faithfully on $F^*\big(X/O_{2'}(X)\big)$.

Finally, these two residual cases are ruled out as follows. Consider first the case that $F^*\big(X/O_{2'}(X)\big)$ is a 2-group. If N contains a four-group U in Thompson's set $\mathfrak{U}_G(2)$,[3] then an elementary argument of Thompson [**Tho68**, Lemma 6.1b], applicable to elements of $\mathfrak{U}_X(2)$ in any group X of characteristic 2 type, yields that $[U, O_{2'}(X)] = 1$, whence $O_2(X) \neq 1$. Then $R = O_2(X)$ serves as the desired subgroup. But the only obstruction to the existence of such a subgroup U would be a 2-local subgroup N_0 of G containing B and containing transvections in its action on one of its normal elementary abelian 2-subgroups. Therefore such an obstruction N_0 exists, and McLaughlin's classification A.6.1 of $\mathbb{F}_2$-linear groups generated by transvections restricts the structure of N_0 to a handful of specific configurations. These are then dealt with by *ad hoc* arguments.

In the complementary case, $F^*\big(X/O_{2'}(X)\big)$ is the product of, say, n components permuted transitively by B. Here the condition in Theorem 5.2.2 about primes q with $m_{2,q}(G) > m_{2,p}(G)$ comes into play for the only time in Chapter 2. (It appears in a few places in Chapter 1 as well.) The automorphism group of a simple $\mathcal{K}$-group G has p-rank at most 1 if G is a p'-group. This well-known fact, and the faithful action of B on $F^*\big(X/O_{2'}(X)\big)$, force $n \geq p^{m-1}$, where $m = m_{2,p}(G)$ is the p-rank of B. An easy consequence is that $m_{2,q}(G) > n - 1 \geq m$ for some odd prime divisor q of $|F^*(X/O_{2'}(X)|$; and so $(\Gamma_2^o)_q$ holds by hypothesis. Thus, for any Sylow q-subgroup Q of G, there is a 2-local subgroup N_1 of G containing $\Gamma^0_{2,Q}(G)$. Taking Q to contain a Sylow q-subgroup of N, one shows easily that $N \leq N_1$, and so $R = O_2(N_1)$ serves as the desired 2-group. The proof of Theorem 5.2.6 is then complete.

Moving around. Recall that the second stage in the proof of Theorem 5.2.2 is to extend control of normalizers of elementary abelian p-groups of G, established

[3] In the N-group paper [**Tho68**, Defn. 2.8], $\mathfrak{U}_G(2)$ is defined whenever Sylow 2-subgroups of G have noncyclic normal abelian subgroups. It is the set of all four-subgroups of G which are central in some Sylow 2-subgroup of G, if this set is nonempty; otherwise it is the set of all four-subgroups of G which are normal in some Sylow 2-subgroup of G.

for subgroups of B of rank at least 3 in the first stage, to all elementary abelian p-subgroups of G of rank 2 that are 2-connected to B. In particular as part of this stage, the conclusion $\Gamma_{3,B}(G) \leq M$ will be strengthened to $\Gamma_{2,B} \leq M$, making up for the lack of balance on B.

The main result (Theorem 2 of [**GL83**], established in Sections 9–13 of Chapter 2) constructs a suitable moving-around functor $\theta^{(2)}$; in particular, note that condition (1) below corresponds to the G-equivariance requirement (I1) in Definition 0.3.16:

THEOREM 5.2.7. *Let B and M satisfy the conditions of Theorem 5.2.6. If Theorem 5.2.2 does not hold, then for each elementary abelian p-subgroup D of G of rank 2, there exists a subgroup $\theta^{(2)}(D)$ of G with the following properties:*

(1) $\theta^{(2)}(D^g) = \theta^{(2)}(D)^g$ for all $g \in G$; and

(2) If $D \leq M$ and D is 2-connected to B in M, then $O_2(M) \leq \theta^{(2)}(D) \leq M$; and furthermore if $O_2(M) \leq T \leq \theta^{(2)}(D)$, then $N_G(T) \leq M$.

The subgroup $\theta^{(2)}(D)$ is actually a 2-group. The construction of the "preferred" candidate Γ for the functor $\theta^{(2)}$ is somewhat technical.

DEFINITION 5.2.8. Let T be any 2-subgroup of G. We recursively define the descending chain $\gamma^1(T) \geq \gamma^2(T) \geq \cdots \geq \gamma^\infty(T)$ of subgroups of T as follows:

$$\gamma^1(T) := T \cap \left(\bigcap_{1 \neq T_0 \text{ char } T} O_2\big(N_G(T_0)\big)\right);$$
$$\gamma^{n+1}(T) := \gamma^1\big(\gamma^n(T)\big); \text{ and}$$
$$\gamma^\infty(T) := \bigcap_{n=1}^{\infty} \gamma^n(T).$$

Let D be an elementary abelian p-subgroup of G of rank 2. We set:

$$\Sigma^*(D) := \langle O_2\big(C_G(d)\big) : d \in D^\# \rangle,$$
$$\Sigma(D) := O_2\big(\Sigma^*(D)\big), \text{ and}$$
$$\Gamma(D) := \gamma^\infty\big(\Sigma(D)\big).$$

The definitions lead easily at 2.9.3 of [**GL83**] to the following properties of $\Gamma(D)$, including the basic equivariance property for part (1) of Theorem 5.2.7:

LEMMA 5.2.9. *Let D be an elementary abelian p-subgroup of G of rank 2, and $g \in G$. Then*

(1) $\Gamma(D^g) = \Gamma(D)^g$;

(2) $\Gamma(D) \leq \Sigma(D)$, and $\Gamma(D) \trianglelefteq \Sigma^*(D)$; and

(3) $\Gamma(D) \leq O_2\big(N_G(T_0)\big)$ for any non-identity characteristic subgroup T_0 of either $\Gamma(D)$ or $\Sigma(D)$.

Hence to establish Theorem 5.2.7, with Γ in the role of "$\theta^{(2)}$", it will suffice to prove:

$\Gamma(D) = O_2(M)$ for each elementary abelian D of rank 2 in M which is 2-connected to B.

For taking D as in part (2) of the Theorem, we have $O_2(M) = \theta^{(2)}(D)$. Also $O_2(M)$ is the only group that satisfies the hypotheses of "T" in 5.2.7, and furthermore we get $N_G\big(O_2(M)\big) = M$, since M is a maximal 2-local in Theorem 5.2.6.

Consider first the case $D \leq B$. It is not obvious from the definition that $\Gamma(D)$ is even nontrivial. However, using both conclusions of Theorem 5.2.6, it is shown that $C_{O_2(M)}(b) \leq O_2\big(C_G(b)\big) \leq M$ for all $b \in B^\#$, leading to $O_2(M) \leq \Sigma^*(D) \leq M$.

So suppose for some $D \leq B$ that $\Gamma(D) \neq O_2(M)$. The proof of 2.9.5 at [**GL83**, p 576] shows that conclusion (3) of Lemma 5.2.9 leads to the existence of a certain proper "characteristic 2-core" $C\big(Y, \Gamma(D)\big)$ in the language of the Local $C(G,T)$ Theorem B.7.3.

The argument goes on to use pushing-up results (e.g. I.28.10) to show that some p-component L of $L_{p'}(M)$ must be a *block* in M, and indeed of a restricted type: namely $L/O_{p'}(L)$ is isomorphic to an alternating group or an orthogonal group over $\mathbf{F}_2$; and the unique nontrivial chief factor of L in $O_2(L)$ is either a natural module, or a 4-dimensional module for $L/O_{p'}(L) \cong A_7$. Then, as noted at [**GL83**, p 572], a contradiction could be obtained immediately, by quoting the Global $C(G,T)$ Theorem B.7.8. However, Gorenstein and Lyons had arrived at this point of the proof while the proof of Global $C(G,T)$-Theorem was still in progress, so they derived a contradiction without it. Their argument is presented in the remainder of Section 2.9 of [**GL83**]. It involves the fine structure of the definition of Γ in Definition 5.2.8—as well as taking $\theta^{(2)}$ to be a slight *ad hoc* variation Γ^* of Γ designed to circumvent the alternating and orthogonal configurations where Γ was ineffective.

The final step in the proof of Theorem 5.2.7 is to prove:

PROPOSITION 5.2.10. *For any E_{p^2}-subgroup D of G which is 2-connected to subgroups of B,*
(1) $O_2(M) \leq \Gamma^*(D)$; and
(2) $N_G(T) \leq M$ for any T such that $O_2(M) \leq T \leq \Gamma^*(D)$.

This is considerably harder to show for general D than for subgroups D of B, because an analogue of Theorem 5.2.6 has not been established for elementary abelian subgroups containing D. As we move along the commuting graph Λ_2, effective substitutes for Theorem 5.2.6 must always be proved for the elementary abelian group of current interest. This requires quite intricate arguments, carried out in the final four sections and 50-odd pages of Chapter 2, and completing the proof of Theorem 5.2.7.

Finally, we indicate (cf. [**GL83**, p 525]) how Theorem 5.2.2 follows as a consequence of Theorem 5.2.7. Let B and M be as in Theorem 5.2.6 (and hence as in Theorem 5.2.7). Let $P \in Syl_p(M)$ with $B \leq P$. As we are assuming in Theorem 5.2.6 that $(\Delta O)_p$ fails, to prove Theorem 5.2.2 we must establish $(\Gamma_2^0)_p$: namely we will show that $P \in Syl_p(G)$ and that $\Gamma_{2,P}^0(G)$ lies in the 2-local M. For the latter, we must show that $N_G(Q) \leq M$ for all $Q \leq P$ with $m_p(Q) \geq 2$ and with $m_p\big(QC_P(Q)\big) \geq 3$; and this suffices for the former as well, since then in particular $N_G(P) \leq M$, so that $P \in Syl_p(G)$.

So fix Q. Fix any E_{p^3}-subgroup $E \leq Q_1 := QC_P(Q)$, and any normal E_{p^2} subgroups $D \trianglelefteq Q$ and $U \trianglelefteq P$; D and U exist by A.1.8. Then $D \trianglelefteq QC_P(Q) = Q_1$, whence by B.4.8, D is 2-connected to E in Q_1; likewise E and B are 2-connected in P to U and thus to each other. Therefore D is 2-connected in P to B.

For any $x \in N_G(Q)$ and integer i, $D^{x^i} \trianglelefteq Q$, so D^{x^i} is also 2-connected to B in P by the previous paragraph. Then by Theorem 5.2.7.2, $O_2(M) \leq \theta^{(2)}(D^{x^i})$. Hence setting:

$$T := \bigcap_{i=1}^{\infty} \theta^{(2)}(D^{x^i}),$$

we have $O_2(M) \leq T \leq O_2(\theta^{(2)}(D))$. Theorem 5.2.7.2 now yields $N_G(T) \leq M$.

But x permutes the D^{x^i}, and hence permutes the terms $\theta^{(2)}(D^{x^i})$ by the equivariance in Theorem 5.2.7.1; thus x normalizes T. As x was arbitrary, we conclude that $N_G(Q) \leq M$. Finally, as we saw, this suffices to complete the proof of Theorem 5.2.2.

Recall from our discussion after the statement of Theorem 5.2.2 that we may assume that $(\Gamma_2^0)_p$ "eventually" fails; so we may choose a prime p with $(\Delta O)_p$ holding, and subject to this condition, with $m := m_{2,p}(G) \geq 4$ as large as possible. Thus, Theorem 5.2.2 has the following immediate corollary:

COROLLARY 5.2.11. *Suppose that the hypotheses of the Pretrichotomy Theorem hold, and that the conclusions (2) and (3) of that theorem both fail. Then there exist a prime $p \in \sigma(G)$ and an integer $m \geq 4$ such that the following conditions hold:*
(1) $m_{2,p}(G) = m$ and $(\Delta O)_p$ holds;
(2) for every odd prime q such that $m_{2,q}(G) > m$, $(\Gamma_2^0)_q$ holds;
(3) for every odd prime q such that $m_{2,q}(G) = m$, either $(\Gamma_2^0)_q$ or $(\Delta O)_q$ holds.

5.3. The Component Analysis (leading to standard type)

With Corollary 5.2.11 providing a good prime p via signalizer analysis, the proof of the Pretrichotomy Theorem 5.1.1 is completed with the following result, which Gorenstein and Lyons establish as Theorem A, in Chapter 1 of Part II of [**GL83**]:

THEOREM 5.3.1. *Assume G is a $\mathcal{K}$-proper simple group of characteristic 2-type with $e(G) \geq 4$, and suppose that conclusions (2) and (3) of the Pretrichotomy Theorem 5.1.1 do not hold.*

Then there is a prime $p \in \sigma(G)$ which satisfies conclusions (1)–(3) of Corollary 5.2.11 and such that one of the following also holds:
(1) G is of $GF(2)$-type and $p = 3$;
(2) G is of standard type with respect to some element of $\mathcal{S}^*(G;p)$.

The proof aims at establishing conclusion (2), by successively developing the various properties in Definition 5.1.4 of a group of standard type – aside from certain exceptional situations in which the $GF(2)$ type conclusion (1) can be verified.

For most of the proof to follow, through the discussion of Proposition 5.3.10,

> *We fix any p satisfying the conclusions of Corollary 5.2.11. (And so $(\Delta O)_p$ holds.)*

Before discussing the proof in detail we make two further remarks:

First: Gorenstein and Lyons study the r-local structure of G only for $r = p$ and $r = 2$. However, other primes q such that $m_{2,q}(G) \geq m_{2,p}(G)$ come into play in obtaining some of the "standard type" conditions—namely, that $L/O_{p'}(L)$ is of Lie type in characteristic 2 (which is proved before quasisimplicity of L), that p splits L, and that under certain conditions p half-splits the neighbors of L. In the configurations where such primes q arise, it is established that $m_{2,q}(G) = m_{2,p}(G)$ and $(\Delta O)_p$ holds. For otherwise $(\Gamma_2^0)_q$ holds by the hypothesis of 5.3.1. Then the existence of a 2-local subgroup containing $\Gamma_{2,Q}^0(G)$ for some $Q \in Syl_q(G)$ is exploited to contradict $(\Delta O)_p$.

REMARK 5.3.2. Our second remark is simple but important; it already in evidence in the N-group paper, and fundamental in the work of Klinger and Mason: Let $B \in \mathcal{B}_{max}(G;p)$. Then B is either maximal with respect to inclusion among all elementary abelian p-subgroups of G, or not. In each case there is a useful conclusion which can be drawn about centralizers of subgroups of B:

In the maximal case, the Bender-Thompson Signalizer Lemma (quoted as [**GL83**, I.20-1], see our Lemma B.1.10) implies that for any B-invariant p'-subgroup T of G and any subgroup $E \leq G$ such that $C_G(E)$ is p-constrained, we in fact have $T_0 := C_T(E) \leq O_{p'}(C_G(E))$.

In the non-maximal case, for any $E \leq B$, $C_G(E)$ has p-rank greater than B, and hence greater than $m_{2,p}(G)$. Consequently $O_{p'}(C_G(E))$ has odd order. Thus for any B-invariant 2-subgroup T of G, we get $T_0 := C_T(E)$ "visible" in the quotient $\overline{C}_E := C_G(E)/O_{p'}(C_G(E))$; i.e. it acts faithfully on $F^*(\overline{C}_E) = E(\overline{C}_E)O_p(\overline{C}_E)$. Thus the structure of T_0 is constrained by the isomorphism types of components in p-local subgroups, as well as by $m_{2,p}(G)$. The latter dependence is through the Thompson Dihedral Lemma B.1.7, which applied to the action of T_0 on $O_p(\overline{C}_E)$ yields that $m_2(\ T_0\ /\ C_{T_0}(O_p(\overline{C}_E))\ ) \leq m_{2,p}(G) + 1 - m_p(E)$. ◊

Obtaining a p-component L. The opening move in the proof of 5.3.1 is as usual to locate a p-component—in fact in the centralizer of an element from the following set, whose definition is relevant to condition (4) in Definition 5.1.4.

DEFINITION 5.3.3. Assume that $B \in \mathcal{B}_{max}(G;p)$. Then B_e is the set of all elements $x \in B^\#$ such that B acts nontrivially on a 2-subgroup of $C_G(x)$ of rank at least 2. ◊

Thus the first main result [**GL83**, Proposition 1.3.2] produces such a p-component, and so corresponds roughly with completing Step 3 of Strategy 1.1.1:

PROPOSITION 5.3.4. *Assume that $B \in \mathcal{B}_{max}(G;p)$. Then one of the following conclusions holds:*

(1) $p = 3$ and G is of $GF(2)$-type; or

(2) there exists $x \in B_e$ such that $L_{p'}(C_G(x)) \neq 1$.

We sketch the proof; assume that both conclusions fail. Asnce $B \in \mathcal{B}_{max}(G;p)$, B is contained in a maximal 2-local subgroup N of G; set $T := O_2(N) = F^*(N)$. Thus $[B, C_T(B_0)] \neq 1$ for some hyperplane B_0 of B. At 1.3.1 it is established that $B_0^\# \subseteq B_e$ for a suitable such hyperplane B_0. Indeed this is trivial unless $p = 3$ and $[B, C_T(B_0)] \cong Q_8$ or 1 for all hyperplanes of B. But in that case it is shown that T is of symplectic type, contrary to the assumed failure of (1).

Next the failure of (2) means that $C_G(x)$ is p-constrained for all $x \in B_e^\#$ (cf. B.1.9), and in particular for all $x \in B_0^\#$. So at 1.3.4.i, B cannot be a maximal elementary abelian subgroup of G (with respect to inclusion); if it were, then by Remark 5.3.2, $C_T(B_0) \leq O_{p'}(C_G(x))$ for all $x \in B_0^\#$, whence $[B, C_T(B_0)] \leq \Delta_G(D)$ for each D of rank 2 in B_0, contrary to our assumption in this section that $(\Delta O)_p$ holds. So B is not maximal; and Remark 5.3.2 yields at 1.3.4.ii that $O_{p'}(C_G(D))$ is of odd order for each $1 \neq D \leq B$. Using this, Gorenstein and Lyons are now able at 1.3.5 to follow part of the argument for 2.2 in Klinger-Mason [**KM75**], which we briefly summarize:[4]

The argument concerns subgroups $E \leq B$ with B/E of rank $k \leq 2$. Suppose for all such E that $L_{p'}(C_G(E)) = 1$ (so that $C_G(E)$ is p-constrained, e.g. B.1.9).

[4]In the language of our discussion of the Klinger-Mason Argument in Remark 2.2.2: we initially obtain just the weaker conclusion that G is not of characteristic p type; but then the Weak Trichotomy Theorem 2.2.1 gives the stronger conclusion that G is of p-component type.

Then 2.2.i of [**KM75**] holds—obviating any need for the the hypothesis of 2.2 that $[B, C_T(E)] \neq 1$. And above we got $O_{p'}\big(C_G(E)\big)$ of odd order, as required early in the second paragraph of the proof in [**KM75**]. Thus we have $m_2\big(C_T(E)\big) \leq k+1$. And in this situation, other arguments in [**KM75**, Sec 2] show that $p = 3$ and G is of $GF(2)$ type—contrary to the assumed failure of (1). This contradiction shows that $L_{p'}\big(C_G(E)\big) \neq 1$ for some E. Now for $x \in E \cap B_0 \subseteq B_e$ it follows from $L_{p'}$-balance in known groups (quoted as I.18.2 there; cf. our 0.5.5) that $L_{p'}\big(C_G(x)\big) \neq 1$. Of course this contradicts the assumed failure of (2), completing the proof.

Hence, since conclusion (1) of Proposition 5.3.4 is conclusion (1) of the Pretrichotomy Theorem 5.1.1, Gorenstein and Lyons may now assume for the remainder of the analysis that conclusion (2) of Proposition 5.3.4 holds. In other words, we obtain an initial approximation to the set of triples $\hat{\mathcal{S}}(G;p)$ defined shortly after the statement of 5.1.2; namely the set:

$$\mathcal{S}(G;p) := \big\{(B, x, L) : B \in \mathcal{B}_{max}(G;p),\ x \in B_e,\ L \text{ is a } p\text{-component of } C_G(x)\big\}$$

is nonempty.

Obtaining "quasi"-quasisimplicity for such p-components. A partial ordering is defined on this set at the start of Section 1.4, based on $L_{p'}$-balance (I.18.2 there, cf. our 0.5.5) and pumpups; and $\mathcal{S}_{max}(G;p)$ is defined to be the set of all maximal elements of $\mathcal{S}(G;p)$ under this ordering. We omit the definition but state the resulting critical property, a kind of $\mathcal{B}_{max}$-localized terminality: if $(B, x, L) \in \mathcal{S}_{max}(G;p)$, then it follows that for any $x^* \in C_B(L/O_{p'}(L))$ and $(B, x^*, L^*) \in \mathcal{S}(G;p)$, the group $L_{p'}(C_L(x^*))$, which covers $L/O_{p'}(L)$, must either cover or centralize $L^*/O_{p'}(L^*)$. The full definition of $\mathcal{S}_{max}(G;p)$ is a certain extension of this to all triples $(B^*, x^*, L^*) \in \mathcal{S}(G;p)$—with B^* not required to equal B.

For brevity we shall write $\mathcal{S}$, $\mathcal{S}^*$, etc., for $\mathcal{S}(G;p)$, $\mathcal{S}^*(G;p)$, etc.
As a further refinement, a subset $\mathcal{S}^o$ is defined as the set of maximal elements of $\mathcal{S}$ with respect to a different and elaborately defined linear ordering. Again without the detailed definition, the effect is that when one chooses $(B, x, L) \in \mathcal{S}^o$, one is *de facto* exercising certain preferences about the isomorphism type of $L/O_{p'}(L)$ and, to a certain extent, the embedding of B in $C_G(x)$. For the sake of exposition, let us say that the larger elements with respect to this ordering are "better". The "best" triples (B, x, L) are essentially those for which $L_{p'}(C_L(E)) \neq 1$ for some subgroup $E \leq B$ of rank 3. Apart from these, the rest of $\mathcal{S}$ is ordered by the isomorphism type of $L/O_{p'}(L)$. The "worst" have $L/O_{p'}(L) \in Chev(p)$. Sporadic groups and groups of p-rank 1 or 2 are also rather bad as well. Thus if $(B, x, L) \in \mathcal{S}^o$, then $L^*/O_{p'}(L^*)$ is at least as bad as $L/O_{p'}(L)$, for all $(B^*, x^*, L^*) \in \mathcal{S}$. Elements (B, x, L) of $\mathcal{S}^o$ are also required to have B normalizing L; and sincei we have $m_{2,p}(G) = m_p(B)$, this turns out always to be possible, and indeed is true for any $(B, x, L) \in \mathcal{S}^o$ (see [**GL83**, I.11-3]), except when $p = 3$ and $L/O_{3'}(L)$ is of 3-rank 2.

The linear ordering defining $\mathcal{S}^o$ contains the partial ordering defining $\mathcal{S}_{max}$. Hence we may define:

$$\mathcal{S}^o_{max} := \mathcal{S}^o \cap \mathcal{S}_{max} \neq \emptyset.$$

Restricting attention to this set, Gorenstein and Lyons are able, in the present branch of the proof of Theorem 5.2.2—corresponding roughly to the triviality of signalizer functors in the Gorenstein-Walter Alternative, cf. Remark 1.1.2—to come close to moving from a p-component to an actual quasisimple component, as in condition (3) in the Definition 2.3.2 of $\mathcal{S}^*$. (That is, they almost complete Step 4 in the language of Strategy 1.1.1.) Specifically, they prove the following result [**GL83**, Proposition 1.4.15]:

PROPOSITION 5.3.5. *If* $(B, x, L) \in \mathcal{S}^o$, *then* $[L, O_{p'}(C_G(x))]$ *has odd order.*

The proof exploits properties of known groups developed in Part I, and uses a variety of arguments, depending on how "bad" $L/O_{p'}(L)$ is, in the sense of the ordering used to define $\mathcal{S}^o$. In the best case, for example, as noted above B has a subgroup $E \cong E_{p^3}$ such that $L_{p'}(C_L(E)) \neq 1$. If $[L, O_{p'}(C_G(x))]$ has even order in that case, then a quick argument using the B_p-property for $\mathcal{K}$-groups (namely [**GL83**, 14-5]) yields that $\Delta_G(D)$ has even order for some hyperplane D of E, and $(\Delta O)_p$ fails, contrary to hypothesis.

Recall that by definition, for any $(B, x, L) \in \mathcal{S}$, the p-element x lies in B_e, so B acts nontrivially on some 2-subgroup T of $C_G(x)$. As a consequence of Proposition 5.3.5, T is "visible" in $C_G(x)/[O_{p'}(C_G(x)), \langle L^{C_G(x)}\rangle]$.

Restricting the centralizer of L. Recall from our overview of standard form, toward the end of Chapter 0, that one requirement in the definition of a standard component L is that $C_G(L)$ should be "essentially cyclic". Proposition 1.5.1 of [**GL83**] provides a first step in that direction, which completes the verification of the conditions defining $\hat{\mathcal{S}}$:

PROPOSITION 5.3.6. *If* $(B, x, L) \in \mathcal{S}^o_{max}$, *then* $C_B(L/O_{p'}(L)) = \langle x \rangle$.

To convey the gist of the argument, we make two simplifying assumptions. Let us pretend, first of all, that p-components are components.

Assume by way of contradiction that $D := C_B(L)$ has rank at least 2. Suppose that a is any element of $D^\#$, so that $L \leq C_G(a)$. Then from $(B, x, L) \in \mathcal{S}^o_{max}$ we get (using the details of the orderings) that L is a component of $C_G(a)$. Our second simplifying assumption, used at one point below, is that L is normal in $C_G(a)$.

Since D centralizes L, it centralizes a Sylow 2-subgroup of L, and so D lies in some 2-local subgroup N of G. If possible, we choose the 2-local N so that $DL \leq N$. Set $R := O_2(N)$. Since G is of characteristic 2 type, D acts faithfully on R. As R is generated by the subgroups $C_R(a)$ with $a \in D^\#$, $R_a := [D, C_R(a)] \neq 1$ for some $a \in D^\#$. But $[D, L] = 1$ and L is normal in $C_G(a)$, so R_a centralizes L. Thus $N_G([D, R_a])$ is a 2-local containing DL. Hence by our choice, $L \leq N$. Then $C_L(R) = L \cap Z(R) \leq O_2(L) = 1$ by Proposition 5.3.5. Now with L as with D before, we find that $S_b := [L, C_R(b)] \neq 1$ for some $b \in D^\#$. Then $S_b = [L, S_b]$. But this contradicts the subnormality of L in $C_G(b)$, as subnormality implies that $S_b \leq L$ and then $S_b \leq [L, O_{p'}(L)] = 1$, a contradiction.

Of course in reality, p-components are not necessarily components, and L is known only to be subnormal rather than normal, so the actual argument needed is more elaborate. Nonetheless, Proposition 5.3.5 suffices to push it through.

REMARK 5.3.7. If nontrivial cores had not been assumed away, then the contradiction to subnormality in the last sentence of the argument would amount to the statement that $L_{p'}(C_L(D))$ does not cover any component of $C_G(b)/O_{p'}(C_G(b))$. A similar situation will arise in proving the existence of neighbors, below.

It is then fairly easy to bootstrap Proposition 5.3.6 on the centralizer in B up to the following statement [**GL83**, 1.5.8] on the centralizer in G, eventually needed to establish condition (4) in the Definition 2.3.2 of $\mathcal{S}^*$:

PROPOSITION 5.3.8. *There is a triple* $(B, x, L) \in \mathcal{S}^o_{max}$ *such that* $C_G(L/O_{p'}(L))$ *has cyclic Sylow p-subgroups.*

Indeed at [**GL83**, 1.5.3], it is shown easily that to prove our Proposition 5.3.8, it suffices to show that:

> B normalizes a non-trivial 2-subgroup of $LO_{p'}(C_G(x))$, for some suitable triple $(B, x, L) \in \mathcal{S}^o_{max}$;

and most of Section 1.5 is devoted to establishing this assertion. Its importance is the consequence that any p-subgroup of $BC_G(L/O_{p'}(L))$ then normalizes a non-trivial 2-subgroup of L. Then if $m_p(C_G(L/O_{p'}(L))) > 1$, B can be replaced by another $B^* \leq C_G(x)$ such that $B^* \leq BC_G(L/O_{p'}(L))$, $(B^*, x, L) \in \mathcal{S}^o_{max}$, and $C_{B^*}(L/O_{p'}(L)) > \langle x \rangle$, contrary to 5.3.6.

As a result of Proposition 5.3.8, we can introduce still another subset of $\mathcal{S}$:

$$\mathcal{S}^*_{max} := \{(B, x, L) \in \mathcal{S}^o_{max} : m_p(C_G(L/O_{p'}(L))) = 1\} .$$

The definitions immediately imply that:

$$\mathcal{S}^*_{max} \subseteq \mathcal{S}^*(G; p).$$

Recall that Theorem 5.3.1.2 asserts the existence of a suitable triple (B, x, L) in $\mathcal{S}^*(G; p)$. Normally the triple that will be produced will be in $\mathcal{S}^*_{max}$. However, on exactly one occasion (Proposition 5.3.11 below), Gorenstein and Lyons make an *ad hoc* escape from an exceptional configuration by verifying Theorem 5.3.1.2 for some element of $\mathcal{S}^*(G; p)$ which may not belong to $\mathcal{S}^*_{max}$.

As a result of Proposition 5.3.6, $Aut(L/O_{p'}(L))$ has p-rank at least as large as $m_p(B) - 1 = m_{2,p}(G) - 1 \geq 3$ for any $(B, x, L) \in \mathcal{S}^*_{max}$, so the isomorphism type of $L/O_{p'}(L)$ avoids many small possibilities with exceptional behavior. Moreover B acts nontrivially on some 2-subgroup of $C_G(x)$ as $x \in B_e$. If $B = \langle x \rangle \times (B \cap L)$, then such a 2-group lies in L, so:

$$m_{2,p}(L) \geq m_p(B) - 1 \geq 3.$$

For example, this inequality applies if $L/O_{p'}(L)$ is a sporadic group, in which case its outer automorphism group has order 1 or 2. This remark immediately cuts the number of sporadic possibilities for $L/O_{p'p}(L)$ by more than half, from 26 to 9 (at [**GL83**, I.10-17]); in fact there are 12 cases to be treated, since some central extensions among the 9 could arise.

Indeed Gorenstein and Lyons are ready to rule out all unwanted isomorphism types for $L/O_{p'p}(L)$.

Reducing to $L \in Chev(2)$**.** The next three sections [**GL83**, Sections 1.6–1.8] are devoted to the proof of the following proposition [**GL83**, 1.8.1], eventually needed for condition (1) of the Definition 5.1.4 of standard type:

PROPOSITION 5.3.9. *Let* $(B, x, L) \in \mathcal{S}^*_{max}$. *Then one of the following holds:*
(1) $p = 3$ *and* G *is of* $GF(2)$*-type; or*
(2) $L/O_{p'}(L)$ *is a finite group of Lie type defined over a field of characteristic* 2.

This is achieved via a case-by-case argument according as $L/O_{p'p}(L)$ is an alternating group, a group of Lie type in characteristic p, a group of Lie type of any other characteristic, or a sporadic group. Of the 12 sporadic possibilities, 9 are treated together. The arguments of course depend on shared properties of groups within each case, and the ordering defining $\mathcal{S}^o$ is of critical importance.

We illustrate by outlining the easiest case, treated at [**GL83**, 1.6.1], in which (1) is shown to hold. There is significant overlap here with Proposition 5.3.4, and much of the proof in this case has already appeared in [**GL83**, Section 1.3].

Thus, suppose that $L/O_{p'}(L)$ is of Lie type over a field of characteristic p. As we had indicated above, $L/O_{p'}(L)$ then ranks lowest in the ordering defining $\mathcal{S}^o$, so all relevant p-components are in $Chev(p)$ as well.

Now groups in $Chev(p)$—with a single exception, requiring yet another *ad hoc* argument—are strongly locally 1-balanced with respect to any maximal elementary abelian p-subgroup of their automorphism groups. In effect this pretty property provides an extension of Remark 5.3.2, making the p-constraint assumption in that remark unnecessary in the maximal case. Therefore if B is a maximal elementary abelian p-subgroup with respect to inclusion, the same argument is available that was used above in the proof of Proposition 5.3.4 to show that (1) holds. Conclusion: Gorenstein and Lyons may assume that $O_{p'}(C_G(b))$ has odd order for all $b \in B^\#$, by Remark 5.3.2.

They then show that:

$$L_{p'}(C_G(E)) = 1 \text{ for all } E \leq B \text{ such that } |B/E| \leq p^2,$$

and reuse the argument in the third paragraph of our discussion of Proposition 5.3.4—without change—to reach a contradiction. The above condition is a consequence of the Borel-Tits Theorem [**BT71**] unless there exists $(B, u, K) \in \mathcal{S}^o$ with $u \in E$ such that every element of E either centralizes $K/O_{p'}(K)$ or induces a non-inner automorphism on it. Using this fact, and the fact that the outer automorphism groups of groups in $Chev(p)$ have cyclic Sylow p-subgroups (again there is one exception), they deduce that $m_p(E) = 2$, so $m_p(B) = 4$. Finally an analysis of B-invariant 2-subgroups of $C_G(x)$ in this minimal case yields a contradiction to $(\Delta O)_p$.

The interesting sporadic case is treated in section 1.8 of [**GL83**]. In the majority of situations here, $\overline{L} := L/O_{p'}(L)$ a large sporadic group containing an extraspecial 2-subgroup R normalized by some $B^* \in \mathcal{B}_{max}(G;p)$ with $x \in B^*$. One gets $p = 3$, or perhaps $\overline{L}$ is the Monster M and $p = 5$. Indeed these cases must be handled carefully, because locally they closely resemble the structure in groups of Lie type over $GF(2)$, which do appear as conclusions in Proposition 5.3.9. Let M be a maximal 2-local subgroup of G containing $N_G(R)$. Gorenstein and Lyons are able to eliminate $p = 5$, and reduce to the situation where $O_2(M)$ is extraspecial, $p = 3$, and G is of $GF(2)$-type. So instead of a contradiction, conclusion (1) holds.

The $Chev(p)$ and sporadic cases of Proposition 5.3.9, and the Klinger-Mason case handled in Proposition 5.3.4, are the only places where the proof of the Pretrichotomy Theorem 5.1.1 reaches conclusion (1): G is of $GF(2)$ type.

Concerning the remaining cases in the proof of Proposition 5.3.9, perhaps the most noteworthy aspect occurs in the case [**GL83**, 1.6.5] where $L/O_{p'}(L) \in Chev(q)$ with $q \neq p$ odd. Here the choice of p (see (2) of Corollary 5.2.11) can be exploited in the argument handling the major subcase where $m_{2,q}(G) > m_p(B) = m_{2,p}(G)$. In that case $(\Gamma_2^0)_q$ holds by hypothesis. This yields the existence of a 2-local subgroup of G covering $C_G(x)/O_{p'}(C_G(x))$, and leads to a contradiction to Proposition 5.3.5. A similar argument will appear shortly in the discussion of Proposition 5.3.10 below.

Showing that L is a component. Gorenstein and Lyons are finally able (recall our discussion of Proposition 5.3.5) to show that p-components in members of $\mathcal{S}^*_{max}$ are components. Thus their key result [**GL83**, 1.9.4] completes Step 4 in the language of Strategy 1.1.1:

PROPOSITION 5.3.10. *If* $(B, x, L) \in \mathcal{S}^*_{max}$, *then* L *is quasisimple.*

The result is a special case of their more general 1.9.2, which is also used later in the analysis of neighbors. The proof again uses the full strength of Theorem 5.2.2, i.e. our choice of the prime p, in the following way: We assume that L is not quasisimple. By Proposition 5.3.5, $[L, O_{p'}(L)]$ is of odd order, and so solvable. hence there exists an odd prime $q \neq p$ for which we get $C_L(O_q(L)) \leq O_{p',Z}(L) = O_{p',p}(L)$. We let Q_0 be a critical subgroup (cf. Lemma B.1.3) of $O_q(L)$, and set $Q_1 := \Omega_1(Q_0)$.

The following claim [**GL83**, 1.9.5]) will enable the proof to be reduced to a single possibility for $L/O_{p'}(L)$:

$$m_{2,q}(G) \leq m_{2,p}(G).$$

Suppose the claim is false; then by the choice of p, $(\Gamma_2^o)_q$ holds. So for a Sylow q-subgroup R of G containing Q_1, $\Gamma^0_{2,R}(G)$ is contained in some maximal 2-local subgroup M of G. Gorenstein and Lyons show there is a rank-3 subgroup $E \leq B$ with $L_{p'}(C_L(E)) \neq 1$. Taking a p-component J of $L_{p'}(C_L(E))$, they find as usual that there is a subgroup D of E of rank 2 with $Q_2 := [J, C_{Q_1}(D)] \neq 1$, and as in the best case of Proposition 5.3.10 they use the B_p-property in $\mathcal{K}$-groups (namely [**GL83**, 14-5]) to deduce that $Q_2 \leq O_{p'}(C_G(a))$ for all $a \in D^{\#}$. But they also show that $m_q(Q_2) \geq 3$, so that for $Q_2 \leq Q_3 \in Syl_q(O_{p'}(C_G(a)))$, they have Q_3 and hence $N_G(Q_3)$ in $\Gamma^0_{2,R}(G) \leq M$. Then $J \leq M$ and by a Frattini argument, M covers $C_G(a)/O_{p'}(C_G(a))$ for all $a \in D^{\#}$. Recall that M is a 2-local subgroup of the group G of characteristic 2 type, so that $F^*(M) = O_2(M)$, and in particular $[J, O_2(M)] \neq 1$. Since D has rank 2, by 0.3.11.2 there exists $a \in D^{\#}$ such that $S := [J, C_{O_2(M)}(a)] \neq 1$, and it follows from the coverage by M that we have $S = [S, J] \leq O_{p'}(C_G(a))$. This leads with some further argument to a contradiction to $(\Delta O)_p$.

Thus the claim holds. But now the claim together with the Thompson Dihedral Lemma B.1.7 yields: If S is an elementary abelian 2-subgroup of L of maximal rank, and S_1 is a complement to $C_S(Q_1)$ in S, then

$$\begin{aligned} m_p(B) &= m_{2,p}(G) \geq m_{2,q}(G) \geq m_2(S_1) - 1 \\ &\geq m_2(L) - m_2(O_{p'}(L)) - 1 \geq m_2(L) - m_2(M(L/O_{p'}(L))) - 1 \end{aligned}$$

where $M(L/O_{p'}(L))$ is the Schur multiplier of $L/O_{p'}(L)$. Now B normalizes a nontrivial 2-subgroup of $C_G(x)$ and so $B/\langle x \rangle$ must do the same in $Aut(L)$, with $m_p(B) \geq 4$. By Proposition 5.3.9 we may assume that $L/O_{p'}(L)$ is a group of Lie type in characteristic 2, where all 2-locals are 2-constrained. It is a bit surprising that there is any L satisfying all these conditions, but inspection of known groups (e.g. [**GL83**, I.10.7]) shows that there is exactly one possibility: $L \cong U_5(2)$, with $p = 3$, and $|B| = 3^4$. This final configuration is dispatched by yet another *ad hoc* argument.

Neighbors and standard type. The proof is now in the home stretch, having achieved rather tight control over the centralizer of the element $x \in B$: namely by Propositions 5.3.8–5.3.10, L is a quasisimple normal subgroup of $C_G(x)$, isomorphic to a group of Lie type in characteristic 2, and $C_G(L)$ has cyclic Sylow p-subgroups.

Thus L is a standard subgroup of G in the sense of Definition 2.3.2; and indeed we have condition (1) from the definition of standard type in Definition 5.1.4, as well as condition (4) since $x \in B_e$ from Proposition 5.3.4. So it remains to establish the other technical conditions in that definition, corresponding roughly to an extension of Step 5 in Strategy 1.1.1; these provide important information later used by Gilman and Griess in their identification of groups of standard type discussed in our Chapter 6.

Restricting the embedding of B in $C_G(x)$. Control over the embedding of B in $N_G(L)$ involves making a good choice of the triple (B, x, L), and in the case of the splitting condition (2) of Definition 5.1.4, even a good choice of p.

We recall from our choice of the prime p that our arguments apply to any prime q in the set:

$$\mu(G) := \{q : m_{2,q}(G) = m_{2,p}(G) \text{ and } (\Delta O)_q \text{ holds}\}.$$

At this point, we locate (and fix) some prime q—which is required to split L, and also to satisfy $m_{2,q}(G) \geq m_{2,p}(G)$. It might be that $q = p$.

In any case, such a prime q may even be chosen in $\mu(G)$. For otherwise the hypothesis of Theorem 5.3.1 implies that $(\Gamma_2^0)_q$ holds. Then a 2-local subgroup of G containing $\Gamma_{2,Q}^0(G)$ for a suitable $Q \in Syl_q(G)$ must contain $C_G(x)$; and this is shown to conflict with $(\Delta O)_p$, by an argument in the style of Proposition 5.3.5.

Replacing p by q, Gorenstein and Lyons restrict the embedding of B as required in conditions (2) and (3) of Definition 5.1.4, except for a few cases, for which standard type can be directly verified for some $(B, x, L) \in \mathcal{S}^*(G; p)$. The result is [**GL83**, Propositions 1.10.1 and 1.11.4]:

PROPOSITION 5.3.11. *One of the following conclusions holds:*
(1) There is a prime $p \in \mu(G)$ and a triple $(B, x, L) \in \mathcal{S}^*_{max}$ such that every element of B induces an inner-diagonal automorphism on $L/Z(L)$, and p splits L; or
(2) Theorem 5.3.1.2 holds, with $p = 3$, and G is of standard type with respect to some $(B, x, L) \in \mathcal{S}^*(G; 3)$ such that $L/Z(L) \cong U_5(2)$, $U_6(2)$, or $\Omega_8^+(2)$.

Properties of neighbors. The final stage of the proof obtains the remaining conditions (5) and (6) of Definition 5.1.4; it establishes the existence of neighbors, and analyzes the structure of the neighbors of a "good" triple (B, x, L)—chosen as in Proposition 5.3.11.1. The key results are 1.12.2, 1.12.3, and 1.12.7 in [**GL83**]:

PROPOSITION 5.3.12. *Suppose that $p \in \mu(G)$, $(B, x, L) \in \mathcal{S}^*_{max}$, every element of B induces an inner-diagonal automorphism on L, and p splits L. Assume that (B, x^*, L^*) is a neighbor of (B, x, L) in G. Then the following hold:*

(1) B normalizes L^*;
(2) L^* is a quasisimple group of Lie type defined over a field of characteristic 2;
(3) either p half-splits L^*, or x induces a nontrivial field automorphism on L^*;
(4) for any $d \in \langle x, x^* \rangle^\#$, $[L_{p'}(C_L(x^*)), O_{p'}(C_G(d))]$ has even order.

Moreover, if D is a subgroup of B of rank 2 and (D, K) is a standard subcomponent of (B, x, L) where $K := L_{p'}(C_L(D))$, then there exists $x^* \in D \setminus \langle x \rangle$ and a neighbor (B, x^*, L^*) of (B, x, L) in G with respect to (D, K).

It is not difficult to verify that B normalizes L^*, using the facts that B has rank $m_{2,p}(G)$ and $C_G(x^*)$ is a $\mathcal{K}$-group. However, the argument does use the technicality about avoiding $U_4(2)$ and A_6, which we had mentioned in our discussion of the definition of standard subcomponents (following the statement of 5.1.2). Then as L^* is a pumpup of the group K of characteristic 2, it is immediate (except in a few small cases) that L^* is also of Lie type in characteristic 2, while the quasisimplicity follows from 1.9.2 in [**GL83**], a more general version of Proposition 5.3.10. In proving the assertion (3) about half-splitting, it is first observed to be impossible that $m_{2,q}(L^*) > m_{2,p}(G)$ for some prime q, using $(\Gamma_2^0)_q$ as before to contradict $(\Delta O)_p$. Thus $e(L^*) \leq m_{2,p}(G) = m_p(B)$. But $C_B(L^*) = \langle x^* \rangle$. These two conditions on L^*, which are essentially an abstract formulation of "half-splitting", yield (3) by a close examination of groups of Lie type. They also explain why the singular case (4) is included in the definition of half-splitting.

Finally in Part (4) there exists, with a few exceptions, a subgroup $E \leq B$ such that $D \leq E \cong E_{p^3}$ and $L_{p'}(C_L(E)) \neq 1$, and one can proceed as in the proof of Proposition 5.3.5 to complete the proof of Proposition 5.3.12.

Thus the final key point is to establish the existence of neighbors. To achieve this, without the availability of an odd prime analogue of Glauberman's Z^*-Theorem, Gorenstein and Lyons must again exploit the fact that G is of characteristic 2 type. Thus, suppose that (D, K) is a standard subcomponent of (B, x, L), but there is no associated neighbor (B, x^*, L^*). Then for every $x^* \in D \setminus \langle x \rangle$, setting $K_{x^*} := \langle K^{L_{p'}(C_G(x^*))} \rangle$, we have that K covers $K_{x^*}/O_{p'}(K_{x^*})$. This configuration is reminiscent of the proof of Proposition 5.3.6 and specifically Remark 5.3.7; it leads in a similar manner to a contradiction to the fact that G is of characteristic 2 type. Namely at 1.12.6 (except in a few cases which are treated separately), there exists an elementary abelian p-subgroup $B^* \leq LB$ such that $B < B^*$ and we have $B^* = (B^* \cap K)N_{B^*}(U)$ for some nonidentity 2-subgroup U of K. Set $N := N_G(U)$. Since $F^*(N) = O_2(N)$, D acts faithfully on $O_2(N)$, and so (using 0.3.11.2 as before) there exists $x^* \in D \setminus \langle x \rangle$ such that $T := [C_{O_2(N)}(x^*), x] \neq 1$. On the other hand, at 1.12.4 the earlier signalizer analysis, using $(\Delta O)_p$, shows that $[K_{x^*}, O_{p'}(K_{x^*})]$ is of odd order. It then follows, using the fact that K covers its pumpups in $C_G(x^*)$ for all $x^* \in D \setminus \langle x \rangle$, that $[K, T] = 1$. But now we have $N_G(T) \geq KN_{B^*}(U) \geq B^* > B$, whence $m_{2,p}(G) \geq m_p(B^*) > m_p(B)$, contrary to the fact that $B \in \mathcal{B}_{max}(G; p)$.

Now with Proposition 5.3.12, Gorenstein and Lyons complete the verification of all the conditions for G to be of standard type with respect to (B, x, L); hence they also complete the proof of Theorem 5.3.1, and with it, the proof of the Pretrichotomy Theorem 5.1.1.

REMARK 5.3.13. Reviewing the steps of the proof of Theorem 5.3.1, Gorenstein and Lyons note that a stronger version of (2) has actually been proved (see [**GL83**, Section II.1.13]). For any $p \in \mu(G)$, let $\mathcal{T}(G; p)$ be the set of all those triples $(B, x, L) \in \mathcal{S}^*_{max}(G; p)$ such that p splits L and B induces inner-diagonal automorphisms on L. The strengthening is in the final clause as follows:

> Unless the special situation 5.3.11.2 of Proposition 5.3.11 holds, some $p \in \mu(G)$ has the property that $\mathcal{T}(G; p) \neq \emptyset$, and further every $p \in \mu(G)$ has the property that every element of $\mathcal{T}(G; p)$ is a standard triple.

The Trichotomy Theorem can therefore be strengthened in the same way. $\diamondsuit$

Is the $GF(2)$ type branch really necessary? After the CFSG, one may observe that the simple groups G of characteristic 2 type with $e(G) \geq 4$ are in fact *all of standard type*—and so already appear in conclusion (2) of the Pretrichotomy Theorem. In particular, many of the groups defined over $\mathbb{F}_2$ can be identified by the work of Gilman-Griess on standard type. Why then did Gorenstein and Lyons retain the groups of $GF(2)$ type as a conclusion in the Trichotomy Theorem? There were a number of considerations:

For one thing, a classification of groups of $GF(2)$ type (which we have stated as Theorem 7.0.1) was already available in the literature and had particular philosophical significance: Over half of the sporadic groups known at that time (and all 26 had by then been discovered) were of $GF(2)$ type, and in particular all the sporadics discovered late in the game shared that property. Moreover the CFSG had been reduced, modulo a few isolated results, to a classification of the groups of characteristic 2 type, so Theorem 7.0.1 provided strong evidence that the number of sporadics was finite, and the classification effort could succeed. Thus the lasting relevance of Theorem 7.0.1 was apparent.

A second motivation was that the argument in [**GL83**] seemed to be greatly simplified by shifting certain difficult cases into the $GF(2)$ type branch. As we mentioned earlier in the course of the chapter, those difficult cases arose at two points: in certain Klinger-Mason type arguments (during the proof of Proposition 5.3.4), as well as in the treatment of certain large sporadic components and $Chev(p)$ type components (during the proof of Proposition 5.3.9).

The difficulties with the Klinger-Mason arguments were less serious. We note first that, because they were using different hypotheses, Gorenstein and Lyons quoted certain arguments, rather than results, from Klinger and Mason [**KM75**]—in which $GF(2)$ type is a possible conclusion. However, in the situation of the Klinger-Mason Dichotomy B.9.1, groups of $GF(2)$ type do not actually arise when we have $e(G) \geq 4$. Finally, B.9.1 just shows that G is not of characteristic p type; so in order to produce the p-component required by Gorenstein and Lyons, some further work would remain (which is fairly routine—see e.g. Theorem 2.2.1).

By contrast, there are various larger sporadic groups G (such as the Monster) which have a smaller sporadic group L occurring as a p-component in the centralizer of a p-element. In some other sporadic groups, Chevalley groups L in (the same!) characteristic p occur in the same context. These groups G do not arise in the Gorenstein-Lyons theorem, since they do not have characteristic 2 type; indeed they have a 2-local with a component—but such 2-locals are not necessarily visible from the 2-locals which arise most naturally. For example they rarely contain a Sylow 2-subgroup of G. Hence the 2-local configurations of these groups G would loom as obstacles in the proof. Gorenstein and Lyons chose to keep their proof shorter, by routing this case into the $GF(2)$ type branch of their trichotomy.

(On the other hand, eight years later, they drafted an unpublished preprint of about 130 additional pages, indicating how to eliminate the sporadic configurations without using the theorem on groups of $GF(2)$ type. The preprint had additional motivations, to be sure: It uses the notions of "even type"—a weakening of characteristic 2 type which allows components of highly restricted isomorphism types, in centralizers of involutions such as those occurring in large sporadic groups. It also uses an analogous notion of "weak p-type" for an odd prime p. It argues that a minimal counterexample G to the CFSG cannot have both these properties if $m_{2,p}(G) \geq 4$, by showing that G would then be isomorphic to one of the larger sporadic groups. The second-generation GLS approach to the CFSG incorporates these ideas in the logic leading to the large sporadic groups.)

Comments on quoted and preliminary results. We have mentioned that Part I of [**GL83**], totaling over 400 pages, and comprising more than half of the total work, is devoted to preliminary results—most concerned with properties of known simple groups. We now give a sketchy description of some of that material. (Gorenstein and Lyons provide a briefer summary at the end of their General Introduction.)

In particular, Part I contains statements of most of the results which are quoted from the literature, so that typically the external references made in Part II can just be to the statement given in Part I. Of course, many of these references are also cited in other parts of the classification of groups of characteristic 2 type, and so have also been mentioned in our other chapters. But we emphasize that the bulk of Part I is devoted to developing properties beyond those quoted from references.

Part I is subdivided into seven chapters and an appendix, each of which we will now quickly summarize.

Part I, Chapter 1: Introduction. After introducing notation in Section 1, there is a review in Section 2 of the elementary theory of p-layers; much of this is quoted from Gorenstein and Walter [**GW75**] [**GW71**]. Sections 3 and 4 review the most basic properties of the known simple groups, particularly the groups of Lie type—primarily quoting Carter [**Car89**], Steinberg [**Ste68b**] [**Ste68a**], and Tits [**Tit63**]. Further developments (notably Lang's Theorem) are quoted from Springer-Steinberg [**SS70**], Seitz's generation work [**Sei82**] (cf. Theorem A.3.4), and the Bourbaki chapters [**Bou68**] on root systems. Section 5 collects material on sporadic groups: For certain groups there are references to unpublished tables of O'Nan (cf. material now appearing in [**GLS98**]), as well as to Conway's lectures [**Con71**]. For the individual groups, there are references to other standard sources in the literature; we will not repeat those here, as they are included in our Appendix Section A.3.

Part I, Chapter 2: Coverings and automorphisms of simple groups. Section 6 recalls the elementary theory of Schur multipliers, and quotes their structure for the simple groups—primarily from Griess [**Gri73, Gri74, Gri80**]. Section 7 develops properties of automorphisms of simple groups, based on the general references indicated for Sections 3–5 above, as well as some more special references such as [**GL75**].

Part I, Chapter 3: Local subgroups of simple groups. Sections 8–17 go well beyond the basics of subgroup structure given in Chapter 1. Much of the material is on groups of Lie type, and builds on the references indicated above, as well as other sources such as [**Die55**], [**Wei55**], [**BGL77**], [**BW77**], and [**BW76**].

The study of local subgroups includes a discussion of parabolic subgroups, and sections on Sylow structure, fusion, subcomponents and their pumpups, splitting primes in the sense of Definition 5.1.3, and standard components in the sense of Definition 2.3.2.

Part I, Chapter 4: Balance and signalizers in known groups. Section 18 recalls basic results on balance and signalizer functors, primarily from Gorenstein and Walter [**GW75**], but also referring to work on nonsolvable signalizer functors as in [**GL77**]. (We discuss signalizer functors and balance in our Section B.3.) Sections 19 and 20 explore the subgroups of the known groups, focusing on the structure of signalizers (p'-subgroups invariant under p-groups) and possible obstructions to balance conditions; the development refers to work such as Seitz generation [**Sei82**] (cf. Theorem A.3.4) and the Bender-Thompson Signalizer Lemma B.1.10. Finally Section 21 establishes an analogue, under suitable hypotheses, of completeness in the Signalizer Functor Theorem 0.3.15.1. There are references to material on balance, and to the standard papers [**Gol72b**] [**Gla76**] on the classical solvable signalizer functor theorem, as well as to McBride's work [**McB82a**] on the nonsolvable case. The section ends by introducing Aschbacher-Goldschmidt signalizer functors, an important generalization usable in the context of weak balance rather than balance itself. The references are [**Asc77a**, Sec 5], based on [**Gol72**].

Part I, Chapter 5: Generation of known groups. Section 22 develops some notation and general properties for generation, based on references mentioned for Chapter 1 and other material such as [**Gol75**]. Section 23 explores generation in known groups, particularly building on results in Seitz's work [**Sei82**] (cf. Theorem A.3.4), but also using material such as [**BGL77**] and [**O'N76a**]. Section 24 develops an aspect crucial for inductive arguments (used also in Aschbacher's $e(G) = 3$ work discussed in Chapter 4), namely the determination of strongly p-embedded subgroups in known groups, with some use of references mentioned earlier. The next two sections explore generation of pumpups and layers in known groups. Section 27 demonstrates roughly that known quasisimple groups are either well-generated, or satisfy good balance properties.

Part I, Chapter 6: Failure of factorization and pushing-up. Section 28 briefly reviews some of the literature on these topics: For failure of factorization, there is reference to Glauberman [**Gla73**] for the p-solvable case, as well as to McLaughlin's Theorem A.6.1 on groups generated by transvections, and to Aschbacher's general discussion in [**Asc82**]. For the theory of pushing-up, there are references to the work of Baumann [**Bau79**] and Glauberman-Niles (as mentioned in [**Gla77**]), from which a couple of other results are quoted as well, and to Campbell's thesis [**Cam79**]; and more generally to Aschbacher's Local $C(G,T)$ Theorem B.7.3.

It is worth noting that while the material in the other chapters of Part I has developed only marginally in the years since the publication of [**GL83**], the subject of its Chapter I.6 has undergone a rich development during that time. A powerful motivation, given special impetus by the paper [**Gol80**] and the subsequent development of the "amalgam method", has been to understand and even classify the finite simple groups of characteristic 2-type solely through the analysis of their 2-local subgroups. Indeed a strategy to accomplish this classification has been in the process of implementation for some years by—among others—U. Meierfrankenfeld, B. Stellmacher, and G. Stroth (see e.g. [**MSS03**] for an overview).

Part I, Chapter 7: Miscellaneous results. These sections are primarily devoted to proving or quoting an assortment of results, some about $\mathcal{K}$-groups and others in the general theory of finite groups. For example, the tables of involution centralizers in [**AS76a**] are quoted, as well as [**FW69**].

Some basic general results are quoted from standard texts and articles (generally classic ones) e.g. for Dickson's Theorem on subgroups of $SL_2(q)$, Theorem B of Hall-Higman [**HH56**], several basic elementary lemmas from the N-group paper of Thompson [**Tho68**], and a theorem of Konvisser on 3-groups [**Kon71**]

Part I, Appendix. Section 34 contains work of N. Burgoyne [**Bur83**]: this appendix tabulates the conjugacy classes and centralizers of inner-diagonal automorphisms of order 3 for the Lie type groups of characteristic 2. As has often been observed, in the analysis of a finite simple group—whether a minimal counterexample to the CFSG or a simple $\mathcal{K}$-group—it is the smallest primes that command the most detailed attention, together with any prime(s) p for which the group is of characteristic p type or nearly so. For example, the centralizers of p-elements, p odd, in groups of Lie type in characteristic 2 show the most exceptional behavior in the case $p = 3$. For these reasons Burgoyne's appendix is especially useful.

Quoted results in Part II. In this part a very small number of results not mentioned in Part I are quoted from the literature, beyond foundational results such as the Bender-Suzuki Theorem and the Z^*-Theorem: notably a transfer theorem of Glauberman [**Gla70**] used in the analysis of one special configuration in Chapter 2 of Part II.

CHAPTER 6

The classification of groups of standard type

In the classification of groups of characteristic 2 type, the next step along the branch of the proof passing through conclusion (II) of the Trichotomy Theorem 5.1.2 for $e(G) \geq 4$ leads to the problem of identifying the groups of *standard type*. In the classification of groups of characteristic 2 type with $e(G) = 3$, Aschbacher dealt with most of the analogous identification problems as the final step in his treatment of those groups.

These standard type problems are the analogues, for centralizers of elements of odd order, of the *standard form* problems for involution centralizers—which in turn arose in the classification of groups of GW type from the Standard Component Theorem 1.8.12. (In the language of Strategy 1.1.1, these are the final identification problems in Step 6.) Recall we saw at the end of Section 1.8 that for groups of *GW* Type, it was necessary to treat the standard form problem for *every* possible quasisimple group as the standard component—and we indicated in the two subsequent sections the very large number of papers in the literature that contain those corresponding treatments.

By contrast, in the problems involving centralizers of standard type in groups of characteristic 2 type, it is only necessary to treat components which are of Lie type in characteristic 2, from condition (1) of Definition 5.1.4. We will see in this chapter that the solutions of problems of this type are contained in just six papers: Gilman and Griess [**GG83**] gave a complete treatment of the standard type problems for $e(G) \geq 4$; and in the case of $e(G) = 3$, the related *odd standard form* problems were handled in Aschbacher [**Asc81b, Asc83a**] and in Finkelstein-Frohardt [**FF84, FF79, FF81b**].

6.1. The Gilman-Griess Theorem on standard type for $e(G) \geq 4$

Recall the definition of standard type that we gave at 5.1.4, with its substantial number of detailed properties obtained in the process of establishing conclusion (II) of the Trichotomy Theorem. That setup provides the basic hypothesis for this section. Without repeating all the important parts of the setup, we mention that if a triple (B, x, L) lies in $S^*(G;p)$, then $x \in G$ is an element of prime order p, $L \in Chev(2)$ is a component of $C_G(x)$, and x lies in the elementary abelian group B. Moreover, B acts nontrivially on some 2-subgroup of $C_G(x)$, and $m_p(B) = m_{2,p}(G) \geq 4$.

The principal theorem treating the Standard Type Case is the Gilman-Griess Standard Type Theorem, the main theorem in [**GG83**]:

THEOREM 6.1.1 (Standard Type Theorem). *Let G be a $\mathcal{K}$-proper finite simple group of characteristic 2 type, with $e(G) \geq 4$. Suppose that G is of standard type with respect to $(B, x, L) \in \mathcal{S}^*(G; p)$ for some odd prime p such that $m_{2,p}(G) \geq 4$.*

Then $G \in Chev(2)$.

The explicit list of groups in $Chev(2)$ that satisfy the hypothesis $e(G) \geq 4$ of Theorem 6.1.1 can be obtained from the list of groups of Lie type in section 0.4 by removing the groups in $Chev(2)$ which have arisen in the Quasithin Theorem 3.0.1 and in the $e(G) = 3$ Theorem 4.0.1. The result includes all groups over any field of order 2^m, $m \geq 1$, of certain types: L_n with $n \geq 8$, U_n with $n \geq 7$, Sp_{2n} with $2n \geq 10$, $\Omega_{2n}^{\pm}$ with $2n \geq 10$, E_6, ${}^2E_6$, E_7, and E_8. The rest of the list consists of all groups over any field of order 2^m, $m \geq 2$, of certain smaller types: namely the types $L_n (n = 5, 6, 7)$, U_6, Sp_8, Ω_8^+, and F_4. The condition $m \geq 2$ is necessary for these smaller types to satisfy the condition $e(G) \geq 4$.

Sketch of the strategy. The overall strategy of the proof is as follows. Fix some triple $(B, x, L) \in \mathcal{S}^*(G; p)$ with respect to which G is of standard type. Thus we have $L \in Chev(2)$. Expand B to B^*, maximal among elementary abelian p-subgroup of G with respect to inclusion. This B^* will play the role of the exponent-p subgroup of a split or semisplit torus of G_0 of maximal rank. Set $A = A_G(B^*)$, the *automizer* of B^*, i.e., the image of $N_G(B^*)$ in $Aut(B^*)$, which is isomorphic to $N_G(B^*)/C_G(B^*)$. The group A should play the role of the Weyl group associated to that torus. Then fix (D, J), a standard subcomponent of L with $x \in D \leq B$ and $D \cong E_{p^2}$, and consider all the "neighboring components" K coming from neighbors (B, z, K) of (B, x, L) with respect to (D, J). Such neighboring components K exist, and each one contributes its own Weyl group (of a torus containing $B^* \cap K$) as a subgroup of A. Now construct a chain of overgroups of L lying in $Chev(2)$, terminating in a group G_0. Namely, throw in one neighboring K at a time, quoting the propositions from Section 3 of [**GG83**] and the Gilman-Griess Presentation Theorem (see 6.1.3 below), to deduce the isomorphism types of the new Weyl group and the new overgroup of L, respectively. Deduce that $G_0 \in Chev(2)$, and that the Weyl group of the torus containing B^* is normal in A. Finally, develop p-local uniqueness properties of $M := N_G(G_0)$ and use them to verify the hypotheses of Holt's Theorem B.7.16, yielding $G = M$, and then $G = G_0$ by simplicity.

We add three remarks. First, it turns out that one or two neighboring components suffice. Second, it is not until simplicity is invoked and the proof is complete that one knows that A is generated by the contributions from L and its neighbors; if just $F^*(G) = G_0 < G$ were true, then A would not be so generated. Finally, it is a great convenience that Gilman and Griess can rule out the possibility that some element of B^* induces a nontrivial field automorphism on some neighboring component K.

Identifying a large Lie-type subgroup G_0. We now expand on the process of defining and identifying G_0. There are two essential ingredients: analyzing the "Weyl group" $A_G(B^*)$, and using the Gilman-Griess Presentation Theorem [**GG83**, Proposition 2.30].

Analyzing the Weyl group. Fix (B, x, L), (D, J), and B^* as just described. The possibilities for the standard component L are known—indeed $L/Z(L) \in Chev(2)$

by (1) of Definition 5.1.4. By inspection of the cases listed in 2.32 (Table B) of [**GG83**], B is of index 1 or p in B^*.

The preliminaries used in the identification of the "Weyl group" $A = A_G(B)$ are established in Sections 3 and 4.

Section 3 contains four major propositions, labeled by the types of Weyl groups being considered: CF, D, E, and A, where CF means "C or F." Each proposition treats the same basic set-up:

> F is a field of characteristic $p \neq 2$, and B^* is an $(n+1)$-dimensional F-vector space, for some $n \geq 3$. An F-basis for B^* is fixed: $\{b_0, b_1, \ldots, b_n\}$.

The purpose in each case is to identify a certain subgroup[1] W of $Aut_F(B^*)$. They set:

$$H := C_W(b_0) \text{ and } H^* := N_W(\langle b_0 \rangle),$$

and make some fairly precise assumptions about the structures of H and H^*; these assumptions are of the form:

> H is the Weyl group of a split or semisplit torus in a group of type T,

where the name of the proposition is Proposition T. (For Proposition CF, the assertion is that H is a Weyl group of type C *or* of type F_4.) Moreover, it is assumed that $H^* < W$. The conclusion in each case is of the form:

> Either $p > 0$ and $O_p(W) \neq 1$; or W is a Weyl group of a certain type; or $n = 3$ and we are in one of a few exceptional cases.

The proof of the propositions proceeds roughly as follows. A short argument shows if $O_{2'}(W) \neq 1$, then $O_p(W) \neq 1$, so we may assume that $O_{2'}(W) = 1$.

Suppose first that H is of type C or D, so that $O_2(H) \neq 1$. Then an eigenvalue argument determines the set of W-conjugates of some $y \in O_2(H)$ having either one or two eigenvalues -1 in its action on B^*. This shows, in general, that we have $N_W(O_2(H^*))/O_2(H^*) \cong S_{n+1}$. Finally, Goldschmidt's 2-Fusion Theorem [**Gol74**] establishes that $O_2(H^*)$ is normal in W, whence W is isomorphic to a Weyl group of type C_{n+1} or D_{n+1}.

Suppose that H is of type A or E. Here it is established that W contains a conjugacy class of odd transpositions in the sense of Fischer and Aschbacher. Then Aschbacher's classification [**Asc72**] of groups generated by odd transpositions is invoked to identify W as a symmetric group or as a Weyl group of type E_6, E_7, or E_8.

Quite a few small cases must be handled by specialized arguments.[2]

These results will be applied with $A_G(B^*)$ in the role of W in Section 5. But first, in Proposition 4.1, Gilman and Griess establish that $O_p(A_G(B^*)) = 1$. For if this fails, they are able to build a Sylow p-subgroup of G, and then to establish by a transfer argument using the Hall-Wielandt Theorem (see e.g. [**Hal76**, Theorem 14.4.2] or [**Har78**, Theorem 9]) that $[G, G] < G$, contradicting the simplicity of G. Then their easy Corollary 4.2 shows that x does not induce a field automorphism on L.

---

[1]We are altering two pieces of Gilman and Griess's notation here, for clarity. In Section 3 they write B instead of B^*, and K for the subgroup W of $Aut(B^*)$. We make the change to avoid confusion with our established notation.

[2]In particular, in some cases where B^* is small the proofs make use of the 2-Rank 2 Theorem and the Sectional 2-Rank 4 Theorem, as well as some other results from the literature.

The Gilman-Griess Presentation Theorem. The proof of this result [**GG83**, Proposition 2.30] is based directly on the fundamental Steinberg relations for a Lie type group (see [**GLS98**, 2.9.1, 2.9.2]). Below we state the version of their presentation given in [**GLS98**, 2.9.9]—where it is derived as a corollary of the Curtis-Tits Presentation Theorem A.5.1, which is in turn based on the Steinberg relations.

We will require the following concept:[3]

DEFINITION 6.1.2. Two groups $M = {}^d\Sigma(q)$ and $M_1 = {}^{d_1}\Sigma_1(q)$ of Lie type over the field $\mathbb{F}_q$ are said to be *compatible over the group* Y if and only if the following conditions hold:

(1) The BN-ranks of M and M_1 are at least 3 and 2, respectively;
(2) M and M_1 are either both twisted, or both untwisted;
(3) The root system $\tilde{\Sigma}_1$ for M_1 can be identified with a proper subsystem, $\hat{\Sigma}_1$, of the root system $\tilde{\Sigma}$ for K, and both are irreducible;
(4) The root subgroup $X_{\hat{\alpha}}$ of M for $\hat{\alpha} \in \hat{\Sigma}_1$ is isomorphic to the root subgroup $X_{\hat{\alpha}}$ of M_1; and
(5) Y is a subgroup of the Weyl group $\tilde{W}$ of $\tilde{\Sigma}$; and for every pair of roots $\tilde{\alpha}$ and $\tilde{\beta}$ from $\tilde{\Sigma}$, there is an element $w \in Y$ such that $w(\{\tilde{\alpha}, \tilde{\beta}\}) \subseteq \hat{\Sigma}_1$.

In this setup, under additional conditions, we can in effect identify M as being canonically generated by conjugates of root subgroups of M_1 under the Weyl group of M. These conjugates obey the appropriate relations by part (5) of the definition above:

THEOREM 6.1.3 (Gilman-Griess Presentation Theorem). *Assume $I = {}^d\Sigma(q)$ and $I_1 = {}^{d_1}\Sigma_1(q)$ be groups of Lie type over the field $\mathbb{F}_q$, compatible over a subgroup Y of the Weyl group $\tilde{W}$ of I. Let $\tilde{W}_1$ be the Weyl group of $\tilde{\Sigma}_1$, identified naturally with a subgroup of $\tilde{W}$. Suppose that R is a group and $\delta : R \to Y$ a surjective homomorphism. By this homomorphism, we regard R as acting on $\tilde{\Sigma}$. Let I_0 be a group containing I_1 and R. Suppose that the following conditions hold:*

(1) $\delta(R \cap I_1) = Y \cap \tilde{W}_1$;
(2) $R \cap I_1$ permutes $\{X_{\hat{\alpha}} : \hat{\alpha} \in \hat{\Sigma}_1\}$ in the same way that it permutes $\tilde{\Sigma}_1$ via δ;
(3) $R \cap I_1$ transitively permutes the roots of $\tilde{\Sigma}_1$ of a given length; and
(4) For each $\hat{\alpha} \in \hat{\Sigma}_1$, the stabilizer $R_{\hat{\alpha}}$ of $\hat{\alpha}$ in R normalizes $X_{\hat{\alpha}}$.

Let $I_1^* := \langle I_1^R \rangle$ be the subgroup of I_0 generated by all R-conjugates of I_1.

Then I_1^* is a quasisimple group of Lie type with $I_1^*/Z(I_1^*) \cong I/Z(I)$.

Defining and identifying G_0. Now let $K_2, \ldots, K_r$ be the components arising from the totality of neighbors of (B, x, L) with respect to the fixed standard subcomponent (D, J) of (B, x, L). Also set $K_1 := L$. Gilman and Griess define:

$$G_0 := \langle K_1 = L, K_2, \ldots, K_r \rangle.$$

and consider the chain $L = K_1 \leq \langle K_1, K_2 \rangle \leq \langle K_1, K_2, K_3 \rangle \leq \cdots \leq G_0$. Notation may be chosen so that if two consecutive terms in this chain are equal, then they equal G_0.

[3]We review notation and terminology for Lie type groups at Remark A.3.1 in the Appendix; for example the superscripts d and d_1 allow for the possibility that the groups may be twisted, as in [**GLS98**, 2.2.4].

Corollary 4.2 applies in fact not just to L but to all the K_i: no element of B^* induces a nontrivial field automorphism on any K_i, $1 \leq i \leq r$. Indeed, as noted in [**GG83**, Remark 5.1], it follows from Section 4 and [**GL83**] that B^* induces inner-diagonal automorphisms on each K_i.

Gilman and Griess now have considerable control over the automizer subgroups $A_i := A_{K_i}(B^*)$. They set:

$$A^* := \langle A_1, \ldots, A_r \rangle.$$

Thus A^* is expected to be $A_G(B^*)$ or at least a very large subgroup of $A_G(B^*)$.

From here, the identifications of A^* and G_0 proceed hand-in-hand, using the results of Sections 3 and 4, and the Gilman-Griess Presentation Theorem 6.1.3, respectively. In [**GG83**, Lemma 5.4], the key condition 6.1.3.4 is established in sufficient generality for the subsequent applications of Theorem 6.1.3.

Note that since p splits L by definition of standard type, it follows that A_1 may be identified with a very large normal subgroup of the group $H = C_K(b_0)$ from Section 3. In particular, as B^* induces inner-diagonal automorphisms on L, it follows, as established in Lemma 2.50, that H is isomorphic to one of the Weyl groups considered in Propositions CF, D, E, and A of Section 3. Moreover, the existence of a neighbor K_2 with $K_2 \not\leq N_G(\langle b_0 \rangle)$ guarantees that $H^* < K$. Since Proposition 4.1 shows that $O_p(A_G(B^*)) = 1$, Gilman and Griess are set to invoke the principal conclusions of the propositions from Section 3.

Specifically, Lemmas 5.6 through 5.12 in [**GG83**] consider the possibilities for pairs A_1,A_2, go on to determine the subgroup $A := \langle A_1, A_2 \rangle$ of $A_G(B^*)$, and then deduce the isomorphism type of $G_1 := \langle K_1, K_2 \rangle$ from Theorem 6.1.3. They then use Section 3 again to determine the possibilities for $\langle A_1, A_2, A_3 \rangle$. Typically, it turns out that this group equals $\langle A_1, A_2 \rangle$, and $G_0 = G_1$. The few cases in which $G_0 \neq G_1$ are identified and worked out in Lemma 5.15 and Proposition 5.16, which specify the structures of $\langle A_1, A_2, A_3 \rangle$ and $\langle G_1, K_3 \rangle$, and thereafter show that $G_0 = \langle G_1, K_3 \rangle = \langle K_1, K_2, K_3 \rangle$.

Finally, Lemma 5.18 verifies that $A^* \trianglelefteq A_G(B^*)$, and Corollary 5.20 records an important consequence for the final step to come:

$$N_G(B) \text{ and } N_G(B^*) \text{ normalize } G_0.$$

The final step: $G = G_0$. Section 6 of [**GG83**], completes the proof of Theorem 6.1.1 with:

THEOREM 6.1.4. $G = G_0$.

To prove Theorem 6.1.4, Gilman and Griess set $M := N_G(G_0)$ and devote the entire section to proving the following result.

PROPOSITION 6.1.5. *There exists an involution $r \in G_0$ such that the following conclusions hold:*

(1) $C_G(r) \leq M$;
(2) r is 2-central in G; and
(3) $r^M = r^G \cap M$.

Proposition 6.1.5 quickly implies Theorem 6.1.4: Conditions (1) and (3) of 6.1.5 say that M is the unique fixed point of r on G/M, so as r is 2-central in G by condition (2), the hypotheses of Holt's Theorem B.7.16 are satisfied. By that result, either $M = G$ or G is isomorphic to an alternating group or a Bender group. Moreover in the last two cases, $E(M)$ is an alternating group or is trivial. But $E(M)$

has the component $G_0 \in Chev(2)$. The small alternating groups that are in $Chev(2)$ do not arise because M has p-rank at least 4. Thus $G = M$, and then $G = G_0$ as G is simple.

In this section of [**GG83**], many of the basic steps depend on properties enjoyed by most but not necessarily all groups $G_0 \in Chev(2)$. Hence as in some of the previous sections, the proofs of many intermediate results require more elaborate arguments in a small number of exceptional situations.

The p-uniqueness properties of M. Gilman and Griess let r be any long-root involution in G_0—or even a short root involution if G_0 is of type C or F. In any case, r is chosen to be 2-central in G_0. Moreover $C_{G_0}(r)$ typically contains a G_0-conjugate of a hyperplane of B^*; $C_{G_0}(r)$ always contains a G_0-conjugate of a subgroup of B^* of co-rank 2 so $C_{G_0}(r)$ contains a noncyclic subgroup of B^*. The strategy for proving Proposition 6.1.5 is to establish p-uniqueness properties of B^* and M which, together with the fact that G has characteristic 2 type, yield the desired conclusions.

As mentioned above, the previous section gives the starting point: M contains $\langle N_G(B), N_G(B^*)\rangle$. The following notion is basic for the p-local analysis.

DEFINITION 6.1.6. Set

$$U(M) = U_G(M) := \{x \in M \ : \ \text{for all } g \in G,\ x^g \in M \iff g \in M\}. \qquad \diamondsuit$$

Observe that $x \in U(M)$ if and only if M is the unique fixed point of x on G/M, or, as in our discussion of (1) and (3) above, $x \in U(M)$ if and only if $x^G \cap M = x^M$ and $C_G(x) \leq M$. Gilman and Griess refer to this condition as "M controlling strong fusion of x in G"; moreover, for any subset $A \subseteq M$, they say that M controls strong fusion of A if and only if M controls strong fusion of each nonidentity element of A. However, as a different notion of "control of strong fusion" is in frequent use in the literature (e.g. [**AG67**, p 243]), we shall use the $U(M)$ terminology to avoid confusion.

REMARK 6.1.7. Trivially, if $x \in U(M)$ and $x \in A \leq M$, then $N_G(A) \leq M$. This argument may be iterated to show that if $x \in A \trianglelefteq\trianglelefteq B \leq G$, then $B \leq M$. In particular if $U(M)$ contains any p-element, then M contains a Sylow p-subgroup of G.

The arguments in Section 6 revolve about finding elements of $U(M)$ of order p and exploiting their existence via Remark 6.1.7. Most importantly, Gilman and Griess find an E such that:

$$E \cong E_{p^2},\ E \leq B^*, \text{ and } E^\# \subseteq U(M).$$

They first show that these conditions hold for $E = D$, where (D, J) is the standard subcomponent of the triple $(B, x, L) \in \mathcal{S}^*(G;p)$ used in the construction of G_0. The definition of G_0 quickly yields that $C_G(D) \leq M$ at 6.2. Using the fact that $F^*(C_G(t))$ is a 2-group for all involutions of J, Gilman and Griess establish at 6.9 that $C_G(d) \leq M$ for all $d \in D^\#$; they show also that $C_G(G_0) = 1$. Likewise at 6.5 they settle an issue that had been left open in [**GL83**], proving that if $d \in D^\#$ and $JO_{p'}(C_G(d)) \trianglelefteq\trianglelefteq C_G(d)$, then J is a component of $C_G(d)$. As for fusion of p-elements, they observe that B^* is weakly closed in a Sylow p-subgroup of G. Consequently $N_G(B^*)$ controls G-fusion in B^*. Moreover, B^* typically contains representatives of all G_0-conjugacy classes of elements of order p. The proof of these facts constitutes the main part of their proof at 6.11 that, with the possibility of three exceptional cases, any $b \in B^*$ such that $C_G(b) \leq M$ is in $U(M)$. At 6.13 they finally show that in all cases, $D^\# \subseteq U(M)$.

However, Gilman and Griess are looking for an E with additional properties with respect to involutions of G_0; most importantly, they want E to centralize some G_0-conjugate of r. Often D centralizes such a conjugate, and then $E = D$ is a good choice. But for certain choices of G_0, including the infinite families $U_8(q)$ and ${}^2E_6(q)$ with p dividing $q-1$, finding E requires more work. Gilman and Griess manage the task, and replace r by a suitable conjugate to obtain:

$$[E, r] = 1.$$

Completing the proof. Any E-invariant 2-subgroup of G is contained in the subgroup $\langle C_G(e) \mid e \in E^\# \rangle \leq M$, because $E^\# \subseteq U(M)$. So we see that M contains $R_t := O_2\big(C_G(t)\big) = F^*\big(C_G(t)\big)$ for any involution t with $t \in C_G(E) = C_M(E)$. On the other hand $E \cap G_0 \neq 1$, since B^* faithfully induces inner-diagonal automorphisms on G_0, and contains every element of order p in $C_G(B^*)$. Consequently:

$$1 \neq [E \cap G_0, R_t] \leq O_2(C_M(t)) \cap G_0 \leq O_2(C_{G_0}(t)).$$

This shows that t cannot centralize G_0, and similarly that t does not induce a field automorphism on G_0, for in those cases $O_2\big(C_{G_0}(t)\big)$ would be trivial. Together with basic facts about $Aut(G_0)$, similar arguments yield that M/G_0 has odd order [**GG83**, Lemma 6.20], reducing 6.1.5 to a proof that $r \in U(M)$.

To prove 6.1.5.1, write $C_{G_0}(r) = R_0 L_0$ where $R_0 = O_2(C_{G_0}(r)) = O_2(P_0)$ for a certain parabolic subgroup P_0 of G_0 whose co-rank in G_0 is usually 1, and always either 1 or 2. Here L_0 is a normal subgroup of a Levi subgroup of P_0 with cyclic factor group, and $J_0 := O^{2'}(P_0) \leq R_0 L_0$. Gilman and Griess compare the two series

$$R_0 \trianglelefteq R_0 J_0 \trianglelefteq C_M(r) \quad \text{and}$$
$$R_r \trianglelefteq R_r L_r \trianglelefteq C_G(r),$$

where $R_r = O_2\big(C_G(r)\big)$ and $L_r = L_{p'}\big(C_G(r)\big)$. The existence of E easily gives that $R_0 = R_r$. A series of $\mathcal{K}$-group lemmas [**GG83**, 3.16, 3.17, 3.22], which in turn use Seitz's generation results [**Sei82**] and a detailed examination of alternating and sporadic groups, establishes that in general, $R_0 J_0$ is a p-component of $C_G(r)$. Apart from the exceptional cases that arise, the fact that $E \cap J_0 \neq 1$ by construction then yields $C_G(r) \leq M$ by Remark 6.1.7, with $C_G(r)$ and $R_0 J_0$ in the roles of B and A.

The proof of Proposition 6.1.5.1 culminates in the discussion in and around Lemmas 6.18 and 6.19 of [**GG83**], by which point Gilman and Griess have found enough p-elements of $U(M)$ to be able to handle all the exceptional cases in their general argument. Finally they prove Proposition 6.1.5.3 at [**GG83**, Lemma 6.21]. Since $|M : G_0|$ is odd, what they must show is that $r^G \cap G_0 = r^M$. Suppose that $g \in G$ and $r^g \in G_0$. As $C_G(r) \leq M$, the group $C_M(r^g)$ is isomorphic to a subgroup of $C_M(r)$. This restricts the possibilities for r^g, and using the Aschbacher-Seitz calculations of centralizers of involutions in groups in $Chev(2)$ [**AS76a**], they are able to show that r^g centralizes some $x \in U(M)$ of order p. But then we have $x \in C_G(r^g) \leq M^g$, so $g \in M$, as required.

This completes the proofs of Proposition 6.1.5 and Theorem 6.1.1.

As we mentioned after Section 5.3, all groups of characteristic 2 type which satisfy $e(G) \geq 4$ in fact are of standard type. Hence all those groups are identified in the Standard Type Theorem 6.1.1.

Comments on quoted results. The introductory part of the paper appeals to the Gorenstein-Lyons Trichotomy work [**GL83**] for motivation and the definition of standard type, and to various sources such as [**Gor80a**] for notation and terminology. There is also mention, for context, of other work on standard form problems, such as that of Finkelstein and Frohardt [**FF84, FF79, FF81b**], along with some expository articles on such problems.

Of course the paper makes heavy appeals to the literature for detailed properties of groups of Lie type, and many such results are quoted in Section 2. The sources most frequently cited are Carter [**Car89**] and Steinberg [**Ste62, Ste63, Ste68b**]. In particular the latter two are quoted at [**GG83**, 2.27] for the Steinberg relations which underlie the Gilman-Griess Presentation Theorem 6.1.3. A number of the early lemmas in Section 2 reference the methods in a paper of Curtis-Kantor-Seitz [**CKS76**]. For properties of twisted groups there are some references to Burgoyne-Griess-Lyons [**BGL77**] and Burgoyne-Williamson [**BW77**]. For algebraic groups, the paper quotes Lang's Theorem (from [**BCC$^+$70**]) as 2.33, and later the Borel-Tits Theorem [**BT71**] (see also [**BW76**]). We had also mentioned the use of the Seitz Generation Theorem A.3.4 (see especially the results around 3.18).

There are references to properties of known groups, such as Schur multipliers (Griess [**Gri73**] and later papers, also [**Sch04**]), and cohomology (Pollatsek [**Pol71**] and D. Robinson [**Rob76**]). In support of tables for properties of sporadic groups there is specific mention of the papers of O'Nan [**O'N76b**] and Lyons [**Lyo72**] on their groups. Tables of involution centralizers in Lie type groups are quoted from Aschbacher and Seitz [**AS76a**].

The preliminary results in Section 3 of the paper make use of a number of standard sources from the literature on classification theorems and methods. We had mentioned the Hall-Wielandt transfer theorem earlier. Recognition theorems quoted (cf. Section A.5 etc) include McLaughlin's Theorem A.6.1, Aschbacher's work [**Asc72**] on odd transpositions, and Timmesfeld's work [**Tim75a**] on root involutions. Classifications quoted include the 2-Rank 2 Theorem 1.4.6 and some early works of Gorenstein and Harada [**GH71b, GH71a, Har75**], as well as a characterization of $Sp_6(2)$ by Yamaki [**Yam69**]. Use is made of Glauberman's Z^*-Theorem B.2.1 as well as a characteristic subgroup of a p-group from [**Gla71**] (although the well-known Glauberman-Thompson Normal p-Complement Theorem [**Gla68**] could also have been used instead of the latter); of Aschbacher's sufficient condition for a strongly embedded subgroup [**Asc73**]; and of Goldschmidt's strongly closed theorem [**Gol74**]. At 3.22 some intermediate results from the Trichotomy paper [**GL83**] are used.

6.2. Odd standard form problems for $e(G) = 3$ (Finkelstein-Frohardt)

In view of the Trichotomy Theorem 5.1.2, we see that the Gilman-Griess Standard Type Theorem 6.1.1 has treated all necessary standard type problems for finite simple groups G of characteristic 2 type with $e(G) \geq 4$. So given the Quasithin Theorem 3.0.1, it only remains to treat any such problems needed for the case $e(G) = 3$.

We recall from Section 2.4 and Chapter 4 that Aschbacher's treatment of groups G with $e(G) = 3$ in [**Asc81b, Asc83a**] involved a branch characterized by information related to the Gorenstein-Lyons notion of standard type, and that local data can be viewed as an odd standard form for G, in the sense of Remark 2.3.3. As we indicated in Chapter 4, Aschbacher himself handled most such problems; but in a few cases, he quoted treatments by Finkelstein and Frohardt in [**FF84, FF79, FF81b**]. In this section we will also discuss those papers.

We briefly recall Aschbacher's setup:

We saw in Section 4.1, in the case in [**Asc81b**] where $p > 3$, that Aschbacher treated the necessary standard form problems using a process similar to that of Gilman and Griess—in particular with an appeal to the Curtis-Tits Theorem A.5.1, and Phan's Theorem (see A.5.2). And we saw in Section 4.2 that a similar approach identified the groups in (23.1) of [**Asc83a**] when $p = 3$.

However the treatment of three particular standard 3-components, in (21.2) of [**Asc83a**], invoked the work of Finkelstein and Frohardt. Aschbacher had verified their hypotheses (given below) by that point: Recall that [**Asc83a**] treats the case where $\sigma = \{3\}$. In particular, we have $p = 3$, with $m_{2,q}(G) < 3$ for $q > 3$. Conditions (b) and (c) below hold, as we saw in 4.1.4—they arise from Aschbacher's definition of the set $\mathcal{S}$, corresponding to the case where the relevant signalizer functor is trivial. Condition (a) is a consequence of the assumption that G is a counterexample to (21.2) of [**Asc83a**], so that $W \not\leq H$. Finally the quasisimplicity of L is (21.3) of [**Asc83a**].[4]

THEOREM 6.2.1 (Finkelstein-Frohardt Theorem). *Let G be a $\mathcal{K}$-proper finite simple group of characteristic 2 type, with $e(G) = 3 > m_{2,p}(G)$ for all primes $p > 3$. Suppose that $b \in G$ is of order 3, and $C := C_G(b)$ has a quasisimple normal subgroup L with $L/Z(L) \cong L_4(2)$, $L_5(2)$, or $Sp_6(2)$. Suppose further that the following conditions hold:*

(a) $\langle b \rangle$ is not weakly closed in C with respect to G;

(b) If $B \in Syl_3(C)$ and $a \in B^\#$, then $[\theta(C_G(a)), B]$ is of odd order;

(c) If B_0 is a hyperplane of B, then $\bigcap_{b \in B_0^\#} \theta(C_G(b))$ is of odd order.

Then $Z(L) = 1$, so that L is simple; and furthermore:

(1) If $L \cong L_4(2)$, then $G \cong L_6(2)$ or $\Omega_8^-(2)$.

(2) If $L \cong L_5(2)$, then $G \cong L_7(2)$.

(3) If $L \cong Sp_6(2)$, then $G \cong Sp_8(2)$ or $F_4(2)$.

Actually the assumption that G is $\mathcal{K}$-proper is only used in treating the case in which $L/Z(L) \cong L_4(2)$, i.e. in [**FF84**]; the other two papers manage without this hypothesis.

Each of the three papers begins with some 3-fusion analysis, which refines the information about the structure of C.

When $L/Z(L) \cong L_4(2)$ or $L_5(2)$, in [**FF84, FF79**] Finkelstein and Frohardt analyze the centralizer of an involution t which is 2-central in L. They establish that $F^*(C_G(t))$ is an extraspecial 2-group, and determine enough further information about $C_G(t)$ to be able to identify G by invoking the $GF(2)$ Type Theorem 7.0.1. This theorem, to be discussed in the following chapter, had of course been established prior to the Finkelstein-Frohardt work. In [**FF79**], the authors in fact refer to a special case established earlier by Suzuki [**Suz69a**] and sufficient

[4]Quasisimplicity should technically have been stated instead just *before* (21.2), since it is required in order to apply Finkelstein-Frohardt there (and the proof of quasisimplicity is independent of that application).

for their purposes; likewise in [**FF84**], they refer to Dempwolff [**Dem73b**] and Smith [**Smi80b**].

When $L/Z(L) \cong Sp_6(2)$, in [**FF81b**] they proceed in a manner analogous to Gilman and Griess, as described in the previous section. In Proposition 3.1 of [**FF81b**], they determine that there are only two possibilities for the structure of $A_G(B^*)$, where B^* is an elementary abelian 3-subgroup of G of rank 4 with a hyperplane lying in C. They then identify a subgroup of G (an analogue of the subgroup G_0 of Gilman-Griess in the previous section) as either $Sp_8(2)$ or $F_4(2)$, using the Curtis-Tits Presentation Theorem A.5.1. As in Gilman-Griess, the final identification of G is accomplished via an appeal to Holt's Theorem.

Comments on quoted results. Naturally the Finkelstein-Frohardt work quotes much the same background literature as that quoted by Gilman-Griess. Since Finkelstein-Frohardt is a level 1 result in our viewpoint, we will not attempt a thorough analysis of that literature—but will instead confine ourselves to a few remarks on certain additional results beyond the background used by Gilman-Griess.

Probably the most significant addition is the above mentioned use of the $GF(2)$ Type Theorem 7.0.1, rather than the Curtis-Tits Theorem A.5.1 or the Steinberg relations, to identify certain groups.

A number of further classification theorems are quoted, such as Walter [**Wal69b**] on groups with abelian Sylow 2-subgroups, as well as Collins [**Col73**] on groups with Sylow 2-subgroups of type $L_3(q)$, and results of Fletcher-Stellmacher-Stewart in [**FSS77**], and D. Page's (unpublished) Ph.D. thesis [**Pag69**]. There are appeals to unpublished lecture notes of Graham Higman [**Hig68, Hig72**] including his well-known theorem that if A_5 acts on a 2-group T and there is some element x of A_5 of order 3 with $C_T(x) = 1$, then T is elementary abelian and the direct sum of irreducible natural $L_2(4)$-modules.

We note that the references to Page's work and Higman's manuscript [**Hig72**] are only used in the proof of a subsidiary result [**FF79**, 2.3], about groups of 3-rank 2, which in turn is only applied to a proper subgroup of G. The application helps to yield a conclusion that would be clear if the proper subgroup were known to be a $\mathcal{K}$-group. Thus, one can sidestep these two unpublished references by adding an assumption to the Finkelstein-Frohardt Theorem 6.2.1 that the group G is $\mathcal{K}$-proper.

Finally, some scattered intermediate results are also used, e.g. from [**FF81a**] and from Finkelstein and Solomon [**FS79b**]; an early result of Feit and Thompson [**FT62**]; a result of Prince on actions of A_5 and A_6 on 2-groups [**Pri77**]; and a transfer result of Yoshida in [**Yos78**].

CHAPTER 7

The classification of groups of $GF(2)$ type

In the classification of groups of characteristic 2 type, conclusion (I) of the Gorenstein-Lyons Trichotomy Theorem 5.1.2 requires the identification of groups with $e(G) \geq 4$ which are also of $GF(2)$ *type*. Such groups are known from the $GF(2)$ Type Theorem 7.0.1, which classifies *all* simple groups of $GF(2)$ type. We describe that work in this chapter, although historically work on groups of $GF(2)$ type had been completed a few years before the proof of the Trichotomy Theorem, essentially by 1978.

Indeed as we mentioned in Chapter 5, Gorenstein and Lyons took advantage of the availability of the $GF(2)$ Type Theorem primarily as a labor-saving device in their analysis: For all the groups of characteristic 2 type with $e(G) \geq 4$ in fact arise in the Standard Type Theorem 6.1.1, corresponding to conclusion (II) of the Trichotomy Theorem. So presumably with additional effort, conclusion (I) of the Trichotomy Theorem (and hence this use of the $GF(2)$ Type Theorem) might have been avoided.

However, the $GF(2)$ Type Theorem *was* part of the original CFSG. Moreover the theorem had a psychological influence on simple group theory during the 1970s: since so many of the sporadic groups being discovered at that time were of $GF(2)$ type, the completion of the $GF(2)$ type problem convinced many group theorists that the classification of *all* simple groups might in fact be within reach.

Since the $GF(2)$ Type Theorem preceded most of the work more directly involved in the classification of the groups of characteristic 2 type, it was naturally established for all finite groups, not just those of characteristic 2 type. Furthermore we emphasize that the work in this chapter does not require the K-group Hypothesis 0.5.4, which was also assumed in most of the work on groups of characteristic 2.

Finally we also mention some other expository articles on this subject: The paper [**Smi80**] is a version of a lecture given at the 1979 AMS Summer Symposium at Santa Cruz. That paper was in turn largely based on the exposition given in several of the original research papers: Timmesfeld's paper [**Tim78a**] reviews (starting at (3.4) of [**Tim78a**]) the history of the problem, up to his fundamental case division. Then Smith, in Section 1 of [**Smi80b**] as well as [**Smi79b**], details the eventual treatment of those cases, with the final result summarized at the end of [**Smi79b**].

In this chapter we will primarily follow [**Smi80**], with suitable additions: for example, in addition to stating the various level 1 results whose union provides the $GF(2)$ Type Theorem, we will also include at least some indication of other papers quoted during the proofs of those results.

Introduction

Recall in 2.3.7 we defined a group G to be of $GF(2)$ type if G possesses an involution centralizer whose generalized Fitting subgroup is a 2-group of symplectic type. The reader may wish to review the example given at (2.3.6), as well as the more extensive introduction given in Section 3 of [**Smi80**]—indicating that the condition arises naturally in the groups of Lie type of types A, D, and E (both ordinary and twisted) defined over the field $\mathbb{F}_2$ of exactly two elements.

Thus those groups provide the generic conclusion in the following theorem[1] classifying the groups of $GF(2)$ type; they appear as conclusion (1) of the theorem, and, as usual, many other groups also manage to arise:

THEOREM 7.0.1 ($GF(2)$ Type Theorem). *Let G be a finite simple group of $GF(2)$ type. Then G is one of*

(1) $L_n(2), n \geq 3$; $U_n(2), n \geq 4$; $\Omega^{\pm}_{2n}(2), n \geq 3$; ${}^3D_4(2)$, $E_6(2)$, ${}^2E_6(2)$, $E_7(2)$, $E_8(2)$;
(2) $L_2(q)$, $q = 2^m \pm 1, m \geq 3$; $L_3(3)$, $U_3(3)$, $L_4(3)$, $U_4(3)$, $G_2(3)$, $\Omega^+_8(3)$;
(3) A_6, A_8, A_9;
(4) One of the following 16 *sporadic groups:*

$$M_{11}, M_{12}, M_{24}, J_2, J_3, J_4, Co_2, Co_1, HS, Suz, He, Fi'_{24}, HN, Th, BM, M.$$

REMARK 7.0.2. We recall that some of the groups in conclusions (2)–(4) are not in fact of characteristic 2 type. Also a number of the smaller groups G in the list satisfy $e(G) \leq 2$, although the $GF(2)$ Type Theorem was not quoted during the proof of the Quasithin Theorem 3.0.1. An analogous statement holds for $e(G) = 3$—though in proving the $e(G) = 3$ Theorem 4.0.1, Aschbacher made use of work of Finkelstein and Frohardt, which (as we mentioned in Section 6.2) does quote the $GF(2)$ Type Theorem to identify the groups $L_6(2)$, $L_7(2)$ and $\Omega^-_8(2)$.

Recall the groups G in conclusion (I) of the Gorenstein-Lyons Trichotomy Theorem 5.1.2 are the groups of characteristic 2 type with $e(G) \geq 4$. Such groups form the sublist of conclusion (1) given by $L_n(2)$ for $n \geq 8$, $U_n(2)$ for $n \geq 7$, $\Omega^{\pm}_{2n}(2)$ for $n \geq 5$, ${}^2E_6(2)$, $E_6(2)$, $E_7(2)$, $E_8(2)$. But as we have mentioned, these groups also arise in the Standard Type Theorem 6.1.1. ◇

In the remainder of the chapter, we will state and discuss the results from the literature which constitute the $GF(2)$ Type Theorem.

We begin by establishing some notational conventions which are fairly standard in that literature:

NOTATION 7.0.3. We let G denote a simple group of $GF(2)$ type. Then by Definition 2.3.7, G contains:

z an involution of G ,
$M := C_G(z)$,
$Q := F^*(M)$,

where Q is a 2-group of symplectic type—see Definition A.1.11. We recall from that definition that Q is a central product

$$Q = E * R, \tag{7.0.4}$$

[1] We mention that Table 1 on p.279 of [**Smi79b**] inadvertently omits $G \cong L_3(3)$ in the row with $G \cong M_{11}$.

where E is either trivial or extraspecial; and R is either cyclic—or dihedral, semidihedral, or generalized quaternion of order at least 16. In particular, it follows that Q has a unique central involution (since this holds in the factors E and R). So as z is central in $C_G(z) = M$, and M contains Q, z must be that involution. As a consequence, (cf. [**Asc77**, 4.2]) z is 2-central in G, so that a Sylow 2-subgroup of M is in fact Sylow in G.

We also recall from Definition A.1.10 that when E is extraspecial, it is of order 2^{2w+1}, where w is called the *width* of E. ◊

We mention that groups of $GF(2)$ type often arise from the following situation: Suppose we assumed instead just that M is maximal in G, and that $Q := F^*(M)$ is of symplectic type. Then M normalizes Q, and hence centralizes the unique involution (say z) of $Z(Q)$. It follows that $M = N_G(Q) = N_G\big(Z(Q)\big) = C_G(z)$ by maximality.

In related terminology, Janko referred to a symplectic-type 2-group Q satisfying the condition $Q = F^*\big(\ N_G\big(Z(Q)\big)\ \big)$ as a *large* 2-subgroup of symplectic type. Indeed Janko and his students and collaborators were among the first to explore the properties of groups of $GF(2)$ type, after Thompson called attention to such groups in [**Tho68**].

We mention than many of the results in this literature do not assume that G is simple. Indeed such results typically include the case where z is weakly closed in M—where Glauberman's Z^*-Theorem B.2.1 forces $z \in Z^*(G)$. But we will simplify our exposition in this chapter by restricting to the case where G is simple, so the conclusion $z \in Z^*(G)$ in those results will not arise. In particular, our discussion can then assume that z is not weakly closed in M; that is:

(7.0.5) We may assume z has a distinct G-conjugate (denoted by a) in M.

This is where the analysis essentially begins in most of the papers.

7.1. Aschbacher's reduction of $GF(2)$ type to the large-extraspecial case

In the generic examples of groups of $GF(2)$ type, (the groups of Lie type over $\mathbb{F}_2$ in conclusion (1) of Theorem 7.0.1) we have $R = 1$ in (7.0.4)—so that $Q = E$ is in fact extraspecial. This is because (cf. [**Smi80**, p. 113]) Q is then a unipotent radical, generated by root groups taken from a set of roots which is "almost abelian"—in the sense that the only root in the set which can be a sum of roots in the set is the root corresponding to the central involution z.

Thus one might hope to reduce the general problem to the case Q extraspecial, by first handling any cases where $R \neq 1$ in (7.0.4). This was accomplished by Aschbacher in [**Asc76a**]; in particular he notes in his introduction that such a reduction had already been achieved in the $GF(2)$ type subcase of Thompson's work [**Tho68**] on N-groups; and in the work of Janko [**Jan72**] and F. Smith [**Smi75**] on groups with all 2-local subgroups solvable, where the reduction was provided by Lundgren [**Lun73**]. Below we state Aschbacher's result [**Asc76a**, Thm 1] in the case of G simple:

THEOREM 7.1.1 (Aschbacher). *Let G be a finite simple group of $GF(2)$ type, with Q as in Notation 7.0.3. Then one of the following holds:*

(1) $G \cong L_2(q), q = 2^m \pm 1, m \geq 3$.

(2) $G \cong M_{11}$, $L_3(3)$, or $U_3(3)$.

(3) $G \cong HS$.

(4) Q is extraspecial (and large), with width $w \geq 2$.

Since conclusions (1)–(3) above appear in conclusions (2) and (4) of the $GF(2)$ Type Theorem 7.0.1, we see that the problem is indeed now reduced to the case of a large extraspecial 2-subgroup Q. We now outline Aschbacher's proof.

We may assume that conclusion (4) fails. If Q is of symplectic type but not extraspecial, then from (7.0.4), either R is of order at least 4, or $R = 1$ and $Q = E$ is extraspecial of width 1, so Q is either dihedral or quaternion of order 8. In any case, Q contains some cyclic subgroup Z of order 4 containing z.

In Aschbacher's initial reductions, the possibilities arising in some non-generic cases are small, in the sense that G is shown to be of low 2-rank. First, the result 4.3 of [**Asc76a**] deals with the case of Z weakly closed in $N_G(Z) \leq M$: Here the preliminary result 2.5 shows that $\langle z^G \rangle$ (which for us is G, as G is simple) has dihedral or semidihedral Sylow 2-subgroups. So Aschbacher can quote the corresponding parts of the 2-Rank 2 Theorem 1.4.6, to conclude that G is $L_2(2^m \pm 1), m \geq 3$, where $Q = R$ is dihedral and equal to M; though he also has a non-simple case $Aut(L_3(3))$ with $Q = R$ semidihedral of order 16. Then a short argument at 4.4, using 4.3 and the preliminary result 2.6, reduces to Q of exponent 4. Next, 4.5 deals with the cases where either $Q = E$ is extraspecial of width 1, or else $Q = R$:[2] one quickly reduces to $Q \cong Q_8$ and $M \cong GL_2(3)$, so that G has semidihedral Sylow 2-subgroups; and again quoting part of 1.4.6, one gets $G \cong M_{11}$ or $L_3(3)$.

Determining $\overline{M}$ via transposition groups. So at 4.6, we are left with the "generic" case where Q is of symplectic type of exponent 4, but not extraspecial. Thus in (7.0.4), we have E extraspecial, and $R = Z(Q)$ cyclic of order 4. We then take R in the role of "Z". The analysis now focuses on the action of M on Q. We establish some more notation, which remains important throughout the chapter:

NOTATION 7.1.2. We continue Notation 7.0.3; in particular w is the width of E. Set

$$\tilde{Q} := Q/Z(Q) \text{ and } \overline{M} := M/Q.$$

Then the commutator map on Q defines a nondegenerate symplectic form on $\tilde{Q}$, and $\overline{M}$ is contained in the corresponding symplectic group $Sp(\tilde{Q})$. In our case, $Z(Q) \cong \mathbf{Z_4}$, but when Q is extraspecial, the power map on Q defines a quadratic form on $\tilde{Q}$, also preserved by $\overline{M}$, and then $\overline{M}$ is a subgroup of $O(\tilde{Q}) \cong O^{\epsilon}_{2w}(2)$. ◇

REMARK 7.1.3 (The approach via transposition-group properties). We briefly interrupt our discussion of Aschbacher's proof, to preview some features of his approach which became standard throughout much of the literature on groups of $GF(2)$ type. In rough summary:

---

[2]The case Q semidihedral is inadvertently omitted in the statement of 4.5 there, though covered by the proof.

The study of conjugates a of z in M as in (7.0.5) leads[3] to involutions $\bar{t}$ of $\overline{M}$ with restricted action on $\tilde{Q}$. And the $\overline{M}$-conjugacy class of $\bar{t}$ inherits, from the symplectic group $Sp_{2w}(2)$ or orthogonal group $O^{\epsilon}_{2w}(2)$ in 7.1.2, the property of being a set of odd transpositions or $\{3,4\}^{+}$-transpositions, in the sense of Section A.6. Then transposition-group theory can be exploited to try to determine the quotient $\overline{M}$. This can give the structure of the involution centralizer $M = C_G(z)$, which in turn should allow the identification of G. (And many of the groups in the conclusion of Theorem 7.0.1 can in fact be recognized via transposition-group properties, e.g. when z^G is a set of $\{3,4\}^{+}$-transpositions in G [**Tim73**].)

In transposition-group theory itself, after the original work of Fischer on 3-transpositions, Aschbacher and Timmesfeld had made many fundamental contributions. So in retrospect it is not surprising that they also established the transposition-group approach to groups of $GF(2)$ type. $\diamondsuit$

We now return to our discussion of Aschbacher's proof in [**Asc76a**]. By the result 4.3 indicated earlier, we may assume that Z is not weakly closed in M; so let t denote a generator of some distinct conjugate Z^g in M. Now note since $M = C_G(z)$ that *Z is in fact a TI-set in G*. Aschbacher exploits the fact that Z is a cyclic TI-set as the key technique for establishing strong restrictions on the action of $\bar{t}$ on $\tilde{Q}$. He first shows at 4.7 that $\langle t \rangle$ intersects Q in $\langle t^2 \rangle$ (and t^2 plays the role of "a" as in (7.0.5)). Hence $\bar{t}$ induces an involution on $\tilde{Q}$. Then at 4.9, he shows that the involution $\bar{t}$ induces either a transvection, or an element of Suzuki type c_2 in the language of [**AS76a**] (namely $[\tilde{Q}, \bar{t}]$ is of dimension 2, and is totally isotropic but not totally singular). As indicated in our preview in Remark 7.1.3, the $\overline{M}$-conjugacy class of such an involution (in either case) is an odd transposition, and Aschbacher's strategy is to use transposition-group theory to study the set $D := (Z^G \cap M) \setminus \{Z\}$ of all G-conjugates of Z in M; eventually he can identify the possibilities for the normal subgroup $\langle \overline{D} \rangle$ of $\overline{M}$ in various cases.

Identifying G from $\overline{M}$. First 4.16 deals with the subcase of 4.9 where some $\bar{t}$ induces a transvection. Here 4.14 had earlier shown that $\overline{D}$ is a set of 3-transpositions in $\overline{M}$. Further argument now show $\overline{M}$ is S_3, with $w = 1$. Consequently G has wreathed Sylow 2-subgroups; so again, part of Theorem 1.4.6 can be quoted, and the present hypotheses now yield $G \cong U_3(3)$.

Thereafter we may assume in 4.9 that all such $\bar{t}$ induce involutions of type c_2. Then 4.19 shows that $\overline{D}$ must be a set of $\{3,5\}$-transpositions in $\overline{M}$; indeed 4.18 gives a restricted list of possibilities for pairs from $\overline{D}$, with w limited to the small values 2, 3, and 5. In 4.23, a situation from 4.18.4 leads to a non-simple case with $F^*(G) \cong U_4(3)$, identified using Aschbacher's Classical Involution Theorem 1.7.5. In 4.26, another situation leads to $\overline{M} \cong S_5$ with $w = 2$; then G is identified as HS by the structure of the involution centralizer M, using Janko-Wong [**JW69**]. The only remaining situation now leads to a numerical contradiction after 4.27, and the proof is complete.

The results in Section 3 of [**Asc76a**] on 3-transpositions use Fischer's Theorem A.6.3; more general results there on odd transpositions use Aschbacher's

[3]To see how such $\bar{t}$ emerge in similar later situations, consider the "a" in (7.3.4): the intersection of the corresponding conjugate Q_a of Q with M determines the "$\bar{t}$" parametrized by the matrix entries denoted by T_i there; and these $\bar{t}$ are root involutions in the orthogonal group $\overline{M}$ there.

work [**Asc72**]. Some of the preliminary results in Section 2 make use of Aschbacher's earlier papers [**Asc75a, Asc76b, Asc75b**] involving the theory of tight embedding and groups of component type, as well as a result from Goldschmidt's product fusion paper [**Gol75**].

We mention that Fred Smith in [**Smi77a**] gave a somewhat more elementary proof of the simple subcase of Theorem 7.1.1. Smith focuses more directly on restricting the structure of Q; his main theorem essentially amounts to a sharpening of Aschbacher's initial reductions: His conclusion (3) for the case $Q = R$, much as in Aschbacher's 4.5, reduces to Q dihedral or semidihedral; and his conclusion (4) for Q symplectic but not extraspecial reduces to R cyclic of order 4, much as in Aschbacher's 4.6—but here with E of width $w \leq 2$ (the bound is obtained in Smith's lemma 3.5 and also after 3.8). In Section 4, Smith is able to deduce Aschbacher's result without needing much of Aschbacher's generic analysis of $\overline{M}$ on $\tilde{Q}$: When E is of width $w \leq 1$, Smith identifies G using the Sectional 2-Rank 4 Theorem 1.5.7. When $w = 2$, he identifies G as HS much as Aschbacher had, using Janko-Wong.

7.2. The treatment of some fundamental extraspecial cases

From here on, we may assume we are in conclusion (4) of Theorem 7.1.1; so G *has a large extraspecial subgroup* Q, of order 2^{2w+1} with width $w \geq 2$. (In particular we have $R = 1$ and $Q = E$ in (7.0.4); so we no longer reserve the letter "E" for that notation.)

It is now convenient, for the purposes of our exposition, to separate out the case $w = 2$, which originally appeared as the intermediate result 4.6.1 inside Aschbacher's more general work [**Asc77**] which we will consider in a moment. (For convenience we include also the case $w = 1$, which can be obtained as the appropriate sublist from conclusions (1)–(3) of Theorem 7.1.1.)

THEOREM 7.2.1. *Let G be a finite simple group with a large extraspecial 2-subgroup Q of width $w \leq 2$. Then G is one of*

$(w = 1)$ $L_2(7) \cong L_3(2), L_2(9) \cong A_6, M_{11}, L_3(3)$
$(w = 2)$ $U_4(2), L_4(3), U_4(3), G_2(3), L_4(2) \cong A_8, A_9.M_{12}, J_2, J_3$.

PROOF. Aschbacher notes as in 7.1.2 that $\overline{M}$ is a subgroup of the orthogonal group $O_4^{\pm}(2)$; and shows then that G is of sectional 2-rank at most 4. So he can use the Gorenstein-Harada Theorem 1.5.7 to determine those G satisfying the present hypothesis. □

So from here on we can assume that Q *has width* $w \geq 3$. (Although we sometimes point out where cases with $w = 1$ or 2 arose in the original papers.)

The case of z weakly closed in Q. We had observed at (7.0.5) that we may assume that z is not weakly closed in M. But z might still be weakly closed in $Q = O_2(M)$; and in fact the unitary groups give the main example of this important general subcase: For if G is unitary we can take z to be a transvection, analogous to the element "t" illustrated in (2.3.6); and now Q will afford the unitary subgroup of the group "U" there. Indeed if we choose our unitary basis with hyperbolic pairs in opposite positions 1,4 and 2,3, then any involution y other than z in Q will have nonzero entries which exhibit (hermitian) symmetry about the "anti-diagonal", because of the unitary restriction; and so will have commutator space of dimension (at least) 2, so that it cannot be G-conjugate to the transvection z—for example:

$$z = \left(\begin{array}{c|cc|c} 1 & & & \\ \hline & 1 & & \\ & & 1 & \\ \hline 1 & & & 1 \end{array}\right) \text{ and } y = \left(\begin{array}{c|cc|c} 1 & & & \\ \hline & 1 & & \\ 1 & & 1 & \\ \hline & 1 & & 1 \end{array}\right). \tag{7.2.2}$$

The situation with z weakly closed in Q was handled by Aschbacher in [**Asc77**]. His introduction mentions that a similar subcase had occured in Section 13 of Thompson's work [**Tho68**] on N-groups. Again we state Aschbacher's result only when G is simple and $w \geq 3$. Furthermore we state it in an extension due to Fred Smith in [**Smi77b**, 1.3], which removed Aschbacher's original assumption of a hypothesis related to the characteristic 2 type hypothesis:

THEOREM 7.2.3 (Aschbacher; F. Smith). *Let G be a finite simple group with a large extraspecial 2-subgroup Q, of width $w \geq 3$. Assume further that z is weakly closed in Q with respect to G.*

Then either G is the unitary group $U_{w+2}(2)$; or $w = 4$ and G is the Conway group Co_2.

The work continues the tranposition-group approach of Remark 7.1.3, which we saw in our discussion above of Aschbacher's earlier paper [**Asc76a**].

Indeed this time the only real preliminary reduction is 4.6, for the case $w = 2$—which we had already separated out above as Theorem 7.2.1; this identifies G as $U_4(2)$ or $L_4(3)$, under our hypothesis that z is weakly closed in Q. (For $w = 1$, add M_{11} and $L_3(3)$.) Thus the main argument in Section 6 assumes that $w \geq 3$.

Determining $\overline{M}$ via $\{3,4\}^+$-transpositions. The result 6.4 establishes the hypotheses of a special minimal situation: Namely as in (7.0.5), there is a distinct conjugate $a = z^g$; and here $a \in M \setminus Q$, as z is weakly closed in Q. Furthermore among subgroups H of G which are minimal subject to $Q \leq H \not\leq M$ and $O_2(H) \neq 1$, we can find an H of the form $\langle Q, Q^g \rangle$, with z^H of order 5.

At the start of Section 5, Aschbacher defines E as the set of such conjugates; and he studies the resulting configurations in Sections 4 and 5, under a more general large-extraspecial hypothesis which does not require z weakly closed in Q. In fact, the a of Section 6 also satisfies $az \notin z^G$; and he defines D as the subset of E with this further property. The result 5.1 then shows that the class z^H gives the 5 singular vectors of an orthogonal space V of dimension 4 and minus type, with H inducing the orthogonal group $\Omega_4^-(2) \cong A_5$. The structure of the intersection $Q \cap Q^g$ also plays a role in the analysis—a feature that will recur in the later literature. Furthermore $\overline{a}$ induces on $\tilde{Q}$ an involution of Suzuki type a_2 in the language of [**AS76a**] (namely $[\tilde{Q}, \overline{a}]$ is totally singular of dimension 2). Now the approach of 7.1.3 reappears: the conjugacy class of type a_2 in $\overline{M}$ is a set of $\{3,4\}^+$-transpositions in the sense of Timmesfeld [**Tim73**]; and Aschbacher then uses transposition-group methods to study the normal subgroup $\overline{L} := \langle \overline{E} \rangle$ of $\overline{M}$. The analysis culminates in the result 5.12, which determines a list of six possible cases for $\overline{L}$ (and in some of the cases, Aschbacher goes on to determine G).

Identifying G from $\overline{M}$. Section 6 then returns to the original hypothesis of z weakly closed in Q; this eliminates the cases (2), (4), and (5) in 5.12: For in the last two cases, the intersection of $D \subseteq z^G$ with $Q^g \cap M$ is larger than $a = z^g$, so that z^g is not weakly closed in Q^g; while in the first case, one checks directly that weak closure fails in the two groups $G \cong Suz$ and Co_1 (which are identified via the involution centralizer M using the work of Patterson-Wong [**PW76**], and Patterson's thesis [**Pat72**] which is now covered by the more modern [**Asc94**, 49.15]). On the other hand, conclusions (1) and (6) of 5.12 are the "small" groups $U_5(2)$ and Co_2 (with $w = 3$) in the conclusion of Theorem 7.2.3; they are identified via the centralizer M, using work of G. Thomas [**Thm70**] (see 5.11) and of F. Smith [**Smi74**].

So at 6.5, we have just the generic conclusion (3) of 5.12, where $\overline{L}/Z(\overline{L}) \cong U_w(2)$ with $w \geq 4$. In particular here $\overline{D}$ is a set of 3-transpositions in $\overline{M}$. Aschbacher is able to show then that z^G is a set of 3-transpositions in G, and uses Fischer's Theorem A.6.3 to conclude that $G \cong U_{w+2}(2)$, completing the proof.

The preliminary results in Section 3 of [**Asc77**] also use Fischer's Theorem for properties of 3-transposition groups; and they quote Timmesfeld's work [**Tim73**] on $\{3,4\}^+$-transpositions, and [**Tim75a**] on root involutions. There are also references to some fairly standard background results such as the Strongly Embedded Theorem 1.3.5, and to various fairly specific results in papers such as [**Asc75a**], [**Tim75b**], [**BGL77**], [**Fra51**], [**Smi76a**, Cor 1].

Removing the characteristic 2 type restriction. Aschbacher's argument at various points (e.g. 5.9) uses the hypothesis that $z \in O_2(C_G(b))$ for all $1 \neq b \in Q$. As z is 2-central in $C_G(b)$, this containment holds if we assume G to be of characteristic 2 type.

So we now briefly discuss Smith's result at 6.5 in [**Smi77b**]—which shows that the condition on b always holds, and so need not be assumed. Smith first establishes the condition at 6.3 in the special case of $b \in Q \cap Q^g$: For if it fails, Smith shows that $C_G(b)$ satisfies a generalized version of the large-extraspecial hypothesis, but with a strong restriction on conjugates of z; and in an independent result 1.3, he shows, using Timmesfeld's work [**Tim75a**] on root involutions, that this restriction leads to the double cover $2J_2$, whereas in the present case the center would be of order 4. Finally at 6.5, he applies the Z^*-Theorem in $C_G(b)$ in order to take "g" in that group; so that $b \in Q \cap Q^g$ and hence 6.3 applies, completing the proof. The references used by Smith are essentially a subset of those used by Aschbacher; he also uses a standard cohomology result of Gaschütz [**GLS96**, 9.26], and standard facts about automorphisms of groups of Lie type, as in [**GLS98**, 2.5.1].

The case of $\overline{M}$ reducible on $\tilde{Q}$. In view of Theorem 7.2.3, from now on (7.0.5) can be strengthened to:

(7.2.4) We may assume that there is some $a = z^g$ lying in $Q \setminus \langle z \rangle$.

Next, the linear groups provide the main example of another important subcase, in which the action of $\overline{M}$ on $\tilde{Q}$ is reducible: For note in (2.3.6) that Q has the subgroups

$$(7.2.5) \qquad A = \left(\begin{array}{c|cc|c} 1 & & & \\ \hline * & 1 & & \\ * & & 1 & \\ \hline * & & & 1 \end{array}\right) \text{ and } B = \left(\begin{array}{c|cc|c} 1 & & & \\ \hline & 1 & & \\ & & 1 & \\ \hline * & * & * & 1 \end{array}\right);$$

generated by transvections, and $\tilde{A}$ and $\tilde{B}$ give a decomposition of $\tilde{Q}$ under the action of $\overline{M} \cong L_2(2)$.

The reducible situation was handled by Dempwolff and S. K. Wong in Theorem B of [**DW78**]. As usual we restrict our statement to the case of G simple, and width $w \geq 3$.

THEOREM 7.2.6 (Dempwolff-Wong). *Let G be a finite simple group with a large extraspecial 2-subgroup Q of width $w \geq 3$. Assume further in Notation 7.1.2 that $\overline{M}$ acts reducibly on $\tilde{Q}$.*

Then either G is the linear group $L_{w+2}(2)$; or $w = 3$ and G is M_{24} or He.

Below we summarize the proof, with an indication of the main further results which are quoted along the way. (For readers interested in deeper dependency analysis, we will also intersperse brief parenthetical remarks about additional papers quoted in the level one references; of course the reader interested only in an outline of the proof can ignore such remarks.)

From our hypothesis that $\overline{M}$ is reducible on $\tilde{Q}$, there is some A normal in M with $\langle z \rangle < A < Q$. Suppose first that A is extraspecial. Then Q is the central product of A with $B := C_Q(A)$, which is also normal in M. In this situation, a result of Fred Smith [**Smi77b**, 1.2] shows that $w = 2$—contrary to our hypothesis that $w \geq 3$. (The actual hypotheses of Dempwolff and Wong in fact allow for $w = 2$, and so they treat that case. We mention also that the proof of Smith's result, in section 4 of [**Smi77b**], uses some of his other papers which we had mentioned earlier—notably [**Smi76a**, Thm 3], and a lemma 1.1 in [**Smi77b**] which is much like [**Smi77a**, 2.4].)

So we now assume instead that A is not extraspecial. Then $\langle z \rangle < Z(A)$, which is also normal in M; so we may replace A by $Z(A)$, to reduce to the case where A is abelian. Here [**DW78**, Thm A] shows that A is elementary of rank $w + 1$, and hence maximal abelian in Q. (We mention that the proof of Theorem A, in section 3 of [**DW78**], uses some work from their earlier paper [**DW77b**], as well as Smith's result [**Smi77b**, 1.2] indicated above; and in implementing the approach of Remark 7.1.3, they use Timmesfeld's root involution work [**Tim75a**] and Aschbacher's result Theorem 7.2.3 above.)

Now that A is maximal abelian in Q, an earlier "Theorem A" from [**DW77b**] applies: The possibility there that z is weakly closed in Q is eliminated by Aschbacher's Theorem 7.2.3—where one checks in each case in the list that M is instead irreducible on $\tilde{Q}$. The only remaining possibility is $\tilde{Q}$ of maximal Witt index, with $\overline{M} \cong L_w(2)$. (We mention that the proof of this earlier Theorem A, in section 3 of [**DW77b**], implements the approach of 7.1.3, using McLaughlin's Theorem A.6.1.)

Having obtained $\overline{M} \cong L_w(2)$, the authors quote an earlier "Theorem B" from [**DW77b**], to determine G generically as $L_{w+2}(2)$; although M_{24} and He also arise when $w = 3$.[4] This completes the deduction of Theorem 7.2.6. (We mention that the proof of this earlier Theorem B, in Section 4 of [**DW77b**], uses Theorem A from [**DW77b**], and yet another "Theorem B" from an earlier paper [**DW77a**]; along with [**Dem73a**] and "well known classification theorems"—including Theorem 7.2.1, Held's paper [**Hel69**], and [**Dem73b**] which characterizes $L_n(2)$ using [**Suz69a**].)

The case of $\overline{M}$ containing $\Omega(\tilde{Q})$. In contrast to the "small" case of $\overline{M}$ reducible, one can consider the case where $\overline{M}$ is nearly the orthogonal group $O^{\epsilon}_{2w}(\tilde{Q})$. Fred Smith in [**Smi77c**, 1.2] showed that the situation can arise only in a few cases of small width w. We now state his result, extended by the identification of the groups arising in those cases, using the characterizations of Patterson-Wong [**PW76**] and Patterson's thesis [**Pat72**] (now covered by [**Asc94**, 49.15]), already mentioned in our discussion of Aschbacher's proof of Theorem 7.2.3.

THEOREM 7.2.7 (F. Smith). *Let G be a finite simple group with a large extraspecial 2-subgroup Q of width $w \geq 3$. Assume further in Notation 7.1.2 that $\overline{M}$ contains $\Omega^{\epsilon}_{2w}(2)$. Then either*
(1) $w = 3$, $\epsilon = -$, and $G \cong Suz$; or
(2) $w = 4$, $\epsilon = +$, and $G \cong Co_1$.

In Smith's proof, preliminary arguments (including one quoted from [**PW76**]) eliminate the small-width case[5] with $w = 3$ and $\epsilon = -$; and show, when $w = 4$ and $\epsilon = +$, that z is not weakly closed in Q—via elementary arguments, since at the time of writing, Theorem 7.2.3 had not yet been published. Thus in Section 4, the real work remaining is to eliminate the cases $w > 4$, and $w = 4$ with $\epsilon = -$. The proof exploits the existence of an element $x \in M$ of order 3 with $[\tilde{Q}, \overline{x}]$ of rank 6 and minus type. At 4.2 it is shown that z can have no G-conjugate in $C_M(x)$; but (in part using the canonical forms for involutions acting on $\tilde{Q}$, much as in [**Smi77b**, 1.1]) the inclusion of $\Omega(\tilde{Q})$ in $\overline{M}$ quickly gives a contradiction.

The proof in Section 4 also uses [**Smi76a**, Thm 2] (essentially a transfer result) to conclude that $\overline{M}$ contains no orthogonal transvection, so that $\overline{M}$ is exactly $\Omega(\tilde{Q})$, rather than $O(\tilde{Q})$; this condition is in turn part of the hypothesis of the preliminary lemmas 2.2, 2.4, 2.6, and 2.7.

7.3. Timmesfeld's reduction to a list of possibilities for $\overline{M}$

We now turn to Timmesfeld's Main Theorem in [**Tim78a**],[6] which brought the large-extraspecial problem within range of its eventual completion.

As background, we mention that those groups appearing as conclusions in Theorem 7.0.1 which do *not* appear in the results discussed so far include one infinite family (namely the orthogonal groups over $\mathbb{F}_2$ in conclusion (1)) as well as certain cases of fairly large width w (including some exceptional groups of Lie type and sporadic groups). While Timmesfeld does not identify the group G in these cases,

---

[4] As small cases from Theorem 7.2.1, one gets A_6, A_9 when $w = 1, 2$, and M_{12} when $w = 2$, with A uniquely determined in Q. When $w \geq 3$, notice that M is uniquely determined by w.

[5] Also, using Theorem 7.2.1, one gets $L_3(2)$, A_6, M_{11}, and $L_3(3)$ when $w = 1$; and J_2 and J_3 when $w = 2$. (Smith also mentions M_{12}, but that group does not seem to satisfy the hypothesis here.)

[6] The short note [**Tim79b**] just fixes a minor problem in (3.9) of the original paper [**Tim78a**].

he does essentially determine the corresponding possibilities for $\overline{M}$. Using this information, others were soon able to complete the determination of G. His analysis of $\overline{M}$ continued (and indeed largely perfected) the transposition-group approach of Remark 7.1.3.

The statement below collects the cases handled so far into conclusion (2), and part of (1); thus the problem is reduced to a treatment of the remaining cases in (1), along with (3)–(8).

THEOREM 7.3.1 (Timmesfeld). *Let G be a finite simple group with a large extraspecial 2-subgroup Q. Adopt Notation 7.1.2, where w is the width of Q. Then one of the following holds:*

(1) $w \le 4$. (Using Theorem 7.2.1, we may in fact assume that $w = 3$ or 4.)

(2) $G \cong L_{w+2}(2)$ or $U_{w+2}(2)$.

(3) $w \ge 6$ is even and $\overline{M} \cong S_3 \times \Omega_w^{\pm}(2)$ or $S_3 \times O_w^{\pm}(2)$.

(4) $w = 6$, $E(\overline{M})$ is quasisimple, and $|O(\overline{M})| \le 3$. Furthermore $\overline{M}$ contains an involution $\bar{t}$ such that $C_{\overline{M}}(\bar{t})$ has a normal subgroup $\overline{R}$ which satisfies $\Phi(O_2(\overline{R})) \le \langle \bar{t} \rangle$ and $\overline{R}/O_2(\overline{R}) \cong L_3(2)$, $\Omega_4^+(2)$, or $\Omega_6^{\pm}(2)$, where $O_2(\overline{R}) = \langle \bar{t} \rangle$ in the last case.

(5) $w = 10$ and $\overline{M} = \overline{M}_0\langle \sigma \rangle$, where $\sigma^2 = 1$ and $\overline{M}_0 \cong L_6(2)$ or $U_6(2)$.

(6) $w = 11$ and $\overline{M} \cong Co_2$.

(7) $w = 12$ and $\overline{M} \cong Co_1$.

(8) $\overline{M}$ is simple and contains an involution $\bar{t}$ such that $\overline{N} := F^*\big(C_{\overline{M}}(\bar{t})\big)$ is extraspecial, and one of the following holds:
(i) $w = 16$, $\overline{N}$ is of width 8, and $C_{\overline{M}}(\bar{t})/\overline{N} \cong S_3 \times \Omega_8^+(2)$;
(ii) $w = 28$, $\overline{N}$ is of width 16, and $C_{\overline{M}}(\bar{t})/\overline{N} \cong \Omega_{12}^+(2)$.

As we will see in due course, conclusion (3) is the generic case, arising in the remaining infinite family $G \cong \Omega_{w+4}(2)$. Conclusions (5) and (8) arise in the larger exceptional groups $E_6(2)$, ${}^2E_6(2)$, $E_7(2)$, and $E_8(2)$. Conclusions (6) and (7) arise in the sporadic groups B and M, while (4) arises in J_4 and Fi'_{24} (beyond the orthogonal cases $\Omega_{10}^{\pm}(2)$ already appearing in (3)). And in conclusion (1), aside from the groups in (2) and in earlier results, we get $w = 4$ with $G \cong {}^3D_4(2)$, $\Omega_8^+(3)$, HN, and Th.

In the remainder of the section, we will outline Timmesfeld's proof. The overall logic sequence appears in Section 14 of [**Tim78a**]; while the more detailed treatment of the various subcases arising in that overall sequence is relegated to earlier sections.

First if z is weakly closed in Q, then the groups from Theorem 7.2.3 (and the relevant groups from 7.2.1) appear in conclusions (1) and (2)—notably $U_{w+2}(2)$ in the latter. So as in (7.2.4) we may assume that there is some $a = z^g \in Q \setminus \langle z \rangle$. We introduce:

NOTATION 7.3.2. Abbreviate Q^g by Q_a, and set $L := Q(Q_a \cap M)$.

Involutions in $\overline{L}$ of type a_2. Timmesfeld's first main reduction is 14.2, which deals with the cases where some element of $\overline{L}$ induces an involution of type a_2 on $\tilde{Q}$. Very roughly, most such examples should in fact arise in conclusion (2): For in $G \cong L_{w+2}(2)$, the element z is a transvection, and for a transvection t in the complement L in Example 0.3.2, the element $\bar{t} \in \overline{M}$ has $[\tilde{Q}, \bar{t}]$ of dimension 2, because of the decomposition in (7.2.5). Similarly in $U_{w+2}(2)$ an examination of a transvection z leads to $\bar{t} \in \overline{M}$ of type a_2 on $\tilde{Q}$. By contrast, in the remaining infinite family, the groups $\Omega^{\pm}_{w+4}(2)$ contain no transvections; rather the element z is of type a_2 in G, and in (7.2.2), the picture in (2.3.6) is replaced (for example in $G \cong \Omega_8^+(2)$) by:

(7.3.3)

$$z = \left(\begin{array}{cc|cccc|cc} 1 & & & & & & & \\ & 1 & & & & & & \\ \hline & & 1 & & & & & \\ & & & 1 & & & & \\ & & & & 1 & & & \\ & & & & & 1 & & \\ \hline 1 & & & & & & 1 & \\ & 1 & & & & & & 1 \end{array}\right) \quad \text{and} \quad Q = \left(\begin{array}{cc|cccc|cc} 1 & & & & & & & \\ & 1 & & & & & & \\ \hline * & * & 1 & & & & & \\ * & * & & 1 & & & & \\ * & * & & & 1 & & & \\ * & * & & & & 1 & & \\ \hline * & * & * & * & * & * & 1 & \\ * & * & * & * & * & * & & 1 \end{array}\right).$$

(with suitable anti-diagonal symmetry restrictions on the entries $*$). Here the doubling of columns and rows in Q implies that a conjugate t, which is of type a_2 in G, will in fact have $[\tilde{Q}, \bar{t}]$ of dimension 4. So we might expect to find that there are no more examples in which t is of type a_2 on $\tilde{Q}$.

Timmesfeld's proof of 14.2 is actually carried out in Section 12 of [**Tim78a**] (and eventually summarized at 12.16). First, the $\overline{M}$-class of any element $\bar{t} \in \overline{L}$ of type a_2 on $\tilde{Q}$ is a set of $\{3,4\}^+$-transpositions in $\overline{M}$ (where 3-transpositions might occur as a degenerate case); this sets up the approach of 7.1.3. Timmesfeld now obtains some strong restrictions: At 12.4, he shows that $N_{\overline{M}}(\overline{L}) = C_{\overline{M}}(\tilde{a})$, and that $C_{\overline{M}}(\overline{L})$ is a 2-group. Then in 12.5, he shows that all such $\bar{t}$ are conjugate in $\overline{M}$. An earlier lemma 12.3 makes a more general analysis of these properties, using Fischer's Theorem and Timmesfeld's work [**Tim73**] on $\{3,4\}^+$-transpositions, obtaining a list of 6 possibilities—which by 12.8 can be applied to describe the subgroup $\overline{X}$ of $\overline{M}$ generated by the class of $\bar{t}$. Cases (6), (4), and (3) of 12.3 are then successively eliminated; the Dempwolff-Wong Theorem 7.2.6 is used at several points to obtain a contradiction (though not yet to identify groups in the conclusion); it is applied in the form of Timmesfeld's Lemma 3.7, which assumes irreducible action of $\overline{M}$ on $\tilde{Q}$. Cases (5) and (2) do lead to some groups of small width; these appear implicitly under $w \leq 4$ in conclusion (1) of 7.3.1, though Timmesfeld describes them somewhat more explicitly. Namely, two cases, with $\overline{M} \cong \Omega_6^-(2)$ or $\Omega_8^+(2)$ and $w = 3$ or 4, appear in Fred Smith's Theorem 7.2.7. In the first, Timmesfeld quotes Patterson-Wong [**PW76**] to identify $G \cong Suz$—but in the second, he stops at the description of $\overline{M}$ (though we saw in 7.2.7 that one can show that $G \cong Co_1$ using Patterson's thesis [**Pat72**], now covered by [**Asc94**, 49.15]). Some further cases with $w = 3$ and $\overline{M} \cong L_3(2)$ are handled implicitly via the assumption of irreducibility in 3.7; more explicitly one could get $G \cong L_5(2)$, M_{24} or He using the Dempwolff-Wong Theorem 7.2.6. Finally at 12.15, there remains only conclusion (1) of 12.3; this is reduced to $\overline{M} \cong L_w(2)$ (using the treatment of the case $w = 3$ to get $w \geq 4$)—and again the Dempwolff-Wong Theorem 7.2.6 in the form of 3.7 identifies $G \cong L_{w+2}(2)$, which of course appears in conclusion (2) of 7.3.1. This completes the proof of Timmesfeld's main reduction in 14.2.

Determining the possibilities for the subgroup $\overline{R}$ of $C_{\overline{M}}(\bar{t})$. As a result of 14.2 in [**Tim78a**]:

we may assume that no element of $\overline{L}$ induces an involution of type a_2 on $\tilde{Q}$. And indeed the bulk of the argument in [**Tim78a**], starting for example with results like 3.9, has already proceeded under this hypothesis. (Usually it is also assumed that $w \geq 3$; as we may since conclusion (1) of 7.3.1 covers all cases with $w \leq 4$.)

The absence of a_2 involutions quickly produces very strong information: Namely the result 3.11 shows that $Q \cap Q_a$ is a maximal abelian subgroup of Q; and also that $C_{\tilde{Q}}(\overline{L}) = \langle \tilde{a} \rangle$ (only $[\overline{L}, \tilde{a}] = 1$ is immediate from the definition in 7.3.2); and that $\overline{L}$ is elementary of rank $w - 1$ and weakly closed in $N_{\overline{M}}(\overline{L})$. The arguments use standard results on root involutions, and at one point Timmesfeld's earlier work [**Tim75b**]. Then at 3.13, it follows from the fact that $\overline{L}$ is weakly closed that M controls fusion in $Q \setminus \langle z \rangle$.

These results supply the basis for the further analysis. It may be illuminating to examine them in the context of the example from (7.3.3). The matrices below exhibit a choice of a and the corresponding form of $Q_a \cap M$:

(7.3.4)

$$\left(\begin{array}{cc|cccc|cc} 1 & & & & & & & \\ & 1 & & & & & & \\ \hline & & 1 & & & & & \\ & & & 1 & & & & \\ & & & & 1 & & & \\ 1 & & & & & 1 & & \\ \hline & & & & & & 1 & \\ & & 1 & & & & & 1 \end{array}\right), \quad \left(\begin{array}{cc|cccc|cc} 1 & & & & & & & \\ T_3 & 1 & & & & & & \\ \hline A_1 & & 1 & & & & & \\ A_2 & & T_2 & 1 & & & & \\ A_3 & & T_1 & & 1 & & & \\ A_4 & & & T_1 & T_2 & 1 & & \\ \hline A_5 & & & & & & 1 & \\ & A_5 & A_4 & A_3 & A_2 & A_1 & T_3 & 1 \end{array}\right);$$

the entries are 0 or 1, and symmetric about the anti-diagonal. Further $Q \cap Q_a$ is of rank 5 and consists of the matrices with $T_i = 0$ for $1 \leq i \leq 3$. Similarly $\overline{L}$ is of rank 3, consisting of the matrices with $T_i \in \{0, 1\}$. Finally taking $T_1 = 1$ and $T_2 = T_3 = 0$ gives a particular choice of $\bar{t} \in \overline{L}$.

Notice in particular that the 4-subgroup of $\overline{L}$ consisting of the matrices in which $T_3 = 0$ is contained in a subgroup $\Omega_4^+(2)$ of $\overline{M}$, determined by the middle four rows and columns; and $\bar{t}$ lies in other conjugates of this 4-group in $\Omega_4^+(2)$, so that $\overline{L}$ is not a TI-set in $\overline{M}$. This behavior is in fact typical.

The next major result is 14.3, which handles certain small cases where $\overline{L}$ is a TI-set in $\overline{M}$; the lemma depends upon work carried out earlier in 4.2. Here the possibilities for the normal closure of $\overline{L}$ in $\overline{M}$ can be determined, using Timmesfeld's work in [**Tim75b**] on weakly closed TI-sets. Most of them are then eliminated—indeed in most of the cases $\overline{M}$ is reducible on $\tilde{Q}$, contradicting 3.7. (More explicitly, we saw that the groups in the Dempwolff-Wong Theorem 7.2.6 arose instead in the earlier analysis, where there are involutions of type a_2.) The remaining configurations have $\overline{M} \cong A_9$ or $L_2(8)$, with $w = 4$ so that they appear in conclusion (1) of 7.3.1. (The corresponding groups $G \cong Th$ or ${}^3D_4(2)$ are identified later in Theorem 7.4.3).

Thus after 14.3, *we may assume that $\overline{L}$ is not a TI-set in $\overline{M}$.* Then at 4.4, Timmesfeld defines the following notation (used in conclusion (4) of 7.3.1):

NOTATION 7.3.5. Fix an involution $\bar{t}$ of $\overline{L}$. Set

$$\overline{R} := \langle \overline{L}^h : h \in M, \bar{t} \in \overline{L}^h \rangle, \quad \tilde{V} := \langle \tilde{a}^h : h \in M, \bar{t} \in \overline{L}^h \rangle, \quad \text{and} \quad \overline{N} := C_{\overline{R}}(\tilde{V}).$$

Note that $C_{\overline{M}}(\bar{t})$ permutes the conjugates of $\overline{L}$ defining $\overline{R}$, and so normalizes both $\overline{R}$ and $\tilde{V}$, and hence also $\overline{N}$. ◇

For example for the $\bar{t}$ in (7.3.4), we have $\overline{N} = \langle \bar{t} \rangle$ and $\overline{R}/\overline{N} \cong \Omega_4^+(2)$.

For certain $\bar{t}$, we may have $\overline{R} = \overline{L}$, but for at least *some* $\bar{t}$, we will have $\overline{L} < \overline{R}$—so we focus on such $\bar{t}$. Timmesfeld's main strategy can now be summarized as follows: Determine the action of $\overline{R}/\overline{N}$ on $\tilde{V}$, and then the structure of the normal subgroup $\overline{R}$ of $C_{\overline{M}}(\bar{t})$. Finally use characterizations by involution centralizers to determine $\overline{M}$ in Theorem 7.3.1. (Sometimes a smaller case of the Theorem itself will be used inductively to provide such a characterization.)

The fundamental first step in the program is provided by Timmesfeld's 4.5.2, which states (now that $\overline{L}$ is not a TI-set) that:

THEOREM 7.3.6. *Suppose that $\overline{L} < \overline{R}$. Then for some m, $\overline{R}/\overline{N}$ is one of $L_m(2)$, $Sp_{2m}(2)$ (including $Sp_4(2)' \cong A_6$), or $\Omega_{2m}^{\pm}(2)$. Further $\tilde{V}$ is the natural module for $\overline{R}/\overline{N}$. (Indeed the proof also shows that $\overline{LN}/\overline{N}$ is the unipotent radical $O_2\big(C_{\overline{R}/\overline{N}}(\tilde{a})\big)$.)*

The proof proceeds by showing that suitable elements of $\overline{L}$ act on $\tilde{V}$ as either transvections or involutions of type a_2, and then uses McLaughlin's Theorem A.6.1 and Timmesfeld's work [**Tim73**] on $\{3,4\}^+$-transpositions to identify $\overline{R}/\overline{N}$.

Treatment of the cases for $\overline{R}/\overline{N}$. The cases indicated in Theorem 7.3.6 are handled in detail (after some preliminary work in Sections 5 and 6) in the course of Sections 7–11 and 13 of [**Tim78a**]. But recall that these more detailed earlier sections are quoted from the overall logic sequence in Section 14.

The case $\overline{N} = \langle \bar{t} \rangle$. The first reduction is 14.4, where the case $\overline{N} = \langle \bar{t} \rangle$ is handled using earlier results summarized at 7.6. Conclusions arising during this step include some smaller-width cases, but also the remaining infinite family, namely the generic case corresponding to the orthogonal groups $\Omega_{w+4}^{\pm}(2)$. We outline the argument.

First, 6.4 and 7.1 reduce the list of Theorem 7.3.6, in the present situation, to $\overline{R}/\overline{N} \cong \Omega_{2m}^{\pm}(2)$ or A_6. The case of A_6 is shown at 7.5 to lead to $w = 4$; thus the corresponding conclusion (2) of 7.6 is included in conclusion (1) of 7.3.1. (In fact in this case $\langle \overline{L}^{\overline{M}} \rangle \cong S_3 \times A_6$; and later Theorem 7.4.3, more specifically [**Smi79b**, 3.2], would show that no simple group arises here.)

Thus we may assume that $\overline{R}/\overline{N} \cong \Omega_{2m}^{\pm}(2)$. At 7.2 it is observed when $m = 2$ that we have $w = 4$; so again conclusion (1) of 7.6 is included in conclusion (1) of 7.3.1. (Later Theorem 7.4.3 identifies G as $\Omega_8^{\pm}(2)$, $\Omega_8^+(3)$, or HN.)

Now we may assume that $m \geq 3$. From $\overline{LN} = O_2\big(C_{\overline{R}/\overline{N}}(\tilde{a})\big)$ one gets $2m = w$. Thus the case $m = 3$ in 7.3.1 gives $w = 6$; and the corresponding conclusion (3) of 7.6 is essentially conclusion (4) of 7.3.1—though here Timmesfeld is implicitly also using 4.7 and 5.5.1 to verify certain details of that conclusion. (Later Theorem 7.4.3 identifies the groups as $\Omega_{12}^{\pm}(2)$ or Fi_{24}'.)

So finally we may assume that $m \geq 4$ (and hence $w \geq 8$). Then 7.3.2 and 7.4 describe the generic case, showing that $\bar{t} \in Z^*(\overline{M})$ and $O(\overline{M}) \neq 1$. These facts can be seen to hold in the example of orthogonal groups by extending (7.3.4) to higher dimensions. The corresponding conclusion (4) of 7.6, as extended in Corollary 7.7, shows that $\overline{M}$ is the direct product of S_3 with $\Omega_w^\pm(2)$ or $O_w^\pm(2)$, as in conclusion (3) of 7.3.1. (Later Theorem 7.4.1 identifies the groups here with $w \geq 4$ as in fact being the orthogonal examples $G \cong \Omega_{w+4}^\pm(2)$.)

The case $\overline{L} \cap \overline{N} = \langle \overline{t} \rangle$. In fact Timmesfeld states his 14.4 as covering not just the case $\overline{N} = \langle \overline{t} \rangle$, but even the case $\overline{L} \cap \overline{N} = \langle \overline{t} \rangle$. This follows using his Lemma 6.5 when $\overline{R}/\overline{N}$ is symplectic or orthogonal (and not $\Omega_4^+(2)$); while when $\overline{R}/\overline{N} \cong L_m(2)$ or $\Omega_4^+(2)$, he implicitly observes (cf. [**Smi79b**, 1.12]) that the proofs of 7.1 and 7.2 in fact establish this stronger statement.

Reducing to $\overline{R}/\overline{N}$ *symplectic or orthogonal.* Since the previous reduction in 14.4 included the remaining infinite family given by the orthogonal groups, only a finite number of further configurations will now emerge as conclusions of Theorem 7.3.1.

At 14.5, Timmesfeld reduces to $\overline{R}/\overline{N}$ symplectic or orthogonal (but not $\Omega_4^+(2)$), by treating the cases $\overline{R}/\overline{N}$ linear or $\Omega_4^+(2)$. In fact these cases lead to only a few possibilities in the conclusion of Theorem 7.3.1; and below we outline the argument.

The case $\overline{R}/\overline{N} \cong \Omega_4^+(2)$ is treated (and indeed essentially eliminated) in a number of results culminating in 13.9. The analysis begins in Section 10, and shows at 10.8 that either $w \leq 6$ or $O_3(\overline{M}) \neq 1$. The case $w = 5$ is eliminated at 10.9. Thus $w \leq 4$ and $w = 6$ give conclusion (i) of 13.9. The case $w \leq 4$ is of course conclusion (1) of 7.3.1; but in fact the argument of 7.2 shows that $w = 4$ and $\overline{N} = \langle \overline{t} \rangle$, contrary to our assumption that $\overline{L} \cap \overline{N} > \langle t \rangle$. The case $w = 6$ is one of the cases in conclusion (4) of 7.3.1; as in the previous reduction, Timmesfeld is implicitly appealing to 4.7 and 5.5.1 to verify certain details of that conclusion. (However later Theorem 7.3.1 shows that $\overline{N} = \langle \overline{t} \rangle$ in this case, contrary to the present hypothesis.) The remaining possibility $O_3(\overline{M}) \neq 1$ in 10.8 gives the hypothesis of Section 13. The analysis from 13.5 onward eliminates case (ii) of 13.4, leaving case (i) of 13.4 as the remaining conclusion (ii) of 13.9. But that in turn is just the generic conclusion (3) of 7.3.1; and indeed the analysis shows that $\overline{N} = \langle \overline{t} \rangle$, again contrary to our hypothesis.

The case $\overline{R}/\overline{N} \cong L_m(2)$ is treated in 11.9; in fact the hypothesis of Section 11 is that $\overline{R}/\overline{N}$ is linear whenever $\overline{L} < \overline{R}$. At 11.1 it is shown that a preimage t of $\overline{t}$ can be taken to be a conjugate of z lying in Q_a, which provides the hypotheses of 6.8. Close analysis eventually yields at 11.9 that either $w = 4$ with $m = 2$, or $w = 6$ with $m = 3$. In the former case conclusion (1) of 7.3.1 holds; although later in Theorem 7.4.3, more specifically [**Smi79b**, 2.9], we find that this case does not arise under the present hypothesis. In the latter case, conclusion (4) of 7.3.1 holds, again implicitly using 4.7 and 5.5.1 to verify certain details of that conclusion. (And Theorem 7.4.3 shows that this case leads to $G \cong J_4$).

Isolating the final conclusions of the Theorem. After 14.5, *we may assume that for some* $\overline{t}$, $\overline{R}/\overline{N}$ *is one of the remaining possibilities in Theorem 7.3.6, namely an orthogonal group (other than* $\Omega_4^+(2)$*) or a symplectic group.* This gives the hypothesis of Section 8 or 9, respectively. We mention also that Lemma 2.1 is now available—which shows that $\overline{N}$ must involve a spin module for $\overline{R}/\overline{N}$; this is a key fact for limiting the possibilities. (And this final branch of the proof will lead to conclusions (1) and (4)–(8) of Theorem 7.3.1.)

The argument through 8.4/9.4 in fact reduces to six/two cases, respectively, for w and $\overline{R}/\overline{N}$. Further 8.5 and 9.5 show that $\overline{N}$ is extraspecial—which suggests the possibility of applying Theorem 7.3.1 inductively.

The results 14.7–14.9 reduce to the case where $F^*(\overline{M})$ is simple, and (modulo one final possibility, eliminated at the end of the section) the extraspecial group $\overline{N}$ is large in $F^*(\overline{M})$; thus induction applies with $C_{\overline{M}}(\overline{t})$, $F^*(\overline{M})$ playsing the roles of "M, G". The reductions here exclude cases (ii) and (iii) in 5.5, where $w = 4$. (In fact they are essentially eliminated, since later Theorem 7.4.3 shows in these cases that $\overline{L}$ is a TI-set or $\overline{N} = \langle \overline{t} \rangle$, contrary to our hypothesis.)

Then 14.10 eliminates cases (1) and (4) of 8.4,[7] as well as case (2) of 9.4: For in those cases, $\overline{N}$ has width 8, so that induction would allow only conclusions (2) or (3) of 7.3.1, with $C_{\overline{M}}(\overline{t})/\overline{N}$ playing the role of "$\overline{M}$". In (2), this quotient is $L_8(2)$ or $U_8(2)$—which does not have the normal subgroup $\overline{R}/\overline{N} \cong \Omega_8^-(2)$, $\Omega_{10}^+(2)$, or $Sp_8(2)$ indicated in those three cases. In (3), we get a similar contradiction—except[8] possibly when that quotient is $S_3 \times \Omega_8^-(2)$ or $S_3 \times O_8^-(2)$, and $\overline{R}/\overline{N} \cong \Omega_8^-(2)$ from case (1) of 8.4: But then by 6.5, the quotient $\overline{N}/\langle \overline{t} \rangle$ is the irreducible 16-dimensional spin module in 2.1 for $\Omega_8^-(2)$. So by Schur's lemma, an element of order 3 in S_3 would have to centralize this quotient, and hence would centralize $\overline{N}$ by coprime action, contrary to the self-centralizing property of $\overline{N} = F^*\big(C_{\overline{M}}(\overline{t})\big)$.

Next 14.12 handles cases (2) and (6) of 8.4, showing that when $\overline{R}/\overline{N} \cong \Omega_6^\pm(2)$ and $w = 10$, we have $F^*(\overline{M}) \cong L_6(2)$ or $U_6(2)$, giving conclusion (5) of 7.3.1. (Later Theorem 7.4.2 shows that G is $E_6(2)$ or ${}^2E_6(2)$.)

Then 14.13 handles case (3) and the $w = 12$ subcase of case (5) of 8.4, as well as the case (1) remaining in 9.4. In the latter two situations, $\overline{N}$ has width 4 with $\overline{R}/\overline{N} \cong \Omega_8^+(2)$ or $Sp_6(2)$, so that $F^*(\overline{M})$ can be identified as Co_1 or Co_2 using the involution-centralizer characterizations of Patterson's thesis [**Pat72**] (now covered by [**Asc94**, 49.15]) and of F. Smith [**Smi74**]. Thus conclusion (6) or (7) of Theorem 7.3.1 holds. (Later Theorem 7.4.3 shows that $G \cong M$ or B here.) Case (3) of 8.4 with $\overline{R}/\overline{N} \cong \Omega_{12}^+(2)$ quickly leads to conclusion (8)(ii) of 7.3.1. (Later Theorem 7.4.2 shows $G \cong E_8(2)$.)

After 14.13, it remains only to treat the subcase of case (5) of 8.4 in which $\overline{N}$ is of width 8 and $\overline{R}/\overline{N} \cong \Omega_8^+(2)$. If $\overline{N}$ is large, then we may apply induction; and as in the argument above for 14.10, only conclusion (3) is possible, and then we have $C_{\overline{M}}(\overline{t})/\overline{N} \cong S_3 \times \Omega_8^+(2)$ (and this time the action of $\Omega_8^+(2)$ on $\overline{N}/\langle \overline{t} \rangle$ is reducible). This quickly leads to conclusion (8)(i) of 7.3.1. (Later Theorem 7.4.2 shows $G \cong E_7(2)$.)

As mentioned earlier, the case where $\overline{N}$ is not large is then eliminated. This completes the proof of Theorem 7.3.1

Timmesfeld's arguments also use some standard background material such as Glauberman's Z^*-Theorem B.2.1, cohomology, Schur multipliers (e.g. [**Gri72**]), Goldschmidt's strongly closed papers [**Gol74, Gol75**], and Shult's fusion theorem—which remained unpublished, but can replaced by [**Gol74**, Cor 3], or Theorem I.8.3 in [**AS04a**].

[7] We mention that "$Sp_6(2)$" in [**Tim78a**, 14.10] is a misprint for "$Sp_8(2)$". Similarly "$\Omega_6^+(2)$" in [**Tim78a**, 14.11,14.12] is a misprint for "$\Omega_6^\pm(2)$".

[8] Here we supply an argument which seems only implicit in the original paper.

7.4. The final treatment of the various cases for $\overline{M}$

Thus it now remains for us to indicate how the description of $\overline{M}$ in conclusions (1) and (3)–(8) of Timmesfeld's Theorem 7.3.1 was translated into an identification of G. That work is largely contained in three papers of Stephen Smith [**Smi80b, Smi80a, Smi79b**], with contributions by other authors which we will indicate along the way.

Orthogonal groups. The generic conclusion (3) in Theorem 7.3.1 was handled in [**Smi80b**]. The statement assumes width $w \geq 4$; the case $w = 2$ is covered by Theorem 7.2.1. As usual, we restrict the statement to the case of G simple.

THEOREM 7.4.1 (S. Smith). *Let G be a finite simple group with a large extraspecial 2-subgroup Q of even width $w \geq 4$ in Notation 7.1.2. Assume further that $\overline{M} \cong S_3 \times \Omega^{\epsilon}_{w}(2)$ or $S_3 \times O^{\epsilon}_{w}(2)$, where $\epsilon = +$ or $-$.*
Then $G \cong \Omega^{\epsilon}_{w+4}(2)$.

The strategy of the proof is to exploit the methods of Timmesfeld in [**Tim78a**], in order to show that z has the fusion pattern of the orthogonal target group; in particular z^G should be a class of $\{3,4\}^+$-transpositions, so that G can be identified (using the structure of the involution centralizer M) via Timmesfeld's work [**Tim73**].

The preliminaries in Section 1 of [**Smi80b**] include the reductions through 14.4 of Timmesfeld [**Tim78a**], described in the previous section; in particular we let t denote a preimage of $\bar{t} \in \overline{L}$ as in 7.3.2, with $\overline{R}/\overline{N} \cong \Omega^{\epsilon}_{w}(2)$ as in 7.3.5. Heavy use is made of the material around 4.5 from Timmesfeld's paper. Also important are Lemmas 1.15 and 1.16, adapted from Section 3 of Fred Smith's paper [**Tim78a**]. Reference is made to some other standard works in the area, such as [**Asc77**] and [**AS76a**].

As a first step, Section 2 analyzes the fusion of z in Q; that is, the possibilities for "a" in (7.2.4). At 2.5, these are characterized as in the target group, via the language of 7.3.5; the description is in terms of the singular vectors in the S_3-conjugates of $\tilde{V}$ (namely $\tilde{V}$, $\tilde{W}$ and $\tilde{W}^t$, where $W := C_Q(V)$). This yields a description of $Q \cap Q_a$ in 2.7, and at 2.8 helps to show that the extension M of Q by $\overline{M}$ is split.

The next step in Section 3 is to show that the fusion of z in M is as expected. Indeed 3.3 shows that for $x \in z^G \cap (M \setminus Q)$, either $\bar{x}$ is in S_3, or $\bar{x}$ is an involution of type a_2 in $\Omega^{\epsilon}_{w}(2)$, $Q \cap Q_x = [Q, x]$, and we may choose a in this group. In particular it follows at 3.4 that $z^G \cap M$ is a set of $\{3,4\}^+$-transpositions. (This fact had been established earlier in unpublished work of Thompson.)

Section 4 is primarily devoted to extending this observation to a proof at 4.4 that z^G is a class of $\{3,4\}^+$-transpositions. The approach is motivated by the fact that the commuting graph on z^G in the target orthogonal group has diameter 2. The crucial result 4.2.1 shows that, given any path $x - z - w - y$ of length 3 in the graph, there is a vertex v defining a path $x - v - y$ of length 2. In other words, the intersection $M_x \cap M_y$ contains a conjugate of z. To approach this, note first that the relationship of x and w in $M_z = M$ can be described using the analysis in Section 3, as can that of z, y in M_w. The resulting cases can then be further analyzed in terms of intersections of the form $Q \cap Q_w$ (as in Section 2) and $Q \cap Q_y$

(as in Section 3). This allows the completion of the proof of 4.2.1;[9] and it then follows that a connected component of the graph has diameter 2. Finally Holt's Theorem B.7.16 is applied at 4.3 to conclude that the graph is in fact connected. So now for *any* pair $x, y \in z^G$, we can embed x and y in some M_v; hence it follows from the work in Section 3 on fusion of z in M that z^G is a class of $\{3,4\}^+$-transpositions in G. Then Timmesfeld's work [**Tim73**] on $\{3,4\}^+$-transpositions can be applied to identify $G \cong \Omega^\epsilon_{w+4}(2)$, completing the proof of Theorem 7.4.1.

Groups of type E over $\mathbb{F}_2$. We now turn to conclusions (5) and (8) in Theorem 7.3.1; the corresponding groups G in the four cases there were identified in:

THEOREM 7.4.2 (Reifart, S. Smith). *Let G be a finite simple group with a large extraspecial 2-subgroup Q in Notation 7.1.2. Assume further that the width w of Q is one of $10, 10, 16, 28$; and that $F^*(\overline{M}) \cong U_6(2)$ or $L_6(2)$, respectively, in the first two cases, or in the last two cases (using Notation 7.3.5) that $C_{\overline{M}}(\overline{t})/\overline{N} \cong S_3 \times \Omega_8^+(2)$ or $\Omega_{12}^+(2)$, respectively.*

Then $G \cong {}^2E_6(2), E_6(2), E_7(2), E_8(2)$, respectively.

The first two cases were treated by Reifart in [**Rei78b, Rei78c**]. His approach initially resembles that of Patterson-Wong [**PW76**] and Patterson [**Pat72**], in the use of the Thompson order formula (e.g. [**GLS96**, 34.1]) to compute the group order $|G|$; this requires detailed calculation of certain parameters for fusion in M. Once $|G|$ is known, values of further fusion parameters show that z^G is a class of $\{3,4\}^+$-transpositions, so that G can, as usual, be identified via Timmesfeld's work [**Tim73**].

All four cases were in fact independently treated, in a shorter and uniform manner, by Smith in [**Smi80a**, Thm A]—where he was able to implement a fairly straightforward extension of the method of [**Smi80b**] discussed above. So below we will indicate only a few points of difference from that earlier exposition.

The initial result Theorem B observes, in the third case, that Theorem 7.4.1 may be applied—so that we may assume $\overline{M} \cong \Omega_{12}^+(2)$ during its proof; and also that, assuming the final result for the third case, we may assume that $\overline{M} \cong E_7(2)$ during the proof of the fourth case.

This time the preliminaries in Section 1 quote those of [**Smi80b**]; including a bit morefrom Timmesfeld's reductions discussed earlier, namely through 14.13 of [**Tim78a**].

In Section 2, this time $\overline{M}' = F^*(\overline{M})$ is a known group of Lie type over $\mathbb{F}_2$; then 2.2 identifies $\tilde{Q}$ as the irreducible module in which $C_{\overline{M}'}(\tilde{a})$ is a parabolic subgroup whose Levi complement is given by $L_3(4), L_3(2) \times L_3(2), L_6(2), E_6(2)$ in the four cases. Now the fusion of z in Q is described at 2.4, in terms of the four suborbits in the rank-4 permutation action of $\overline{M}$ on $\tilde{a}^{\overline{M}}$. Furthermore 2.5 shows that a non-singular vector of $\tilde{V}$ determines a TI-set in $\overline{M}$. This is all the fusion information within Q that will be needed later.

In Section 3, it is shown that for $x \in z^G \cap (M \setminus Q)$, $\overline{x}$ is 2-central in $\overline{M}'$; again $Q \cap Q_x = [Q, x]$, and we can take a in this group. As before, it follows that $z^G \cap M$ is a set of $\{3,4\}^+$-transpositions.

[9]We mention that a special argument needed for the case $\overline{R}/\overline{N} \cong \Omega_4^+(2)$ was inadvertently omitted in [**Smi80b**, 4.2.1.c]; it can be found in the later paper [**Smi81**, p. 420].

In Section 4, the rest of the proof follows the outline discussed for [**Smi80b**]: in particular the analogue of 4.2.1 (involving information from Sections 2 and 3) shows that the commuting graph on z^G has diameter 2, with Holt's Theorem B.7.16 used for connectedness. Hence z^G is a class of $\{3,4\}^+$-transpositions, with G identified by [**Tim73**].

In addition to the references used in the earlier paper [**Smi80b**], there are also appeals in [**Smi80a**] to standard references on groups of Lie type (e.g. [**Car89**]) and their modular representations. An application in 2.2 of the unpublished Cooperstein-Mason work [**CM80**] can now be replaced by reference to Guralnick-Malle [**GM02, GM04**]; another mention of [**CM80**] in 2.3 can be replaced by [**AS76a**].

The remaining cases for certain specified widths. Finally we turn to conclusions (1), (4), (6), and (7) in Theorem 7.3.1; here the groups were identified in Smith [**Smi79b**]

THEOREM 7.4.3 (S. Smith). *Let G be a finite simple group with a large extraspecial 2-subgroup Q of width w in Notation 7.1.2.*

(1) If $w = 3$, then $G \cong L_5(2)$, $U_5(2)$, M_{24}, He, or Suz;
if $w = 4$, then $G \cong L_6(2)$, $U_6(2)$, $\Omega_8^{\pm}(2)$, ${}^3D_4(2)$, $\Omega_8^+(3)$, Co_2, Co_1, Th, or HN.

(2) If $w = 6$, then $G \cong L_8(2)$, $U_8(2)$, $\Omega_{10}^{\pm}(2)$, J_4, or Fi_{24}'.

(3) If $w = 11$ and $\overline{M} \cong Co_2$, then $G \cong BM$;
if $w = 12$ and $\overline{M} \cong Co_1$, then $G \cong M$.

We first discuss case (3), since it can be handled comparatively easily. The proof in Section 4 of [**Smi79b**] was suggested by Stroth and Reifart: in overview, the idea is to pass to the centralizer $C := C_G(b)$ of a *non*-2-central involution b—constructing C via earlier results from large-extraspecial theory; and then to identify G using a characterization in the literature via C (rather than via $M = C_G(z)$ for the 2-central involution z).

We can discuss the two subcases of (3) in parallel. First we choose a suitable involution $b \in Q$ which is *not* a G-conjugate of z; we set $C := C_G(b)$, and then set $C^* := C/\langle b \rangle$. Then $C_Q(b)^*$ is a large extraspecial subgroup of C^*—of width 10 or 11 in the two subcases; the role of "$\overline{M}$" is played by $C_M(b)^*/C_Q(b)^* \cong U_6(2).2$ or Co_2. Using Theorem 7.4.2 in the first case, and assuming the final result of the first case in the proof of the second case, we conclude that $C^* \cong {}^2E_6(2).2$ or BM, respectively. In the first case, we obtain $G \cong BM$ by a result of Stroth [**Str76**]; in the second, we obtain $G \cong M$ from work of Griess [**Gri76**] based on the structure of the centralizers of both b and z. We also emphasize that these "isomorphisms" (as is the case for certain other groups in the conclusions) really depend on the solution of some uniqueness problems that were not completed until several years later; results from this period instead used the language of a "group of type M" (etc.)—this assumption included a great deal of local information (such as that in the tables of [**GLS98**, 5.3]), on the basis of which the uniqueness of the groups was later proved. (References for the existence and uniqueness of these sporadic groups can be found in our table of such information for all the sporadic groups given in Section A.3 in the Appendix.)

Thus we turn to conclusions (1) and (2) of 7.4.3, and in particular to cases with width $w = 3$, 4, and 6. This requires an exhaustive analysis using Timmesfeld's description of $\overline{M}$, which we now outline.

Initial reductions. Much as in [**Smi80a**], the preliminaries in Section 1 are essentially reproduced from [**Smi80b**], with further argument where needed. Again Timmesfeld's reductions through 14.13 of [**Tim78a**] are essentially quoted; but this time, with explicit listing of the groups as they arise. We outline that process below; much of it in fact reviews the relevant parts of results discussed earlier.

First when z is weakly closed in Q, Theorem 7.2.3 produces $G \cong U_5(2)$, $U_6(2)$, Co_2, and $U_8(2)$ in conclusions (1) and (2) above. Next when $\overline{M}$ is reducible on $\tilde{Q}$, Theorem 7.2.6 produces $L_5(2)$, M_{24}, He, $L_6(2)$, and $L_8(2)$. Then 1.9 extends Timmesfeld's reductions at [**Tim78a**, 14.2] (more precisely at 12.13 and 12.14) in the case where $\overline{L}$ contains an involution of type a_2; recall in this case that Theorem 7.2.7 gives $G \cong Suz$ or Co_1.

Next 1.11 gives a similar extension of Timmesfeld's reduction at [**Tim78a**, 14.3] in the case where $\overline{L}$ is a TI-set: From 4.2 we have $\langle \overline{L}^{\overline{M}} \rangle \cong L_2(8)$ or A_9. In the case of $L_2(8)$, Reifart in [**Rei78a**] shows that $G \cong {}^3D_4(2)$, using only a comparatively elementary analysis of fusion (not requiring the computation of the Thompson order formula) in order to conclude that z^G is a class of root involutions, so that G can be identified via Timmesfeld's work [**Tim75a**]. (A short independent proof, again producing root involutions, is given at [**Smi79b**, 3.1].) The other case leads to $\overline{M} \cong A_9$, so that G can be identified as Th by a result of Parrott [**Par77**].

After 1.11, there is some $\bar{t} \in \overline{L}$ with $\overline{L} < \overline{R}$ in the notation of 7.3.5. Then a list is given in 1.17 of the possibilities[10] for $\overline{R}/\overline{N}$ which remain to be analyzed. This makes use of Timmesfeld's reductions at [**Tim78a**, 14.4] and thereafter, from the original list of cases at his 4.5. In particular, the case with $w = 4$ and $\langle \overline{L}^{\overline{M}} \rangle \cong S_3 \times A_6$ is eliminated (with the details given at [**Smi79b**, 3.2]), so that only orthogonal and linear possibilities for $\overline{R}/\overline{N}$ remain. Furthermore when $w = 3$, it is shown that no further groups can arise. Thus cases (1) and (2) of Theorem 7.4.3 are now reduced to $w = 4$ and 6.

Note that we are not *yet* applying Theorem 7.4.1 to produce orthogonal groups; instead they will appear in Section 2.

A final reduction at 1.18 produces some refinements of Timmesfeld's discussion at [**Tim78a**, 5.5] of $E(\overline{M})$: When $E(\overline{M}) = 1$, 7.3.1 says that $w = 4$ and $O(\overline{M})$ is an elementary 3-group of rank at most 4, and it is shown that $\overline{M}$ is solvable, and $O(\overline{M})$ has rank 3 or 4; the argument at one point quotes the work of Lundgren-Wong [**LW76**] on large extraspecial 2-subgroups where $\overline{M}$ is solvable. Next when $E(\overline{M})$ has a single component, $O(\overline{M})$ is of order 1 or 3. Finally when $E(\overline{M})$ has more than one component, Timmesfeld's configuration [**Tim78a**, 5.5.2] with $w = 4$ and $\overline{M} \cong A_5 \wr \mathbb{Z}_2$ is shown (with details at [**Smi79b**, 3.3]) to lead to $G \cong HN$. The argument uses the Reifart-Stroth method indicated earlier for conclusion (3) of 7.4.3: In that notation, for suitable $b \in Q$ not conjugate to z, $O^2(C_M(b)^*)$ is a large symplectic but not extraspecial 2-subgroup: namely a maximal subgroup of $C_Q(b)^*$ given by the central product of a cyclic group of

[10]We mention that $\Omega_6^+(\tilde{V})$ in the statement of [**Smi79b**, 1.17] is a misprint for $\Omega_6^{\pm}(\tilde{V})$ arising in the proof.

order 4 with an extraspecial subgroup of width 2. Then $C_G(b)^* \cong HS.2$ using Theorem 7.1.1. This allows us to identify $G \cong HN$ by Harada [**Har76**].

Section 1 concludes with 1.19 and 1.20, which use the Gorenstein-Harada Theorem 1.5.7 to determine the possibilities for certain subgroups of $\overline{M}$ which can be shown to have sectional 2-rank at most 4.

Section 2 then carries out the analysis of the various cases from 1.17.

The case $\overline{R}/\overline{N} \cong \Omega_w^{\pm}(2)$. The largest possibility for $\overline{R}/\overline{N}$ in 1.17 is $\Omega_w^{\pm}(2)$. The first few results assume at 2.1 that this case occurs for some $\bar{t}$; then 2.2 shows that $\overline{N} = \langle \bar{t} \rangle$. But here since $w = 4$ or 6, no other information is provided by [**Tim78a**, 7.6], so that further analysis is required to determine $\overline{M}$.

First 2.4 covers the case where $\bar{t}$ is 2-central in $\overline{M}$. In the subcase of 1.18 where $E(\overline{M})$ is quasisimple, 1.19 and 1.20 are used to eliminate $\overline{R}/\overline{N} \cong \Omega_4^+(2)$, and to show in the other cases that $\overline{M} \cong S_3 \times \Omega_w^{\epsilon}(2)$ or $S_3 \times O_w^{\epsilon}(2)$—so that we get $G \cong \Omega_{w+4}^{\epsilon}(2)$ by Theorem 7.4.1. Otherwise 1.18 says that $\overline{M}$ is solvable (so $\overline{R}/\overline{N}$ can only be $\Omega_4^+(2)$), with $O(\overline{M})$ an elementary 3-group of rank 3 or 4. In the rank-3 case, further argument (again using Lundgren-Wong) shows that we get the same structure for $\overline{M}$ as in the case above, and hence get $G \cong \Omega_8^+(2)$ by 7.4.1. So the rest of the proof is devoted to treating the case where $O(\overline{M})$ is of rank 4. The structure of $\overline{M}$ is determined, with $|\overline{M}| = 3^4 2^3$, again referencing Lundgren-Wong. At this point, one can obtain a subgroup isomorphic to $SL_2(3)$ subnormal in the involution centralizer M, so G can be identified as $\Omega_8^+(3)$ using the Classical Involution Theorem 1.7.5.

Thus 2.5 turns to the case where $\bar{t}$ is not 2-central in $\overline{M}$. It is fairly easy to eliminate the case $w = 4$; again arguments from Lundgren-Wong are helpful. So $w = 6$, and now a rather detailed analysis ensues. Eventually $\overline{R}/\overline{N} \cong \Omega_6^+(2)$ is eliminated, and it is shown that $\overline{M} = E(\overline{M})\langle \bar{t} \rangle$ with $E(\overline{M})$ a 3-fold covering group of $U_4(3)$. Then G is identified as Fi_{24}' by work of Parrott [**Par81**].

The case $\overline{R}/\overline{N} \cong L_{w/2}(2)$. Starting at 2.6, the next few results assume $\overline{R}/\overline{N}$ is not isomorphic to $\Omega_w^{\pm}(2)$ for any other choice of $\bar{t}$, but the case $L_{w/2}(2)$ does occur. Another rather detailed analysis ensues; much of the argument is similar to that for $\Omega_w^{\pm}(2)$. This time the study of $\overline{M}$ makes use of Fischer's Theorem A.6.3 and root involution theory [**Tim75a**]. At 2.9, the case $w = 4$ is eliminated, again using Lundgren-Wong when $\overline{M}$ is solvable. So $w = 6$, and there is a fairly easy reduction to the case in which $\bar{t}$ is not 2-central in $\overline{M}$. At this point, the analysis becomes even more strikingly parallel to the earlier argument in 2.5. This time it emerges that $\overline{M}$ is an extension of $3M_{22}$ by $\langle \bar{t} \rangle$; and G can be identified as J_4 by Janko [**Jan76**].

And now the end is in sight: The case $w = 4$ is complete, since we have treated both possibilities for $\overline{R}/\overline{N}$ in 1.17. When $w = 6$, it remains to eliminate the case where $\overline{R}/\overline{N}$ is one of the other possibilities in 1.17, namely $L_2(2)$ and $\Omega_4^+(2)$ (and in fact the latter must occur for some $\bar{t}$). Analysis of this situation leads to a contradiction at 2.10, completing the proof of Theorem 7.4.3.

Finally that result in turn completes the proof of the $GF(2)$ Type Theorem 7.0.1.

7.5. Chapter appendix: The classification of groups of $GF(2^n)$ type

Most groups of $GF(2)$ type are of Lie type over the field of order 2. In groups over fields of order $2^n > 2$, the nontrivial Cartan subgroups in long-root SL_2-subgroups lie in a Sylow 2-normalizer of G, and have a strong regulating effect on all 2-local subgroups.

During the classification of the groups of characteristic 2 type, it was realized that it would be useful to have an analogous classification of groups with a large special subgroup, whose center is a root group—over an arbitrary field $\mathbb{F}_{2^n}$ of characteristic 2. This led to the corresponding theory of groups *of* $GF(2^n)$ *type* described in this section. Some of its applications by Aschbacher are indicated in our discussion in Chapter 4 and 8; cf. also the references such as [**Asc78b**] on p. 119 of [**Tim78b**].

In implementing this extended theory, the role of Timmesfeld's work [**Tim73**] on $\{3,4\}^+$-transpositions is now largely played by his corresponding work [**Tim75a**] on root involutions. More generally, many of the arguments from the theory of groups of $GF(2)$ type can be carried through for groups of $GF(2^n)$ type; indeed typically the arguments are easier, roughly because of the nontrivial Cartan groups noted above.

Consequently we can give a briefer sketch of the treatment of groups of $GF(2^n)$ type, largely consisting of reference to our discussion of groups of $GF(2)$ type, with occasional deviations as needed. Furthermore we will include fewer details in our discussion of groups of $GF(2^n)$ type, since the result is at level 1 in our viewpoint. As usual, the reader can also find some exposition on groups of $GF(2^n)$ type in the introductory sections of the original papers, especially [**Smi81**] which completed the classification.

Statement of the result. Recall that root groups in a group of Lie type are TI-sets. In the case of a group of $GF(2)$ type, our candidate $\langle z \rangle$ for a root group is of order 2, and hence is automatically a TI-set. In the case of groups of $GF(2^n)$-type, we will assume that a certain subgroup "B"—which will usually turn out to be a root group in our group G—is TI. Hence the group "M" is no longer the centralizer of z, but the normalizer of B. In contrast to the case of groups of $GF(2)$ type, we make no assumption on the structure of Q, such as the symplectic-type condition; but analogous restrictions will soon emerge during the analysis. These considerations suggest:

DEFINITION 7.5.1 ($GF(2^n)$ type). A finite group G is said to be *of* $GF(2^n)$ *type* if there is a maximal 2-local subgroup M satisfying $F^*(M) = O_2(M)$ (we denote this subgroup by Q), such that a maximal elementary abelian normal subgroup B of M is a TI-set in G.

Notice the maximality of M implies that $M = N_G(B)$. By analogy with the discussion after 7.0.3, the literature sometimes refers to B as a *large* TI-subgroup.

Timmesfeld introduced this concept in [**Tim78b**]. His initial result (Theorem 1) did not in fact require the assumptions that M is maximal and B is of exponent 2. Theorem 1 in turn suggested Theorem 2, which did assume M to be maximal; Timmesfeld observed after Theorem 1 that the additional assumption that B be elementary abelian can always be added. Indeed the latter form of the statement is used in most applications. So we have included the condition that B

be elementary abelian in Definition 7.5.1, and hence in our statement below. As usual we restrict to the case of G simple, so the conclusion in Timmesfeld's original work with $G > \langle B^G \rangle \cong L_3(4)$ and $m(B) = 2$ (where $B \cong \mathbb{Z}_4 \times \mathbb{Z}_4$ can arise) does not appear in our statement. In fact we use the slightly more explicit form of the statement given in [**Tim81**]. Timmesfeld's original result determines the branches below, giving just the first sentence in each conclusion; in the second sentence, we indicate the groups G which arise—these identifications were given in later papers by various authors, which we will be discussing shortly. (We are not sure if this more complete statement, including the explicit list of simple groups of $GF(2^n)$ type, has in fact been given in the literature; most statements from that period, such as [**Asc83a**, 3.10] and [**Smi81**, Corollary], just conclude that "G is known".)

THEOREM 7.5.2 ($GF(2^n)$ Type Theorem). *Let G be a finite simple group of $GF(2^n)$ type, with B, Q, and M as in Definition 7.5.1; set $q := |B|$. Then one of the following holds:*

(1) $m(B) = 1$, and Q is of symplectic type (and large). Hence G is in the list of Theorem 7.0.1.

(2) $m(B) \geq 2$, and B is weakly closed in Q. Indeed either

(i) B is not weakly closed in M,
and if G is of characteristic 2 type, then $G \cong U_m(q) (m \geq 4)$; or

*(ii) B is weakly closed in M; hence by [**Tim75b**], G is one of*
$L_m(r), U_3(r), Sz(r)($ for some $m \geq 2,$ and $r = 2^s, s \geq 1)$;
A_6, A_7, A_8, A_9; M_{22}, M_{23}, M_{24}.

(3) $m(B) \geq 2$, G-conjugates of B generate Q, and Q is special (and large) with center B. Indeed G is one of $\Omega_6^-(q)$, $\Omega_{2m}^{\pm}(q) (m \geq 4)$, $G_2(q)$, ${}^3D_4(q)$, ${}^2E_6(q)$, $E_6(q)$, $E_7(q)$, or $E_8(q)$.

We now briefly discuss Timmesfeld's arguments in [**Tim78b**].

Some of the preliminary results in Section 2 use facts from Timmesfeld's earlier work [**Tim75b, Tim77**] on weakly closed abelian TI-sets, and from Aschbacher's paper [**Asc76b**] on tight embedding. We also mention that a number of the papers in this area, including the present paper of Timmesfeld at 4.8, make use of result of G. Higman on splitting of modules for $SL_2(2^n)$, and an analogous result of Martineau [**Mar72**] for $Sz(2^n)$; Higman's result appeared in the unpublished [**Hig68**], but a similar result can be found as G.1.3 in [**AS04a**].

Section 3 presents the proof of Theorem 1, which assumes that B is abelian, but not necessarily that B is elementary abelian or that M is maximal. It is shown that if B is neither cyclic nor elementary, then B is $\mathbb{Z}_4 \times \mathbb{Z}_4$, $\langle B^G \rangle$ is $L_3(4)$, and G is not simple. The proof uses the theory of root involutions [**Tim75a**].

The proof of Theorem 2 (that is, of the first sentences in the conclusions of 7.5.2 above) is carried out in Section 4. First 4.1 deals with the case that B is cyclic. The hypothesis of 7.5.1 forces $|B| = 2$, but Timmesfeld's result is more general, assuming that B is maximal abelian normal, but not necessarily elementary abelian. It is shown (partly using arguments from Aschbacher [**Asc76a**]) that either Q is of symplectic type, so that conclusion (1) holds; or that $\Omega_1(B) \leq Z^*(G)$—giving a degenerate case of conclusion (2), which does not arise under our assumption that G is simple. Thus we may assume that B is noncyclic. Now Timmesfeld uses Theorem 1 to reduce to the case B elementary, so that now $m(B) \geq 2$. Indeed we may assume that B is not weakly closed in Q, as otherwise conclusion (2) holds; hence $B < P$, where P is the weak closure of B in Q, and it follows fairly easily that P is special with center B. Thus to establish conclusion (3), it only remains to show that $P = Q$. So Timmesfeld assumes that $P < Q$; this amounts to assuming that $B < C_Q(P) =: T$, since in fact $Q = PT$. Furthermore T must also be special with center B, a fact used crucially at 4.7: essentially starting to show that $B^G \cap M \subseteq Q$. Indeed a bit more is established at 4.10 and 4.13,[11] using the preliminaries above (and standard results such as McLaughlin's Theorem A.6.1). The situation now appears as a very special subcase [**Tim75a**, Cor B] of the theory of root involutions—a subcase in which no group G arises; more precisely, three families in the list of that Corollary are of $GF(2^n)$ type, but in fact they fail to satisfy $T > B$. This contradiction completes the proof of Theorem 2 (that is, of the initial statements in the conclusions of 7.5.2).

It remains to establish the final statements in the conclusions of 7.5.2, by identifying the groups which can arise. Of course in conclusion (1) where $m(B) = 1$, this was already done in the $GF(2)$ Type Theorem 7.0.1. In the other cases where $m(B) \geq 2$, the process largely consisted of adapting the arguments from the analysis of groups of $GF(2)$ type to the larger root group B in a group of $GF(2^n)$ type.

The treatment of B weakly closed in Q. Conclusion (2) where B is weakly closed in Q is the analogue of the earlier situation of Aschbacher and Fred Smith in Theorem 7.2.3; the corresponding classification appeared in Timmesfeld [**Tim81**, Cor 2]. Our outline below will largely parallel the introductory discussion on p. 2 of that paper.

First if B is weakly closed in M, then G is determined as in case (ii) of conclusion (2) by Timmesfeld's earlier work [**Tim75b**] on weakly closed TI-subgroups. Note when $G \cong L_m(r)$ with $m \geq 3$ that B is not just a root group, but the unipotent radical of the maximal parabolic stabilizing a point or hyperplane of the defining module, so that $B^\#$ consists of transvections.

So for the rest of this subsection *we assume that B is not weakly closed in M*. Thus M contains G-conjugates of B distinct from B—but these conjugates do not lie in Q, as we are in conclusion (2). As we are in case (i) of conclusion (2), *we also assume that G is of characteristic* 2 *type*.

The identification of G as $U_m(q)$ in Sections 6 and 7 of the paper is based on the study of certain set $\mathcal{H}$ of 2-local subgroups which do *not* lie in M. So the analysis from Section 3 onward assumes $\mathcal{H}$ is nonempty, where $\mathcal{H}$ consists of those overgroups H of $C_Q(B)$ such that $O_2(H) \neq 1$, but $H \not\leq M$. Indeed the case where $\mathcal{H}$ is empty is the alternative to $G \cong U_m(q)$, in the analogue of conclusion (2)(i) in Timmesfeld's more general Theorem 1.

Thus the hypothesis that G is of characteristic 2 type is actually used only in Section 8: to deduce Corollary 2 from Theorem 1, by showing that $\mathcal{H}$ must be nonempty under this assumption. The argument assumes that $\mathcal{H}$ is empty; and then the theory of pushing-up (cf. Section B.7) can be applied. We see in particular that M is the unique maximal 2-local over $C_Q(B)$. It follows that a Sylow 2-subgroup T of M over $C_Q(B)$ is in fact Sylow in G; then we get $C(G,T) \leq M < G$ in the language of Definition B.7.1. Timmesfeld in fact refers just to Aschbacher's paper [**Asc81a**] on the Local $C(G,T)$-Theorem B.7.3 (which only applies to groups

[11] The statement "By (4.5) it suffices" in [**Tim78b**, 4.13] also implicitly uses the TI-hypothesis; cf. [**Smi81**, 1.12], where (1.4)(b) plays the role of Timmesfeld's (4.5).

with $F^*(G) = O_2(G)$), along with a preprint [**Smi**] of Fred Smith which remained unpublished—Smith's results were later generalized by Foote in [**Foo82**]. Since the underlying argument is left implicit, we briefly sketch it here: First, one could nowadays in fact obtain a quicker contradiction, by applying the Global $C(G,T)$-Theorem B.7.8)—which was not yet available when Timmesfeld was writing: just observe that the groups listed in [**Foo80**, Cor 1] do not satisfy the hypotheses of case (2)(i). However, we instead sketch Timmesfeld's more elementary argument: First the groups in the Strongly Embedded Theorem 1.3.5 have B weakly closed in M and so do not arise here. Thus it follows from Theorems 2 and 3 of [**Asc81a**] that there is an $\mathcal{X}$-block X contained in a unique maximal subgroup N of G; and indeed X is normal in N by [**Foo82**, Thm C]. Finally we can take N to be M: the proof of this fact was apparently inadvertently omitted in [**Tim81**], but it can be deduced from [**Asc78a**] and [**Sol81**]: Again checking that the groups listed in those papers do not satisfy our present hypotheses, we are left with the alternative that $N_T(S) \leq N$, where S is Sylow in $N_G(O^2(X)) = N$; thus S is Sylow in G, so we can take $S = T \geq C_Q(B)$, and hence $N = M$. With this situation in hand, Timmesfeld quickly shows that $C_G(X)$ is tightly embedded in G, so that he can apply Solomon-Timmesfeld [**ST79**]; this leads to B weakly closed in M, contrary to case (i) of conclusion (2). Thus we have obtained a contradiction from assuming $\mathcal{H}$ is empty.

So from Section 3 onward, $\mathcal{H}$ *is nonempty*. We let H denote a member of $\mathcal{H}$, and embark upon the identification of G as $U_m(q)$. The result 3.3 quotes an earlier paper [**Tim79a**] of Timmesfeld, which shows in this situation that $B \leq O_2(H)$; in addition to the usual references, that paper uses Aschbacher's work [**Asc72**] on odd transpositions, and a paper of Stroth [**Str78**] on an extension of the theory of groups of $GF(2)$ type, as well as lemmas for example from [**Gol74**] and [**Asc81e**]. Then 3.3 is used crucially in 3.4, to show that B is of root type in the sense of [**Tim77**]; then it follows at 3.5 that B is weakly closed not just in Q but also in fact in $P := C_Q(B)$. Then 3.6 shows in $\tilde{H} := H/C_H(\langle B^H \rangle)$ that $\tilde{P}$ is a weakly closed TI-set. This makes possible the application of [**Tim79a**], to determine at 3.7 a list of possiblities for $\tilde{H}/Z(\tilde{H})$ (assuming H is generated by the conjugacy class of P). The analysis in Section 4 establishes at 4.5 the further restriction that a "minimal" such H, generated by two conjugates of B, satisfies $\tilde{H} \cong L_2(q^2)$, acting as $\Omega_4^-(q)$ on $\langle B^H \rangle$; this is the analogue of Aschbacher's [**Asc77**, 6.4; cf. 4.8.2]. Furthermore 4.6 shows that $P = Q$, and this group is special with center B. Section 5 then considers pairs of such minimal choices of H over B, culminating in 5.13 which shows essentially that $\overline{B}^g$ determines a class of odd transpositions in $\overline{M} := M/Q$—this is the analogue of Aschbacher's reduction at his 6.6 to 3-transpositions in case (3) of [**Asc77**, 5.12]. Next 6.1.2 shows that Q has order q^{1+2w}, where $w \geq 2$ is an analogue of the width of an extraspecial group. And 6.2 is an analogue of Aschbacher's analysis of $\overline{Q^g \cap M}$, leading to the analogue 6.3 of [**Asc77**, 6.5]—which shows that the class of $\overline{B}^g$ generates $L_2(q)$ when $w = 2$, and $U_w(q)$ for $w \geq 3$.

The cases $w = 2, 3$ are shown at 6.5 to lead to $G \cong U_4(q), U_5(q)$; the groups are identified by Aschbacher's work [**Asc72**] on odd transpositions. (In the analogous small cases for $q = 2$ in [**Asc77**], Aschbacher had used some earlier classification results to identify G).

Finally at 7.8 it is shown for $w \geq 4$ that $G \cong U_{w+2}(q)$. The argument follows the logic in Aschbacher [**Asc77**] fairly closely—showing now that B defines a class of odd transpositions in G (where of course Aschbacher had just 3-transpositions), so that G is again identified via [**Asc72**].

This completes the proof of Timmesfeld's Corollary 2; that is, of the detailed statements in conclusion (2) of Theorem 7.5.2.

The treatment of large special Q. So we turn to the remaining case (namely conclusion (3) of Theorem 7.5.2), where B is not weakly closed in Q—indeed Q is generated by conjugates of B, and is special with center B.

In his note [**Tim79c**], Timmesfeld provided further information on Q: he showed (via an elementary proof) that it is a central product of Sylow 2-subgroups of $L_3(q)$. In particular, it follows that $|Q| = q^{1+2w}$ for the minimal number w of such factors—so that w is an analogue of the width of an extraspecial group.

Determining the cases for $\overline{M}$. Then Stroth in [**Str80**] established the analogue of Timmesfeld's Theorem 7.3.1: namely he obtained a corresponding list of possibilities for $\overline{M}$.

REMARK 7.5.3 (Stroth's list of conclusions in [**Str80**]). For the convenience of the reader, we compare and contrast some details of the two lists.

First, certain of the cases for $\overline{M}$ in Timmesfeld's list in Theorem 7.3.1 will not arise for Stroth: For example, because Stroth has $q > 2$, he obtains only configurations corresponding to groups of Lie type over $\mathbb{F}_q$; thus the sporadic examples with $q = 2$ in (4), (6), and (7) of 7.3.1 will not arise. Also, Stroth assumes that B is not weakly closed in Q; and in a unitary group $U_n(q)$, the maximal elementary abelian normal subgroup B is $Z(Q)$, which *is* weakly closed—so the unitary groups in (2) of 7.3.1 will not arise.

But aside from these exclusions, Stroth can expect essentially the remaining cases from 7.3.1—though he organizes them somwhat differently. Below we give that list from [**Str80**, p 461]; however, for expository purposes we will make certain simplifications, such as writing $\overline{L}$, $\bar{t}$, $\overline{N}$, $\overline{R}$, for the analogues in Stroth of the structures with those names in our statement of 7.3.1:

(1) $w = 1$, and Q is isomorphic to a Sylow of $L_3(q)$.
(2) $w = 2$, and $\langle \overline{L}^{\overline{M}} \rangle \cong L_2(q)$.
(3) $w = 4$. and $\langle \overline{L}^{\overline{M}} \rangle \cong L_2(q^3)$.
(4) $\overline{L}$ is not a TI-set in $\overline{M}$, and for suitable $\bar{t}$ we get one of:
  (i) $w = 4$, $|N| = q$, $R/N \cong \Omega_4^{\pm}(q)$.
  (ii) $w = n$, $|N| = q$, $R/N \cong \Omega_n^{\pm}(q)$, $\langle \overline{L}^{\overline{M}} \rangle \cong L_2(q) \times \Omega_n^{\pm}(q)$.
  (iii) $w = 10$, $E(\overline{M}) = \langle \overline{L}^{\overline{M}} \rangle \cong U_6(q)$ or $L_6(q)$.
  (iv) $w = 16$, $|N| = q^{17}$, $R/N \cong \Omega_8^{+}(q)$.
  (v) $w = 28$, $|N| = q^{33}$, $R/N \cong \Omega_{12}^{+}(q)$.
  (vi) $\langle \overline{L}^{\overline{M}} \rangle \cong L_n(q)$, with $\tilde{Q}$ the sum[12] of a natural and a dual module.

To compare the two lists more directly: First, Stroth's cases (1), (2), (3), and (4)(i) above are a refinement of Timmesfeld's less explicit case $w \leq 4$ in (1) of 7.3.1.[13] Next, the four cases (ii)–(v) in (4) above are the exact analogues of (3), (5), (8)(i),

[12]We mention that the "$\otimes$" in [**Str80**] is a misprint for "$\oplus$".

[13]We mention that (2) now also contains the local structure for $G \cong G_2(q)$—which did not arise for Timmesfeld, since $G_2(2)$ is not simple. (Instead the commutator group $G_2(2)' \cong U_3(3)$ had appeared earlier in Theorem 7.1.1, where Q is symplectic but not extraspecial.)

and (8)(ii) in 7.3.1. Finally (4)(vi) above corresponds to the linear groups in (2) of 7.3.1. (But notice it can only arise when G is a non-simple extension of a linear group by a graph automorphism—since here B is not $Z(Q)$, but the group of all transvections with a given center or axis, which would be weakly closed in Q in the absence of that automorphism.) ◇

At the outset of [**Str80**], Stroth mentions one reason why his proofs for $q > 2$ are often easier than in Timmesfeld's case for $q = 2$: since B is not weakly closed in Q, the result [**Tim78b**, 2.18] provides a useful element of order $q - 1$ acting on B—but that element is trivial in the analogous result [**Tim78a**, 3.8] for $q = 2$. (In the examples, this element is in a Cartan subgroup acting on the root group B.) There are many parallels among the preliminary results: for example Stroth's 2.1, 3.3, 3.5, and 3.7 correspond to Timmesfeld's 2.1, 3.10, 3.11, and 4.2. Here and elsewhere, Stroth uses work of Beisiegel in [**Bei77**] on "semi-extraspecial" p-groups; there are also some references to the unpublished work of Cooperstein-Mason [**CM80**], which can be replaced by the work of Guralnick and Malle [**GM02, GM04**].

In particular (as we assumed in our notation in Remark 7.5.3), there is an analogue of the group $\overline{L}$ in Notation 7.3.2; and when $\overline{L}$ is not a TI-set in $\overline{M}$, there are analogues of $\overline{R}$ and $\overline{N}$ in Notation 7.3.5. Stroth's 3.10 is the analogue of Timmesfeld's crucial [**Tim78a**, 4.5], giving a list of possibilities for $\overline{R}/\overline{N}$. The proof uses various of the references standard in the area (notably on root involutions), as well as Harada [**Har68**].

The main part of the proof appears in Stroth's Section 12, closely following the sequence in Section 14 of [**Tim78a**]; for comparison we will refer to our discussion of the latter work in Section 7.3.

The case $w = 1$ is essentially Stroth's conclusion (1) in Remark 7.5.3, while the case $w = 2$ leads at 12.3 to conclusion (2) there; so $w \geq 3$ is assumed in the other results in the section (and it is also required elsewhere, for example in Section 11). The case where $\overline{L}$ is a TI-set in $\overline{M}$ is shown at 3.7 to lead to conclusion (3) (the configuration in ${}^3D_4(q)$); this is the analogue of 14.3 (and 4.2) in Timmesfeld. Thus we now assume that $\overline{L}$ is not a TI-set; so all remaining cases will appear as subcases of Stroth's conclusion (4). The case of an involution of type a_2 in $\overline{L}$ is shown at 11.11 to lead to conclusion (4)(vi), the configuration corresponding to a non-simple extension of a linear group by a graph automorphism. This argument is a simpler analogue of Timmesfeld's 14.2 (based on 12.16): first, sporadic configurations for $q = 2$ do not arise here; and second, an analysis corresponding to the Dempwolff-Wong Theorem 7.2.6 (where $\overline{M}$ is reducibile on $\tilde{Q}$) is not required here—since, as we saw, the simple linear groups do not arise.

Thus we now assume there is no involution of type a_2 in $\overline{L}$. This makes possible various results starting at 3.3;[14] furthermore $\overline{R}/\overline{N}$ must be one of the possibilities in 3.10. The case where $\overline{N}$ has order q is the analogue of $\overline{N} = \langle \overline{t} \rangle$, as in Timmesfeld's 14.4; in Stroth's 6.5, this is shown to lead to conclusions (4)(i) or (4)(ii) in Remark 7.5.3—and again the argument is easier, because of the absence of sporadic configurations. The analogue of Timmesfeld's 14.5 is provided in Stroth's 12.1—via 10.8[15] and 9.8, which eliminate the cases where $\overline{R}/\overline{N}$ is linear

[14]A hypothesis of no a_2 involutions is apparently implicit in the results 3.6–3.10 in [**Str80**], following 3.5 where it is explicitly stated.

[15]The "10.9" in [**Str80**, 12.1] is a misprint for "10.8".

or $\Omega_4^+(q)$ (with $|\overline{N}| > q$); again the exceptional cases for $q = 2$ in Timmesfeld do not arise. Now the remaining possibilities in 3.10 are $\overline{R}/\overline{N}$ symplectic or orthogonal (and not $\Omega_4^+(q)$). The case where $\overline{R}/\overline{N}$ is symplectic is eliminated later in Stroth's 12.1, corresponding to parts of Timmesfeld's 14.10 and 14.13: for in the two configuration in Stroth's 8.3 (similar to Timmesfeld's 9.4) one can apply induction to $C_{\overline{M}}(\bar{t})$, where the "width" is 4 or 8—the former is even more easily eliminated than in Timmesfeld, due to the existence of fewer conclusions for Stroth with $w = 4$. Thus $\overline{R}/\overline{N}$ must be orthogonal. The possibilities in Stroth's 7.8 correspond to those in Timmesfeld's 8.4. The cases with $\overline{R}/\overline{N} \cong \Omega_8^-(q)$ or $\Omega_{10}^+(q)$ are similarly eliminated by induction still later in Stroth's 12.1, as in Timmesfeld's 14.10. Induction also eliminates $\overline{R}/\overline{N} \cong \Omega_8^+(q)$ with $w = 12$, as in Timmesfeld's 14.13—this elimination seems to have been inadvertently omitted in Stroth's text. The remaining three possibilities in 7.8 now lead to subconclusions (iii)–(v) in Stroth's conclusion (4). And this completes the proof.

Identifying G in the cases for $\overline{M}$. The cases from Stroth, indicated in Remark 7.5.3 above, were handled by Smith in [**Smi81**]; this involved generalizing to $q > 2$ the techniques for $q = 2$ in [**Smi80b, Smi80a**]—those from [**Smi79b**] were not really needed. Again we will refer for comparison to our treatment of these earlier papers in Section 7.4, and will confine ourselves to a discussion of various points of difference.

Of course the preliminaries in [**Smi81**] quote a good deal of work from the earlier papers, especially Stroth [**Str80**]; thus below we will maintain the notation from our discussion above of Stroth's argument. We mention that 1.8 shows in addition (when $\overline{L}$ is not a TI-set and contains no a_2 involutions) that there is an involution $\bar{t} \in \overline{L}$, which, together with $\overline{R}/\overline{N}$, uniquely determines one of the cases (i)–(v) in Stroth's conclusion (4) in Remark 7.5.3. And as in Stroth: results from Beisiegel [**Bei77**] are used at various points, and references to the unpublished Cooperstein-Mason work [**CM80**] can be replaced by Guralnick-Malle [**GM02, GM04**].

The main argument is outlined in Section 2. First 2.1 (via 1.12) treats Stroth's conclusion (1), where $w = 1$: here root involution theory [**Tim75a**, Cor B] is used to show that G is a non-simple extension of $F^*(G) \cong L_3(q)$ containing a graph automorphism. Thus after 2.1, *we may assume that* $w \geq 2$. Next it is observed at 2.2 that Stroth's conclusion (2) (where $w = 2$) splits into two subcases, according to whether $\overline{L}$ does, or does not, contain an involution of type a_2. The bulk of Section 2 then gives a sketch of the proof. The emphasis is on the parallel treatment of various cases via the approach of [**Smi80b, Smi80a**]; thus 2.3 through 2.19 give a statement of the intermediate results needed—with details of the proof in the various cases postponed to later sections. For comparison with our earlier discussion in Section 7.4, we summarize these via:

2.3, 2.4	—the structure of $\overline{M}$
2.5–2.9	—fusion of B in Q
2.10, 2.11	—treatment of some small cases
2.12, 2.13	—fusion of B in M: $z^G \cap M$ is a set of root involutions
2.14–2.19	—z^G is a class of root involutions, identifying G

In particular we mention that again Holt's Theorem B.7.16 is used at 2.17 to establish that the graph on z^G is connected. Section 2 closes with a few technical lemmas needed in the process.

Section 3 treats the case where there are involutions of type a_2 in $\overline{L}$: We saw that this includes a subcase of Stroth's conclusion (2) with $w = 2$, as well as his conclusion (4)(vi), where $w \geq 3$ and $\overline{M}$ contains $SL_w(q)$ along with a graph automorphism. And notice since B is not weakly closed in Q that the treatment can begin at the situation analogous to 1.5 in [**Smi80b**], and proceed through the analogue of 1.9 there. But now instead of the approach of Dempwolff-Wong in the proof of 7.2.6, a shortened form of the approach of Section 2 is available: The fusion analysis through 2.12 shows that G is not simple, using the Thompson Transfer Lemma B.2.9. Then $F^*(G)$ can be identified as $L_{w+2}(q)$ as in 3.1, since here "V_1" is a unipotent radical weakly closed, as in conclusion (2)(ii) of 7.5.2.[16]

Thus after Section 3, *we may assume that* $\overline{L}$ *contains no involution of type* a_2; as usual this imposes various important restrictions dating back to Timmesfeld's analysis in [**Tim78a**]. Then Section 4 treats the case where $\overline{L}$ is a TI-set; this is essentially the analogue of 1.11 in [**Smi80b**]. Here we have the remaining subcase of Stroth's conclusion (2) where $w = 2$; and also of his conclusion (3) with $w = 4$ and $\langle \overline{L}^{\overline{M}} \rangle \cong L_2(q^3)$. Again only an abbreviated version of the approach of Section 2 is needed: namely fusion analysis through 2.10, showing that $z^G \cap M \subset Q$. As observed at 2.11, this condition allows the use of Corollary B in Timmesfeld's work [**Tim75a**] on root involutions, to identify G as $G_2(q)$ or ${}^3D_4(q)$.

So after Section 4, *we may also assume that* $\overline{L}$ *is not a* TI*-set in* $\overline{M}$; this gives us the restricted list of possibilities for $\overline{R}/\overline{N}$ indicated at 1.8. Section 5 now treats Stroth's conclusions (4)(i) and (4)(ii), where $|\overline{N}| = q$ and $\overline{R}/\overline{N} \cong \Omega^{\epsilon}_w(q)$ with $w \geq 4$. This argument very closely parallels the treatment of the orthogonal groups in [**Smi80b**], and indeed for the most part is easier. In particular, it uses the approach of Section 2—through 2.19 where z^G is shown to be a class of root involutions, at which point G is identified as $\Omega^{\epsilon}_{w+4}(q)$ using [**Tim75a**].

Finally in Section 6, we are left just with Stroth's cases (iii)–(v) in his conclusion (4)—namely the special configurations with $w = 10$, 16, or 28. Here the argument closely parallels the treatment of groups of type E in [**Smi80a**]. Indeed at 2.3, just as in that earlier paper, in the orthogonal case, one can identify $\overline{M}$ as $\Omega^+_{12}(q)$ when $w = 16$; then the result for $w = 16$ is used to identify $\overline{M}$ as $E_7(q)$ when $w = 28$. The fusion analysis also follows [**Smi80a**], using the approach of Section 2. Again z^G is shown to be a class of root involutions, so that G is identified as ${}^2E_6(q)$, $E_6(q)$, $E_7(q)$, or $E_8(q)$ by [**Tim75a**].

This completes the work of [**Smi81**], and hence the treatment of conclusion (3) in the $GF(2^n)$ Type Theorem 7.5.2.

[16] We mention that the "[20]" in [**Smi81**, 3.1] is a misprint for "[26]", namely Timmesfeld's treatment [**Tim81**] of the weakly closed case.

CHAPTER 8

The final contradiction: Eliminating the Uniqueness Case

In the classification of groups of characteristic 2 type, the Preuniqueness Case for $\sigma(G)$ is defined (see 2.3.8) as a subcase of the general case for $e(G) \geq 3$ by the following condition:

For every prime $p \in \sigma(G)$ and Sylow p-subgroup P of G, there is a maximal 2-local subgroup M of G such that $\Gamma^0_{2,P}(G) \leq M$.

This is the common statement of conclusion (3) of the Gorenstein-Lyons Pretrichotomy Theorem 5.1.1 for $e(G) \geq 4$, and of conclusion (5) of the $e(G) = 3$ Theorem 4.0.1 of Aschbacher.

Thus to complete the characteristic 2 type classification, we must show that no groups appear in the Preuniqueness Case.

Recall (from 2.3.9 and 2.4.1) that for a group G of characteristic 2 type such that $e(G) \geq 3$, $\sigma(G)$ is defined as the first nonempty set of odd primes p on the following list:

(1) $\{p : m_{2,p}(G) \geq 4\}$
(2) $\{p : m_{2,p}(G) = 3 \text{ and } p \geq 7\}$
(3) $\max\{p : m_{2,p}(G) = 3\}$

When $\sigma(G)$ is defined by the third case, $\sigma(G) = \{3\}$ or $\{5\}$.

The first step toward eliminating the Preuniqueness Case for $\sigma(G)$ is to show that it in fact leads to the considerably stronger Uniqueness Case for $\sigma(G)$ of Definition 2.7.2—in which (among other things) the full 2-generated core $\Gamma_{2,P}(G)$ lies in M. Indeed usually even $\Gamma_{1,P}(G) \leq M$; that is, M is strongly p-embedded in G. Obtaining this is a bootstrapping task: from the assumed initial property $\Gamma_{2,P}(G) \leq M$ of the 2-local subgroup M, through stronger and stronger p-uniqueness properties, and also stronger 2-uniqueness properties of M, to the initial goal of M essentially being strongly p-embedded in G.[1]

The Uniqueness Case for $\sigma(G)$ appears as conclusion (III) in the full Trichotomy Theorem 5.1.2, and we repeat it here for convenience. It asserts for every $p \in \sigma(G)$ and Sylow p-subgroup P of G that there is a maximal 2-local subgroup M of G with the following two properties:

(1) M is almost strongly p-embedded in G; and
(2) For any 2-local subgroup H of G, $m_p(H \cap M) \geq 2$ implies $H \leq M$.

For convenience we recall the definition (see 2.7.1) of "almost strongly p-embedded" for an odd prime divisor p of $|G|$:

[1] The theme of bootstrapping between p-uniqueness and 2-uniqueness is then implicitly continued in Aschbacher's subsequent elimination of the Uniqueness Case: for example, Aschbacher repeatedly exploits 2-uniqueness via the Global $C(G,T)$ Theorem B.7.8.

The subgroup M of G is almost strongly p-embedded in G if and only if M contains $\Gamma_{2,P}(G)$ for some Sylow p-subgroup P of G, and one of the following conditions holds:

(1) M is strongly p-embedded in G, i.e., $\Gamma_{1,P}(G) \leq M$.
(2) $p \geq 5$, P is abelian, M is solvable, and the following hold:
  (a) $P \geq P_0$ of order p with P_0 weakly closed in P with respect to G and $C := C_G(P_0) \not\leq M$.
  (b) $C_G(P_1) \leq M$ for all subgroups P_1 of order p in P except P_0.
  (c) $E(C) \not\leq M$, $E(C) \cong L_2(p^n)$ for some $n \geq 2$, and $C_C(E(C))$ has cyclic Sylow p-subgroups.
  (d) $O_{p'}(M)P_0$ is a Frobenius group with kernel $O_2(M)$ and complement $O_{p'}(C)P_0$.
(3) $p = 3$, $P \cong \mathbb{Z}_3 \wr \mathbb{Z}_3$, M is solvable, and the following hold:
  (a) $\Gamma_{1,J(P)}(G) \leq M$, but $C := C_G(P_0) \not\leq M$ where $P \geq P_0$ is of order 3 with $P_0 \not\leq J(P)$.
  (b) Let $W := O_{2'}\big(O_{3'}(C)\big)$. Then M covers C/W and $F(W) \cap M = 1$.
  (c) Let T be a P-invariant Sylow 2-subgroup of $O_{3'}(M)$. Then $C_T(P_0)$ is isomorphic to a central product $Q_8 * Z_{2^n}$ for some $n \geq 1$.

This first step was accomplished by Gorenstein and Lyons (building on an earlier collaboration with Aschbacher in [**AGL81**]) in Theorem C of their Trichotomy Memoir [**GL83**]—which unlike Theorems A and B of that paper, treats not just groups G such that $e(G) \geq 4$, but also groups G with $e(G) = 3$. We state the result in the form:

THEOREM 8.0.1 (The Preuniqueness-implies-Uniqueness Theorem). *Let G be a K-proper simple group of characteristic 2 type with $e(G) \geq 3$. If G satisfies the Preuniqueness Case for $\sigma(G)$, then G in fact satisfies the Uniqueness Case for $\sigma(G)$.*

We have chosen to discuss this result in Section 8.1 of this chapter, rather than in Chapter 5 on the Trichotomy Theorem—since the methods involved are more closely related to those in Aschbacher's treatment [**Asc83b, Asc83c**] of the Uniqueness Case, than to the signalizer functor and standard form arguments used in the proof of the Trichotomy Theorem.

The huge second and final step after the Preuniqueness-implies-Uniqueness Theorem 8.0.1 was accomplished by the work of Aschbacher in [**Asc83b, Asc83c**] which eliminated the Uniqueness Case:

THEOREM 8.0.2 (Uniqueness Case Theorem). *No finite group G having characteristic 2 type and $e(G) \geq 3$ can satisfy the Uniqueness Case for $\sigma(G)$.*

The bulk of this chapter will be devoted to an outline of Aschbacher's proof of this result.

We begin by defining a subset of the notation used in the papers indicated above.

NOTATION 8.0.3. Assume first that we are in the Preuniqueness Case 2.3.8. Then for each prime $p \in \sigma(G)$ and $P \in Syl_p(G)$, there is a maximal 2-local subgroup M of G containing P, such that:

$$\Gamma^0_{2,P}(G) \leq M.$$

We will abbreviate such a choice of p and M by "$p \in \sigma(M)$".[2]

Assume next that we are in the Uniqueness Case. Then for "$p \in \sigma(M)$" we have the stronger condition from Definition 2.7.1 that M is almost strongly p-embedded in G. In particular:

$$\Gamma_{2,P}(G) \leq M;$$

and usually M is even strongly p-embedded in G—except in cases (2) and (3) of 2.7.1. ◊

REMARK 8.0.4 (Uniqueness theorems). To reduce to the Uniqueness Case in 8.0.1, we must show that $\Gamma_{2,P}(G) \leq M$. Then many arguments in the elimination of the Uniqueness Case in 8.0.2 focus on some p-subgroup R of P, with the aim of showing that $N_G(R) \leq M$. More generally, we may wish to show that $H \leq M$, for some 2-local subgroup H of G which contains some specific p-subgroup R of P.

Often one shows that $H \leq M$ by proving that M is the unique maximal 2-local subgroup of G containing R. Results in this spirit are called *uniqueness theorems* on p. 386 of [**Asc83b**]. In much of the literature (and in particular in [**Asc83b, Asc83c**]), this assertion is expressed in the form $\mathcal{M}(R) = \{M\}$—where $\mathcal{M}$ denotes the set of maximal 2-local subgroups of G, and $\mathcal{M}(R)$ consists of those members of $\mathcal{M}$ containing R. In the quasithin work [**AS04a, AS04b**] that appeared much later, the notation $M = !\mathcal{M}(R)$ was used for this assertion, and we follow that convention here.

We mention that uniqueness theorems are often established using pushing-up techniques; cf. Remark B.7.7. ◊

8.1. Prelude: From the Preuniqueness Case to the Uniqueness Case

Before discussing the Preuniqueness-implies-Uniqueness Theorem 8.0.1, we first record some preliminary uniqueness theorems that were established expressly for the proof of that result.

Uniqueness theorems of Aschbacher-Gorenstein-Lyons. The bootstrapping of uniqueness properties involves proving some intermediate implications, between certain p-local uniqueness properties of M and certain 2-local uniqueness properties. Results are available in both directions, and the proof of the Preuniqueness-implies-Uniqueness Theorem zigzags from a p-local property to a 2-local property to a stronger p-local property, and so on.

A number of uniqueness theorems of the first type, i.e., that p-local uniqueness properties imply 2-local uniqueness properties (in the sense of Remark 8.0.4), which are needed in the proof both of the Preuniqueness-Implies-Uniqueness Theorem 8.0.1 and of the Uniqueness Case Theorem 8.0.2, are in fact established in a joint paper [**AGL81**] of Aschbacher, Gorenstein, and Lyons, which was written as a preliminary to their separate papers on those two Theorems.

The results of the second type, namely that 2-local uniqueness properties imply p-local uniqueness properties, are contained in Chapter II.3 of [**GL83**], which is written to cover the case $e(G) = 3$ as well as $e(G) \geq 4$. This chapter, with crucial references to [**AGL81**], thus provides the zigzag proof of the Preuniqueness-implies-Uniqueness Theorem.

[2]In fact, we will normally use the particular letter "p" *only* for a prime in $\sigma(M)$—though we try to always specify this at each use. We mention that in [**Asc83b, Asc83c**], this convention for p is often used implicitly.

As these preliminary results are at level 1 from our viewpoint, we will not discuss them in full detail; however, they are sufficiently important in establishing the level 0 results 8.0.1 and 8.0.2 that we will provide at least a brief description in this subsection. In fact, the results in [**AGL81**] were originally heading for publication separately in the $e(G) = 3$ and $e(G) \geq 4$ papers [**Asc81b, Asc83a, GL83**] until Aschbacher, Gorenstein and Lyons, seeing the similarity of their results, decided to join forces.

Results developed for the Preuniqueness-Implies-Uniqueness proof. We indicate several of these preliminary results, within a rough outline of the overall deduction of the Uniqueness Case.

Theorem A of [**AGL81**] treats 2-local subgroups H containing a large p-subgroup of M; we state it in the form:

THEOREM 8.1.1. *Assume the Preuniqueness Case, with $p \in \sigma(M)$ as in Notation 8.0.3. If H is a 2-local subgroup of G with $m_p(H \cap M) \geq 3$, then $H \leq M$.*

So for a p-subgroup $R \leq M$ with $m_p(R) \geq 3$, we have $M = !\mathcal{M}(R)$ in the language of 8.0.4.

This result is quoted as Theorem D1 on p. 12 of [**GL83**]: as we will see below, it has numerous applications in Chapter II.3 of that paper, which establishes the Preuniqueness-implies-Uniqueness Theorem 8.0.1; for example it is used crucially in Section II.3.2, to show that, for each $P \in Syl_p(G)$, the condition $\Gamma^0_{2,P}(G) \leq M$ implies in fact that $\Gamma_{2,P}(G) \leq M$—the first step in obtaining the Uniqueness Case.

As the process of verifying the remaining hypotheses of the Uniqueness Case continues in [**GL83**], Gorenstein and Lyons later make use of Theorem B of [**AGL81**], which assumes that $\Gamma_{2,P}(G) \leq M$. When $3 \in \sigma(G)$, Theorem B also assumes some of the hypotheses of the Uniqueness Case: namely that with respect to $p = 3$, M satisfies either (1) or a variant of (3) in Definition 2.7.1.

Theorem B is an analogue of Theorem 8.1.1, in the sense that the hypothesis of $m_p(H \cap M) \geq 3$ is replaced by $m_p(H \cap M) \geq 2$. Notice that in the situation of Theorem 8.1.1, $H \cap M$ contains a subgroup $R \cong E_{p^3}$, and $\Gamma_{2,R}(G) \leq M$ by the Preuniqueness Case. In Theorem B, a similar role is played by some $R \cong E_{p^2}$ for which it is explicitly assumed that $\Gamma_{1,R}(G) \leq M$. In addition as just noted, an extra assumption is needed when $p = 3$. Here is the statement of Theorem B of [**GL83**]:

THEOREM 8.1.2. *Assume the Preuniqueness Case, and $M \in \mathcal{M}$ with $\sigma(M) \neq \emptyset$, such that for all $p \in \sigma(M)$, we have $\Gamma_{2,P}(G) \leq M$ for $P \in Syl_p(M)$. When $p = 3$, assume further that M contains $N_G(Q)$ for all $1 \neq Q \leq P$ with $m_3(QC_P(Q)) \geq 3$;*[3] *and indeed when M is nonsolvable, that M is strongly 3-embedded in G.*

If H is a 2-local subgroup of G containing a noncyclic p-subgroup R of M with $\Gamma_{1,R}(G) \leq M$, then also $H \leq M$; that is, $M = !\mathcal{M}(R)$.

This result is quoted as Theorem D2 on p. 12 of [**GL83**]. It strengthens the *initial* statement of Theorem 8.1.1 by requiring only that $m_p(H \cap M) \geq 2$.

Eventually in [**GL83**], enough cyclic p-subgroups of M are shown to have their normalizers contained in M so that the hypothesis above that $\Gamma_{1,R}(G) \leq M$ can

[3] This extends the usual condition $\Gamma^0_{2,P}(G) \leq M$ by also allowing the possibility of Q cyclic.

be removed; this in fact gives condition (2) of the Uniqueness Case. That is, the following theorem is established [4] in [**GL83**]:

THEOREM 8.1.3. *Assume the Preuniqueness Case with $M \in \mathcal{M}$ and $p \in \sigma(M)$. If H is a 2-local subgroup of G with $m_p(H \cap M) \geq 2$, then $H \leq M$.*

So for R a noncyclic p-subgroup of M, we have $M = !\mathcal{M}(R)$.

This is the second of the two defining properties of the uniqueness case in [**GL83**, p 9]. It is quoted as Theorem E in [**AGL81**]; and as 6.1 in [**Asc83b**], where it is used frequently in arguments eliminating the Uniqueness Case.

And Chapter II.3 of [**GL83**] is primarily devoted to establishing the first defining property of the uniqueness case, namely that M is almost strongly p-embedded in G; which therefore completes the proof of the Preuniqueness-implies-Uniqueness Theorem.

We will comment more fully below on these various steps; but first we zigzag back to [**AGL81**], for two more results that will be used to help rule out the Uniqueness Case in [**Asc83b, Asc83c**].

Further results from [**AGL81**], *for the Uniqueness Case Theorem 8.0.2.* As in Notation 8.0.3, we assume for each $p \in \sigma(M)$ that M is almost strongly p-embedded in G.

First, Theorem C of [**AGL81**] extends Theorem 8.1.2, so that its conclusion also applies to *all* odd primes q such that $m_q(M) \geq 2$—not just those in $\sigma(M)$; the result is quoted as 6.3 in [**Asc83b**]:

THEOREM 8.1.4. *Assume the Uniqueness Case, let $M \in \mathcal{M}$ with $\sigma(M) \neq \emptyset$, and let q be an odd prime.*

If H is a 2-local subgroup of G containing a noncyclic q-subgroup Q of M with $\Gamma_{1,Q}(G) \leq M$, then also $H \leq M$; that is, $M = !\mathcal{M}(Q)$.

Finally Theorem D of [**AGL81**], quoted as 6.2 in [**Asc83b**], shows that M contains a Sylow 2-normalizer of G:

THEOREM 8.1.5. *Assume the Uniqueness Case, and let $M \in \mathcal{M}$ with $\sigma(M) \neq \emptyset$. Then for $T \in Syl_2(M)$, we have $N_G(T) \leq M$. In particular, $T \in Syl_2(G)$.*

Comments on the proofs of Theorems A–D in [**AGL81**]. An outline of the proofs of Theorems A–D of [**AGL81**] is provided on pp. 338–339 of [**AGL81**]. Large parts of the proofs of Theorems A–C are fairly similar, so that substantial parts of the analysis can be carried out essentially simultaneously.

The proofs consider a suitable counterexample $H \not\leq M$. Since H is a 2-local and M is a maximal 2-local, $H \cap M$ is a proper subgroup of both H and M.

In each case, H possesses a proper subgroup C that by hypothesis contains the normalizers of many p-subgroups of H (such as $\Gamma^0_{2,Q}(H)$ for $Q \in Syl_p(H)$) such that $C \leq H \cap M < H$. This is called a *nongeneration* condition on p. 338 of [**AGL81**]. It yields $O_{p'}(H)R \leq H \cap M$, with $R \in Syl_p(H)$; and since $H \cap M < H$, it also yields that $L_{p'}(H) \not\leq M$. Then results on generation of known groups, such as those of Seitz for groups of Lie type [**Sei82**], lead to a very restricted list of possibilities for H. In almost all of these, and indeed in all cases for

[4] In fact (as we discuss toward the end of the section) this is obtained at II.3.4.9 of [**GL83**] in a generic subcase, and at the end of Section II.3.6 in the remaining case.

which we have $m_p(H \cap M) \geq 3$, any component of $H/O_{p'}(H)$ is strongly locally 1-balanced,[5] like Chevalley groups in characteristic p; as a consequence, when the strong local balance condition holds, $O_{p'}(H)$ is the unique maximal R-invariant p'-subgroup of H, and hence of $H \cap M$. This is useful in locating $O_{p'}(H)$ and $O_2(H) = O_2\big(O_{p'}(H)\big)$ as subgroups in M.

Another consequence of the nongeneration of H is obtained at 9.8 of [**AGL81**]: we usually have $S := O_2(H) = O_2(H \cap M)$, and at 9.8 the few configurations which are exceptions to this statement are specified. This condition leads to properties called *nonfactorization* on p. 339 of [**AGL81**]—the term is used in the sense of the failure of factorization appearing in pushing-up. Namely, in the spirit of the Local $C(G,T)$ Theorem B.7.3, typically we have:

$$S \trianglelefteq C(M,S) = H \cap M < M;$$

though of course here S need not be Sylow in M. (Cf. also Remark B.7.7.) The obstructions which are forced by such a proper containment—playing a role analogous to that of χ-blocks in the Local $C(G,T)$-Theorem B.7.3—are given the name *weak S-blocks*. At 10.1 of [**AGL81**], there is a description of weak S-blocks, leading to a quite explicit description of the structure of M if such blocks exist.

In particular, the typical weak S-blocks are components K of $M/O_2(M)$ lying in $Chev(2)$ or isomorphic to alternating groups of even degree. If $K \in Chev(2)$, then $H \cap K$ is either a parabolic subgroup, or a symplectic group (with K an orthogonal group and $O_2(H \cap M)$ containing an element acting as a transvection on K). If $K \cong A_{2n}$, then $(H \cap K)/\big(O_2(H) \cap K\big)$ has at most two components, and they are alternating groups. On the other hand from the structure of H, the section $(H \cap M)/O_{p'}(H)$ typically has the structure of a strongly p-embedded subgroup of some simple $\mathcal{K}$-group K, or the normalizer of such a subgroup in some overgroup of K in $Aut(K)$. As a result, one can usually observe that the embeddings of $H \cap M$ in H and M are clearly incompatible, which is (generically, at any rate) the contradiction completing the proof of the theorem in question.

The proof of Theorem D of [**AGL81**] is accomplished mainly by pushing-up, making use of the power provided by Theorems A–C. Let $p \in \sigma(M)$. By Theorems A and B of [**AGL81**] and by [**GL83**, Theorem E], M is almost strongly p-embedded in G. If M is not actually strongly p-embedded in G, then its structure is very limited by definition, and with Thompson factorization one easily finds a characteristic subgroup C of a Sylow 2-subgroup T of M such that $N_M(C)$ contains an E_{p^2}-subgroup R with $\Gamma_{1,R}(G) \leq M$. Thus by Theorem B, $N_G(C) \leq M$, yielding the conclusion of Theorem D. So we may assume that M is strongly p-embedded in G; again by Theorem B, the conclusion of Theorem D holds if $m_p\big(N_G(C)\big) \geq 2$ for some characteristic subgroup C of T. Moreover, any $p \in \sigma(M)$ will do. In Section 15 of [**AGL81**], this approach is shown to complete the proof of Theorem D, except in the case that $O^{p'}(M)/O_2\big(O^{p'}(M)\big) \cong U_4(2^n)$, $2^n \equiv -1 \pmod{p}$. This final case is treated in Section 16 by weak closure methods—such as are discussed in our Section B.8.

Here is a quick sketch of the layout of [**AGL81**]: Sections 3–7 review properties of $\mathcal{K}$-groups. Sections 8–10 provide preliminaries on groups of characteristic 2 type, as well as general properties of pairs H,M of subgroups of G, such that $H \cap M$ is embedded in H and M as above, and the theory of weak S-blocks. Section 11 treats

[5] We discuss balance conditions in Section B.3.

the generic case for Theorems A and B, while Section 12 does the same for Theorem C; then Sections 13 and 14 similarly treat exceptional cases for those results. Sections 15 and 16 then treat the generic and exceptional cases for Theorem D.

The literature cited in the papers includes fairly standard references, typically mentioned already in earlier chapters and in the Appendix, on topics such as properties of known groups (particularly groups of Lie type), pushing-up (including failure of factorization), and weak closure. Among these are the Local $C(G,T)$-Theorem B.7.3, Baumann's Lemma B.7.5 B.7.5, and a variant of the Glauberman-Niles Theorem B.7.6 from Campbell's Ph.D. thesis [**Cam79**]. There is also reference to Goldschmidt's work on trivalent graphs [**Gol80**] (cf. our discussion of amalgams in Section 3.3).

The proof of 8.13 in [**AGL81**] quotes 20.12 from Aschbacher's work [**Asc83a**] on the $e(G) = 3$ Theorem 4.0.1.

Comments on the proof of Theorem C in [**GL83**]. The proof of Theorem C of [**GL83**], namely the Preuniqueness-Implies-Uniqueness Theorem 8.0.1, is carried out in Chapter 3 of Part II of that work.[6] Our exposition here will largely parallel the introduction contained in Section 3.1 of [**GL83**].

We first make some general remarks. To start, fix a maximal 2-local M and a prime $p \in \sigma(M)$; we aim to prove that M is almost strongly p-embedded in G, and that $M = !\mathcal{M}(R)$ for any noncyclic p-subgroup $R \leq M$—the two conditions of the Uniqueness Case in Definition 2.7.2.

The structure of the 2-local subgroup M, together with the uniqueness theorems from [**AGL81**], exert significant control over the p-local structure of G, especially the structure of normalizers of p-subgroups of p-rank at most 2, which is the focus of this chapter, at least p-locally.

However, local analysis sometimes reaches beyond the primes 2 and p to "guest" odd prime divisors f of $|M|$. Since we are in the Preuniqueness Case, Theorems 8.1.1 and 8.1.2, and any partial results so far proved in [**GL83**], may be applied *to any prime* $f \in \sigma(G)$. This observation is critically important for this chapter. For any $f \in \sigma(G)$, it yields at a minimum that there is a maximal 2-local M_f of G with properties relative to f that are analogous to those of M relative to p. Moreover if $f \in \sigma(M)$, then $M_f = M$ by Theorem 8.1.1.

The analysis is complicated considerably by the possible existence of E_{p^2}-subgroups B of M such that $m_p(C_G(B)) = 2$, i.e., such that B contains every element of order p in its centralizer in G. Section (II.)3.6, the longest and most complex section of Chapter II.3 of [**GL83**], deals with this case. A number of the results in that section either establish or use uniqueness properties of primes $f \notin \{2, p\}$, for example at 3.6.5, 3.6.12, 3.6.13, and 3.6.20.

REMARK 8.1.6. An important $\mathcal{K}$-group lemma [**GL83**, I.29-1][7] classifies all those $\mathcal{K}$-groups X such that $m_p(X) \geq 3$, but X possesses an E_{p^2}-subgroup B such that $m_p(C_G(B)) = 2$. Of these the most demanding (when they occur as overgroups X of B in G, as in the previous paragraph) are: the wreathed configuration

[6]As in Chapter 5, our references for chapters in [**GL83**] will typically be abbreviated, by omitting the prefix "II"; though we may sometimes use notation such as II.3 or I.28 to distinguish references within the different Parts.

[7]The alternative (c) stated in the conclusion of this lemma is a red herring: it does not actually occur but was erroneously included.

(see also [**GL83**, I.6-9]), where $X/O_{p'}(X) \cong K \wr Z_p$, with K simple of p-rank 1; and the "PGL/PGU-configuration", where $X/O_{p'}(X) \geq L \cong SL_p(q)$ or $L_p(q)$, with $q \equiv 1 \pmod p$, and $X/O_{p'}(X) \cong PGL_p(q)$ (or a similar configuration involving $PGU_p(q)$, with $q \equiv -1 \pmod p$). Easier are the remaining cases, in which X is p-constrained or $E(X/O_{p'}(X))$ is in $Chev(p)$—in fact $E(X/O_{p'}(X)) \cong B_2(p)$, $G_2(p)$ or $L_2(p^p)$. ♢

The full 2-generated core. The first step toward establishing the Uniqueness Case is to show in Proposition 3.2.1 that for $M \in \mathcal{M}$ and $p \in \sigma(M)$, we can use the hypothesis $\Gamma^0_{2,P}(G) \leq M$ of the Preuniqueness Case to show that:

$$\Gamma_{2,P}(G) \leq M.$$

The proof proceeds by contradiction, so there exists some noncyclic $D \leq P$ such that $N_G(D) \not\leq M$. Since $\Gamma^0_{2,P}(G) \leq M$, $m_p(DC_P(D)) = 2$, and in particular $\Lambda_2(P)$ is disconnected (cf. B.4.8). Now D is chosen maximal with respect to a number of properties; and detailed analysis of D-signalizers, i.e., D-invariant p'-subgroups of G, using the Bender-Thompson Signalizer Lemma B.1.10 and examination of the wreathed and PGL configurations above, leads to a contradiction.

In more detail, the cases $|N_G(D)|_p = |P|$ and $|N_G(D)|_p < |P|$ are considered separately:

In the first case, it is enough to show that all D-signalizers lie in M. For then, the subgroup X that they generate contains $O_2(M)$ and so satisfies $O_2(X) \neq 1$. As $P \leq N_G(D) \leq N_G(O_2(X))$, it follows that $N_G(D) \leq N_G(O_2(X)) \leq M$ by Theorem 8.1.1. To show that all D-signalizers are in M, a subgroup $U \leq D$ with $U \trianglelefteq M$ and $U \cong E_{p^2}$ is used; all D-signalizers are U-signalizers, and to show that they lie in M, one is helped by the fact that $m_p(C_M(U)) \geq 3$.

In the second case, namely when $|N_G(D)|_p < |P|$, a quick reduction is made to the case $D \cong E_{p^2}$, $Z(P) < D$, $N_0 := N_G(Z(P))$ is p-constrained and lies in M, and $Aut_G(D)$ contains $SL_2(p)$. This time, D-invariant 2-groups are investigated. It is shown first at 3.2.13 that any $T \in \mathcal{U}^*_M(D;2)$ (the set of maximal $2'$-subgroups of M invariant under D) is a Sylow 2-subgroup of the P-invariant subgroup $O_{p'}(M)O_{p'}(N_0)$ of M, so $N_G(T) \leq M$ by Theorem 8.1.1 and $T \in \mathcal{U}^*_G(D;2)$. Second, a transitivity theorem 3.2.15 à la Thompson is also proved: namely $C_G(D)$ permutes $\mathcal{U}^*_G(D;2)$ transitively. This leads to $N_G(D) \leq C_G(D)N_G(T) \leq M$, a contradiction.

In establishing 3.2.15, the fact that G is of characteristic 2 type helps to compensate for the fact that $m_p(D)$ is only 2, whereas Thompson transitivity normally would require p-rank 3.[8] A certain 2-local H containing D and obstructing the usual transitivity argument is selected, maximizing the size of a D-signalizer in $H \cap M$. The obstruction occurs because $H \not\leq M$, and indeed because H has a p-component K not contained in M. Using Theorem 8.1.1, it is shown that $m_p(H) = 2$. Then Thompson factorization (and the maximality) yields the contradiction $K \leq M$, unless $p = 3$ and $K/O_{3'}(K) \cong A_7$. In this final case, a pushing-up argument of Aschbacher and Chermak is needed to get a contradiction and complete the proof that $\Gamma_{2,P}(G) \leq M$.

---

[8]Namely, one can show that for a pair R, T of maximal D-invariant 2-subgroups of G such that $R \not\leq M$, $T \leq M$, and $R \cap M = R \cap T$, there is $x \in D \setminus Z$ with nontrivial fixed points on both $N_R(R \cap T)/(R \cap T)$ and $N_T(R \cap T)/(R \cap T)$. This is because $N_G(Z) \leq M$, D acts faithfully on R and T, and T is invariant under P—in which the subgroups of D of order p other than Z are all fused to one another.

Most of the arguments for $\Gamma_{2,P}(G) \le M$ are comparatively elementary—using for example properties of $\mathcal{K}$-groups, including the list in the preliminary result I.24.4 of strongly p-embedded subgroups in known groups. But as noted, crucial use is made of Theorem 8.1.1; and the final argument uses pushing-up results quoted in Section I.28, such as the Local $C(G,T)$ Theorem B.7.3 (their I.28.3).

Obstructions to strong p-embedding; the set $\mathcal{P}(M)$. Now whenever we can prove $\Gamma_{1,P}(G) \le M$, then we have case (1) of almost strongly p-embedded in Definition 2.7.1; which gives condition (1) for the Uniqueness Case 2.7.2. Moreover, if we can prove that $\Gamma_{1,P}(G) \le M$ for every $p \in \sigma(M)$, then the special hypothesis of Theorem 8.1.2 is satisfied with respect to $p = 3$, and that theorem implies that any 2-local containing a noncyclic p-subgroup of R lies in M; this gives the remaining condition (2) of the Uniqueness Case, and we are done.

Therefore from this point on, we may assume that we have chosen $p \in \sigma(M)$ such that

$$\Gamma_{1,P}(G) \not\le M, \tag{8.1.7}$$

and it remains to establish the more technical conditions in the exceptional cases (2) or (3) of Definition 2.7.1 of almost strongly p-embedded (as well as condition (2) of the Uniqueness Case 2.7.2). The choice of p will be refined further in the final section [**GL83**, II.3.6].

Since $\Gamma_{1,P}(G) \not\le M$, the analysis will focus on "obstructions" $x \in M$ of order p, having the property that $C_G(x) \not\le M$. Let:

$$\mathcal{P}(M) := \{x \in M : x^p = 1 \ne x,\ m_p(C_G(x)) \ge 3,\ C_G(x) \not\le M\};$$

namely, the set of such elements with $m_p(C_G(x)) \ge 3$. The strategy subdivides into the cases where $\mathcal{P}(M)$ is nonempty, or empty—so that in the latter case, for each obstructing x, we have $m_p(C_G(x)) = 2$.

The generic case. Sections 3.3–3.5 treat the case where:

$$\mathcal{P}(M) \ne \emptyset.$$

An initial result 3.3.1 establishes basic properties of any $x \in \mathcal{P}(M)$, including a short list of possibilities for the (unique!) p-component L_x of $C_G(x)$. We have that $L_x \not\le M$. Again the arguments use Theorem 8.1.1, and the list at I.24.4.

We have $O_{p'}(C_G(x)) \le M$ and $m_p(C_{C_G(x)}(L_x/O_{p'}(L_x))) = 1$, because we have $m_p(C_G(x)) \ge 3$, and the preuniqueness condition $\Gamma^0_{2,P}(G) \le M$ holds. So M does not cover $L_xO_{p'}(C_G(x))/O_{p'}(C_G(x))$. Then Theorem 8.1.1 and a Frattini argument give the contradiction $C_G(x) \le M$—unless $O_{p'}(C_G(x))$ has odd order. This conclusion and the fact that $F^*(M) = O_2(M)$ yield at 3.3.3 that in any subgroup $D \le M$ with $D \cong E_{p^2}$, there are at least two subgroups of order p not G-conjugate to $\langle x\rangle$. A detailed analysis ensues of p-fusion and of 2-subgroups of G invariant under E_{p^2}-subgroups of M. This successively eliminates most cases from 3.3.1—yielding in particular at 3.4.1 that $p \ge 5$, and reducing at 3.4.10 to the single possibility $L_x/O_{p'}(L_x) \cong L_2(p^n)$. During this process, which involves further applications of Theorem 8.1.1, condition (2) of the Uniqueness Case is established for p at 3.4.9, making use of the preliminary uniqueness result Theorem 8.1.2. Again some of the arguments use results quoted from Section I.28, on failure of factorization (e.g. B.6.8, their I.28.1) and pushing-up (e.g. the Glauberman-Niles Theorem B.7.6, their I.28.3).

So as $p \ge 5$, it remains to show that condition (2) for almost strongly p-embedded in Definition 2.7.1 holds. This is accomplished in Section 3.5 with straightforward arguments. First, 3.5.1 shows that the p-component L_x must in fact be quasisimple by showing in the contrary case that there is a prime divisor f of $|O_{p'}(L_x)|$ such that $f \in \sigma(M)$ and $m_f\big(O_{p'}(L_x)\big) \geq 3$, which in turn yields the contradiction $L_x \leq M$ by the hypothesis of the Preuniqueness Case. With this information in hand, 3.5.3 shows that M is solvable, and provides details of its structure. Then 3.5.7 makes use of this structure, along with other facts established earlier, to complete the verification of condition (2) in Definition 2.7.1. Thus using (8.1.7), the conclusion of Theorem 8.0.1 holds in the generic case, i.e. for primes p such that either M is strongly p-embedded in G or $\mathcal{P}(M)$ is nonempty.

The small-rank case. Hence we may now assume that:

$$\mathcal{P}(M) \text{ is empty}$$

and so by (8.1.7), there exists some $x \in M$ of order p, which satisfies $C_G(x) \not\leq M$ and $m_p\big(C_G(x)\big) = 2$.

The argument in this case is considerably more technical. The analysis is made for the largest prime $p \in \sigma(G)$ for which the conclusions of the Uniqueness Case fail. It is not until 3.6.27, very near the end, that a contradiction is obtained when $p > 3$.

An E_{p^2}-subgroup $B \leq M$ is fixed with $x \in B$. There is no loss in assuming that $x \in P$, and then $B = \langle x \rangle \times Z$, where $Z := \Omega_1\big(Z(P)\big)$. As $\mathcal{P}(M) = \emptyset$, we have $N_G(Z) \leq M$, and the subgroups of B of order p other than Z are fused in $N_P(B)$. Moreover, B contains every element of order p in $C_G(B)$.

The first step is 3.6.2, which uses the condition $m_p\big(C_M(x)\big) = 2$ to show that either M is p-constrained, or a p-component of M is restricted to a short list (see Remark 8.1.6 above). This information is used in a long analysis of B-signalizers and their normalizers. At 3.6.5 it is shown that any B-invariant subgroup $1 \neq F \leq M$ of odd p'-order satisfies $N_G(F) \leq M$, unless the PGL/PGU-configuration of Remark 8.1.6 occurs in M. It is in the proof of 3.6.5 that the maximality of p is used. Then at 3.6.14 and 3.6.21 it is shown that M contains any B-invariant 2-subgroup S such that $S \cap M \neq 1$, and M contains any 2-local H containing both B and a nontrivial B-invariant 2-subgroup of $H \cap M$. These signalizer and uniqueness results lead, in 3.6.22, to a proof that, when $p > 3$, M covers $C_G(x)/O_{2'}(O_{p'}(C_G(x)))$; hence $O_{2'}(O_{p'}(C_G(x)))$ does not lie in M, and in particular is nontrivial—and then this subgroup is exploited to eliminate most cases from 3.6.2, culminating in a proof that $p = 3$ at 3.6.27. Again the arguments use the list of strongly p-embedded subgroups in known groups (this time in the form of I.24.1 rather than I.24.4), as well as Theorem 8.1.1; also some specialized results from Section I.29 are often applied. In addition, there is more systematic use of pushing-up results from Section I.28, including as usual the Local $C(G,T)$ Theorem B.7.3; but also a ZJ-factorization result at I.28.6, and details of pushing-up $L_3(2)$ from Campbell's thesis [**Cam79**] quoted at I.28.12 and I.28.13, which are covered by the later more general treatment of Meierfrankenfeld and Stellmacher [**MS93**].

So now we are reduced to the case $p = 3$ and $e(G) = 3$, with 3 the unique prime in $\sigma(G)$ by definition of $\sigma(G)$. Therefore the various details of condition (3) in Definition 2.7.1 need to be verified; some of them are already in place. First, 3.6.32 establishes fairly easily that M is solvable, and that $P \cong \mathbb{Z}_3 \text{ wr } \mathbb{Z}_3$. And then at 3.6.33 it is observed that this in fact completes the necessary verification. (In addition to the results 3.6.27 and 3.6.32 indicated there, relevant remarks also include 3.6.22.i and the introduction to Section 3.6 on p. 690 with $\langle x \rangle$ in the role of "P_0".) At the end of the section, condition (2) of the Uniqueness Case is also easily established using 3.6.14 and Theorem 8.1.2. Hence the Uniqueness Case conditions hold for those primes p satisfying this final case in which $\mathcal{P}(M)$ is nonempty, completing the proof of Theorem 8.0.1.

8.2. Introduction: General strategy using weak closure and uniqueness theorems

The first section of [**Asc83b**] contains a brief introduction to the proof of the Uniqueness Case Theorem 8.0.2, which we will reproduce in part here.

The role in the CFSG. First as background on p. 384, Aschbacher records the Trichotomy Theorem 5.1.2 for $e(G) \geq 3$. Recall there is some discussion of the Trichotomy Theorem for $e(G) = 3$ in the latter part of Section 2.4, since that result is at best implicit in [**Asc81b, Asc83a**]. In particular, conclusion (5) in 4.0.1 leads by Theorem 8.0.1 to the Uniqueness Case appearing as conclusion (III) in 5.1.2. Aschbacher then quotes the literature for the treatment of conclusions (I) and (II) of 5.1.2 (note for $e(G) = 3$ that he includes in (I) the groups of $GF(2^n)$ type, as opposed to just $GF(2)$ type). Of course we have described the corresponding results in our Chapters 6 and 7 (including Section 7.5).

Thus the classification of simple groups of characteristic 2 type with $e(G) \geq 3$ is reduced to the elimination of the Uniqueness Case, as in conclusion (III) of 5.1.2. And in view of the Quasithin Theorem 3.0.1 (which was not yet available when Aschbacher was writing), the elimination of the Uniqueness Case also completes the classification of all simple groups of characteristic 2 type. Indeed the elimination of the Uniqueness Case even completes the Classification of the Finite Simple Groups (Theorem 0.1.1)—in view of the Dichotomy Theorem 0.3.10 and the GW Type Classification Theorem 1.0.1.

We mention also that Stroth has produced a preprint, described in [**Str96**], which gives an alternate approach to treating the Uniqueness Case—largely replacing Aschbacher's use of weak closure with arguments using the modern amalgam method (cf. the discussion of amalgams within our Section 3.3). It is expected that a suitable version of this treatment will constitute a future volume in the GLS series, eliminating the GLS version of the Uniqueness Case.

An overview of the approach. Aschbacher begins by recalling[9] the Definition 2.7.2 of the Uniqueness Case.

A choice of 2*-locals* M *and* H. Recall for G in the Uniqueness Case, that, for each prime p in $\sigma(G)$, G possesses a maximal 2-local M which is almost strongly p-embedded in the sense of Definition 2.7.1. Whenever possible, we will want to make a fairly specific choice of M; to this end, Aschbacher defines:

NOTATION 8.2.1. Let $\mathcal{M}(p)$ be the set of almost strongly p-embedded maximal 2-locals of G.

A member $M \in \mathcal{M}(p)$ satisfies one of the conditions (1)–(3) in Definition 2.7.1. We write $p \in \sigma^i(G)$ if some M satisfies the condition numbered (i); thus $\sigma(G)$ is the union of the $\sigma^i(G)$.

---

[9] We mention that there is an obvious misprint at [**Asc83b**, p 385] where in the usual definition of $\Gamma_{k,P}(G)$, "$> k$" should read "$\geq k$".

Furthermore an M in $\mathcal{M}(p)$ might also lie in some other $\mathcal{M}(q)$; so we write $\sigma(M)$ for the set of primes q for which $M \in \mathcal{M}(q)$, and again $\sigma(M)$ is the union of corresponding subsets $\sigma^i(M)$.

The generic case of the argument corresponds to the situation $\sigma(M) = \sigma^1(M)$—namely where M is strongly p-embedded in G for all relevant primes p. The exceptional cases, where condition (2) or (3) of 2.7.1 might hold for some prime, are carried along for much of the proof; but typically they cause little complication. (Indeed the reduction to $\sigma(M) = \sigma^1(M)$ is accomplished in the main branches of the argument at 12.19 (for Sections 16–19), 22.8, and 23.6.)

To deal with the possibility of the exceptional cases, Aschbacher makes a choice of M (and the corresponding possibilities for $p \in \sigma(M)$) from a certain subcollection[10] $M\Sigma$—we will not reproduce the exact definition here, but just comment informally that, if possible, p is chosen as 3, and otherwise M is chosen with condition (2) if possible; leaving only the generic case of $\sigma(M) = \sigma^1(M)$ when these choices are not possible. ◊

We choose $T \in Syl_2(M)$. Applying the preliminary uniqueness result Theorem 8.1.5, we have $N_G(T) \leq M$ and in particular $T \in Syl_2(G)$.

Next, since the goal of Theorem 8.0.2 is to show that no group satisfies the Uniqueness Case, in the background we always have the principle that:

(8.2.2) no *known* group (that is, no $\mathcal{K}$-group) satisfies the Uniqueness Case.

(This can be checked using the list of strongly p-embedded subgroups in known groups appearing at [**GL83**, Sec I.24]. In [**Asc83b, Asc83c**], this fact is often used implicitly—typically to conclude that G is not one of the known groups appearing in one of the various classification theorems in the literature. We often indicate these, and other implicit references, via remarks contained within parentheses.)

For example, Aschbacher observes at [**Asc83b**, p 386] that:

M is not the only maximal 2-local containing T.

REMARK 8.2.3. Here the proof applies the Global $C(G,T)$ Theorem B.7.8—Aschbacher in fact supplies references to a list of papers which establish that theorem. Namely the hypothesis of the $C(G,T)$ Theorem must fail (since by (8.2.2), our counterexample G cannot be one of the groups in its conclusion); that is, for the Sylow 2-subgroup T of M, we must have $C(G,T) \not\leq M < G$. Thus some 2-local subgroup H, among those generating $C(G,T)$, has $T \leq H \not\leq M$—and furthermore we have $F^*(H) = O_2(H)$ as G is of characteristic 2 type.

(See also the parallel discussion at Remark 3.4.3 in the proof of the Quasithin Theorem—where we cannot quote Theorem B.7.8, since there G is not of characteristic 2 type, and so we must argue directly in order to show that there are 2-locals over T outside M.) ◊

Thus in addition to the maximal 2-local M chosen above, Aschbacher can also choose some H in the following set (though he does not introduce this specific notation until much later, at the start of Section 20):

[10]From Section 6 of [**Asc83b**] onward, the notation for this set changes to a calligraphic form $\mathcal{M}\Sigma$.

DEFINITION 8.2.4. Let $\mathcal{H}$ denote[11] the set of $H \leq G$ with $T \leq H \not\leq M$ and $F^*(H) = O_2(H)$. ◇

A strategy for obtaining a contradiction. First for contrast, we recall from Section 3.1 that in the Thompson strategy used in the Quasithin Theorem 3.0.1, the group G was identified as $\langle M, H \rangle$. But here, we must instead show that no group G arises, and so we must produce a contradiction; this will be accomplished at the end of the argument (and indeed elsewhere along the way in eliminating various subcases) by essentially forcing $H \leq M$—contrary to the original choice of H and M.

The conclusion $H \leq M$ is in turn typically obtained by interweaving weak closure arguments with uniqueness theorems.

We discuss these techniques in more detail in Appendix Sections B.7 and B.8. In a typical implementation of the strategy, weak closure arguments produce a factorization of H (cf. Proposition B.8.6) of the form:

$$H = C_H(C_j)N_H(W_i), \tag{8.2.5}$$

where C_j and W_i are certain nontrivial subgroups of T (cf. Definition B.8.1). But we can also hope to establish uniqueness theorems of the form:

$$M = !\mathcal{M}(N_M(W_i)) = !\mathcal{M}(N_M(X)). \tag{8.2.6}$$

for suitable $X \leq C_j$. Then $C_H(C_j) \leq C_G(X) \leq N_G(X)$, so $C_H(C_j)$ lies in the unique maximal 2-local subgroup M over $N_M(X)$ by (8.2.6). Similarly we get $N_H(W_i) \leq N_G(W_i) \leq !\mathcal{M}(N_M(W_i)) = M$. Then the factorization (8.2.5) says that $H \leq M$, for the desired contradiction.

Some details of the strategy. We conclude our introduction with a slightly more detailed discussion of the proof.

Remark B.8.4 discusses the relevance to weak closure arguments of the parameter $n(H)$—which,[12] in the important special case where the structure of the 2-local H is dominated by a group of Lie type defined over the field $\mathbb{F}_{2^n}$, is roughly the $\log_2$ (namely n) of the order of the field.

In particular, specialized weak closure arguments are often required to deal with the smaller case $n(H) \leq 2$—for example, $n(H) = 1$ when H is solvable. Sections 16–19 handle these cases (though with considerable preliminary work done in Sections 12–15). The work culminates in Theorem 19.30, which essentially reduces to the case $n(H) \geq 2$ (and indeed often $n(H) \geq 3$) for all relevant subgroups H. In eliminating subgroups H with $n(H)$ small, the hardest situations require fairly ad-hoc weak closure arguments in Section 16. The analysis culminates in Theorem 16.69, which gives the bound $m(M, V) > 3$ for the parameter m in B.8.2—defined on the $\mathbb{F}_2M$-module V used to define (see Definition B.8.1) the groups C_j and W_i of (8.2.5). Thereafter this bound can be used in some more simple weak closure results designed for the case $n(H) \leq 2$—cf. our Theorem B.8.3. In particular, the bound on m is incorporated into the hypothesis for the frequently-used 12.21 of [**Asc83c**], which corresponds to Proposition B.8.6, as discussed in Remark 8.4.4.

[11]depending implicitly on M and T—cf. the notation $\mathcal{H}_*(T, M)$ of Definition 3.1.2 used in the Quasithin Theorem.

[12]We emphasize that the notation $n(H)$ was developed later for use in [**AS04a, AS04b**]; for simplicity of exposition, we also adopt it here in our discussion of the Uniqueness Case—it provides a reasonably accurate approximation to the more complicated language used in [**Asc83b, Asc83c**].

After the reduction to larger values of $n(H)$, Sections 20–23 treat this remaining generic case: Roughly speaking, we can now force H and M to resemble parabolic subgroups in a group of Lie type of characteristic 2. To begin, we can define a suitable nontrivial subgroup B of a Cartan subgroup of H, which permutes with T. Since BT is solvable, we have $n(BT) = 1$—so that BT is not a candidate for "H", and hence $B \leq M$. In particular, we have produced a partial analogue of the amalgam "(M, BD, H)" in the Generic Case of the Quasithin Theorem discussed in Section 3.6. But here we instead use the embedding of B in H and M to show that H and M are incompatible, leading to a contradiction. For example, B is inverted by an element of the Weyl group in H, which does not lie in M, so that $N_G(B) \not\leq M$. This in turn guarantees that there are no uniqueness theorems of the form $M = !\mathcal{M}(X)$, for any subgroup X of $N_M(B)$. Since we do have such results for many p-subgroups X when $p \in \sigma(M)$, we can typically expect B to be a $\sigma(M)'$-group. Such restrictions lead to stronger weak closure arguments, giving contradictions in the remaining subcases, which involve a certain module U for M.

8.3. Preliminary results and the weak closure setup

Following our convention in other chapters, we will move as quickly as possible to the main part of the proof. So in this section, we will give only a general overview of most of the initial material quoted from the literature, and the preliminary results—making up much of Sections 2–15 of [**Asc83b, Asc83c**]. Then at the end of the chapter, we will include a few more specific remarks on quoted material.

However, in the latter part of this section, we will go into more detail on some of the material from Section 12, which provides fundamental background for the use of weak closure methods throughout the main proof.

The preliminary sections. Section 2 contains notation, and Sections 3–6 (and most of 10) contain results quoted from the literature—many on properties of known groups.

We begin with some of the results from the literature quoted in Section 6:

First 6.1–6.3 are the preliminary uniqueness results of Aschbacher-Gorenstein-Lyons, which we discussed earlier as Theorems 8.1.3, 8.1.5, and 8.1.4. These theorems are of the form $M = !\mathcal{M}(X)$, and several subsequent results derive various consequences, notably 6.6. The arguments will also make use of a weaker notion of uniqueness (which can be compared with the similar set $U(M)$ appearing in Definition 6.1.6):

DEFINITION 8.3.1. We say $X \in \mathcal{U}(M)$ if M is the unique conjugate in M^G over X. ◇

Sufficient conditions for membership in $\mathcal{U}(M)$ are considered in 6.4. (These are sometimes used implicitly, typically for a p-group X when M is strongly p-embedded in G—for example when the proof has reduced to the generic situation, namely where $\sigma(M) = \sigma^1(M)$.)

We also mention that 6.39 is essentially the $GF(2^n)$ Type Theorem 7.5.2, while 6.40 is essentially the Global $C(G,T)$ Theorem B.7.8; these results are quoted at various points, and some of the appeals are particularly significant—e.g. the appeal to the $C(G,T)$ Theorem discussed earlier in Remark 8.2.3.

We return to the overall layout of the initial sections.

Sections 7 and 9 develop various properties of members of the set $\mathcal{H}$ appearing in Definition 8.2.4—along with some techniques for proving uniqueness theorems applicable to such subgroups, including a description of the obstructions to those theorems. Section 8 is devoted to a preliminary analysis of the possibilities for members of the set $\mathcal{C}(M)$ of $2'$-components of M.

Sections 11 and 13–15 are devoted to some "pre-emptive" treatment of certain specialized configurations. For example, 11.12 restricts the weak closure parameter $r(M, V)$ in certain situations. The results in Sections 13 and 14 primarily reduce the list of possibilities for members of $\mathcal{C}(M)$; this theme is continued in Section 15, which also obtains some further information on weak closure.

The setup for weak closure arguments. A number of fundamental lemmas on weak closure from [**Asc81e**] appear early in Section 12. Some of that material is discussed here in Section B.8; for example, the factorizations in [**Asc81e**, 6.11] appear in our Proposition B.8.6, while in [**Asc83b**], they appear in 12.21—except that they are strengthened via an appeal to 11.1 of [**Asc81e**] (which we state as Theorem B.8.3).

But Section 12 also contains some material proved under special hypotheses, such as 12.4 and 12.5. While this material is primarily used in Sections 16–19 to deal with those cases where $n(H)$ is small, some of the issues involved (especially in the argument for 12.8) are at least implicitly also relevant to weak closure arguments elsewhere—particularly for the structure of the module V for M chosen in Definition B.8.1. So we begin with a discussion of some of this material.

REMARK 8.3.2 (The significance of k for weak closure). Arguments in the main part of the proof often subdivide into two cases, which nonetheless can typically be treated in parallel; these are distinguished via the parameter:

$$k := \begin{cases} 1 & \text{if } M \text{ is strongly 3-embedded} \\ 2 & \text{otherwise} \end{cases}$$

This convention is first defined within Hypothesis 12.5, and is used in the subsequent sections treating that Hypothesis; but the convention is also used even after that, in Sections 20–23.

Furthermore the parameter k will arise in weak closure arguments, relative to the generation parameter $n(H)$ for $H \in \mathcal{H}$: Hypothesis 12.5, mainly treated in Sections 16–19, assumes roughly that $n(H) \leq k$ (Indeed it is essentially shown that $n(H) = k$, for certain choices of H, in 12.6). Then k, in its role as $n(H)$, will appear in factorizations of the form (8.4.3)—that is, when in (8.2.5) we specialize "j" to the form "$i + k$". ◊

Some consequences of Theorem 8.1.3. We will use the informal term "module" to mean a nontrivial elementary abelian 2-subgroup V which is normal in M. We also set:

$$\overline{M} := M/C_M(V).$$

As at (B.6.1) we have a Frattini factorization

$$M = C_M(V)N_M\big(C_T(V)\big); \tag{8.3.3}$$

(Such Frattini factorizations are often quoted implicitly in later arguments.) So we see that:

$$\overline{M} = \overline{N_M\big(C_T(V)\big)}.$$

Thus $N_M(C_T(V))$ "contains" the action of $\overline{M}$ on V; and so it might be reasonable to hope that $N_M(C_T(V))$ could serve as a smaller, more convenient substitute for M. Indeed this is roughly the motivation for establishing the uniqueness condition that $M = !\mathcal{M}(N_M(C_T(V)))$.

Furthermore we can expect that condition to hold at least some of the time: (The following observation does not seem to be made explicitly in [**Asc83b, Asc83c**]; but is it used implicitly at many points.)

LEMMA 8.3.4. *Assume $p \in \sigma(M)$, with V as above. Then either*
(a) $m_p(N_M(C_T(V))) \geq 2$, so that $M = !\mathcal{M}(N_M(C_T(V)))$, or
(b) $m_p(C_M(V)) \geq 2$, so that $M = !\mathcal{M}(C_M(V))$.

PROOF. Since $m_p(M) \geq 3$ by definition of $\sigma(M)$, at least one of the factors in (8.3.3) has p-rank at least 2. Then the uniqueness assertion follows from Theorem 8.1.3. □

REMARK 8.3.5 (The significance of $M = !\mathcal{M}(N_M(C_T(V)))$ for weak closure). For weak closure arguments, the optimal situation occurs when we can choose V satisfying $M = !\mathcal{M}(N_M(C_T(V)))$, as in case (a) of Lemma 8.3.4. For example, that condition is hypothesis (a) in the fundamental Theorem B.8.3 (namely 11.1 in [**Asc81e**])—where it is applied in pushing-up arguments in the proof of that theorem.

In fact this condition is even built into the hypothesis of the frequently-used result 12.21 of [**Asc83c**]. (And the verification of this hypothesis is often left implicit, in many of the later applications of 12.21.) That result is a consequence of Theorem B.8.3 and 6.11 of [**Asc81e**], which appears here in the form of Proposition B.8.6.

The alternatives for V. Naturally we will also need to consider the situation when case (a) of Lemma 8.3.4 does not hold for any V; then of course case (b) holds for all modules V. In fact 12.7 and 12.8 in [**Asc83c**] establish further properties in this situation; and this development (along with some related arguments in Section 22) depends on the existence of certain TI-sets, which is essentially implicit on p. 563 of [**Asc81e**] (but seemingly the argument is not explicit there or in [**Asc83c**]):

LEMMA 8.3.6. *Assume $m_p(C_M(V)) \geq 2$ for $p \in \sigma(M)$, so $M = !\mathcal{M}(C_M(V))$. Then V is a TI-set in G.*

In particular this holds if $m_p(\overline{M}) \leq 1$; and this condition in turn holds when either $m(V) \leq 3$, or $M \neq !\mathcal{M}(N_M(C_T(V)))$.

PROOF. The hypothesis in the first statement is just case (b) of Lemma 8.3.4, which includes the uniqueness assertion. From this it follows, since we know that $C_M(V) \leq C_G(v)$ for each $v \in V^\#$, that also $M = !\mathcal{M}(C_G(v))$. Hence if $v^g \in V$, also $M = !\mathcal{M}(C_G(v^g))$—while conjugating by g, we get $M^g = !\mathcal{M}(C_G(v^g))$. Thus $M = M^g$, and hence $g \in M$—so that V is indeed a TI-set.

For the next statement, we saw in proving 8.3.4 that $m_p(\overline{M}) \leq 1$ suffices to give $m_p(C_M(V)) \geq 2$. Furthermore if $m(V) \leq 3$ then $\overline{M}$ is a subgroup of $L_3(2)$, so that $m_p(\overline{M}) \leq 1$. Finally, if $M \neq !\mathcal{M}(N_M(C_T(V)))$, then $m_p(N_M(C_T(V))) \leq 1$ by 8.3.4.a, whence $m_p(\overline{M}) \leq 1$ by (8.3.3). □

We next discuss the argument for 12.8 in [**Asc83c**], which proves that Hypothesis 12.3 is satisfied when Hypothesis 12.4 holds—namely that we can choose V with one of the two possible properties given in Hypotheses 12.1 and 12.2, which are given below. (We will postpone most of our discussion of Hypotheses 12.4 and 12.5 to the next section.)

REMARK 8.3.7 (The alternatives $\mathcal{W}$ or $\mathcal{V}_0$ for V). We observe first that there is always some choice of V which is not a TI-set. To establish this observation, the argument in 12.8 of [**Asc83c**] now makes a significant application[13] of 6.39, namely the $GF(2^n)$ Type Theorem 7.5.2: because (using 8.2.2) G is not one of the known groups in the conclusion of the $GF(2^n)$ Type Theorem, the hypothesis of that theorem must fail—and hence a maximal module V cannot be a TI-set. So *we now restrict attention to modules V which are not TI-sets.*

We prefer (especially in failure of factorization arguments) that V should be "2-reduced": that is, $O_2(\overline{M}) = 1$—which we can often expect, cf. [**AS04a**, B.2.14]. So if possible, we choose V from the set $\mathcal{W}(M)$ of modules satisfying this condition, together with the additional conditions that $M = !\mathcal{M}(N_M(C_T(V)))$ and $m(V) > 2$, which hold by Lemma 8.3.6 since V is not a TI-set. This is the choice assumed in Hypothesis 12.1.

To deal with the cases where this choice is not possible, *we now assume further that $\mathcal{W}(M)$ is empty.* Hence for our choice of V, we have $O_2(\overline{M}) \neq 1$, from our appeal to 8.3.6 above. Thus any module which *is* 2-reduced must be a TI-set; in particular by [**AS04a**, B.2.14], any module of the form $\langle X^M \rangle$, where X is a nontrivial subgroup of $Z := \Omega_1(Z(T))$, is a TI-set.

We let V_0 denote a maximal proper submodule of V. Observe that $V_0 > 1$, as V_0 contains $\langle X^M \rangle$, where $X := Z \cap V \neq 1$. We now choose V minimal subject to not being a TI-set; hence V_0 is a TI-set and $\overline{M}$ is irreducible on $\tilde{V} := V/V_0$, so as $O_2(\overline{M})$ has a nontrivial fixed space on $\tilde{V}$, we conclude that $O_2(\overline{M})$ centralizes $\tilde{V}$, and hence $O_2(C_{\overline{M}}(\tilde{V})) \neq 1$.

We also get $V = \langle v^M \rangle$ for each $v \in V \backslash V_0$: For if $V_1 := \langle v^M \rangle < V$, by maximality of V_0 we would get $V = V_0 + V_1$, with V_1 also a TI-set by minimality of V; and this contradicts 12.14.[14] In particular, this is the place where the argument for 12.8 makes use (implicitly) of Hypothesis 12.4—to which we will return in the next section.

The conditions in the paragraph above define the pairs (V, V_0) in the set $\mathcal{V}_0(M)$, and this is the choice assumed in Hypothesis 12.2, along with the property that for $z \in Z^\#$, we have $C_G(z) \leq M$, but $C_G(z) \not\leq M^g$ for $g \in G \setminus M$. (This follows from the fact that $\langle Z^M \rangle$ is a TI-set.)

Weak closure arguments are somewhat more complicated in the case where we have $(V, V_0) \in \mathcal{V}_0(M)$; so in most of our exposition, we will focus instead on the case $V \in \mathcal{W}(M)$ of Hypothesis 12.1. $\diamondsuit$

---

[13]The result 6.39 is also applied in the proof of the more specialized technical result 8.39.

[14]Thus 12.14 should be stated immediately after 12.7. Note that the proof of 12.14 only uses 12.6, plus parts (1) and (2) of 12.7—which in turn use only Hypothesis 12.4, but not 12.5. Furthermore the reference to 11.1 in the proof of 12.14 does not require the hypothesis for Section 11 that V is in $\mathcal{W}$ or $\mathcal{V}_0$. (And we also mention that the argument for "$[U_2, W] = 1$" in 12.14 implicitly uses 7.1 of [**Asc81e**] inside U_2U^g, in order to then apply $m(U) \geq m(U_2)$.) On the other hand, the reference to 11.15 in the proof of 12.7.3 does require that choice of V, i.e. essentially Hypothesis 12.3—so that 12.7.3 should in fact be stated instead just *after* 12.8, which establishes 12.3.

The choice of V in either $\mathcal{W}(M)$ or $\mathcal{V}_0(M)$ is required during much of Sections 11–19, as well for many weak closure arguments quoted from [**Asc81e**].

8.4. The treatment of small $n(H)$

We regard the main part of the proof as beginning in Section 16. The hypotheses assumed there are implicitly motivated by the following remarks, which actually appear later, just before Theorem 19.30, where this branch of the main argument culminates:

In the first half of the proof, carried out in Sections 16–19, we wish to eliminate the case where some $H \in \mathcal{H}$ roughly satisfies $n(H) \leq k$, where k is defined in Remark 8.3.2, and so in particular $n(H) \leq 2$. The actual assumption is more technical, and is given earlier in Hypothesis 12.4; it restricts the possible composition factors of H. But then the argument in 12.8, which we described in Remark 8.3.7, shows that Hypothesis 12.4 implies Hypothesis 12.3, and hence also Hypothesis 12.5—supplying a module V which is suitable for the application of weak closure arguments.

Thus Sections 16–19 assume Hypothesis 12.5, with the aim of producing a contradiction—which eventually emerges after 19.29. Hence we begin our exposition with a somewhat informal and simplified discussion of that hypothesis.

We consider members H of $\mathcal{H}$. The possible composition factors of H are restricted, using preliminary results such as Theorems 8.1.3 and 8.1.4; see 20.1 in [**Asc83c**] for example. Hypothesis 12.5 assumes further restrictions:

HYPOTHESIS 8.4.1 (A summary of Hypothesis 12.5 of [**Asc83c**]). *Hypothesis 12.4 of* [**Asc83c**] *assumes there exists $H \in \mathcal{H}$ which possesses no composition factor from a certain list of simple groups; for simplicity we will not reproduce the list*[15] *here; instead we summarize by saying that, for the most part, they are the groups L defined over larger fields of even order—namely with $n(L) > k$.*

Next Hypothesis 12.5 extends Hypothesis 2.4 (as we may, using 12.8—see Remark 8.3.7) by adding Hypothesis 12.3: Namely we may choose V to satisfy Hypothesis 12.1 or 12.2 in 8.3.7. Thus we choose V to be a minimal member of $\mathcal{W}(M)$, if $\mathcal{W}(M)$ is nonempty, and otherwise make the choice in Hypothesis 12.2; that is $(V_0, V) \in \mathcal{V}_0(M)$, and in addition for each $z \in Z^{\#}$, we have $C_G(z) \leq M$, but $C_G(z) \not\leq M^g$ for $g \in G \setminus M$.

When $V \in \mathcal{W}(M)$ we set $V_0 = 1$. In either case, we investigate the action of M on the module defined by $\tilde{V} := V/V_0$, and we now write $\overline{M} := M/C_M(\tilde{V})$. (In the case $V_0 > 1$, this differs from our earlier bar convention for $M/C_M(V)$—which we will now denote instead by $\hat{M}$.) ◇

The module V appearing in Hypothesis 8.4.1 is the module whose generalized weak closures W_i, in the sense of Definition B.8.1, will be exploited using weak closure analysis, usually via factorizations of the form (8.4.3).

The list of composition factors for H excluded by Hypothesis 12.4 plays a significant role in the proof of Theorem 8.8 in [**Asc81e**], which in turn is used for example in the proof of 12.7.3.

The later part of Section 12 develops a number of preliminary results under the assumption of Hypothesis 8.4.1: for example giving some fairly general weak

[15] But we mention that in the list of 12.4, the subscript 8 in $\Omega_8^-(4)$ was inadvertently misprinted as "∞".

closure bounds in 12.10, with more specific situations described in 12.23 and 12.26. And 12.19 reduces to the generic situation:

$$\sigma(M) = \sigma^1(M),$$

so that M is strongly p-embedded in G for each $p \in \sigma(M)$.

(The weak closure result 12.21 does not require Hypothesis 8.4.1, and is also used in arguments after Section 19; we discuss it at Remark 8.4.4 as well as various later points.)

Reducing to a "large" action of M on V. After recalling the setup of Hypothesis 8.4.1, the goal of the remainder of Section 16 is in effect to eliminate those subcases with $n(H) \leq k$ which require fairly ad-hoc weak closure arguments. The analysis culminates in Theorem 16.69, which provides a lower bound on the parameter $m(\overline{M}, \tilde{V})$ of Remark B.8.2: (This bound constitutes a crucial part of the hypothesis for more generic weak closure results such as 12.21, which are especially useful when $n(H) \leq 2$—as we will indicate at Remark 8.4.4, after first discussing the proof of 16.69 below.)

THEOREM 8.4.2. *Assume Hypothesis 8.4.1 with $m(V_0) \leq 2$.*
Then $m(\overline{M}, \tilde{V}) > 3$.

Thus the effective hypothesis for Section 16 consists of Hypothesis 8.4.1, along with the conditions that $m(V_0) \leq 2$ and $m(\overline{M}, \tilde{V}) \leq 3$.

In particular, here we do *not* start with a useful lower bound on $m(\overline{M}, \tilde{V})$—which would in turn also guarantee (cf. Theorem B.8.3) bounds on the parameter $r(M, V)$, appearing in the hypothesis of generic weak closure results such as 12.21. Hence the process will instead make heavy use of various more primitive weak closure results established in Section 12, or quoted from [**Asc81e**, Sec 6]: these arguments involve a more careful study of $r(M, V)$, and replace bounds on $m(\overline{M}, \tilde{V})$ with an analysis of more fundamental obstructions to factorizations given by the sets denoted by $\mathcal{O}_n^j$ in [**Asc81e**]. (We omit those definitions here.)

To begin, a preliminary result 12.15 is invoked to verify the hypothesis for another specialized result, established at [**Asc82**, 13.6] for use in the present situation: this gives a list of possibilities for a certain subgroup $L = O^2(L)$ of M (essentially a $2'$-component), and its action on $\tilde{W} := [\tilde{V}, L]$. Furthermore the hypotheses of Section 15 are verified, and results in that section give further information on the possibilities for the action of $\overline{M}$ on $\tilde{V}$, including restrictions on the number of M-conjugates of L. Of course this information leads to further restriction of the weak closure parameters associated to M and V in Remark B.8.2.

The first main step is Theorem 16.7, which methodically reduces the proof to a list of eight cases for $\overline{L}$ and $\tilde{V}$; in each, $\sigma(M)$ consists of a single prime $p \leq 7$. Furthermore in all these cases, one gets $r(M, V) > k$. This makes possible the use at 16.24 of the values 0,k in the roles of "i,$i+k$" in [**Asc81e**, 6.9]—which discusses the primitive obstructions $\mathcal{O}_{i+k}^j$ to factorizations in which the "j" in (8.2.5) is specialized to $i + k$; namely when the goal is to show:

$$H = C_H(C_{i+k})N_H(W_i). \tag{8.4.3}$$

Indeed this is the usual form for factorizations in the remainder of the proof of the Uniqueness Case Theorem 8.0.2.

The rest of Section 16 then eliminates the possibilities for L from 16.7 in a case-by-case analysis, to complete the proof of Theorem 8.4.2. Again the arguments use weak closure results, now also including 16.24, focusing on the obstructions $\mathcal{O}^j_{i+k}$ rather than on $m(\overline{M}, \tilde{V})$.

REMARK 8.4.4 (The relevance of $m(\overline{M}, \tilde{V}) > 3$ for 12.21 etc.). We briefly examine how Theorem 8.4.2 enables smoother weak closure arguments when we have $n(H) = k \leq 2$.

Assume we are in the first (and more favorable) case for V in Hypothesis 8.4.1: namely where $V_0 = 1$ and $M = !\mathcal{M}(N_M(C_T(V)))$. The latter restriction is hypothesis (a) for Theorem B.8.3. Since $V_0 = 1$, we may apply 8.4.2 to conclude that $m(\overline{M}, V) > 3$—giving hypothesis (b) for that Theorem, which then gives the important bound $r(G, V) \geq m(\overline{M}, V)$.

Hence in fact $s(G, V) = m(\overline{M}, V) > 3$, which in turn supplies the hypotheses of results such as Proposition B.8.6, which are specially suited for dealing with the case $n(H) = k \leq 2$: Indeed in this situation, we can use either 0 or 1 in the role of "i", in the inequality $i + k < s(G, V)$ in that hypothesis; which leads to weak closure factorizations of the form (8.4.3). Essentially this result appears as 12.21 in [**Asc83c**],[16] and we state a version as Proposition B.8.6; it is applied at a number of later points. For example, applications of 12.23 in [**Asc83c**] depend on 12.21 when $V_0 = 1$. We comment later at Remark 8.5.16 on the potential usefulness of having *two* factorizations, corresponding to the two values of i above.

(Our exposition emphasizes the easier case in Hypothesis 8.4.1 where $V_0 = 1$; the second case where $V_0 > 1$ appeals instead to results such as 12.18 and 12.22 in [**Asc83c**] for purposes of weak closure.) ◊

Determining possible $2'$-components L of M. Section 17 begins the general analysis of Hypothesis 8.4.1. (And Theorem 8.4.2 can now be applied in those subcases with $m(V_0) \leq 2$.)

The main goal continues a theme from Section 16, namely the determination of possibilities for $2'$-components $L \in \mathcal{C}(M)$, and restrictions on weak closure parameters. The initial work (in Theorems 17.6, 17.17, and 17.33) establishes restrictions on L taken from the major families of known simple groups. This analysis then culminates at Theorem 17.42, reducing to five cases which correspond to certain cases where $X := \langle L^M \rangle$ is of Lie rank 2 in characteristic 2.

The subsequent argument, leading up to 17.64, makes further reductions in case (1) of 17.42; in particular Theorem 8.4.2 is used (along with detailed properties of L obtained during the analysis) to get version (a) of the hypothesis for the preliminary result 13.6. (Cf. also our discussion of Hypothesis 13.1 for Section 13, in Remark B.10.1—except note that here $m_p(S(p)) \leq 1$ comes instead from 17.57.) Then 13.6 produces just two possible very specific structures for X, in terms of factors which are of Lie types L_2 and U_3 in characteristic 2. (We mention also that along the way, the proof of 17.45 uses some specialized results from Section 14.)

After this it is possible at 17.65 to find L of order divisible by $p \in \sigma(M)$—in a certain set $\mathcal{L}_0^*(p)$, which in effect makes a choice of L with respect to an ordering of the five types in 17.42 (refined by 17.64) which do happen to arise. Then 17.66 shows for such L that in most cases $M = !\mathcal{M}(X)$, and even $M = !\mathcal{M}(L)$, after which 17.67 gives further restrictions when $V_0 > 1$.

---

[16]There are several misprints in the statement of 12.21 given in [**Asc83c**]: in the hypothesis, "M" should be "N", and "$C_T(V)$" should be "$N_N(C_T(V))$"; and in conclusion (2), "$j+1$" should be "$j+i$".

The arguments use some of the analysis of members of $\mathcal{C}(M)$ from Section 8, and of subgroups $H \not\leq M$ from Sections 7 and 9. They also methodically use various specialized weak closure results from Section 12 involving the value of $m(\overline{M}, \tilde{V})$—most frequently 12.23, 12.26, and 12.10, but also occasionally 12.21 as described in Remark 8.4.4; as well as some specialized results from [**Asc81e**]. There is also use of results from Sections 10 and 11 on noncyclic p-subgroups X which permute with T, and satisfy conditions such as $\Gamma_{1,X}(G) \leq M$.

Eliminating $m(V_0) \leq 1$. Section 18 adds the assumption $m(V_0) \leq 1$ to Hypothesis 8.4.1. In particular Theorem 8.4.2 is then available (cf. Remark 8.4.4). A choice of p, L, X is made as in 17.65, and in the discussion above.

The initial analysis reduces (at 18.13) to just three possibilities for $\overline{X}$ and $\tilde{V}$, which give lower bounds at 18.16 on $r(G, V)$. Then there is a reduction to a single possibility, for which $C_G(C_2) \leq M$ (at 18.22 and 18.23). But 12.10 gives a factorization of the form (8.4.3) with $0, 2$ in the roles of "i, k", so that $H \not\leq M$ forces $N_G(W_0) \not\leq M$. Then closer study of some $A := V^g \leq M$ with $A \not\leq C_G(V)$ leads to a contradiction to the bound in 18.16. Thus Theorem 18.24 observes that we have reduced Hypothesis 8.4.1 to the case $m(V_0) > 1$.

The arguments continue to use specialized weak closure results from Section 12 (including 12.23 and 12.10 as before) as well as from [**Asc81e**]; but now frequently the restrictions in 17.67 fail, reducing to $V_0 = 1$—and then much more methodical use can be made of 12.21 as in Remark 8.4.4. There are also references to the uniqueness result 17.66, and to the preliminary work from Section 15.

Eliminating the remaining case $m(V_0) > 1$. Thus Section 19 continues Hypothesis 8.4.1, now under the assumption $m(V_0) > 1$ from the reduction in Theorem 18.24. In particular the restrictions in 17.67 apply to a choice of p, L, X made as in 17.65.

The initial results reduce the possibilities for this choice to a sublist at 19.10; use is made of weak closure theorems involving a TI-set such as V_0, notably 12.7 and 12.24, as well as material from Sections 7 and 9 of [**Asc81e**].

Making use of the list in 19.10, a Cartan subgroup A is chosen, and we construct $B := A\langle t \rangle$ for an involution t inverting A. Theorem 19.14 shows that we get $M = !\mathcal{M}(Y)$ for subgroups of the form $Y = BO_2(Y)$ with $O_2(Y) \neq 1$. The proof of this uniqueness theorem as usual involves pushing-up arguments, notably from Section 9 of [**Asc81d**].

At 19.22 and 19.23 an A_0 of prime order r is chosen in A, to guarantee that we have $V_0 \leq H_0 := C_G(A_0) \not\leq M$. And since $B \leq N_G(A_0)$, it follows from $H_0 \not\leq M$ and 19.14 that $O_2(H_0) = 1$, so by 0.3.1 we get $V_0 \not\leq C_{H_0}(F^*(H_0)) = Z(F^*(H_0))$. But as V_0 is noncyclic, from 0.3.11.2 we get that $O_{2'}(H_0)$ is generated by subgroups centralized by involutions v in V—while $C_G(v) = M = N_G(V_0)$ from Hypothesis 8.4.1. Then it follows (at 19.24.1) that V_0 centralizes $O_{2'}(H_0)$, and hence V_0 cannot centralize $E(H_0)$. Indeed after 19.28, it is observed that there must be a component K of H_0 on which V_0 acts faithfully. In particular $K \not\leq M = N_G(V_0)$, since then V_0 would centralize the component K.

However, results from Section 10 now deliver (at 19.29) a short list of possibilities for K. In each case, K is generated by a Sylow 2-group lying in M, along with odd locals which lie in M by various uniqueness results. So $K \leq M$, contrary to the previous paragraph.

Thus Hypothesis 8.4.1 has led to a contradiction, establishing Theorem 19.30, which we state informally as:

THEOREM 8.4.5. *We may assume henceforth that all $H \in \mathcal{H}$ satisfy $n(H) > k$.*

Recall from Remark 8.3.2 that $k = 1$ when M is strongly 3-embedded in G, and otherwise $k = 2$. Furthermore as $n(H) > k$, there is a composition factor of H which is of Lie type of characteristic 2 and Lie rank at most 2, from our discussion of Hypothesis 8.4.1.

8.5. The treatment of large $n(H)$

We now enter the more generic phase of the argument: Since $n(H) > k$ by Theorem 8.4.5, H resembles a 2-local in a group of Lie type of characteristic 2. Using this fact and the embedding of $H \cap M$ in H and M, we can hope to show that M also resembles a parabolic.

But since in the end we must show that there is no group G generated by M and H, there should be incompatibilities—which we will exploit using weak closure factorizations of H, to obtain a contradiction.

We will provide a somewhat more detailed discussion of this generic branch of the argument—in particular this will allow us to see weak closure in action. (In addition, since some of the proofs in [**Asc83c**] are obscured by misprints, we will also provide—mainly in the Appendix—further details in a number of the arguments.)

First, Sections 20 and 21 provide some general background for the generic situation.

A structure theory for H**.** To begin, the result 20.1 recalls some basic properties of groups H in $\mathcal{H}$. In particular there are upper bounds on $m_p(H \cap M)$, coming from 6.6—which in turn is based on the uniqueness theorem 6.3, discussed earlier as Theorem 8.1.4. Furthermore 20.1.4 indicates an important consequence of Theorem 8.4.5: since a solvable group X has $n(X) = 1 < k$, it is not in $\mathcal{H}$, and hence:

(8.5.1) $$O_\infty(H) \leq H \cap M < H.$$

The set $\mathcal{L}(H)$ *of* $2'$*-components not in* M. Given 8.5.1, the following collection of subgroups becomes the focus of interest:

DEFINITION 8.5.2. Let $\mathcal{L}$ denote the set of T-invariant subgroups L of G such that $L = L^\infty$, $LT \in \mathcal{H}$, and $L/O_2(L)$ is the product of a T-orbit of components. We let $\mathcal{L}^*$ and $\mathcal{L}_*$ denote the set of such subgroups which are maximal or minimal under inclusion, respectively. Given $H \in \mathcal{H}$, we write $\mathcal{L}(H)$ for the set of those $L \in \mathcal{L}$ which are *subnormal* in H. ◇

REMARK 8.5.3. A similar collection of subgroups appears in the analysis of quasithin groups of even characteristic, and an even more extensive local theory is developed to prove the Quasithin Theorem 3.0.1, primarily in Sections A.3 of [**AS04a**] and 1.2 of [**AS04b**]; that material in discussed in Sections 3.2 and 3.5. Fairly similar notation is used there, e.g. "$\mathcal{L}(G,T)$".

In fact the restrictions in 20.1 lead to the following structure theorem (at 20.4):

(8.5.4) $H = \langle \mathcal{L}(H) \rangle (H \cap M)$, so in particular $\mathcal{L}(H)$ is nonempty.

Furthermore 20.5 shows that if H contains a member L of $\mathcal{L}$, then $L \leq J$ for some member J of $\mathcal{L}(H)$. This motivates the study of the poset $\mathcal{L}$, ordered by inclusion: for example if L is a subgroup of H in $\mathcal{L}^*$, then L is in $\mathcal{L}(H)$—and so in particular *L is subnormal in H.*

The restrictions in 20.1 also lead at 20.7 to a list of possibile cases for LT where $L \in \mathcal{L}$. As in our discussion of Hypothesis 8.4.1, the groups $L/O_2(L)$ are primarily groups of Lie type of Lie rank 1 or 2, over fields of even order—including pairs of rank-1 groups interchanged by T. Usually $n(L) > 2$, since $n(H) > k$ by Theorem 8.4.5.

From the list in 20.7, one can determine possible inclusions among members of $\mathcal{L}$. This leads to an important uniqueness theorem (at 20.8); given $L \in \mathcal{L}^*$, we set $X := \langle \mathcal{L}(N_G(L)) \rangle$ and $N := N_G(X)$:

(8.5.5) $$\text{Then } N \in \mathcal{H} \text{ and } N = !\mathcal{M}(LT).$$

REMARK 8.5.6. Note that L is a product of $2'$-components of X, and hence also of N; so in particular, L is subnormal in N. (This fact is often used implicitly later.) ◊

REMARK 8.5.7. The property in (8.5.5) is similar to uniqueness theorems used in the proof of the Quasithin Theorem such as [**AS04b**, 1.2.7.3], which we recorded at (3.2.6). But one difference here is that the unique maximal 2-local overgroup is not our "primary" maximal 2-local M, but instead a second maximal 2-local N in the Thompson strategy.

Later N sometimes serves in the role which had been played by "M", in weak closure arguments such as those described in Remark 8.4.4: for example at Remark 8.5.23, we will consider weak closures of a module V normal in N. In particular, the condition $N = !\mathcal{M}(LT)$ will lead to $N = !\mathcal{M}(N_N(C_T(V)))$, which is analogous to the uniqueness condition in Lemma 8.3.4.a. ◊

One can also determine the minimal members in the list of 20.7—obtaining at 20.10 a list of possibilities for members L_0 of $\mathcal{L}_*$. In particular L_0 is either of Lie rank 1, or the product of two groups of rank 1 interchanged by T, or a group $(S)L_3(2^n)$ or $Sp_4(2^n)$ of rank 2, with T nontrivial on the Dynkin diagram. Furthermore one checks that[17]

(8.5.8) $$n := n(L_0) > 2 \geq k,$$

essentially as a consequence of the bound $n(H) > k$ in Theorem 8.4.5. The members L_0 of $\mathcal{L}_*$ are roughly analogous to the members of $\mathcal{H}_*(T, M)$ in Definition 3.1.2, used in the proof of the Quasithin Theorem, which are abstract minimal parabolics (in the sense of Remark B.7.6). Indeed here the unique proper T-invariant parabolic subgroup of L_0 is the Borel subgroup $N_{L_0}(T \cap L_0)$. This has the consequence that

(8.5.9) If $L_0T = XY$ for parabolics X, Y with X proper, then $L_0T = Y$,

which is sometimes used implicitly later when we have weak closure factorizations.

---

[17]Our convention will be to fix the meaning of "n" as $n(L_0)$; this differs slightly from the usage late in [**Asc83c**], where n is used differently at several points. In particular it is used for $m(U_1)$ in 23.2 (cf. Remark 8.5.24) and hence for $m(Z)$ after 23.4; but notice that this "usually" agrees with our convention of taking $n(L_0)$ as n—e.g. in all but one case in 23.5.

Choice of a subgroup B_0 from a Cartan subgroup in $L_0 \cap M$. Using the list in 20.10, it is possible (at 20.11) to choose a certain subgroup B_0 of a Cartan subgroup in $N_{L_0}(T \cap L_0)$. In particular B_0 permutes with T, and is inverted by an element of the Weyl group of L_0, not lying in M. Furthermore the primes dividing the order of B_0 are those dividing $2^n - 1$. For simplicity of exposition, we summarize only some of the properties of B_0:

REMARK 8.5.10. In the generic case of 20.11, B_0 can be taken in the set $\mathcal{B}(L_0)$, consisting of those B_0 which are cyclic, with order the product of distinct odd primes.

However in the exceptional cases, B_0 must be taken in a more general set $\mathcal{B}_1(L_0)$, consisting of the abelian groups such that $B_0 = BC$ for some $B \in \mathcal{B}(L_0)$ and, for each $q \in \pi(C)$, C has elementary abelian Sylow q-subgroups of order q^2. We will largely ignore this more complicated exceptional situation in our exposition.

Since $N_{L_0}(T \cap L_0)$ is solvable, we see that:

$$B_0 \leq L_0 \cap M,$$

in view of (8.5.1). ♢

Section 20 concludes with some further reductions in a few special cases.

The arguments in Section 20 make use of Theorem 8.4.5 and uniqueness results such as 6.6. But there is some reference to results in Sections 3, 4, 7, and 8; and the proof of 20.13.2 quotes 6.40, which is essentially the Global $C(G,T)$ Theorem B.7.8.

Analysis of subgroups of B_0. Section 21 establishes further properties of the standard setup given in Remark 8.5.10, where we have a choice of $L_0 \in \mathcal{L}_*$ and $B_0 \in \mathcal{B}_1(L_0)$. It considers certain subgroups B of B_0 in $\mathcal{B}_1(L_0)$, In most situations we can just take $B = B_0$, but other subgroups are sometimes used, for example at 22.14.

Now $B \leq L_0 \cap M$ (from Remark 8.5.10), and we can hope to show roughly that M resembles a parabolic in which B is contained in a Cartan subgroup of M. In particular we have some information about primes dividing the order of M; for example we can ask whether these primes are in $\sigma(M)$. Also B has p-rank at most 2 for any prime p from the list in 20.10.

Restrictions related to uniqueness conditions. First 21.2 observes that as B_0 is inverted in $L_0 \setminus M$, so is any $1 \neq A \leq B$; it follows that $N_G(A) \not\leq M$.

REMARK 8.5.11. For $p \in \sigma(M)$, and at least some p-subgroups R of M, we have $N_G(R) \leq M$; so from applications of 21.2, we can often show that $R \cap B = 1$, and even that B is a $\sigma(M)'$-group. For example this will hold whenever we are in the generic situation $\sigma(M) = \sigma^1(M)$; and indeed (e.g. in Lemma 8.5.28.2) even in the process of obtaining that situation. ♢

We mention one fairly immediate application of 21.2. Namely 21.10 shows that if X is a q-subgroup normalizing A, and in the set $\mathcal{U}(M)$ of Definition 8.3.1, then A is also inverted in M.

And 21.1 observes roughly that if $A \in \mathcal{U}(M)$, then often we get $M = !\mathcal{M}(Y)$, where $Y := \langle A, T \rangle$. It then follows that $X \not\leq N_G(T)$ using the Global $C(G,T)$ Theorem B.7.8 in the form of 6.40 (and (8.2.2)).

Such restrictions are applied later at various points of the argument, to roughly describe q-subgroups for primes q not in $\sigma(M)$, refining the possibilities for the groups L_0 and M.

In this direction, the properties of B are next used to describe the possibilities for $2'$-components in $\mathcal{C}(M)$, which we will denote by K.[18] Since we have not yet focused on a particular module for M, we re-define our standard bar convention to be:

$$\overline{M} := M/O_2(M),$$

so that $\overline{K}$ is then a component of $\overline{M}$. First, it is shown at 21.5 that B acts on K, and then 21.3 shows that B is usually faithful on $\overline{K}$.[19] Then the choices for $\overline{K}$ are successively restricted in 21.6–21.15, culminating in the list of possibilities in 21.16—with $|K^M| \leq 2$ by 21.17. We will not reproduce the list of possibilities for $\overline{K}$ here—but we do mention that usually $\overline{K}$ is of Lie type and rank at most 2, defined over a field of order 2^m, where $m := n(K) \geq 2$. The arguments primarily use properties of known groups, for example from Section 10, as well as results from Sections 6–8; also the reference to 12.25 involves another appeal to the Global $C(G,T)$ Theorem B.7.8, in the form of 6.40.

REMARK 8.5.12. As just indicated, our convention will be to write m for $n(K)$; recall from (8.5.8) that we are already writing n for $n(L_0)$. This differs from the convention in [**Asc83c**], where often n is used for both purposes. However since we have $B \leq L_0 \cap M$, we will often be able to show that n divides m. In particular, most possibilities avoid the situation of 21.3, in that B is faithful on $\overline{K}$—then we try to show that B induces *inner* automorphisms on K. $\diamondsuit$

The remainder of Section 21 is primarily devoted to two technical results 21.21 and 21.22, which we now discuss for background.

Consequences when $F(\overline{M})$ is cyclic. Each component of $\overline{M}$ centralizes $F(\overline{M})$ and lies in $\overline{M}^{\infty}$.

Indeed later it will sometimes be possible to show that $\overline{M}^{\infty}$ centralizes $F(\overline{M})$. Then by 21.18, which uses 21.17, $\overline{M}^{\infty} = E(\overline{M})$. So we now study the situation where $[\overline{M}^{\infty}, F(\overline{M})] = 1$, as in Hypothesis 21.20, which is summarized in Hypothesis 8.5.14 below. But first we note:

REMARK 8.5.13. Notice that $[\overline{M}^{\infty}, F(\overline{M})] = 1$ when $F(\overline{M})$ is cyclic; this is the case that will arise most frequently in applications. Indeed by 21.23.2, it will hold even when just $F([\overline{M}, \overline{M}])$ is cyclic.[20]

Furthermore if $[\overline{M}^{\infty}, F(\overline{M})] = 1$, then using 0.3.1 and setting $C := C_M\big(E(\overline{M})\big)$, we have $\overline{C}^{\infty} \leq Z\big(F^*(\overline{M})\big)$, so that C is solvable. $\diamondsuit$

Now we turn to a somewhat informal discussion of Hypothesis 21.20:

[18] Our notational convention will be to continue to use L for a member of $\mathcal{L}$; but from here on, to instead use K for a member of $\mathcal{C}(M)$. We mention that this differs from the usage in [**Asc83c**], where L is used for both purposes.

[19] There is a misprint in the statement of 21.3, which should also in fact allow L to be the product of the M-orbit of $2'$-components of M, as in 8.69.

[20] We mention a misprint in 21.23: conclusion (1) should say that $\overline{C} \cap [\overline{M}, \overline{M}]$ lies in $F([\overline{M}, \overline{M}])$ and so is cyclic—and conclusion (3) should say that $\overline{P} \leq \overline{C} \cap [\overline{M}, \overline{M}]$, and so is cyclic. Further $[\overline{C}, \overline{C}]$ should be replaced by $\overline{C} \cap [\overline{M}, \overline{M}]$ in the proof.

HYPOTHESIS 8.5.14 (A summary of Hypothesis 21.20 in [**Asc83c**]). *Assume:*
(1) $[\overline{M}^{\infty}, F(\overline{M})] = 1$.
(2) Each $K \in \mathcal{C}(M)$ *is normal in* M*, and certain possibilities for* K *from the list of 21.16 are excluded.*
(3) Recall that $B \in \mathcal{B}_1(L_0)$*, for some* $L_0 \in \mathcal{L}_*$*. For each* $p \in \sigma(M)$*,* B *is normalized by a Sylow* p*-group of* $\overline{C} := C_M\big(E(\overline{M})\big)$*. (Notice that this is vacuous when* $\overline{C}$ *is a* $\sigma(M)'$*-group.)*
(4) If the Sylow subgroup in (3) is noncyclic, then B *is not inverted in* M*.* $\diamondsuit$

We next record 21.21 as:

LEMMA 8.5.15. *Assume Hypothesis 8.5.14. For each prime* $p \in \sigma(M)$*, we set* $\overline{K}(p) := O^{p'}\big(E(\overline{M})\big)$ *and* $\overline{S}(p) := O^{p'}\big(C_{\overline{M}}(E(\overline{M})\big)$*. Then:*
(1) $\overline{M}^{\infty} = \overline{K}(p)$*.*
(2) $e(G) = 3$*.*
(3) Either $\overline{K}(p) \cong L_2(2^m) \times L_2(2^m)$ *with* $S(p)$ *of* p*-rank* 1*, or* $\overline{K}(p) \cong U_3(2^m) \times L_2(2^m)$ *with* $\overline{S}(p) = 1$*.*
(4) m *is a power of* 2*, and* p *divides* $2^m + 1$*.*
(5) $|B|$ *divides* $2^m - 1$*.*

In Remark B.10.1 in the Appendix, we discuss the proof of the lemma, in order to shed light on how the various conditions in Hypothesis 8.5.14 are applied.

A lemma for exploiting a pair of weak closure factorizations. Lemma 8.5.15 will be used crucially later; but it is also applied immediately, in establishing the important technical result 21.22, which we state below as Lemma 8.5.17. We first supply some motivation for the lemma.

REMARK 8.5.16. At several points (e.g. the proof of 23.1—see our discussion at Remarks 8.5.23 and 8.4.4), the situation will arise where two different values of "i" (typically 0 and 1) can be used in weak closure factorizations of H, as in (8.4.3). Furthermore for each i, H is contained in the second factor $N_G(W_i) =: H_i$—so that $H \leq H_0 \cap H_1$. This is the situation considered in Lemma 8.5.17.

In this situation, in order to obtain the contradiction $H \leq M$, we can hope to show (as in the conclusion of the Lemma) that $H_0 \cap H_1 \leq M$. This supplies an alternative to proving $M = !\mathcal{M}(H_i)$, as is often done elsewhere.

Furthermore the hypothesis of the Lemma is stated in terms of factorizations of certain subgroups X of M, which are essentially solvable, in that $n(X) = 1$. We typically obtain that hypothesis from a weak closure factorization of the form (8.4.3), with $i = 0$ and $k = 1$. $\diamondsuit$

We now give our somewhat informal statement of 21.22:

LEMMA 8.5.17. *Assume that* $H_0, H_1 \in \mathcal{H}$ *and* $X = (X \cap H_0)(X \cap H_1)$ *for each* $T \leq X \leq M$ *for which* $n(X) = 1$*. Then* $H_0 \cap H_1 \leq M$*.*

Again in the Appendix, Remark B.10.2 discusses the proof. As a brief summary: Using uniqueness theorems from Section 21, we find in (B.10.4) that X is a $\sigma(M)'$-group. This in turn establishes hypothesis (1) of Hypothesis 8.5.14, while hypotheses (3) and (4) are satisfied vacuously. But this fact also rules out the conclusions of Lemma 8.5.15. Hence hypothesis (2) of 8.5.14 is not satisfied, so that some $K \in \mathcal{C}(M)$ is among the groups excluded there. These possibilities are then successively eliminated, again using uniqueness theorems.

The final case division. With preliminary results in hand, we are ready to return to the main part of the proof. We continue to assume some choice of L_0 and $B \in \mathcal{B}_1(L_0)$. We now define a module U for M, and return to our more usual form of the bar convention; namely we set:

$$Z := \Omega_1(Z(T)),\ U := \langle Z^M \rangle, \text{ and } \overline{M} := M/C_M(U).$$

Since $F^*(M) = O_2(M)$, it is standard (see e.g. Remark 8.3.7) that U is elementary abelian, and also 2-reduced: that is, $O_2(\overline{M}) = 1$, so $O_2(M) \leq C_M(U)$. The proof will consider first the case where U is not a TI-set in G, and finally the case where U is a TI-set.

Notice that this differs from our more typical notational convention such as in Sections 16–19, where V was a module for M, and U was a module for H: indeed later "V" will denote a module for H, for example in weak closure arguments as in Remark 8.5.23.

The case where U is not a TI-set. Since U is 2-reduced, but not a TI-set, it follows (using Lemma 8.3.6) that $M = !\mathcal{M}(N_M(C_T(U)))$, and U is in the set $\mathcal{W}(M)$ of Remark 8.3.7; this gives the optimal situation for weak closure arguments as mentioned in Remark 8.3.5. In particular, we have the hypothesis of Section 11. Now 22.1 quotes results from that section, with U in the role of "V", showing for example that $\sigma^3(M)$ is empty.

The choice of B centralizing Z. Since U is not a TI-set, we might expect that some $U_0 \leq U$ should satisfy $N_G(U_0) \not\leq M$. Indeed Theorem 22.2 shows that

$$C_G(Z) \not\leq M. \tag{8.5.18}$$

The proof uses pushing-up techniques, such as the Global $C(G,T)$ Theorem B.7.8 (in the usual form of 6.40, but also using 12.25, and 3.39 which quotes a result from Campbell's thesis [**Cam79**]). In summary, from the possibilities in 21.16 for components $\overline{K}$ of $\overline{M}$, we reduce to a single case, including action on a certain FF-module F for $\overline{K}$. Further L_0 is determined from 20.10—but here Thompson factorization does *not* fail, forcing L_0 to normalize F. This leads to the contradiction $L_0 \leq !\mathcal{M}(N_G(F)) = M$. The uniqueness result comes from Theorem 8.1.3, applicable since $m_p(N_M(F)) > 1$ from earlier in the proof.

In view of (8.5.18), we can choose $L_0 \leq C_G(Z)$, so $B \in \mathcal{B}_1(L_0)$ centralizes Z. Also, we can now replace U by some irreducible M-submodule V: For V is generated by M-conjugates of $V \cap Z \neq 1$, so that V is also 2-reduced (again cf. Remark 8.3.7); while (8.5.18) gives $C_G(V \cap Z) \not\leq M$, and hence V is also not a TI-set. So we get $V \in \mathcal{W}(M)$, again using 8.3.6. Hence for the rest of our discussion of Section 22, $\overline{M}$ will denote $M/C_M(V)$ instead of $M/C_M(U)$.

In particular we again have the hypotheses of Section 11, now with V rather than U in the role of "V". Next 11.2 is used to show $\sigma^2(M)$ is empty—the setup described in 6.12 and 6.13, so that at 22.8, we reduce to the generic situation of:

$$\sigma(M) = \sigma^1(M),$$

and hence M is strongly p-embedded in G for each $p \in \sigma(M)$. This fact tends to be used implicitly thereafter; for example the proof of 22.11 uses 7.7—which quotes 6.9, and that lemma in turn requires $\sigma(M) = \sigma^1(M)$ as a hypothesis.

Restricting possibilities for $K \in \mathcal{C}(M)$. The middle step in the argument is to reduce at 22.13 to a short list of possibilities for members of $\mathcal{C}(M)$. Some of the groups listed in 22.13 are among the cases excluded in Hypothesis 8.5.14.2; but when such components are not present, we want to establish, in alternative (4) of 22.13, that Hypothesis 8.5.14 holds.

To facilitate weak closure during that verification, the result 22.9 establishes that $m(\overline{M}, V) \leq n(L_0) + 1$. The argument involves 12.21 in [**Asc83c**], which we have been emphasizing: but since it closely parallels the application of 12.21 in the proof of 23.1, which we discuss in Remark 8.5.23, we restrict ourselves here to only a brief summary of the argument: Assuming the failure of this bound gives hypothesis (b) of Theorem B.8.3, since $n = n(L_0) > 2$ by (8.5.8); while hypothesis (a) (namely that $M = !\mathcal{M}(N_M(C_T(V)))$ holds since $V \in \mathcal{W}(M)$. Then we use the factorizations in Proposition B.8.6 to get first that $L_0T \leq H_0 \cap H_1$ for suitable H_i; and second, to establish that $X = (X \cap H_0)(X \cap H_1)$ for suitable subgroups X of M. Then by Lemma 8.5.17, $H_0 \cap H_1 \leq M$, contradicting $L_0 \not\leq M$ in the definition of $\mathcal{L}$.

The verification of Hypothesis 8.5.14 begins more explicitly at 22.11, where the bound from 22.9 is applied in 5.16, which, together with results from Sections 20 and 21, leads to a contradiction—establishing that $F([\overline{M}, \overline{M}])$ is cyclic, and centralized by B. This provides the hypothesis for 21.23: so at 22.12, setting $M^* := M/O_2(M)$ and $C := C_M(E(M^*))$, we get that $C^* \cap [M^*, M^*]$ is cyclic[21] and $(M^*)^\infty = E(M^*)$. Notice in particular that $(M^*)^\infty$ centralizes $F(M^*)$, as required for Hypothesis 8.5.14.1 (cf. our discussion at Remark 8.5.13). And 22.12 also shows that B centralizes a conjugate of a Sylow p-group of C for $p \in \sigma(M)$, and that Sylow group is cyclic; this gives hypothesis (3) of 8.5.14, and makes hypothesis (4) vacuous.

Thus when hypothesis (2) of 8.5.14 also holds, we may apply Lemma 8.5.15, to obtain alternative (4) of 22.13. Otherwise some $K \in \mathcal{C}(M)$ is among the cases excluded in 8.5.14.2; these are described in alternatives (1)–(3) of 22.13, including various restrictions from Section 21 on the list in 21.16.

Exploiting $B \leq C_G(Z)$ for a contradiction. We now come to the final phase of the argument.

First the list from 22.13 is reduced at 22.17 to essentially a single possibility; indeed M contains the product K_1K_2 of two members of $\mathcal{C}(M)$ interchanged by T, with $\overline{K}_i \cong L_2(2^m)$, for some $m := n(K)$, with just a few choices for the module V. Further for $p \in \sigma(M)$, we have $m_p(\overline{M}) = m_p(\overline{K}_1\overline{K}_2) = 2$ by 22.18. As $m_p(M) \geq 3$, we have $m_p(C_M(V)) \geq 1$. Further since V is not a TI-set, in fact $m_p(C_M(V)) = 1$ by Lemma 8.3.6. We get $m_p(M) = 3 = m_p(G) = e(G)$ since we have $p \in \sigma(M)$.

Recall that $\sigma(M) = \sigma^1(M)$ by 22.8, so $p \notin \pi(B)$ by Remark 8.5.11. We will obtain a contradiction from this fact.

In order to do so, we first pin down the structure of L_0 and a suitable $L \in \mathcal{L}^*$. As L_0 centralizes Z, there is some maximal 2-local $H \neq M$ containing $C_G(Z \cap V)$, and in particular contains L_0. We saw that p divides the order of $C_M(V)$—and hence it must also divide the order[22] of $C_M(Z \cap V) \leq H \cap M$. In this situation, 20.14 and 21.16 describe the groups $O^{p'}(H) =: L \in \mathcal{L}^*$ and $L_1 := O^{p'}(H \cap M)$. Namely by 20.14, $L/O_2(L) \cong (S)L_s(2^n)$, and $L_1/O_2(L_1) \cong SL_{s-1}(2^n)$, for some $s \geq 3$. Furthermore L_1 centralizes V, since it centralizes $Z \cap V$, while $C_{K_1K_2}(Z \cap V)$ is a p'-subgroup. Thus $L_1 \in \mathcal{C}(M)$, so by 21.16, $s = 3$. A preliminary result 8.2

[21] Adjust conclusions (1) and (3) of 22.12 for the misprints noted earlier in the corresponding parts of 21.23.

[22] Here [**Asc83c**] only mentions $C_G(Z)$, but seemingly this is a misprint for $C_G(Z \cap V)$.

gives that M is strongly 3-embedded (even if 3 is not in $\sigma(M)$), which in particular makes possible the application of 20.12 to give m odd.

We turn finally to $B \in \mathcal{B}_1(L_0)$. Here B is cyclic, and we can embed B in a rank-2 torus of L, contained in $L \cap M$ by (8.5.1). So we get $A := B_1 \times B$ in M, with $B \cong B_1 \leq L_1$ and $B_1 \leq C_M(V)$ centralizes each $\overline{K}_i$. Let $r \in \pi(B)$, so $r \notin \sigma(M)$ as mentioned above. Indeed we have $N_G(B) \not\leq M$ by 21.2, so M is not strongly r-embedded in G—and hence we get $m_r(M) < e(G) = 3$. Next we obtain $m_3(C_M(A_r)) \leq 1$ for the rank-2 Sylow r-subgroup A_r of A—since otherwise 7.6 and 6.14 imply that M is strongly r-embedded in G. Therefore B acts nontrivially on some $\overline{K}_i$, and $m = n(K_i)$ is odd (as otherwise an element of order 3 in a Cartan group of each $\overline{K}_i$ is centralized). Furthermore B induces *inner* automorphisms, as B centralizes $Z \cap V$. Thus r divides the orders of $\overline{K}_i$ and $\overline{L}_1$, so $m_r(M) \geq 3$, contrary to an earlier observation.

Thus Section 22 culminates in Theorem 22.19, which shows that $U = \langle Z^M \rangle$ is a TI-set in G.

The final case where U is a TI-set. This case is handled in Section 23 of [**Asc83c**]. We mention that this section often makes implicit reference to various consequences of U being a TI-set; we will collect some of these here at the outset. For example, the following general type of argument is made explicitly during the proof of 23.2, but is used implicitly elsewhere: Since $L_0 \not\leq M$ with $L_0/O_2(L_0)$ semisimple, while $C_G(u) \leq M$ for involution u in U, we get

(8.5.19) For $V := \langle Z^{L_0} \rangle$, we have $C_{L_0T}(V) = O_2(L_0T) = C_T(V)$.

Also for the weak closure parameter $r(G,U)$, by the TI-property we now have:

(8.5.20) $$r(G,U) = m(U).$$

By 23.2, $m(U) \geq n$, which can sometimes reduce reliance on results like Theorem B.8.3 for obtaining lower bounds on $r(G,U)$. And sometimes various properties of TI-sets from the literature are implicitly quoted, such as 7.1 of [**Asc81e**], which includes:

(8.5.21) As U is a TI-set normal in T, it is central in its weak closure $W_0(T,U)$.

As U is a TI-set, in contrast to Section 22, we cannot quote Lemma 8.3.6 to get the uniqueness theorem $M = !\mathcal{M}(N_M(C_T(U)))$. But instead of trying to mimic 12.8 to get a member of $\mathcal{V}_0(M)$ as in Remark 8.3.7, we will instead sometimes use a uniqueness result for a maximal 2-local N over L_0T, rather than for M—as indicated in Remark 8.5.23. In addition, there are weak closure results in Section 8 of [**Asc81e**] which apply to the special case where U is a TI-set.

NOTATION 8.5.22. At the beginning of Section 23, several notational conventions are established in lemmas 23.1 and 23.2, involving *choices* that can be made, in view of the development in Sections 20 and 21:

We embed L_0 in some maximal $L \in \mathcal{L}^*$.

By (8.5.5) there is a maximal 2-local $N \neq M$ such that $N = !\mathcal{M}(LT)$.[23] $\diamondsuit$

We will provide a somewhat more extensive discussion of Section 23, and again we will supply even more detailed arguments in Appendix Section B.10.

[23]We mention that there is a misprint in [**Asc83c**]. where N is defined to be $!\mathcal{M}(L)$.

Restrictions on $L \le H$. The analysis begins with some results on the structure of H developed in Section 20, which use the fact that U is TI.

First 23.1 establishes a fairly standard weak closure bound—indeed the statement and proof are close analogues of 22.9. We assume we are given nontrivial modules $V = \langle (Z \cap V)^N \rangle$ for N and $W = \langle (Z \cap W)^{L_0} \rangle$ for L_0.[24] The goal is to show that:

$$m(Aut_N(V), V) \le a(Aut_{L_0T}(W), W) + 1 \le n(L_0) + 1.$$

The second inequality is essentially immediate. The proof of the first involves results on weak closure (especially 12.21 in [**Asc83c**]) which we have been emphasizing at various points: so below we will sketch the logic:

REMARK 8.5.23 (Some features of the proof of 23.1). The role played by Theorem 8.4.2 in Remark 8.4.4 is now played instead by the failure of the first bound above: in particular we have $m(Aut_N(V), V) > 2$, giving hypothesis (b) of Theorem B.8.3.

But to achieve hypothesis (a) of that Theorem, we will now (in contrast to Remark 8.4.4 as well as the proof of 22.9) obtain a uniqueness result not as usual for M, but instead for the maximal 2-local N above LT. For arguing as in (8.5.19): L does not centralize $Z \cap V$, and then $C_{LT}(V) = O_2(LT) = C_T(V)$ from 20.7. Hence we have $LT \le N_N(C_T(V))$, and so $N = !\mathcal{M}(LT) = !\mathcal{M}(N_N(C_T(V)))$ using 8.5.22.

Thus we can apply Theorem B.8.3, with V, N in the roles of "V, M": this with failure of the first bound above gives $s(G, V) = m(Aut_N(V), V) > k + 1$, where we have set $k := a(Aut_{L_0T}(W), W)$. Then for both the values $i := 0$ and 1, we have the relation $i < s(G, V) - k$. We use these bounds in turn as hypothesis for Lemma B.8.5 (that is, part (1) of 12.21 in [**Asc83c**], instead of the later parts of that result appearing in Proposition B.8.6)—with W in the role of "U". From this we obtain $W_i \le C_T(W)$, along with $C_T(W) = O_2(L_0T)$ as in (8.5.19). Then W_i is weakly closed in $O_2(L_0T)$, and hence $L_0T \le N_G(W_i)$. Since $N_G(W_i) \le N_G(Z(W_i))$, we have $L_0T \le H_0 \cap H_1$, where $H_0 := N_G(W_0)$ and $H_1 := N_G(Z(W_1))$.

Now we can apply Lemma 8.5.17: Recall that $s(G, V) = m(Aut_N(V), V) > 2$; and we now replace L_0T by X satisfying $T \le X \le M$ with $n(X) = 1$, and i, k by $0, 1$. So again we have the bound $s(G, V) > 1 = i + k$, which this time we use as the hypothesis for Proposition B.8.6, to get $X = C_X(C_1)N_X(W_0)$. As $Z(W_1) \le C_1$, we may replace the first term by $C_X(Z(W_1))$ and then even by $N_X(Z(W_1)) = H_1$; so we get the condition $X = (X \cap H_1)(X \cap H_0)$ required for the hypothesis of Lemma 8.5.17—which now delivers $H_0 \cap H_1 \le M$, contradicting $L_0 \not\le M$ in the definition of $\mathcal{L}$. ◊

At 23.2, we choose $V := \langle Z^N \rangle$, and set $N^* := N/C_N(V)$; we obtain strong restrictions on V and L. In the argument, the bound in 23.1 is used (along with results in Section 20, and $n(L_0) > 2$ in (8.5.8)) to verify the hypotheses of 14.5 in [**Asc82**]. In particular, there is implicit use here of the fact in Remark 8.5.6 that L is a $2'$-component of N. The proof also shows that $L \trianglelefteq N$. Summarizing:

REMARK 8.5.24 (Some features of 23.2 in [**Asc83c**]). The list in 20.7 is reduced to a sublist of 11 possibilities for $L^* = Aut_L(V)$. These are all small-rank classical matrix groups of characteristic 2, with $n(L)$ "usually" equal to $n(L_0) = n$—that is,

[24]This corrects a misprint in [**Asc83c**] where W is defined to be $\langle Z^{L_0} \rangle$; the proof is unchanged.

in all but one case, which we de-emphasize for simplicity of exposition. Furthermore in each case, V is the natural module for L^*. It follows that $N_1 := N \cap M$ contains the (maximal) parabolic of L stabilizing a root space in V, which "usually" has rank n—and this root space is given by $U_1 := \langle Z^{N_1} \rangle \leq \langle Z^M \rangle = U$, with $N_1/C_{N_1}(U_1)$ of odd order (induced by the relevant part of a Cartan subgroup in $L_0 \cap M$) acting irreducibly on U_1. ◊

Reduction to $U = Z$ and consequences for M. Attention now turns to the structure of M, where strong restrictions follow from the proof that $U = Z$ in 23.3 and 23.4.

The initial step in 23.3 shows U is the root subgroup U_1—determined by Z in the module V for L as in Remark 8.5.24 above. We regard this initial reduction as particularly significant, and so will discuss it in some detail. To lead up to it, recall from 8.5.24 that $N \cap M$ is the stabilizer in N of U_1. Then 23.3 shows that M is the stabilizer in G of U_1. In particular U_1, although ostensibly determined by L and N, is in fact determined only by M, independent of L and L_0. In addition, the final contradiction for the Uniqueness Case at the end of Section 23 traces back naturally to 23.3; and in particular to a contradiction of the form $L_0T \leq H_0 \cap H_1 \leq M$—which as we will see is based on using Lemma 8.5.17 late in the proof. So this contradiction provides a good representative of various weak closure arguments used elsewhere in the elimination of the Uniqueness Case. To proceed, we state 23.3 as:

(8.5.25) $$U = U_1.$$

The proof given in [**Asc83c**] is rather terse, and indeed parts of it contain misprints. So in the Appendix we expand on that argument (which involves some fairly typical weak closure arguments) at Remark B.10.5. As a brief summary: Assuming that $U_1 < U$, for $i = 0, 1$ we can establish that L_0T normalizes the generalized weak closure W_i of U in T. Thus $L_0T \leq H_0 \cap H_1$, where we set $H_0 := N_G(W_0)$ and $H_1 := N_G(Z(W_1)) \geq N_G(W_1)$. Using Lemma 8.5.17 we get the contradiction $H_0 \cap H_1 \leq M$, much as in the final paragraph of Remark 8.5.23; except that showing that $X = (X \cap H_0)(X \cap H_1)$ when $n(X) = 1$ requires a somewhat more detailed analysis.

This initial reduction $U = U_1$ i (8.5.25) now leads more easily to the final reduction at 23.4:

(8.5.26) $$U = Z, \text{ so } Z = \Omega_1(Z(T)) \text{ is a TI-set which is normal in } M.$$

Again in the Appendix (at Remark B.10.6) we provide some comments on the transfer argument used in the proof, which applies a fusion result of McBride, that was never in fact published in that original form (and so we quote a modern replacement).

REMARK 8.5.27. We comment briefly on the significance of the reduction in (8.5.26) to $U = Z$: From Remark 8.5.24, $L_0 \cap M$ induces an irreducible action of odd order on the root space U_1 in V_1, containing the action of $B \in \mathcal{B}_1(L_0)$. Since $U_1 = U$ by (8.5.25), this is in turn the action of $\overline{B}$ in $\overline{M} = M/C_M(U)$. But $U = Z$ by (8.5.26), so $T \leq C_M(U)$ and hence:

$\overline{M}$ also has odd order.

Indeed by taking an appropriate subgroup C of a Cartan subgroup of $L_0 \cap M$, $\overline{M}$ even contains the overgroup $\overline{C}$ of $\overline{B}$, which is regular on the nonzero vectors of $U_1 = U = Z$.

These conditions in fact have a further consequence, namely:

$\overline{C}$ is normal in $\overline{M}$, with $\overline{M}/\overline{C}$ cyclic of order dividing $m(Z)$.

The normality of $\overline{C}$ follows from the standard Lemma A.1.12 in [**AS04a**], and then the final statement follows using the Schur's lemma argument made for "A" in Remark B.10.6). Of course our usual choice of $\overline{B}$ as in Remark 8.5.10 is given by $\Omega_1(\overline{C})$. (The above structure is sometimes used implicitly in later arguments in [**Asc83c**].) ♢

Initial restrictions on $F^*(\overline{M})$. Since $M/C_M(U)$ has odd order by Remark 8.5.27, the focus of the analysis now shifts to $C_M(U)$. Thus we re-define our standard bar convention, so as to work in the larger quotient:

$$\overline{M} := M/O_2(M).$$

In particular this replaces our earlier use of this bar convention for $M/C_M(U)$. We also now usually write Z for U, and $C_M(Z)$ for $C_M(U)$. From now on in Section 23, we mainly investigate $F^*(\overline{M})$.

As a preliminary, 23.5 collects some consequences for B of the reductions achieved so far. For example from the list of cases for L in 8.5.24, we see that B can be chosen in the subset $\mathcal{B}(L_0)$ of $\mathcal{B}_1(L_0)$, so that in particular B is cyclic. Furthermore "usually" $n = n(L_0)$ is the rank of $U_1 = Z$, and then B is faithful on Z.

In particular, we single out the property in 23.5.4, which is used later at various points. One of these applications is made right away, for the reduction at 23.6 to the generic situation given by $\sigma(M) = \sigma^1(M)$, which we also add here.

LEMMA 8.5.28. *(1) For* $b \in B \backslash C_B(Z)$, *and* X *a nontrivial* q*-subgroup of* $C_M(b)$ *for some prime* q, *we have* $X \notin \mathcal{U}(M)$.

(2) We have $\sigma(M) = \sigma^1(M)$. *Thus for* $p \in \sigma(M)$, M *is strongly* p*-embedded in* G. *Further for* b *as in (1),* $Aut_B(Z)$ *and* $C_M(b)$ *are* $\sigma(M)'$*-groups.*

PROOF. Since $Aut_M(Z) = M/C_M(Z)$ is of odd order, $Aut_B(Z)$ is not inverted in M, and hence neither is B, so (1) follows using 21.10. In particular, if b were a p-element for $p \in \sigma^1(M)$, we would have $\langle b \rangle$ itself in $\mathcal{U}(M)$; this contradiction shows that $Aut_B(Z)$ is a $\sigma^1(M)'$-group. Using this, it is not hard to eliminate the exceptional cases (2) and (3) of Definition 2.7.1, so that we get $\sigma(M) = \sigma^1(M)$—again in the Appendix we provide further details of the argument, at Remark B.10.8. Then the remaining statements in (2) follow from (1). □

During the analysis of $F^*(\overline{M})$ in the remainder of Section 23, there will eventually be a reduction to the case $F(\overline{M})$ cyclic.

REMARK 8.5.29 (Certain uniqueness theorems in 23.7). So it will be necessary to show that Sylow q-groups of $F(\overline{M})$ are cyclic, with the case $q \notin \sigma(M)$ of particular interest. Some uniqueness results applicable to such q-groups are established at 23.7. In fact they are stated in somewhat greater generality, but for expository purposes, we will adopt this oversimplified description. We mention that the proof uses the uniqueness theorem 7.7. Furthermore the analogues of the statements in 23.7 hold also for $p \in \sigma(M)$—even in the cyclic case—since M is strongly p-embedded in G by Lemma 8.5.28.2. ♢

The analysis will also include a determination of the possibilities for the members of $\mathcal{C}(M)$, whose images are the components of $\overline{M}$. So to begin, the list in 21.16 of possibilities for components of $\overline{M}$ is reduced to a sublist at 23.8, essentially of rank-1 groups.

And then the list of 11 possibilities for $L \in \mathcal{L}^*$ from Remark 8.5.24 is reduced to just four cases at 23.9—and in particular it is shown that either $X := (L \cap M)^\infty$ is trivial, or $X \in \mathcal{C}(M)$.

These reductions use results from Sections 4 and 6–8, and the Global $C(G,T)$ Theorem B.7.8 (in the usual form of 6.40).

The eventual proof that $F(\overline{M})$ is cyclic in effect produces the situation described earlier in Hypothesis 8.5.14. The final argument at the end of Section 23 will essentially contradict $F(\overline{M})$ cyclic—to prepare for that contradiction (and others along the way), 23.10 observes that the situation in Lemma 8.5.15 does not occur. We state this as:

LEMMA 8.5.30. *Assume $F(\overline{M})$ is cyclic. Then there is some $p \in \sigma(M)$ such that B does not centralize $O_p(\overline{M})$. In particular $O_p(\overline{M}) \neq 1$, so $F(\overline{M})$ is not a $\sigma(M)'$-group.*

We sketch the proof: As $F(\overline{M})$ is cyclic, it is centralized by $\overline{M}^\infty$; this establishes Hypothesis 8.5.14.1. Furthermore the list of 23.8 provides the restrictions on members of $\mathcal{C}(M)$ in 8.5.14.2. Now we assume that B centralizes $O_p(\overline{M})$ for each $p \in \sigma(M)$. Notice then that $O_p(\overline{M}) = 1$ by 8.5.28.2, so this assumption is equivalent to the assumption that $F(\overline{M})$ is a $\sigma(M)'$-group. Then using 21.23 (with the correction of misprints we mentioned earlier) we see that $O_p(\overline{M})$ is a Sylow p-subgroup of $C_{\overline{M}}(E(\overline{M}))$—and so is centralized by B, giving 8.5.14.3. Furthermore that Sylow group is cyclic, so that 8.5.14.4 is vacuously satisfied. (But even if the Sylow group were noncyclic, B is not inverted in M, as we noted in the proof of 8.5.28.1.) Thus we may apply Lemma 8.5.15, and parts (3) and (4) of that result gives two components of $\overline{M}$ defined over fields $\mathbb{F}_{2^{m_i}}$, with m_i a power of 2. But since we have two components, case (1) of 23.8 must hold, so that instead m_i is odd.

More detailed analysis of $F^*(\overline{M})$. In the remainder of Section 23, we essentially exploit the relationship in 23.9 between the subgroups of some H containing $L \in \mathcal{L}^*$, and the $2'$-components of M in 23.8.

The next few results reduce the four possibilities for L in 23.9.2 to a single case.

The first of these reductions in Theorem 23.11 eliminates $L/O_2(L) \cong SL_3(2^n)$. At 23.14, the argument uses Lemma 8.5.17 (which we have used elsewhere, cf. Remark 8.5.23). Then 23.21 provides a contradiction to Lemma 8.5.30; the use of the uniqueness results discussed in Remark 8.5.29 set up that application.

Next 23.22 quickly eliminates $L/O_2(L) \cong \Omega_4^+(2^n)$. The proof of this result does not invoke Lemma 8.5.30, but it is similar to the proof of that lemma, and produces a contradiction to Lemma 8.5.15; again one uses 21.23 in the verification of Hypothesis 8.5.14.

Finally Theorem 23.23 eliminates $L/O_2(L) \cong \Omega_4^-(2^n)$, completing the reduction to the case $L/O_2(L) \cong L_2(2^n)$—with V the natural module for L as in Remark 8.5.24. At 23.27 there are uses of Remark 8.5.29 and Lemma 8.5.30, much as in the proof of Theorem 23.11; in addition, the proof also uses at 23.25 a standard variation (cf. [**GLS96**, 26.16]) on the usual Thompson factorization (cf. (B.6.5)).

Next recall that $N = !\mathcal{M}(LT)$. Theorem 23.30 uses this uniqueness theorem to establish that $N = !\mathcal{M}(Y)$ for suitable subgroups Y of N containing B and an element of L inverting B. The proof uses techniques of pushing-up, notably from Sections 9 and 10 of [**Asc81d**], and also results on failure of factorization, e.g. from McClurg's thesis [**McC82**], as well as preliminary results from Section 8.

We now turn from N to $\overline{M}$; recall we will eventually show $F(\overline{M})$ is cyclic, in order to contradict Lemma 8.5.30.

Theorem 23.30 is used at 23.37 to prune the earlier list of cases for $K \in \mathcal{C}(M)$ in 23.8 (and improved at 23.29) to the single possibility of $\overline{K} \cong L_2(q)$ for q a prime. (Aside from the cases where $q = 5, 7$, this is promising, since we expect instead a group of Lie type in characteristic 2.) The proof uses Lemma 8.5.30 to handle the case where $F(\overline{M})$ is cyclic, while the proof in the noncyclic case appeals to Remark 8.5.29, to contradict Lemma 8.5.28.1.

Recall the module $V = \langle Z^N \rangle$ for N is defined just before Remark 8.5.24 essentially from $L \in \mathcal{L}^*$. We can now (in the spirit of 23.9) examine the embedding of V in M: In particular we set:

$$E := \langle V^M \rangle,$$

and apply some final elementary weak closure analysis—with E in the role of "V" in Definition B.8.1: namely 23.41 shows that:

$$V \text{ centralizes } W_1 := W_1(T, E), \text{ and hence } V \leq C_1 := C_1(T, E).$$

We remark that the proof uses 23.40, which requires $M = O^p(M)$ for $p \in \sigma(M)$—this follows from the simplicity of G via control of p-fusion by M (this is standard e.g. by B.2.7; compare also with the role of 2-fusion at the end of the proof in Remark B.10.6). Furthermore 23.40 also uses the structure of $\hat{M} = M/C_M(Z)$ in Remark 8.5.27, to get $[\hat{M}, \hat{M}] \leq F(\hat{M})$. In the proof of 23.41, if V were to fail to centralize W_1, then V would be contained in some conjugate M^g of M, such that for some conjugate K^g of K, $m(Aut_V(K^g/O_2(K^g))) = n$—but $n > 2$, whereas the automorphism group of $L_2(q)$ in 23.37 has 2-rank 2.

The weak closure arguments culminate at 23.42, which shows that:

$$X := O_{2,F}(M) \leq N.$$

For 23.41 shows that W_1 is weakly closed in $C_T(V)$, which equals $O_2(LT)$ as in (8.5.19)—so that $N_G(W_1) \leq !\mathcal{M}(LT) = N$. Also $C_G(C_1) \leq C_G(V) \leq N$. However, E is generated by conjugates of the TI-set Z, so $m(Aut_M(E), E) \geq n > 2$, and hence $s(M, E) \geq n$ using 23.40. It now follows from Proposition B.8.6, with $E, X, 0, 1$ in the roles of "V, H, i, k", that $X = C_X(C_1)N_X(W_1)$. Hence $X \leq N$, since we saw above that the two factors lie in N. (We mention that the reference to 12.21 in the proof in [**Asc83c**] seems to be a misprint: as that would require hypothesis (a) of Theorem B.8.3, namely $M = !\mathcal{M}(N_M(C_T(E)))$; this is why we instead used Proposition B.8.6 above.)

So at 23.43, we can finally deduce the desired properties of $\overline{X} = F(\overline{M})$: Assume that $q \in \pi(\overline{X})$ and $Q \in Syl_q(X)$. Then we get $N = O_2(X)N_N(Q)$ by a Frattini argument. Since $O_2(X) \leq T \leq M$ but we have $N \not\leq M$, we get $N_N(Q) \not\leq M$. If $q \in \sigma(M)$, then $N_G(Q) \leq M$, since M is strongly q-embedded by Lemma 8.5.28.2; so $\overline{X}$ is a $\sigma(M)'$-group. Furthermore if Q were noncyclic, then as $q \notin \sigma(M)$, we could apply Remark 8.5.29 (namely 23.7.2 with $O_2(X)Q$ in the role of "P") to get $Q \cap N = 1$—giving the contradiction $Q = 1$, since $O_{2,F}(M) \leq N$ by 23.42 above. Thus Q is cyclic, and we conclude that $\overline{X} = F(\overline{M})$ is a cyclic $\sigma(M)'$-group.

The final contradiction. But this contradicts Lemma 8.5.30, which says that if $F(\overline{M})$ is cyclic, then $F(\overline{M})$ is not a $\sigma(M)'$-group. This final contradiction completes the proof of the Uniqueness Case Theorem 8.0.2.

Some final comments on quoted results. Timmesfeld's work [**Tim75b**] on weakly closed TI-sets is used at many places in the treatment, particularly in the proof of 7.1; that result gives general properties of subgroups generated by TI-sets.

Our Section 7.5 (and more generally Chapter 7) includes the references quoted at 6.39—that is, for the $GF(2^n)$ Type Theorem 7.5.2.

Our Section B.7 includes the references quoted at 6.40—that is, for results related to the Global $C(G,T)$ Theorem B.7.8. Other pushing-up results are quoted from sections 8–10 of [**Asc81d**].

A number of specialized facts about $\mathbb{F}_2$-representations for various known simple groups are established in [**Asc82**], specifically for use in the treatment of the Uniqueness Case. Various references on representations of known groups are included among those indicated in our Section A.4. Other references for facts about modules include James' paper [**Jam73**] on the 2-modular irreducibles for the Mathieu groups, and Wales' paper [**Wal69a**] on the embedding of J_2 in $G_2(4)$.

Similarly many references for properties of known simple groups are included among those indicated in our Section A.3. In particular there is some use of the theory of fundamental subgroups for groups of Lie type in odd characteristic, from [**Asc77a, Asc77b**] and [**Asc80b**]. Also quoted is the list of strongly p-embedded subgroups in $\mathcal{K}$-groups determined at 24.1 in the Gorenstein–Lyons trichotomy work [**GL83**]. There are various references for properties of individual sporadic groups; again these are included among those listed in Section A.3. There are various appeals to Seitz's generation paper [**Sei82**] (on when $\Gamma_{E,1}(G) \neq G$ for G a group of Lie type and E a noncyclic elementary abelian subgroup of $Aut(G)$). But in most cases, some alternate argument is sketched, so it is not clear if these appeals are actually necessary.

Certain technical results are quoted from the papers [**Asc81b, Asc83a**] on the case $e(G) = 3$.

Also in [**Asc82**], there are appeals to Patterson's thesis [**Pat72**], Janko's paper [**Jan76**] on J_4, and a paper [**Smi74**] of F. Smith; these appeals can presumably be replaced by more modern references; the first at least is covered by [**Asc94**, 49.15]. Probably one can also dispense with some of the other specialized appeals above. There are appeals to Fischer's unpublished paper on 3-transposition groups (whose initial sections did appear as [**Fis71**]), about the structure of the Fischer groups; likely Aschbacher's book [**Asc97**] on 3-transpositions would now cover most of these appeals. (We state Fischer's Theorem as A.6.3, but in fact cite Aschbacher [**Asc97**] for the proof.)

Appendices

APPENDIX A

Some background material related to simple groups

A.1. Preliminaries: Some notation and results from general group theory

We will collect here, for the convenience of the reader, some standard notation, as well as statements of a number of well-known elementary results which we use frequently. We typically refer to the second volume [**GLS96**] of the GLS series for their proofs.

We recall that G is a *semidirect product* if $G = AB$ for a normal subgroup A, where $A \cap B = 1$; and a *direct product* if furthermore A and B centralize each other.

DEFINITION A.1.1. We say that G is a *central product* of A and B (and we write $G = A * B$), if $G = AB$ with $[A, B] = 1$.

We say G is a *wreath product* of A and B, and write $G = A \text{ wr } B$, if G is an extension of a direct product of a set Ω of copies of A, with a subgroup B transitively permuting Ω (in some specified way) and $N_B(A) = C_B(A)$.

Note that, in both cases, G is not determined up to isomorphism without additional information—in which case we usually say "the" central product etc. $\diamondsuit$

DEFINITION A.1.2 (TI-subgroup). A subgroup H of a group G is a *TI-subgroup* (short for *trivial intersection* subgroup) if $H \cap H^g = 1$ for all $g \in G - N_G(H)$. $\diamondsuit$

An important example is provided by:

DEFINITION A.1.3 (Frobenius group). A *Frobenius group* is a transitive permutation group G in which the one-point stabilizer G_α is nontrivial, but the two-point stabilizers $G_{\alpha,\beta}$ are trivial. It follows that G_α is a TI-subgroup of G, since $G_\alpha \cap G_\alpha^g = G_{\alpha,\alpha^g}$. (See e.g. [**GLS96**, 9.9])

If K_0 denotes the set of elements of G fixing no points, then $K := K_0 \cup \{1\}$ is a normal subgroup of G ([**GLS96**, 9.10])—nilpotent by Thompson's celebrated result [**GLS96**, 9.11.i]—called the *(Frobenius) kernel*; and G is the semidirect product ([**GLS96**, 9.10]) of K with a point stabilizer G_α, which may be called a *Frobenius complement*. $\diamondsuit$

Notation related to p-groups and p-local analysis. We typically write p for some prime.

NOTATION A.1.4. Let G be a finite group, and p a prime; then $Syl_p(G)$ denotes the set of all Sylow p-subgroups of G.

A *p-local subgroup* of G is the normalizer $N_G(P)$ of some nontrivial p-subgroup P of G. $\diamondsuit$

LEMMA A.1.5. *Let Y denote a normal subgroup of X.*
(1) If $S \in Syl_p(X)$, then $S \cap Y \in Syl_p(Y)$.
(2) (Frattini Argument) If $T \in Syl_p(Y)$, then $X = YN_X(T)$.
In particular $X = YN_X(S \cap Y)$.

PROOF. E.g. [**GLS96**, 4.4]. □

DEFINITION A.1.6 (p-rank and sectional p-rank). Let G be a finite group, and p some prime. The *p-rank* $m_p(G)$ of G is the largest rank of an elementary abelian p-subgroup of G. The *sectional p-rank* of G is the largest p-rank of an elementary abelian p-section of G. Equivalently (e.g. [**Asc00**, (23.1),(23.2)]), G has sectional p-rank at most k, if every p-subgroup of G is generated by at most k elements. ♢

For the definitions of standard classes of 2-groups of small 2-rank, such as *dihedral*, *semi-dihedral* (also called quasi-dihedral), *quaternion* and *generalized quaternion*, and *wreathed*, see e.g. [**Gor80a**, pp. 27, 191, 29, 486].

LEMMA A.1.7. *For odd p, a p-group of p-rank 1 is cyclic.*
A 2-group of 2-rank 1 is either cyclic or generalized quaternion (and so has a unique involution).

PROOF. E.g. [**GLS96**, 10.1]. □

LEMMA A.1.8. *A p-group P of p-rank at least 2 has a normal elementary subgroup of rank 2, unless $p = 2$ and P is dihedral or semidihedral.*

PROOF. E.g. [**GLS96**, 10.11]. □

LEMMA A.1.9. *If the 2-group S is either cyclic, dihedral of order at least 8, semidihedral, or generalized quaternion of order at least 16, then the automorphism group $Aut(S)$ is a 2-group.*

PROOF. For S cyclic, see e.g. [**GLS96**, 10.2.iii]. Otherwise, S has a unique cyclic subgroup of index 2, so the assertion follows from the cyclic case using coprime action [**GLS96**, 11.5]. □

DEFINITION A.1.10. The p-group P is *special* if $P' = Z(P) = \Phi(P)$, and *extraspecial* if furthmore this subgroup is of order p. It follows (e.g. [**GLS96**, 10.4]) that P is of order p^{2w+1}, where w is the *width* of P. ♢

DEFINITION A.1.11 (symplectic type). A p-group T is said to be of *symplectic type* if every characteristic (elementary) abelian subgroup is cyclic. A well-known theorem of P. Hall (see e.g. [**GLS96**, 10.3]) asserts that T is a central product $E * R$; where either $E = 1$ or E is extraspecial (of exponent p when p is odd), and where either R is cyclic, or (when $p = 2$) R is dihedral, semidihedral, or generalized quaternion of order at least 16. ♢

DEFINITION A.1.12. Let p be a prime, and G a finite group. Then $O_p(G)$ is the largest normal p-subgroup of G, and $O^p(G)$ is the smallest normal subgroup of G with p-power index. Similarly $O_{p,q}(G)$ is the preimage in G of $O_q\big(G/O_p(G)\big)$.

More generally if π is a set of primes, we can define $O_\pi(G)$ and $O^\pi(G)$ to be the largest normal π-subgroup, and smallest normal subgroup of index divisible only by primes in π, in G. In particular, writing p' for the set of primes other than p, $O^{p'}(G)$ is generated by all p-elements of G.

The subgroup $O_{2'}(G)$ is sometimes called the *core* of G, and denoted by $O(G)$. ♢

DEFINITION A.1.13. The preimage in G of $Z\big(G/O_{2'}(G)\big)$ is denoted by $Z^*(G)$. ♢

Glauberman's important Z^*-Theorem will be stated in the next chapter as B.2.1.

We mention P. Hall's generalization of Sylow's theorem for solvable groups:

THEOREM A.1.14 (Hall subgroups of solvable groups). *Assume G is solvable. Then for each subset π of the set of primes dividing $|G|$, there is a* Hall π-subgroup G_π, *of order $|G|_\pi$; Hall π-subgroups are all conjugate in G, and any π-subgroup of G is contained in a Hall π-subgroup.*

PROOF. E.g. [**GLS94**, 1.8.2]. □

DEFINITION A.1.15. We say that a group G is *p-nilpotent* if $G = O_{p'}(G)P$ for some (and hence for every) $P \in Syl_p(G)$. In this situation we say that $O_{p'}(G)$ is a *normal p-complement*. ♢

DEFINITION A.1.16 (Fitting subgroup). The *Fitting subgroup* $F(G)$ of a group G is the largest nilpotent normal subgroup of G. Equivalently, $F(G)$ is the product of the subgroups $O_p(G)$ for all primes p. ♢

Notation related to components and the generalized Fitting subgroup. We briefly sketch some concepts involved in Bender's important generalization of the Fitting subgroup. (We had previewed some of this material at the start of Section 0.3.)

DEFINITION A.1.17. A group S is *subnormal* in G if $S \trianglelefteq S_1 \trianglelefteq S_2 \ldots \trianglelefteq G$.

A group L is *quasisimple* if $L = [L, L]$ and $L/Z(L)$ is simple.

A *component* of G is a quasisimple subgroup L which is subnormal in G. ♢

Originally Gorenstein and Walter focused on components of $G/O_{2'}(G)$:

DEFINITION A.1.18 (p-components and the p-layer). Let G be a finite group.

A *p-component* L of G is a subnormal subgroup of G such that $L = O^{p'}(L)$ and $L/O_{p'}(L)$ is quasisimple. Thus the image of L is a component in $G/O_{p'}(G)$. (If also $O_{p'}(L) \leq Z(L)$, then L itself is quasisimple, and so is a component of G itself. Conversely if $p = 2$, then every component of G is a quasisimple 2-component of G, since the odd-order group $O_{2'}(G)$ is solvable. However, if p is odd, then G may well have components which lie in $O_{p'}(G)$ and so do not determine p-components.)

The *p-layer* of G, denoted by $L_{p'}(G)$, is the product of all p-components of G.

Similar definitions can be made for sets π of primes. For example, a *$2'$-component* covers a component of $G/O_2(G)$. ♢

Bender's fundamental construction used the components of G:

DEFINITION A.1.19 ($E(G)$ and $F^*(G)$). Let G be a finite group.

We write $E(G)$ to denote the product of all components of G.

The (central) product $F^*(G) := E(G)F(G)$ is the *generalized Fitting subgroup* of G. ♢

REMARK A.1.20 (Properties of components). See e.g. [**GLS96**, 3.5] for various properties of components. The key property is that distinct components of G commute, so that $E(G)$ is a central product of its quasisimple components. ♢

The generalized Fitting subgroup has an analogue of the self-centralizing property (e.g. [**Asc00**, 3.10]) which the usual Fitting subgroup $F(G)$ possesses when G is a solvable group. Indeed earlier we stated, as Theorem 0.3.1 in our Introduction, Bender's fundamental generalization:

$$C_G\big(F^*(G)\big) = Z\big(F^*(G)\big)$$

for arbitrary finite groups G.

We conclude the section with a result from representation theory:

REMARK A.1.21 (Clifford's Theorem). There are several standard versions of Clifford's Theorem, see e.g. [**GLS96**, 9.7, 32.16]. For the applications in our exposition, probably the following suffices: Assume G is irreducible on the vector space V, and N is a normal subgroup of G. Then N is completely reducible on V; and G/N transitively permutes the N-isotypic summands of V. $\diamondsuit$

A.2. Notation for the simple groups

In this section, we will mainly just indicate the standard names for the simple groups.

For further background on the simple groups, we will typically reference the third volume [**GLS98**] of the GLS series. We also mention Chapter 2 of Gorenstein's introduction [**Gor82**].

Alternating groups. We write A_n for the alternating group on n letters.

We write S_n for the symmetric group; in the literature one also sees Σ_n and Sym_n.

Groups of Lie type. We sometimes use the notation $Chev(p)$ (as in [**GLS94**, p. 81]) to denote the groups of Lie type defined over fields of characteristic p. (Note that this usage includes not only the usual ("untwisted") Chevalley groups, but also the twisted types.)

REMARK A.2.1 (The families of groups of Lie type). The simple groups of Lie type are subdivided into families. We tabulate them below; q denotes a power of a prime p.

Lie type	Notation	Variants	Name
			Chevalley (untwisted) groups
			(classical matrix groups—untwisted)
$A_n(q)$	$L_{n+1}(q)$	$PSL_{n+1}(q)$	linear
$B_n(q)$	$\Omega_{2n+1}(q)$		orthogonal
$C_n(q)$	$PSp_{2n}(q)$		symplectic
$D_n(q)$	$P\Omega_{2n}^{+}(q)$		orthogonal
			(exceptional groups)
$G_2(q)$			
$F_4(q)$			
$E_6(q)$			
$E_7(q)$			
$E_8(q)$			
			twisted groups
${}^2A_n(q)$	$U_{n+1}(q)$	$PSU_{n+1}(q)$	unitary (also classical matrix)
${}^2B_2(2^{2m+1})$	$Sz(q)$		Suzuki, twisted B_2
${}^2D_n(q)$	$P\Omega_{2n}^{-}(q)$		orthogonal (also classical matrix)
${}^3D_4(q)$			triality-twisted D_4
${}^2E_6(q)$			twisted E_6
${}^2F_4(2^{2m+1})$			twisted F_4 (Ree, Tits)
${}^2G_2(3^{2m+1})$	$R(q)$		Ree, twisted G_2

$\diamondsuit$

Our notation is essentially that used in GLS [**GLS98**] and in Carter [**Car89**]; there are minor variations from that used in the Atlas [**CCN$^+$85**].

The notation $O_{2n}^{\epsilon}(q)$ and $\Omega_{2n}^{\epsilon}(q)$, for a sign $\epsilon = \pm 1$, is sometimes used to consider simultaneously the two corresponding families of orthogonal groups. Similarly $L_n^{\epsilon}(q)$ is sometimes used to consider the linear groups $L_n^{+}(q) := L_n(q)$ simultaneously with the unitary groups $L_n^{-}(q) := U_n(q)$.

Sporadic groups. The 26 sporadic groups are usually just known by the names of their discoverers (perhaps with a subscript). Again we will present some information in tabular form.

Notation	Variants	Name
M_{11}		(Mathieu)
M_{12}		(Mathieu)
M_{22}		(Mathieu)
M_{23}		(Mathieu)
M_{24}		(Mathieu)
Co_1	$Co.1$, $\cdot 1$	(Conway)
Co_2	$Co.2$, $\cdot 2$	(Conway)
Co_3	$Co.3$, $\cdot 3$	(Conway)
J_1		(Janko)
J_2	HJ, HaJ	(Hall)-Janko
J_3		(Janko)
J_4		(Janko)
Fi_{22}	F_{22}, $M(22)$	(Fischer)
Fi_{23}	F_{23}, $M(23)$	(Fischer)
Fi'_{24}	F'_{24}, $M(24)'$	(Fischer)
HN	Har, F_5	Harada(-Norton)
He	HHM, F_7	Held(-Higman-McKay)
HS	HiS	Higman-Sims
Ly	LyS	Lyons(-Sims)
McL	Mc	McLaughlin
$O'N$	ON	O'Nan
Ru		Rudvalis
Suz		Suzuki
Th	F_3	Thompson
BM	B, F_2	Baby Monster (Fischer)
M	FG, F_1	Monster, Friendly Giant (Fischer-Griess)

A.3. Properties of simple groups and K-groups

Inductive arguments in the CFSG often make use of fairly detailed properties of K-groups—and in particular of the known simple groups. A good modern general reference for many of these properties is [**GLS98**]. In this section, we give an overview of some of the most-frequently used properties, primarily by referring to sources in the literature. Less frequently-cited references will typically be mentioned where they arise in the discussion of the main theorems in the classification of groups of characteristic 2 type.

First a preliminary remark concerning existence and uniqueness for the various types of simple groups: For the alternating groups, existence is trivial and there are easy proofs of uniqueness subject to a variety of properties. For the Lie type groups, existence and uniqueness are implicit in e.g. [**GLS98**, Sec 2.9]; existence is more explicit in Chevalley's construction described in [**Car89**, Secs 4.4, 13.4], and uniqueness is discussed in Section 12.1 of Carter. For the sporadic groups, we will indicate individual references for existence and uniqueness in the relevant subsection below.

We also mention that various papers quote the list of strongly p-embedded subgroups in $\mathcal{K}$-groups determined at 24.1 in the Gorenstein–Lyons trichotomy work [**GL83**]; this also now appears at [**GLS98**, pp 383–385].

Groups of Lie type. Recall there was some discussion of this material in Section 0.4. Our fundamental reference is Chapters 2–4 of [**GLS98**], along with Carter [**Car89**].

REMARK A.3.1 (A review of certain features of groups of Lie type). The finite *groups of Lie type* (see e.g. [**GLS98**, Sec. 2.2] for fundamental definitions) are analogues of the Lie groups of analysis; in particular they are matrix groups—but defined over finite fields. For the untwisted (Chevalley) groups, this is clear from the *Chevalley construction* ([**Car89**, Sec. 4.4]), while the twisted groups arise as fixed points of certain automorphisms of Chevalley groups ([**Car89**, Sec. 13.4]; cf. [**GLS98**, 2.2.4]).

The *Lie type* of a Chevalley group G is the type of the underlying complex simple Lie algebra appearing in the Chevalley construction; this algebra determines the *Lie rank* (or BN-rank) of G, its *Dynkin diagram*, and its *Weyl group*. For a twisted group G, the superscript in Remark A.2.1 gives the order of the relevant automorphism σ of the ordinary Chevalley group $\overline{G}$ from which we obtain the finite group $G = O^{p'}\left(C_{\overline{G}}(\sigma)\right)$ (for p the characteristic prime); this superscript is also the order of an associated automorphism τ of the underlying Dynkin diagram. Then identifying nodes fused by the automorphism τ, we determine the rank, diagram and Weyl group of G; cf. [**Car89**, Sec. 13.3]. Sometimes these are called the twisted rank, twisted diagram, and twisted Weyl group—to distinguish from other ("untwisted") objects that arise naturally in an alternative approach to groups of Lie type, which we describe next:

All the groups of Lie type can be regarded as fixed points of suitable endomorphisms of infinite simple algebraic groups, defined over the algebraic closure of a finite field of characteristic p; see especially Chapter 1 of [**GLS98**]. In this approach, a finite Chevalley group G and any of its twisted analogues descend from the same algebraic group—whose rank, diagram, and Weyl group can be called the untwisted rank, diagram, and Weyl group of G or its analogue. The differences between G and its twisted analogues arise from different choices of the endomorphism whose fixed points form the finite group. Correspondingly in a group G of Lie type of characteristic p, the p-elements are *unipotent elements*, and the p'-elements are *semisimple elements*; and centralizers of semisimple elements are obtained from the corresponding centralizers in the algebraic group, which are *reductive* groups.

Chapter 3 of [**GLS98**] describes results related to unipotent subgroups: Examples of such subgroups are provided by *root subgroups* (for the notion of *opposite* root groups see e.g. [**Car89**, p 280]). The normalizer of a maximal unipotent subgroup U is a *Borel subgroup* B. We then have $B = UH$ where H is a *Cartan subgroup*, namely a *(split) torus* (other p'-subgroups can come from *non-split tori*.) The overgroups of the Borel subgroups give the important class of p-local subgroups called *parabolic subgroups*. Recall that Lemma 0.4.1 gives some other important properties of parabolics. In more detail, a parabolic P has a *Levi decomposition* $P = VL$, in which $V = O_p(P) \leq U$ is the *unipotent radical* of P, and L is a *Levi complement*. The Borel-Tits Theorem (e.g. [**GLS98**, Sec 3.1]) implies that every p-local subgroup M of a finite group G of Lie type satisfies $M \leq P$ and $O_p(M) \leq O_p(P)$ for some parabolic P of G.

A Borel subgroup B, along with a monomial subgroup N of G (usually[1] we get $N = N_G(H)$ as the normalizer of some Cartan subgroup H of B), determines

[1]This can fail over very small fields, as H can be degenerate in such cases.

the *BN-pair* or *Tits system* of G. This notion provides a useful axiomatic approach for studying groups of Lie type; for example one can see how G and suitable subgroups are generated (see e.g. [**GLS98**, Sec 3.3]). The collection of all parabolic subgroups, along with their inclusions, provides the important combinatorial structure called the *Tits building* for G (cf. [**Car89**, Sec 15.5].)

Chapter 4 of [**GLS98**] discusses automorphisms of a group G of Lie type. In general such an automorphism is a product of four basic types: inner automorphisms; *diagonal* automorphisms—essentially arising from elements of a torus H which fall outside the commutator subgroup $[G,G]$; *field* automorphisms, corresponding to a Galois automorphism of the underlying field of definition of G; and *graph (or diagram)* automorphisms, defined using a symmetry of the underlying Dynkin diagram. These automorphisms and their centralizers are important in the study of the structure of non-p-local subgroups of G. $\diamondsuit$

REMARK A.3.2 (Isomorphisms among Lie types). The groups for distinct Lie types are usually not isomorphic, except in certain small cases: First we observe that $B_2(q) = \Omega_5(q) \cong Sp_4(q) = C_2(q)$; indeed in characteristic 2, more generally $B_n(q) = \Omega_{2n+1}(q) \cong Sp_{2n}(q) = C_n(q)$ for all n. Additional isomorphisms include $D_3(q) = \Omega_6^+(q) \cong L_4(q) = A_3(q)$, and also $\Omega_6^-(q) = {}^2D_3(q) \cong U_4(q) = {}^2A_3(q)$ as well as ${}^2D_2(q) = \Omega_4^-(q) \cong L_2(q^2)$. See also Remark A.3.6 for some isomorphisms not arising from Lie theory. $\diamondsuit$

REMARK A.3.3 (Failure of simplicity in some small cases). The groups of Lie type, as constructed in [**Car89**], are simple—except in certain small cases: First the groups $L_2(2)$, $L_2(3)$, $U_3(2)$, and $Sz(2) = {}^2B_2(2)$ are solvable. Also $\Omega_4^+(q)$ is the commuting product of two $SL_2(q)$ subgroups and so is not simple, and indeed is solvable for $q = 2, 3$. Finally $Sp_4(2) = B_2(2)$, $G_2(2)$, ${}^2F_4(2)$, and ${}^2G_2(3)$ are not simple, but have a simple subgroup of index equal to the associated characteristic (namely 2 or 3). These exceptions should be kept in mind when reading statements of classifications such as the CFSG theorem 0.1.1. $\diamondsuit$

Specialized facts about subgroup structure in groups of Lie type. It is useful to know *all* subgroups of some small groups of Lie type: The subgroups of $SL_2(q)$ are given in Dickson's Theorem; see Dickson [**Dic58**] or e.g. Huppert [**Hup67**, II.8.27]). Some of Aschbacher's papers reference Bloom [**Blo67**] for the subgroups of $GL_3(q)$.

Many papers reference Aschbacher–Seitz [**AS76a**] for involution centralizers in groups of Lie type in characteristic 2. Detailed information on involution centralizers in groups of Lie type in odd characteristic was worked out by Burgoyne and Williamson in [**BW77**] and is available in Chapter 4 of [**GLS98**].

Various papers also reference Seitz's "generation" paper [**Sei82**] on when we can have $\Gamma_{1,E}(G) < G$—for G a group of Lie type, and E a noncyclic elementary abelian subgroup of $Aut(G)$. Here is one version of a basic result there which many papers quote:

THEOREM A.3.4 (Seitz Generation Theorem). *Let K be a finite simple group of Lie type defined over a field of characteristic r. Let $p \neq r$ be an odd prime, and let E be a noncyclic elementary abelian p-subgroup of $Aut(K)$. Then either:*

$$K = \langle C_K(F) : 1 < F \leq E \rangle,$$

or one of the following holds:

(1) $|E| = 5^2$ *and* $K = {}^2F_4(2)'$ *or* $Sz(2^5)$*; or*

(2) $|E| = 3^2$ *and* $K = L_2(8)$, $L_3(4)$, $U_4(2)$, $Sp_4(2)'$, $Sp_6(2)$, $Sp_8(2)$, $G_2(2)'$, *or* ${}^2F_4(2)'$.

□

A corollary of this theorem extends the result to classify all cases where:

$$K > \langle C_K(F) : |E/F| = p \rangle.$$

A proof of these fundamental generation theorems, as well as extensions by Lyons to the case where G is a group of Lie type over a field of characteristic 2, is given in Chapter 7 of [**GLS98**].

For groups of Lie type in odd characteristic, the theory of *fundamental subgroups* (namely certain $SL_2(q)$-subgroups) is developed in [**Asc77a, Asc77b**] and in [**Asc80b**].

Sporadic groups. Many facts about the structure of sporadic groups are contained in the Atlas [**CCN$^+$85**] (though without proofs), and in Section 5.3 of [**GLS98**] with some proofs; see also the references given in [**GLS98**], especially [**Asc97**] for the Fischer groups.

For more details on some particular sporadics (especially the Mathieu and Conway groups), see Conway's lectures [**Con71**] and Aschbacher's book [**Asc94**]; the latter supplies some proofs not given in the former or in the Atlas.

We now list some other references for the subgroup structure of the sporadic groups. This list includes those references most frequently quoted in the papers discussed in this volume. In particular we indicate references for existence and uniqueness, though many other papers could also be listed. For many of the groups, these problems are covered by a more recent systematic approach due to Aschbacher [**Asc94**].

M_{11}: existence and uniqueneness [**Con69**] [**Asc94**]
M_{12}: existence and uniqueneness [**Con69**] [**Asc94**]
M_{22}: existence and uniqueneness [**Con69**] [**Asc94**]
M_{23}: existence and uniqueneness [**Con69**] [**Asc94**]
M_{24}: existence and uniqueneness [**Con69**] [**Asc94**]
Co_3: existence [**Con69**], uniqueness [**Fen73**], maximal subgroups [**Fin73**]
Co_2: existence [**Con69**], uniqueness [**Smi74**]
Co_1: existence [**Con69**], uniqueness [**Pat72**] [**Asc94**]
J_1: existence and uniqueness [**Jan66**]
J_2: [**Jan69**], existence and uniqueness [**HW68**] [**Asc94**], maximal subgroups [**FR73**]
J_3: [**Jan69**], existence and uniqueness [**HM69**], maximal subgroups [**FR74**]
J_4: [**Jan76**], existence [**Nor80**], uniqueness [**AS91**]
HS: existence [**HS68**], uniqueness [**JW69**], maximal subgroups [**Mag71**]
He: [**Hel69**], existence and uniqueness: Higman–McKay (unpublished), see also [**Asc94**]
Ly: [**Lyo72**], existence and uniqueness [**Sim73**] [**AS91**]
McL: existence [**McL69a**], uniqueness [**AS91**], maximal subgroups [**Fin73**]
$O'N$: existence [**O'N76b**] refers to Sims (unpublished)—see also [**CCN**$^{+}$**85**, p 132]), uniqueness [**And80**]
Ru: existence [**CW73**], uniqueness [**Dem74**]
Suz: existence [**Suz69b**], uniqueness [**Asc94**]
Fi_{22}: existence [**Fis71**], uniqueness [**Asc97**]
Fi_{23}: existence [**Fis71**], uniqueness [**Asc97**]
Fi'_{24}: existence [**Fis71**], uniqueness [**Asc97**]
HN: uniqueness [**Har76**], existence [**Gri82**]
Th: Thompson (1974 Sapporo lecture), existence [**Smi76b**], [**Gri82**] uniqueness [**Rei76**]
BM: existence and uniqueness [**LS77**], [**Gri82**]
M: [**Gri76**], existence [**Gri82**], uniqueness [**GMS89**]

Various papers reference the table of involution centralizers in [**AS76b**], and other facts about 2-local structure in [**AS81**]; these in turn depend on further references for the individual groups, such as those just indicated.

Schur multipliers of simple groups. When G is perfect with trivial center (and in particular when G is simple), the *Schur multiplier* is the center of the largest perfect central extension of G. For the theory of the Schur multiplier of an arbitrary group G, see e.g. [**GLS98**, Sec 5.1].

For the alternating group A_n ($n = 5$, $n \geq 8$), the Schur multiplier is of order 2; it is realized via a standard double cover (meaning an extension over a center of order 2), which we denote by $2A_n$. For the exceptional values $n = 6, 7$, the multiplier has order 6; so there is also a triple cover (i.e. with center of order 3), which we denote by $3A_6$ or $3A_7$.

For the groups of Lie type, the multiplier is "usually" explained by the Lie theory—cf. [**GLS98**, Table 6.1.2]. (The extension by the full generic multiplier is the *universal* form of the group.)[2] For the exceptional cases, and for the sporadic groups, we summarize via:

REMARK A.3.5 (Schur multipliers). The Schur multipliers of the finite simple groups not determined by Steinberg using Lie theory were determined by various authors, notably Griess in [**Gri72, Gri80, Gri85**]. Tables of multipliers for the simple groups, along with a useful expository discussion, are given in [**GLS98**, Sec 6.1]. ◊

Isomorphisms among simple groups. Of course for the different names indicated above, the corresponding groups are usually not isomorphic. However, we indicated the exceptional isomorphisms among groups of Lie type of the same characteristic earlier in Remark A.3.2; and we now record the additional cases, as part of a discussion of groups of GW-type:

REMARK A.3.6 (Some cases which are unusual from the point of view of groups of GW type). We indicate some groups which have unexpected behavior from the point of view of the main theorem 1.0.1 for groups of GW type, sometimes because of exceptional isomorphisms.

We mentioned after the definition 0.3.8 of groups of component type that the generic examples are groups of Lie type in odd characteristic, and alternating groups. But the following members of those families are of 2-rank at least 3 but *not* of GW-type; indeed they are of characteristic 2 type (and hence not of component type from the definitions), sometimes because of exceptional isomorphisms:

$$A_8 \cong L_4(2), PSp_4(3) \cong U_4(2), U_4(3), G_2(3).$$

These isomorphisms show that the first two groups are of Lie type in characteristic 2. Also $U_4(3)$ is a "big" subgroup of the group $U_6(2)$ of characteristic 2.

Furthermore some groups of Lie type which are of 2-rank at most 2, and hence of GW type, are also of characteristic 2 type. Among these are certain cases which (again from the indicated isomorphisms) are even of Lie type in characteristic 2:

$$A_5 \cong L_2(5) \cong L_2(4), A_6 \cong L_2(9) \cong Sp_4(2)', L_2(7) \cong L_3(2),$$

$$U_3(3) \cong G_2(2)', {}^2G_2(3)' \cong L_2(8), U_3(4),$$

as well as $L_2(p)$ for p a Fermat or Mersenne prime, and M_{11}. Of course these groups do not arise when we consider groups which are of characteristic 2 type but have 2-rank at least 3. ◊

A.4. Properties of representations of simple groups

Arguments related to 2-local structure in the CFSG often make use of fairly detailed properties of certain $\mathbb{F}_2$-modules V for various almost-simple K-groups—typically "small" modules such as FF-modules.

---

[2]Indeed for groups G of Lie type in characteristic p, the universal version gives the p'-part of the Schur multiplier, with the single exception of $G = Sp_4(2)'$.

Weight theory for groups of Lie type. Various papers in the classification of groups of characteristic 2 type, and especially papers such as [**GM02, GM04**] and [**Asc82**], which are quoted during the proof of the classification, make use of the highest-weight theory for irreducible modules V in the natural characteristic 2 for groups G in $Chev(2)$. One source for this material is Section 2.8 of [**GLS98**], where in particular the Steinberg Tensor Product Theorem is listed as Theorem 2.8.5; other sources include [**Ste68b**] and Section 9 of [**Cur71**].

Cohomology. There are occasional references to facts about module cohomology, typically 1- or 2-cohomology of various small modules. For Lie type groups, several papers quote Jones–Parshall [**JP76**]. Also a number of specific facts are established in Section 2 of Timmesfeld's extraspecial paper [**Tim78a**]; and other references are quoted in that and other papers in the $GF(2)$ type classification, discussed in Chapter 7. There may well be more recent sources for such facts.

Parameters for modules. We mention in Sections B.6 and B.8 that the methods of failure of factorization and weak closure make use of certain parameters for various (usually "small") modules. These parameters are discussed especially in [**Asc82**], which establishes estimates on $m(G, V)$ and $a(G, V)$ for suitable $\mathbb{F}_2G$-modules V. Here $m(G, V)$ is (cf. Remark B.8.2) the minimum codimension of $C_V(t)$ for t an involution in G; while $a(G, V)$ is the maximum k with $\mathcal{A}_k(G, V)$ nonempty, where $\mathcal{A}_k(G, V)$ consists of those nontrivial elementary abelian 2-subgroups A of G such that $C_V(A) = C_V(B)$ for all $B \leq A$ with $m(A/B) < k$. Both parameters are critical in weak closure theory; and the estimates require the weak closure theory itself, as well as facts on subgroup structure and irreducible representations as mentioned earlier in Sections A.3 and A.4. For example, to bound $m\big(V/C_V(t)\big)$ for an involution t in an almost-simple group G, it suffices to show that either there is a big p-subgroup (p an odd prime) $P = [P, t]$ of G; or, that G is generated by a small number of conjugates of t, and then that $m(V)$ is bounded below, often using Landazuri-Seitz [**LS74**] (though seemingly that paper itself is not directly quoted in the proofs of level 0 results for characteristic 2 type).

Other results on representations. Some proofs quote James [**Jam73**] for the 2-modular irreducibles for the Mathieu groups. Wales' paper [**Wal69a**] on the embedding of J_2 in $G_2(4)$ studies the natural module for $G_2(4)$.

A.5. Recognition theorems for identifying simple groups

The proof of a classification theorem will usually finish by quoting (or proving) a suitable *recognition theorem*—namely once we have built up enough information (usually about local subgroups) in our so-far unknown group G, we then invoke a result which shows that this information is sufficient to identify G up to isomorphism as a particular simple group.

Some general references. There is a discussion of the general topic of recognition theorems in Chapter 3 of Gorenstein's introductory volume [**Gor82**]. This includes sections on alternating and sporadic groups; and on Lie type groups. In a moment we will expand a little on the subject of presentation theorems for the groups of Lie type.

Often a recognition theorem will be just a variant of a uniqueness proof; and indeed for sporadic groups (and more generally for groups of $GF(2)$ type as in our

Chapter 7), the approach to uniqueness developed by Aschbacher in [**Asc94**] comes as close as any to a uniform strategy.

The Curtis-Tits and Phan presentation theorems for groups of Lie type. Since most simple groups are of Lie type, recognition theorems for Lie type groups are needed throughout the CFSG. And (as we see especially in Chapter 6 on standard type), the most fundamental result is the presentation of Curtis-Tits and the variant by Phan.

It was observed independently by Tits [**Tit62**] and by Curtis [**Cur65**] (using the Steinberg relations [**GLS98**, 2.9.1, 2.9.2]) that a universal quasisimple group G of Lie type with BN-rank at least 3 can be characterized as the universal completion of an amalgam (cf. 3.3.14) of rank 1 Levi complements. This presentation became the standard method for the identification of groups of Lie type. We state it as in [**GLS98**, 2.9.3]:

THEOREM A.5.1 (Curtis-Tits Theorem). *Let K be the universal version of a finite group of Lie type having BN-rank at least 3 with root system $\hat{\Sigma}$, fundamental system $\hat{\Pi}$, and root groups $X_{\hat{\alpha}}$ for $\hat{\alpha} \in \hat{\Sigma}$ (See* [**GLS98**, 2.3.1, 2.3.6] *for notation).*

For each subset $J \subseteq \hat{\Pi}$, let K_J be the subgroup of K generated by all root subgroups $X_{\hat{\alpha}}$, $\pm\hat{\alpha} \in J$. Let D be the set of all subsets of $\hat{\Pi}$ with at most 2 elements, and partially order D by inclusion.

Then $\{K_J : J \in D\}$ is a defining K-amalgam.

The terminology of defining K-amalgam from [**GLS96**, Sec 29] just says here that K is presented by generating subgroups K_J with $|J| = 2$, subject to relations given by the multiplication tables of the K_J, together with those relation from their intersections given by K_I with $|I| = 1$.

REMARK A.5.2 (Phan's theorem for unitary groups). The statement above of the Curtis-Tits Theorem includes the twisted groups. Originally, it was only stated for untwisted groups; and it was usually quoted only for those untwisted groups. And so in order to handle the unitary groups, Phan in [**Pha77a, Pha77b**] established a result which uses a slightly different hypothesis.

By way of background: in the untwisted case, say with field of definition $\mathbb{F}_q$, for $|I| = 1$ we get $K_I \cong SL_2(q)$ for the group generated by a pair $X_{\pm\hat{\alpha}}$ of "opposite" root groups. However in the unitary group $SU_n(q)$ (with $n \geq 4$), Phan instead takes generating subgroups $K \cong SU_2(q)$. For n even, these do arise from a pair $X_{\pm\hat{\alpha}}$—but for n odd, they arise only from $\Omega_1(X_{\pm\hat{\alpha}})$ (and so represent only a part of a Levi complement). Furthermore an "adjacent" pair of such subgroups will generate a subgroup $SU_3(q)$ (whereas in the Curtis-Tits viewpoint, a pair of $SL_2(q)$ subgroups would generate a group of BN-rank 2). Finally when $q = 3$, Phan needs to assume that three adjacent subgroups generate $SU_4(3)$. He then shows that these generators and relators present $SU_n(q)$.

Since some details in Phan's argument were suspect, we instead refer here to a revised, full proof given later by Bennett and Shpectorov in [**BS04**].

Let $\mathcal{L}(q)$ be a Chevalley group where the simple Lie algebra $\mathcal{L}$ has a single root length. The Curtis-Tits presentation of $\mathcal{L}(q)$ is a subset of the Curtis-Tits presentation of the overlying algebraic group $\overline{G}$. If one descends from $\overline{G}$ to the finite case differently, including the automorphism -1 on the root system of $\mathcal{L}$, then one obtains a putative presentation of Phan type for ${}^2\mathcal{L}(q)$ (if the Weyl group of G does not contain -1) or the group $\mathcal{L}(q)$ itself (if the Weyl group of G does contain -1). Here "Phan type" means that two adjacent $SL_2(q)$'s generate $SU_3(q)$. A comprehensive study of Phan type recognition theorems has been made by Gramlich in his Darmstadt Habilitationsschrift [**Gra04**]. ◇

Another important corollary of the Steinberg presentations is the Gilman-Griess Theorem 6.1.1. This is a key tool for the identification of the groups of Lie type defined over a field of characteristic 2, for which $e(G) \geq 4$.

References for some specific recognition theorems. We saw in Section 1.4 that some early classification theorems made use of the Brauer-Suzuki-Wall Theorem [**BSW58**]. We mention that their work in turn uses a paper of Zassenhaus, namely *Abh. Math. Sem. Hamburg* **11** (1936), 17–40.

A number of recognition theorems used in the proof of the Quasithin Theorem appear in Sections I.4, I.5, and J of [**AS04a**]. The proof of the Uniqueness Case Theorem uses a characterization of Co_2 by F. Smith in [**Smi74**], which may now be covered by more recent results. Also various papers (for example in the proof of the $GF(2)$ Type Theorem) use the characterization of Co_1 in Patterson's thesis [**Pat72**]; this is now covered by [**Asc94**, 49.15].

A.6. Transvection groups and transposition-group theory

In this section, we briefly review a number of results which identify certain groups via elements with "minimal action".

Transvection groups. If G acts on a vector space V, the nontrivial unipotent elements t which move a minimal number of points are the *transvections*; that is, the unipotent elements t with $[V, t]$ of dimension 1. Thus it is useful to be able to identify subgroups generated by transvections.

The most frequently quoted case is when V is an elementary abelian 2-group, regarded as a vector space over $\mathbb{F}_2$. The result appeared in McLaughlin [**McL69b**]; and we give the statement as in [**Gor82**, 4.177]:

THEOREM A.6.1 (McLaughlin's theorem [**McL69b**]). *Let V be an irreducible module of dimension n over $\mathbb{F}_2$ for a group G generated by transvections. Then G is one of $SL_n(2)$, $Sp_n(2)$, $SO_n^{\pm}(2)$, S_{n+1}, or S_{n+2}.*

3-transposition groups. For permutation groups, the elements with minimal action are the transpositions. Fischer discovered his first three sporadic groups in the process of trying to characterize the symmetric groups via the possible orders of products of transpositions. Namely he defined:

DEFINITION A.6.2 (3-transpositions). A set of involutions in G is called a set of 3-*transpositions* if the product of any pair has order 1, 2, or 3. A group generated by a conjugacy class of 3-transpositions is called a 3-*transposition group*. ◇

Clearly the symmetric groups have this property—but so for example do all the Weyl groups with one root length. So Fischer added some further restrictions, still trying to characterize the symmetric groups; but he also obtained his sporadic groups as conclusions in the result usually called Fischer's 3-transposition theorem. In fact Fischer published only part [**Fis71**] of the proof of his theorem, so we instead refer here to a full proof published later by Aschbacher in [**Asc97**].

THEOREM A.6.3 (Fischer's theorem on 3-transposition groups). *Assume G is generated by a conjugacy class of 3-transpositions; and that $Z(G) = 1$ with $[G, G]$ simple. Then G is one of S_n, $Sp_n(2)$, $U_n(2)$, $O_n^{\pm}(2)$, $PO_n^{\mu,\pi}(3)$, Fi_{22}, Fi_{23}, or Fi_{24}.*

Here μ and π are suitable signs, parametrizing projective orthogonal groups over $\mathbb{F}_3$, generated by reflections.

In the proof of the Uniqueness Case Theorem, there are appeals to Fischer's unpublished paper (which would have been the sequel to [**Fis71**]) about the structure of the Fischer groups; now [**Asc97**] covers these appeals.

Other transposition-type characterizations. Several extensions of the notion of 3-transpositions were subsequently studied.

The definition of a set of *odd transpositions* is obtained by replacing the integer 3 in Definition A.6.2 by the set of odd integers. The corresponding groups were determined by Aschbacher in [**Asc72**]. We summarize by saying that the groups in the conclusion include the symplectic, unitary, and orthogonal groups—but now over any field $\mathbb{F}_{2^m}$ of characteristic 2 (not just $\mathbb{F}_2$). In addition there are orthogonal groups over $\mathbb{F}_5$, Suzuki groups, and certain wreath products of groups $L_2(q)$.

Timmesfeld then used the transposition-viewpoint to study more general groups of Lie type in characteristic 2.

The first step was again to consider just the groups over the field $\mathbb{F}_2$. Here the relevant notion is that of $\{3,4\}^+$*-transpositions*: this extends Definition A.6.2 by allowing order 4 as well as 3—but with the restriction that when xy is of order 4, the square $(xy)^2$ is also in the conjugacy class. In [**Tim73**], he produced the characterization of the additional groups over $\mathbb{F}_2$. Assuming that a product of order 4 does exist, one obtains essentially the groups of Lie type with single bonds in the Dynkin diagram (A_n, D_n, E_n) as well as G_2.

Then to treat the same groups, but over all fields of characteristic 2, Timmesfeld introduced the notion of *root involutions*, by again replacing 3 by all odd integers in the definition of $\{3,4\}^+$-transpositions. Timmesfeld obtained the corresponding classification in [**Tim75a**].

Finally, one motivation for classifying the groups of $GF(2^n)$ type (as in Section 7.5) was to obtain a treatment which also includes the remaining Lie type groups—notably those whose Dynkin diagram contains a double bond.

All these results were often used in the role of recognition theorems in the CFSG. (But typically their proofs in turn depended on the Curtis-Tits or Phan presentation theorems.)

APPENDIX B

Overview of some techniques used in the classification

For material on general techniques, the reader may also wish to refer to Chapter 4 of Gorenstein's introduction [**Gor82**]. (However, note that our material on properties of $\mathcal{K}$-groups, corresponding to Gorenstein's Sections 4.14 and 4.15, has been located in Section A.3.)

In addition, Section 0.3 contains a number of basic general results on the generalized Fitting subgroup $F^*(G)$, p-components, and p-layers, not all of which are reproduced here.

B.1. Coprime action

There are many standard results about the action of a p'-group on a p-group; in the literature these are sometimes referred to only implicitly, or just by using the general term "coprime action". One source for such material is Section 11 of [**GLS96**]. For the convenience of the reader, we will record here a few of the statements which we use frequently. (Indeed we previewed some features of coprime action at Lemma 0.3.11, which is repeated below as Lemma B.1.8.)

One of the most important results on coprime action is Thompson's $A \times B$ Lemma. (E.g. [**Asc00**, 24.2] or [**GLS96**, 11.7].)

THEOREM B.1.1 (Thompson's $A \times B$ Lemma). *Let P be a p-group admitting the action of $A \times B$, where A is a p-group and B is a p'-group centralizing $C_P(A)$. Then B centralizes P.* □

A corollary ([**GLS96**, 11.8]) is in turn a key ingredient for the Critical Subgroup Lemma below.

DEFINITION B.1.2 (critical subgroup). A *critical subgroup* of a p-group P is a characteristic subgroup of P on which each p'-automorphism of G acts faithfully.

Thompson's Critical Subgroup Lemma guarantees the existence of a critical subgroup of small nilpotence class; thereby reducing many questions about coprime action to linear (or bilinear) algebra. Here is one version of the lemma:

LEMMA B.1.3 (Thompson's Critical Subgroup Lemma). *Let P be a finite p-group. Then P has a critical subgroup Q with the following properties:*
(1) Q has nilpotence class at most 2;
(2) $[Q, Q] = \Phi(Q)$. □

See e.g. [**GLS96**, 11.11] or [**Gor80a**, 5.3.11] for fuller statements.

Another important application of the $A \times B$ Lemma is the global-implies-local property for characteristic p ([**Asc00**, 31.16.2]) (we had previewed this at 0.3.12):

LEMMA B.1.4 (p-locals in groups of characteristic p). *Assume $F^*(G) = O_p(G)$. Then $F^*(H) = O_p(H)$ for every p-local subgroup H of G.* □

REMARK B.1.5. For a p-subgroup P of a group G, $F^*\big(N_G(P)\big)$ is a p-group if and only if $F^*\big(C_G(P)\big)$ is a p-group. Indeed $O^p(F^*\big(C_G(P)\big)) \leq O^p(F^*\big(N_G(P)\big))$, and $O^p(F^*\big(N_G(P)\big)) = E\big(N_G(P)\big)O_{p'}(F\big(N_G(P)\big)) \leq C_G(P)$ gives the reverse inclusion. ♢

This remark leads to a reduction which seems to be standard in the literature:

LEMMA B.1.6. *Let $H := N_G(P)$ for a p-subgroup P of G. Assume for some a of order p in $Z(P)$ that $F^*\big(C_G(a)\big) = O_p\big(C_G(a)\big)$. Then also $F^*(H) = O_p(H)$.*

Hence to show that G is of characteristic p type, it suffices to show for all elements a of order p in G that $F^*\big(C_G(a)\big) = O_p\big(C_G(a)\big)$.

PROOF. By hypothesis, $P \leq C_G(a) =: K$ and $F^*(K) = O_p(K)$. Then by Lemma B.1.4, $F^*\big(N_K(P)\big)$ is a p-group, and hence by Remark B.1.5, so is $F^*\big(C_K(P)\big)$. But $C_K(P) = C_G(P)$, and so again using that Remark, $F^*\big(N_G(P)\big) = F^*(H)$ is a p-group. □

Next, the Thompson Dihedral Lemma describes the minimal solvable groups X with elementary abelian Sylow 2-subgroups and $O_2(X) = 1$: X is the direct product of dihedral groups of twice odd order. This is 5.34 in Thompson's N-group paper [**Tho68**] (see also [**GLS96**, 24.1]); it is quoted at various points in our discussion (and in particular it plays a crucial role in the Klinger-Mason Method, in later Section B.9).

LEMMA B.1.7 (Thompson's Dihedral Lemma). *If p is an odd prime and E is an elementary abelian 2-group of rank $m \geq 1$ acting faithfully on the p-group P, then P contains an E-invariant subgroup Q such that $QE = D_1 \times \cdots \times D_m$, where each D_i is a dihedral group of order $2p$.*

Here is our re-statement of a result previewed earlier at 0.3.11:

LEMMA B.1.8 (see 18.7.4 in [**Asc00**] or 11.13 in [**GLS96**]). *Let P be a p-subgroup of G, and set $\overline{G} := G/O_{p'}(G)$. Then*

(1) $N_{\overline{G}}(\overline{P}) = \overline{N_G(P)}$.

(2) If P is abelian and normalizes a p'-subgroup X, then:

$$X = \big\langle C_X(Q) : Q \leq P,\ P/Q \text{ cyclic}\big\rangle.$$

Below we review some of the many equivalent definitions of the notion of p-constraint.

LEMMA B.1.9. *Let X be a finite group, choose any $R \in Syl_p\big(O_{p',p}(X)\big)$, and set $\overline{X} := X/O_{p'}(X)$. Then the following properties are equivalent; in which case X is said to be* p-constrained*:*

(1) $F^*(\overline{X}) = O_p(\overline{X})$;

(2) $C_{\overline{X}}(\overline{R}) = Z(\overline{R})$;

(3) $C_X(R) \leq O_{p',p}(X)$; and

(4) $L_{p'}(X) = 1$.

PROOF. (Sketch) Use the self-centralizing property 0.3.1 of $F^*(\overline{H})$, Coprime Action B.1.8.1, and the fact that $\overline{L_{p'}(X)} = E(\overline{X})$ (cf. Definition 0.3.7). □

The Bender-Thompson Signalizer Lemma, an application of the Critical Subgroup Lemma, appeared in [**Ben67**]; a fuller statement can be found as [**GLS96**, 23.3].

LEMMA B.1.10 (Bender-Thompson Signalizer Lemma). *Suppose that p is an odd prime and G is p-constrained. Let B be maximal among elementary abelian p-subgroups of G with respect to inclusion, or more generally a p-subgroup containing every element of order p in $C_G(B)$.*

Then every B-invariant p'-subgroup of G lies in $O_{p'}(G)$.

B.2. Fusion and transfer

In this section, typically p denotes a prime, and P denotes a Sylow p-subgroup of a finite group G.

Fusion and control of fusion. The term *p-fusion* (or just *fusion* when p is clear) refers to the G-conjugacy of elements or subsets of a Sylow p-subgroup P of G. The study of fusion is important throughout local group theory; for example Glauberman's Z^*-Theorem [**Gla66**] gives a structural consequence of a restriction on fusion:

THEOREM B.2.1 (The Z^*-Theorem). *Let G be a finite group with $O_{2'}(G) = 1$. If z is an involution of G which commutes with none of its distinct G-conjugates, then $z \in Z(G)$.* □

In view of the Sylow theorems, we can normally study fusion of p-elements or p-subgroups by studying their fusion to and from extremal conjugates:

DEFINITION B.2.2. If y is an element of $P \in Syl_p(G)$ we say y is *extremal* in P if $C_P(y)$ is Sylow in $C_G(y)$. Likewise a subgroup $Q \leq P$ is extremal in P if $N_P(Q)$ is Sylow in $N_G(Q)$.

It follows from Sylow's Theorem that any element x of P has some G-conjugate y which is extremal in P; indeed it is possible to choose $g \in G$ with $x^g = y$ and $C_P(x)^g \leq C_P(y)$. A similar statement holds for subgroups. See e.g. [**GLS96**, p. 92].

In order to study fusion in G, it is extremely useful to know when a collection of subgroups—or even a single subgroup—*controls* p-fusion, in the sense that any conjugation of subsets of P can be effected via a sequence of conjugations within those controlling subgroups. See e.g. [**GLS96**, Sec 16] for more on this topic; we mention a few basic results:

REMARK B.2.3 (Burnside's Fusion Theorem). When P is abelian, the single subgroup $N_G(P)$ controls fusion by Burnside's Fusion Theorem (see e.g. [**GLS96**, 16.2]).

There are generalizations (e.g. in [**GLS96**, Sec 16]) involving suitable control by the normalizer of a subgroup which is weakly (or strongly) closed in P. We say A is *weakly closed* in B with respect to C, if $A^c \leq B$ for $c \in C$ implies $A^c = A$; and A is *strongly closed*, if for all $c \in C$ we have $A^c \cap B \leq A$. ◊

Indeed when P is abelian, P is the *only* member of the following standard collection, which in general controls fusion for any pair G, P:

DEFINITION B.2.4. The *Alperin-Goldschmidt conjugation family* (see for example [**GLS96**, 16.1]) consists of the normalizers $N_G(D)$ of the members of the set $\mathcal{D}$ of nontrivial subgroups D of P such that:

(1) $N_P(D) \in Syl_p\big(N_G(D)\big)$ (that is, D is extremal in P);

(2) $C_G(D) \leq O_{p',p}\big(N_G(D)\big)$ (in particular, $N_G(D)$ is p-constrained);[1] and

(3) $O_{p',p}\big(N_G(D)\big) = O_{p'}\big(N_G(D)\big) \times D$.

The set $\mathcal{D}$ itself can also be referred to as the Alperin-Goldschmidt conjugation family. ◊

When P has rank greater than 1, we have the following elementary fact:

LEMMA B.2.5. *Suppose that G has p-rank at least 2. If $p = 2$, suppose also that a Sylow 2-subgroup of G is not semidihedral. Let N be a p-local subgroup of G, and let $D \in Syl_p\big(O_{p',p}(N)\big)$. Suppose that N is p-constrained.*

Then D has rank at least 2, as does any member of the Alperin-Goldschmidt conjugation family $\mathcal{D}$.

PROOF. Note that if $D \in \mathcal{D}$, then from Definition B.2.4, setting $N := N_G(D)$, we see that N is p-constrained and $D \in Syl_p\big(O_{p',p}(N)\big)$; so the second conclusion follows from the first.

To prove the first, we assume that $m_p(D) = 1$ and obtain a contradiction.

Let $P \in Syl_p(G)$ with $N_P(D) \in Syl_p(N)$. Then as N is p-constrained by hypothesis, $C_P(D) \leq D$ and $DC_N(D) \leq O_{p',p}(N) = O_{p'}(N)D$. Set $\overline{N} := N/O_{p',p}(N)$; then $O_p(\overline{N}) = 1$. Since P has rank greater than 1 by hypothesis, $D < P$ and hence also $D < N_P(D)$. So as $D \in Syl_p(O_{p'p}(N))$, we have $\overline{N_P(D)} \neq 1$. Then since $C_N(D) \leq O_{p'}(N)D$, $\overline{N}$ is isomorphic to a nontrivial subgroup of the outer automorphism group $\mathrm{Out}(D) = \mathrm{Aut}(D)/\mathrm{Inn}(D)$.

If D is cyclic, then $\mathrm{Aut}(D) = \mathrm{Out}(D)$ is abelian, and so has a normal Sylow p-subgroup, contrary to $O_p(\overline{N}) = 1$. Hence as $m_p(D) = 1$, by A.1.7, $p = 2$ and D is generalized quaternion. If $|D| > 8$, then $\mathrm{Aut}(D)$ is a 2-group by A.1.9, whence $\overline{N}$ is a 2-group, again contrary to $O_2(\overline{N}) = 1$. It follows that $|D| = 8$, and now from $O_2(\overline{N}) = 1$ we get $\overline{N} \cong Out(D) \cong S_3$; and since $C_P(D) \leq D$, $N_P(D)$ is semidihedral or quaternion of order 16, and hence has no normal 4-subgroup.

Suppose that P has a normal 4-subgroup U. Then $[U, D] \leq U \cap Z(P)$, while we have $Z(P) \leq C_P(D) \leq D$, whence $U \leq N_P(D)$. But by the previous paragraph, $N_P(D)$ has no normal 4-subgroup. Hence P has no normal 4-subgroup, whence P is dihedral or semidihedral by A.1.8. As P contains the quaternion group D, P is in fact semidihedral—but this contradicts our hypothesis, so the proof is complete. □

We now follow [**GLS96**, 16.1] in stating the basic result:

THEOREM B.2.6 (Alperin-Goldschmidt Fusion Theorem). *The family $\mathcal{D}$ (the Alperin-Goldschmidt conjugation family) controls p-fusion in G; that is, given subsets $A, B \subseteq P$ with $A^g = B$ for $g \in G$, there is a corresponding sequence of subsets $A =: A_1, A_2, \ldots, A_n := B$ of P, and members $D_i \in \mathcal{D}$ with $x_i \in N_G(D_i)$ for $1 \leq i \leq n-1$, such that:*

(1) $g = cx_1x_2 \ldots x_{n-1}$ for some $c \in C_G(A)$;

(2) $A_i \leq D_i$; and

(3) $A_i^{x_i} = A_{i+1}$.

---

[1] We mention that D is also *p-radical* and *p-centric*, as defined in the topological literature.

COROLLARY B.2.7. *If P has rank at least 2 and is not a semidihedral 2-group, then $\Gamma_{2,P}(G)$ controls G-fusion in P.*

PROOF. By Lemma B.2.5, any member D of the Alperin-Goldschmidt conjugation family has rank at lest 2. Thus $N_G(D) \leq \Gamma_{2,P}(G)$ for all such D, so that $\Gamma_{2,P}(G)$ controls fusion by Theorem B.2.6. □

Transfer. In general, the *transfer* homomorphisms from G into abelian p-groups computes the largest homomorphic image of a group G which is an abelian p-group.

If $N \leq G$ and N contains a Sylow p-subgroup of G, we say that N *controls p-transfer in G* if and only if the largest abelian p-quotients of G and N are isomorphic (or, equivalently, have the same order).

Control of p-fusion implies control of p-transfer, though not conversely; so transfer is often treated alongside fusion—see e.g. the latter part of Section 16 of [**GLS96**] as a background reference for this topic.

REMARK B.2.8 (Burnside's Transfer Theorem). For example, a corollary of Burnside's fusion theorem B.2.3 is Burnside's transfer theorem (cf. [**GLS96**, 16.5]), showing when P is abelian that $P \cap G' = P \cap N_G(P)'$.

A further corollary, for $p = 2$, is the standard fact that P cyclic implies (since $\mathrm{Aut}(P)$ is then a 2-group by Lemma A.1.9) that G is 2-nilpotent (cf. [**Gor80a**, 7.6.1] and [**GLS96**, 16.7]). ◊

The most frequently-used transfer results in 2-local group theory are the following lemma of Thompson and its variants; we record the generalized version given in [**GLS96**, 15.16]:

LEMMA B.2.9 ((Generalized) Thompson Transfer Lemma). *Assume $P \in Syl_2(G)$ is the semidirect product of a normal subgroup Q and a cyclic subgroup A.*

If $G = O^2(G)$, then the involution u of A has a conjugate in Q, which may be chosen extremal in P.

This has the following direct consequence, for example:

LEMMA B.2.10. *Suppose that $G = O^2(G)$ and $m_2(G) \geq 3$. Then $m_2\big(C_G(t)\big) \geq 3$ for every involution $t \in G$.*

PROOF. Let $P \in Syl_2(G)$ with $t \in P$. Since $m_2(P) \geq 3$ by hypothesis, P has a normal four-group U by A.1.8, so that $|P : Q| \leq 2$ where $Q := C_P(U)$. Furthermore $m_2(Q) \geq 3$ (e.g. B.4.8). For an involution $y \in Q$, $C_Q(y)$ contains $U\langle y \rangle$, and equals Q if $y \in U$, so $m_2\big(C_Q(y)\big) \geq 3$. On the other hand if $t \notin Q$, then t has a G-conjugate y in Q by Thompson Transfer B.2.9. In any case $m_2\big(C_G(t)\big) \geq 3$. □

There are many other transfer results in the literature. Some of the papers in the CFSG use results of Yoshida [**Yos78**], which generalize earlier theorems of P. Hall and Wielandt.

REMARK B.2.11. The remaining sections of this chapter discuss techniques more specifically related to classification problems in local group theory. Most of these are also discussed in Chapter 4 of Gorenstein's introduction [**Gor82**]. ◊

B.3. Signalizer functor methods and balance

Our basic reference is [**GLS96**, Sections 19–23], but see also [**Gor83**, 4.3]. In this section we introduce various definitions, with the aim of correcting and making more precise the current terminology of signalizer functor theory. Some of the basic ideas have already appeared in our Introduction, starting at Definition 0.3.14.

The local analysis of signalizers originated in the Feit-Thompson Odd Order Paper [**FT63**], where the term *A-signalizer in G* refers to an A-invariant subgroup W of G such that $A \cap W = 1$. Here G is a minimal simple group of odd order. The idea is applied to an abelian subgroup A of G such that $A \trianglelefteq P$ for some $P \in Syl_p(G)$ and $A = C_P(A)$; these properties imply that any A-signalizer in G is a p'-group. The Thompson Transitivity Theorem [**FT63**, 16.1], only available if $m_p(A) \geq 3$, is a key tool for Feit and Thompson's study of such signalizers for a fixed A. They eventually show that there is a unique maximal A-signalizer W_A in G; moreover, if $W_A \neq 1$, then $N_G(W_A)$ is the unique maximal subgroup of G containing A. In particular $N_G(W_A)$ contains $\Gamma_{1,A}(G)$, so that $N_G(W_A)$ is a uniqueness subgroup in the language of Example 1.3.1.

Later, Gorenstein and Walter had the idea of assembling certain A-invariant odd-order subgroups of a simple group G into an odd-order subgroup when A is an elementary abelian 2-subgroup of G, again of rank at least 3. Signalizer functors were first defined in [**Gor69b**]. The purpose was to study the cores (largest odd order normal subgroups) of involution centralizers[2] in G. Thus, their first object of study was the subgroup-valued mapping which is sometimes called "$O_{2'}$", namely $a \mapsto O_{2'}\big(C_G(a)\big)$, $a \in A^{\#}$. The word "functor" was sometimes used in this context, not with a categorical subtext, but rather to convey the notion of equivariance under conjugation by some subgroup of G (a condition that could of course be phrased in categorical language).

During the late 1960s and early 1970s, Gorenstein, Walter, and others developed the basic tools that are now collectively referred to as the "Signalizer Functor Method". These are the subject of this section and the next.

Subgroup functors, signalizer functors, and A-signalizer functors. We begin with some basic definitions.

DEFINITION B.3.1. Let G be a group and $\mathcal{S}$ a set of subgroups of G. A *subgroup functor from $\mathcal{S}$ to G* is a function Θ on $\mathcal{S}$ such that for each $S \in \mathcal{S}$, $\Theta(S)$ is a subgroup of G.

When $S \in \mathcal{S}$ and $S = \langle x \rangle$ is cyclic, we usually write $\Theta(x)$ for $\Theta(S)$.

If X is a subgroup of G normalizing the set $\mathcal{S}$, then we say that Θ is *X-equivariant* iff:

$$\Theta(S^x) = \Theta(S)^x \text{ for all } S \in \mathcal{S} \text{ and all } x \in X.$$

Finally, if $\mathcal{S}$ consists of elementary abelian p-subgroups of some fixed rank k, then we say that Θ is a subgroup functor *of rank k*. ◊

When $\mathcal{S}$ consists of subgroups of various sizes, it is sometimes natural to consider two categories, whose objects are the elements of $\mathcal{S}$ (resp. the set of all subgroups of G), and whose mappings are all inclusion mappings in both cases; and to restrict attention to the case that Θ is a contravariant functor. An example is

[2]Prior to signalizer functors, Brauer's modular character theory (and the theory of the principal 2-block of G in particular) was the only systematic way to study these cores.

defined by $\Theta(S) := C_G(S)$. However, for our purposes here, this point of view is not necessary.

Signalizer functors are subgroup functors satisfying some of the following conditions:

DEFINITION B.3.2. Let Θ be a subgroup functor from $\mathcal{S}$ to G, as in Definition B.3.1. Then we say that[3]

(1) Θ is *coprime* iff $\gcd(|\Theta(S)|, S') = 1$ for all $S, S' \in \mathcal{S}$;
(2) Θ is *central* iff $\Theta(S) \leq C_G(S)$ for all $S \in \mathcal{S}$;
(3) Θ is *locally constant* iff $\Theta(S) = \Theta(S')$ for all $S, S' \in \mathcal{S}$ such that $[S, S'] = 1$;
(4) Θ is *balanced* iff $\Theta(S) \cap C_G(S') \leq \Theta(S')$ for all $S, S' \in \mathcal{S}$ such that $[S, S'] = 1$;
(5) Θ is *solvable* iff $\Theta(S)$ is solvable for all $S \in \mathcal{S}$; and
(6) Θ has the *$\mathcal{K}$-property* iff all simple sections of $\Theta(S)$ are $\mathcal{K}$-groups for $S \in \mathcal{S}$. $\diamondsuit$

If we assume that Θ is central, and interchange S and S' in the balance condition, we get an equivalent formulation of balance, having the symmetry that the "balance" terminology hints at. Namely we see that Θ is balanced if and only if

$$\text{(B.3.3)} \quad \Theta(S) \cap C_G(S') = \Theta(S') \cap C_G(S) \text{ for all } S, S' \in \mathcal{S} \text{ such that } [S, S'] = 1.$$

Now we can define the fundamental concept of an "A-signalizer functor". The reader will do well to keep in mind the basic case $m_p(A) \geq 3$, $k = 1$.

DEFINITION B.3.4. Let k be a positive integer and let A be an elementary abelian p-subgroup of G of rank at least $k + 2$. An *A-signalizer functor on G of rank k* is a subgroup functor Θ from the set $\mathcal{E}_k(A)$ of all subgroups of A of order p^k to G, such that:

Θ is coprime, central, balanced, and A-equivariant.

Since A is abelian, note that the property $[S, S'] = 1$ holds automatically for all $S, S' \in \mathcal{E}_k(A)$—e.g. in the balance condition (4) of Definition B.3.2, or when we wish to verify the locally constant condition (3) there. Also the A-equivariance shows that A *normalizes* $\Theta(S)$ for each $S \in \mathcal{E}_k(A)$; that is, equivariance is equivalent to "invariance" here. $\diamondsuit$

As Θ is also coprime, the theory of coprime action gives strong tools to analyze the action of A on $\Theta(S)$. At the heart of it all we will see the hyperplane generation property of Lemma B.1.8.2.

Definition B.3.4, with $k = 1$, is the standard definition of a "signalizer functor"; our extension to $k > 1$ is a bit unorthodox. However, for lack of a better word, the term "signalizer functor" has come to be applied more broadly and sometimes inaccurately, for example to describe the *completion* of an A-signalizer functor—which is not central, but *is* locally constant. To make this usage legitimate, we introduce the following definitions, which[4] are useful in showing that certain subgroup functors are locally-constant—while postponing an analysis of the completion of subgroup functors until (B.3.12) in the next subsection:

[3]We mention that these various conditions may be unrelated—for example, the locally constant condition tends to lead away from the centrality condition.

[4]Thus, while any A-signalizer functor is necessarily a signalizer functor, the converse is far from true. For the sake of accuracy we are making a significant distinction between a signalizer functor in general and an A-signalizer functor.

DEFINITION B.3.5. Let $\mathcal{S}$ be a set of p-subgroups of G. A *signalizer functor from* $\mathcal{S}$ *to* G is simply a coprime subgroup functor from $\mathcal{S}$ to G.

Given an A-signalizer functor of rank k on G, we define[5]

$$\Theta^+(D) := \langle \Theta(S) \,:\, S \in \mathcal{E}_k(D) \rangle, \text{ for } D \in \mathcal{E}_{k+1}(A),$$

as a subgroup functor from $\mathcal{E}_{k+1}(A)$ to G. $\diamondsuit$

One of the key results of signalizer functor theory is the following assertion:

THEOREM B.3.6 (A-Signalizer Functors Lead to Locally Constant Functors). *Assume that A is an elementary abelian p-subgroup of G, and Θ is an A-signalizer functor of rank k on G, which has the $\mathcal{K}$-property. Then*

(1) Θ^+ is a locally constant signalizer functor from $\mathcal{E}_{k+1}(A)$ to G.

(2) If X is any subgroup of G such that Θ is X-equivariant, then Θ^+ is also X-equivariant.

Most parts of the proof, given in this paragraph, are easy—and in particular, do not yet use the $\mathcal{K}$-hypothesis: First (2) is immediate from the definitions. Similarly Θ^+ is a subgroup functor by construction. We next verify that $\Theta^+(D) = \Theta^+(D_1)$ for any $D, D_1 \in \mathcal{E}_{k+1}(A)$. Namely take any $S \leq D$ with $m_p(S) = k$. Since Θ is an A-signalizer functor of rank k, $\Theta(S)$ is defined for this S and is an A-invariant p'-group, as Θ is A-invariant and coprime. In particular D_1 normalizes $\Theta(S)$, so using coprime action and balance,

$$\begin{aligned}\Theta(S) \overset{B.1.8.2}{=} \langle C_{\Theta(S)}(S_1) \,:\, |D_1 : S_1| = p \rangle &\overset{B.3.2.4}{\leq} \langle \Theta(S_1) \,:\, S_1 \in \mathcal{E}_k(D_1) \rangle \\ &\overset{B.3.5}{=} \Theta^+(D_1).\end{aligned}$$

Letting S vary over $\mathcal{E}_k(D)$, we get $\Theta^+(D) \leq \Theta^+(D_1)$; and then equality holds by symmetry between D and D_1, so that Θ^+ is locally constant as desired.

However, to complete the proof, we must also verify that $\Theta^+(D)$ is a p'-group. This property is highly nontrivial to establish: in fact we will obtain it a little later, by quoting the completeness property which is the conclusion of the Signalizer Functor Theorem B.3.13. We will develop these concepts in the following subsection.

Completeness and the Signalizer Functor Theorem. An initial version of the Signalizer Functor Theorem appears as 0.3.15, but in this section we provide a deeper discussion of the result, with highlights appearing in B.3.13 and B.3.25.

This theorem emerged historically in an expanding sequence of subcases considered by various authors, over a period spanning almost ten years. The combined result is called the Signalizer Functor Theorem. To state it, we need one further definition, due to Goldschmidt [**Gol72b**].

DEFINITION B.3.7. Let A be an elementary abelian p-group and let Θ be an A-signalizer functor on G of rank k. Then a Θ-*signalizer* is an A-invariant subgroup W of G such that

$$C_W(S) \leq \Theta(S) \text{ for all } S \in \mathcal{E}_k(A).$$

We will see below in Remark B.3.8 that W is necessarily a p'-group.

[5]Often in the literature, and indeed later in Section B.5, one sees $\Theta^+(D)$ given the name W_D, or some other name with subscript "D". We have chosen to avoid the use of the notation W_D, and instead work with $\Theta^+(D)$ and make the connection with the completion $\overline{\Theta}$ later at (B.3.12).

Moreover if Θ is a *solvable* A-signalizer functor, meaning that $\Theta(S)$ is solvable for all $S \in \mathcal{E}_k(A)$, then any Θ-signalizer is in addition required to be solvable.

We say that Θ is *complete* if and only if the set of all Θ-signalizers has a unique maximal member with respect to inclusion. Thus in the solvable case, this maximal member is required to be solvable. ◇

REMARK B.3.8. Note that the condition $C_W(S) \leq \Theta(S)$ in B.3.7 for a Θ-signalizer W is equivalent to

$$C_W(S) = W \cap \Theta(S), \tag{B.3.9}$$

since the reverse inclusion holds as $\Theta(S) \leq C_G(S)$ by centrality (B.3.2.2) in the Definition B.3.4 of an A-signalizer functor.

Notice also that we can write the balance condition for Θ in B.3.2.4 in the form "$C_{\Theta(S)}(S') \leq \Theta(S')$ for all $S' \in \mathcal{E}_k(A)$": and so (taking $\Theta(S)$, S' in the roles of "W, S") we see balance is equivalent to the statement that for each $S \in \mathcal{E}_k(A)$, the subgroup $\Theta(S)$ is a Θ-signalizer.

Next if W is any Θ-signalizer, then W is a p'-group: For otherwise, A normalizes some Sylow p-subgroup $P \neq 1$ of W, whence $1 \neq Z(AP) \cap P \leq C_W(S) \leq \Theta(S)$ for any $S \in \mathcal{E}_k(A)$—contradicting our hypothesis that Θ is coprime.

We can now show that completeness of Θ is equivalent to the statement that the subgroup $\overline{\Theta}$ defined by[6]

$$\overline{\Theta} := \langle \Theta(S) \, : \, S \in \mathcal{E}_k(A) \rangle \tag{B.3.10}$$

is a Θ-signalizer: For we saw just above that each $\Theta(S)$ is a Θ-signalizer, so that $\overline{\Theta}$ is contained in the unique maximal Θ-signalizer W. Hence as W is a p'-group, it follows from B.1.8.2 that W is generated by the $C_W(S)$ for $S \in \mathcal{E}_k(A)$, which lie in $\Theta(S)$ by Definition B.3.7. Hence $W \leq \overline{\Theta}$. In particular as any Θ-signalizer is a p'-group, we get

$$\text{If } \Theta \text{ is complete, then } \overline{\Theta} \text{ is a } p'\text{-group.} \tag{B.3.11}$$

(But the converse may not hold if $C_{\overline{\Theta}}(S) > \Theta(S)$.)

Our final remark does not require that Θ be complete. From Theorem B.3.6 and its proof, Θ^+ is locally constant and for $S \in \mathcal{E}_k(A)$ and $S \leq D \in \mathcal{E}_{k+1}(A)$, $\Theta(S) = \Theta^+(D)$. Therefore

$$\overline{\Theta} \overset{(B.3.10)}{=} \langle \Theta^+(D) \, : \, D \in \mathcal{E}_{k+1}(A) \rangle \overset{B.3.6}{=} \Theta^+(D_0) \text{ for any } D_0 \in \mathcal{E}_{k+1}(A). \tag{B.3.12}$$

Thus when Θ is in fact complete, we can regard both the group $\overline{\Theta}$, and the functor Θ^+ on $\mathcal{E}_{k+1}(A)$ whose common value is $\overline{\Theta}$, as "the completion" of Θ. ◇

Now recall that to complete the proof of Theorem B.3.6, it only remains to show that $\Theta^+(D)$ is a p'-group. In view of (B.3.11) and (B.3.12), it suffices to show that Θ is complete. But this is precisely the conclusion of the Signalizer Functor Theorem B.3.13 below. (In contrast to the other parts of the proof of Theorem B.3.6, the argument for completeness does require the $\mathcal{K}$-hypothesis.)

The main developers of this theorem were Gorenstein in [**Gor69b**], Goldschmidt in [**Gol72a, Gol72b**], Bender in [**Ben75**], Glauberman in [**Gla76**], and McBride in [**McB82b, McB82a**].

[6]Here we are abusing notation slightly, in using $\overline{\Theta}$ to denote a subgroup rather than a functor; but we will see at (B.3.12) that we could in effect identify the subgroup $\overline{\Theta}$ with the locally constant functor Θ^+ having this subgroup as its (constant) value.

THEOREM B.3.13 (Signalizer Functor Theorem). *Assume that A is an elementary abelian p-subgroup of G and Θ an A-signalizer functor of rank $k \geq 1$ on G. If Θ has the $\mathcal{K}$-property, then Θ is complete.*

REMARK B.3.14. An alternative proof of the theorem in the solvable case was given by Aschbacher in [**Asc00**].

Of the above authors, all but McBride assumed that Θ is solvable. Of course if Θ is solvable—in particular if $p = 2$—then Θ automatically has the $\mathcal{K}$-property. For his part, McBride made very economical use of the $\mathcal{K}$-group hypothesis, using just four elegant properties of coprime actions on nonabelian simple $\mathcal{K}$-groups to push his arguments through. We call these properties the K^*-conditions; they are specified in [**McB82a**, Hypotheses A and B, p. 216]. Thus McBride's $\mathcal{K}$-group hypothesis was no more than that all simple sections of each $\Theta(D)$ satisfy the $\mathcal{K}^*$-conditions. Moreover, he proved that the $\mathcal{K}^*$-conditions are inherited by the completion $\overline{\Theta}$.

As just noted, McBride's Theorem is not needed when $p = 2$, i.e., in our Odd Case, where the groups of GW type are classified. However in the Even Case, it was used by Gorenstein-Lyons in dealing with the subcase $e(G) \geq 4$ (e.g. in [**GL83**, Sec 21]), and at 14.6 of Aschbacher's work [**Asc83a**] on the subcase $e(G) = 3$. ◊

REMARK B.3.15 (An inductive argument for $k > 1$). The papers which together yield the Signalizer Functor Theorem B.3.13 actually state the result only for the case $k = 1$. But there are standard arguments e.g. [**Gor71**] that can be assembled to complete a proof of the theorem by induction. We now indicate such a proof:

So we assume that the base step $k = 1$ has been handled by the Classical Signalizer Functor Theorem. For the inductive step, we are given an A-signalizer functor Θ of rank $k > 1$, and we must show that Θ is complete. We may assume the result for ranks smaller than k, including 1.

We will approach the completion $\overline{\Theta}$, by constructing an A-signalizer functor Ψ of rank $k - 1$, which will have the same completion $\overline{\Psi} = \overline{\Theta}$; then we will be able to apply induction to Ψ.

Namely we define

(B.3.16) $$\Psi(E) := \langle \Theta(S) \,:\, E \leq S \in \mathcal{E}_k(A) \rangle \text{ for each } E \in \mathcal{E}_{k-1}(A).$$

We see using the definition in B.3.10 that

(B.3.17) $$\overline{\Psi} = \overline{\Theta}$$

as desired. The A-invariance of Θ in Definition B.3.4 gives the A-invariance of Ψ. Similarly as $C_G(S) \leq C_G(E)$ and Θ is central, Ψ is central. So it remains to show that Ψ is coprime and balanced.

To show that Ψ is coprime, we must show each $\Psi(E)$ is a p'-group. We decompose $A = E \times F_E$, and we will approach $\Psi(E)$ by constructing an F_E-signalizer functor Φ_E of rank 1—which will have the completion $\overline{\Phi}_E = \Psi(E)$, and to which we will be able to apply induction:

Namely we define Φ_E on $\mathcal{E}_1(F_E)$ by

(B.3.18) $$\Phi_E(R) := \Theta(ER) \text{ for } R \in \mathcal{E}_1(F_E).$$

Since any $S \in \mathcal{E}_k(A)$ containing E is of the form $E \times R$ for some $R \in \mathcal{E}_1(F_E)$,

$$\overline{\Phi}_E \overset{B.3.10}{=} \langle \Phi_E(R) \,:\, R \in \mathcal{E}_1(F_E) \rangle = \langle \Theta(S) \,:\, E \leq S \in \mathcal{E}_k(A) \rangle \overset{(B.3.16)}{=} \Psi(E)$$

as desired (and in particular, $\overline{\Phi_E}$ is independent of the choice of F_E). Now we argue much as for Φ above: as $C_G(ER) \leq C_G(R)$ and Θ is central, Φ_E is central. The A-invariance of Θ gives F_E-invariance of Φ_E. As Θ is coprime, so is Φ_E. Finally as Θ is central, $\Phi_E(R) \leq C_G(ER) \leq C_G(E)$ for each R; using this along with the balance condition for Θ, we get

$$\begin{aligned} \Phi_E(R) \cap C_G(R') \overset{\text{(B.3.18)}}{=} \Theta(ER) \cap C_G(ER') &\overset{\text{(B.3.3)}}{=} \Theta(ER') \cap C_G(ER) \\ &\overset{\text{(B.3.18)}}{=} \Phi_E(R') \cap C_G(R), \end{aligned}$$

so that Φ_E is also balanced. Thus Φ_E is indeed an F_E-signalizer functor of rank 1.

Now we conclude from the Classical Signalizer Functor Theorem that Φ_E is complete; then by (B.3.11), $\overline{\Phi_E}$ is a p'-group. Since we saw $\overline{\Phi_E} = \Psi(E)$, this completes the proof that Ψ is coprime.

We next prove that Ψ is balanced. First recall that completeness of Φ_E says that $\overline{\Phi_E} = \Psi(E)$ is a maximal Φ_E-signalizer, so

$$\text{(B.3.19)} \quad C_{\Psi(E)}(R) \overset{\text{(B.3.9)}}{=} \Psi(E) \cap \Phi_E(R) \overset{\text{(B.3.18)}}{=} \Psi(E) \cap \Theta(ER) \overset{\text{(B.3.16)}}{=} \Theta(ER)$$

for all $R \in \mathcal{E}_1(F_E)$; and indeed for all $R \in \mathcal{E}_1(A)$ outside E, since we saw above that $\overline{\Phi_E} = \Psi(E)$ is independent of the choice of F_E.

Next consider any $E' \in \mathcal{E}_{k-1}(A)$. If $E' = E$, the balance condition for E and E' is automatic. So we may assume that $E' \neq E$; hence E' is generated by the members R of the set defined by $\mathcal{E}_1(E';E) := \mathcal{E}_1(E') \setminus \mathcal{E}_1(E' \cap E)$. When we compute with the left-hand side of condition (B.3.3), we see

$$C_{\Psi(E)}(E') = \bigcap_{R \in \mathcal{E}_1(E';E)} C_{\Psi(E)}(R) \overset{\text{(B.3.19)}}{=} \bigcap_{R \in \mathcal{E}_1(E';E)} \Theta(ER).$$

Now each term $\Theta(ER)$ in the right-hand intersection lies in $C_G(E) \leq C_G(R')$, for R' in the analogous set $\mathcal{E}_1(E;E')$; and the intersection of these terms lies in $C_G(E')$ using the left-hand side, so we conclude

$$C_{\Psi(E)}(E') = \bigcap_{\substack{R \in \mathcal{E}_1(E';E) \\ R' \in \mathcal{E}_1(E;E')}} \left(\Theta(ER) \cap C_G(E'R') \right).$$

Since Θ is balanced, each term in the right-hand intersection is symmetric in E and E' by (B.3.3); hence the left-hand side is also symmetric in E and E', giving the balance condition (B.3.3) for Ψ. This completes the proof that Ψ is an A-signalizer functor of rank $k-1$.

Hence Ψ is complete by induction, so that $\overline{\Psi}$ is the maximal Ψ-signalizer; and by (B.3.17), we get $\overline{\Theta} = \overline{\Psi}$.

We now prove that Θ is complete by verifying that $\overline{\Theta}$ is a Θ-signalizer, as in Definition B.3.7. Let $S \in \mathcal{E}_k(A)$, and choose any hyperplane E of S. We write $S = E \times R_E$ with $R_E \leq F_E$ as above. We use the completeness of Ψ and (B.3.19) in computing:

$$\begin{aligned} C_{\overline{\Theta}}(S) &\overset{\text{(B.3.17)}}{=} C_{\overline{\Psi}}(ER_E) &&= C_{\overline{\Psi}}(E) \cap C_G(R_E) \\ &\overset{\text{(B.3.9)}}{=} (\overline{\Psi} \cap \Psi(E)) \cap C_G(R_E) &&\overset{\text{B.3.10}}{=} C_{\Psi(E)}(R_E) \\ &\overset{\text{(B.3.19)}}{=} \Theta(ER_E) &&= \Theta(S). \end{aligned}$$

Thus $\overline{\Theta}$ is indeed a Θ-signalizer, so Θ is complete.

Note also that if Θ is solvable, then $\overline{\Phi}_E$ and $\overline{\Psi}$ are also solvable by induction. Indeed, a similar remark is needed in the nonsolvable case. There one needs to replace the $\mathcal{K}$-property in the hypothesis by the weaker $\mathcal{K}^*$-property B.3.14, and add to the conclusion that $\overline{\Theta}$ satisfies the $\mathcal{K}^*$-conditions. Then McBride's stronger theorem described in B.3.14 fits into the inductive framework above, to yield that $\overline{\Phi}_E$ and $\overline{\Theta}$ satisfy the $\mathcal{K}^*$-conditions.

This completes the inductive step for $k > 1$ in our proof of Theorem B.3.13. $\diamondsuit$

The types of balance, and an application. The CFSG saw the development of increasingly sophisticated subgroup functors. The goal was always to construct a locally constant, equivariant, coprime functor on $\mathcal{E}_k^p(G)$ for some k, whose nontriviality represented the obstruction to some p-local property. Then if the property failed, the functor could be used to construct a uniqueness subgroup. (See e.g. B.4.2.)

Balance, k-balance, and local balance. The prototype of a potential A-signalizer functor of rank 1 on the elementary abelian p-subgroup A is $\Theta := O_{p'}$, defined by

$$\Theta(a) = \Theta(\langle a \rangle) := O_{p'}\big(C_G(a)\big) \text{ for all } a \in A^{\#}.$$

This functor is obviously coprime, central, and A-equivariant. Hence it is an A-signalizer functor if and only if the balance condition for $O_{p'}$ holds:

$$\text{(B.3.20)} \qquad O_{p'}\big(C_G(a)\big) \cap C_G(b) \le O_{p'}\big(C_G(b)\big) \text{ for all } a, b \in A^{\#}.$$

We say that G *is balanced, or* 1*-balanced, with respect to* A if and only if (B.3.20) holds.

We have seen in (0.3.21) that G is balanced with respect to A in the absence of p-components in centralizers of elements of $A^{\#}$. In fact, failure of (B.3.20) always has a local consequence;[7] it entails the existence of a *locally unbalanced* p-*component* L of $C_G(b)$ for some $b \in A^{\#}$. This means that for the quasisimple group $\overline{L} := L/O_{p'}(L)$, there is $c \in \mathrm{Aut}(\overline{L})$ of order p and a p'-subgroup $W \le \mathrm{Aut}(\overline{L})$ such that

$$\text{(B.3.21)} \qquad 1 \ne W \le O_{p'}\big(C_{Inn(\overline{L})W}(c)\big).$$

Failure of local balance is fairly common among quasisimple $\mathcal{K}$-groups, a fact that necessitated the introduction of higher (and weaker) forms of balance. These forms involve the subgroups

$$\text{(B.3.22)} \qquad \Delta_G(D) := \bigcap_{d \in D^{\#}} O_{p'}\big(C_G(d)\big),$$

defined for any elementary abelian p-subgroup D of a group G. For any $k \ge 1$, the group G is said to be *k-balanced with respect to the elementary abelian p-subgroup A* if and only if

$$\text{(B.3.23)} \qquad \Delta_G(D) \cap C_G(b) \le O_{p'}\big(C_G(b)\big) \text{ for every } b \in A^{\#} \text{ and } D \in \mathcal{E}_k(A),$$

a condition that immediately implies that $\Delta_G(D)$ is an A-signalizer functor of rank k. Moreover, failure of k-balance leads to a *locally k-unbalanced* p-component

---

[7] For the more elementary notion of L-balance, see 0.5.5, or the discussion of Section I.18 of **[GL83]** using the language of $L_{p'}$-balance.

of $C_G(b)$ for some $b \in A^{\#}$; the analogue of (B.3.21) is that for some $E \in \mathcal{E}_k^p(\mathrm{Aut}(\overline{L}))$ and some $W \le \mathrm{Aut}(\overline{L})$,

(B.3.24) $$1 \ne W \le \Delta_{\mathrm{Inn}(\overline{L})W}(E).$$

Local 2-balance fails much less frequently for quasisimple $\mathcal{K}$-groups than local balance does; and local k-balance for $k \ge 3$ fails only for coverings of the alternating groups and elementary abelian p-subgroups E (in (B.3.24)) with a regular orbit.

The main application. Many applications of signalizer functors are covered by the following theorem. Recall that $\Lambda_n^p(G)$ is the graph whose vertices are the elementary abelian subgroups of G of order p^n, with edges joining only those pairs of such subgroups that commute elementwise. The second conclusion of the theorem makes a connectedness assumption on the graph $\Lambda_n^p(G)$; we shall explore the failure of the connectedness assumption in the next section.

THEOREM B.3.25. *Let G be a finite group and p a prime such that $O_{p'}(G) = 1$. Let k be a positive integer and assume that $m_p(G) \ge k+2$. Assume also that*

(a) G is k-balanced with respect to every elementary abelian p-subgroup of G of rank $k+2$;

(b) If p is odd, then all p-local subgroups of G are $\mathcal{K}$-groups.

Let Λ_0 be a connected component of $\Lambda_{k+1}^p(G)$ with more than one vertex. Then the following hold for all vertices A of Λ_0:

(1) $N_G(\Lambda_0)$ normalizes the p'-group $\langle \Delta_G(D) : A/D \text{ cyclic}\rangle$; and

(2) If Λ_0 is G-invariant, then $\Delta_G(D) = 1$ for all $D \le A$ with A/D cyclic.

(If $k = 1$, read $O_{p'}(C_G(D))$ for $\Delta_G(D)$.)

PROOF. For any D as in the theorem, set $\Theta(D) := \Delta_G(D)$. Let $\mathcal{A}$ be the set of all elementary abelian p-subgroups A of G of rank at least $k+2$ such that all (equivalently, some) subgroups of A of rank $k+1$ lie in Λ_0. Let $\mathcal{A}_k$ and $\mathcal{A}_{k+1}$ be the set of all subgroups of rank k (resp., rank $k+1$) of members of $\mathcal{A}$. Clearly $\mathcal{A}_k$ and $\mathcal{A}_{k+1}$ are $N_G(\Lambda_0)$-invariant. From the k-balance hypothesis, we see that Θ defines an A-signalizer functor of rank k for any $A \in \mathcal{A}$. Suppose that $B, C \in \mathcal{A}_{k+1}$ and $[B, C] = 1$. Then either $B = C$ or $BC \in \mathcal{A}$. Hence as Λ_0 is connected, its vertex set is $\mathcal{A}_{k+1}$. Moreover using Theorem B.3.6 with BC in the role of "A" if $B \ne C$, we get that $\Theta^+(B) = \Theta^+(C)$. Thus Θ^+ is locally constant on Λ_0. As Λ_0 is connected, Θ^+ is constant on its vertex set $\mathcal{A}_{k+1}$. Clearly Θ is $N_G(\Lambda_0)$-equivariant on $\mathcal{A}_k$, so Θ^+ is $N_G(\Lambda_0)$-equivariant on $\mathcal{A}_{k+1}$. Hence $N_G(\Lambda_0)$ normalizes $\Theta^+(B)$ for all $B \in \mathcal{A}_{k+1}$. Now by the Signalizer Functor Theorem B.3.13 (the reason for our $\mathcal{K}$-group hypothesis), $\Theta^+(B)$ is a p'-group, proving (1). Moreover if Λ_0 is G-invariant, then $\Theta^+(B) \le O_{p'}(G) = 1$ for all $B \in \mathcal{A}_{k+1}$. Thus for any hyperplane D of B, $\Delta_G(D) = \Theta(D) = 1$ by definition of $\Theta^+(B)$. As B was arbitrary, (2) is proved. □

The balance hypothesis is only needed for subgroups of members of $\mathcal{A}$. It is necessarily satisfied if all p-local subgroups are p-constrained, or if all components of $C_G(a)/O_{p'}(C_G(a))$ are locally k-balanced for all $a \in A$ and all $A \in \mathcal{A}$.

Weak balance and $k + \frac{1}{2}$-balanced functors. Aschbacher and Goldschmidt improved the above theorem, constructing a locally constant signalizer functor of rank $k+1$ in the presence only of "weak" k-balance and (ordinary) $k+1$-balance.

Weak k-balance means that

$$\text{(B.3.26)} \qquad \Delta_G(D) \cap [C_G(b), A] \leq O_{p'}(C_G(b)) \text{ for all } b \in A^{\#} \text{ and } D \in \mathcal{E}_k(A),$$

a condition that makes $\Delta^*_{k,A}(D) := \Delta_G(D)O_{p'}(C_G(A))$ an A-signalizer functor of rank k. The obstructions to weak k-balance are again certain p-components L as in (B.3.24), now with the extra requirements that A normalize L and $W = [W, A]$. Unfortunately, the dependence of $\Delta^*_{k,A}(D)$ on A makes equivariance a problem if one attempts to extend $(\Delta^*_{k,A})^+$ to $E_{p^{k+1}}$-subgroups of G not contained in A. However, the functor α on $\mathcal{E}_{k+1}(A)$ defined by

$$\text{(B.3.27)} \qquad \alpha(E) := \langle\ [\Delta_G(D), E]\ :\ D \in \mathcal{E}_k(E)\ \rangle\ \Delta_G(E)$$

turns out to be locally constant, and its definition, intrinsic to E, guarantees equivariance. Moreover, it turns out that $\alpha(E) \leq \overline{\Delta^*_{k,A}}$, so the completeness of $\Delta^*_{k,A}$ makes α coprime. Hence, just as in the balanced case above, if $\Lambda^p_{k+1}(G)$ is connected, then α is trivial.

The functor α is often called the "$(k + \frac{1}{2})$-balanced"-functor. We have noted the variants of it used by Aschbacher in the $e(G) = 3$ subcase of the Even Case, at 4.1.5 and 4.2.6. It is particularly prominent as well in Aschbacher's Classical Involution Theorem [**Asc77a, Asc77b**], discussed in our Section 1.7.

B.4. Connectivity in commuting graphs and i-generated cores

This section provides further details in support of certain statements in Sections 1.3 and 1.5. While this material is not really required for the elementary exposition of Chapter 1, in the chapters thereafter it should provide useful background for various developments which involve the relationship between the commuting graph $\Lambda_2(G)$ and the 2-generated core. Recall that in the previous section we noted the relevance of the completeness of a signalizer functor to connected components in the commuting graph.

For more details, we refer the reader to the discussion in Section 22 of [**GLS96**], and also to Section 2.18 of [**GLS99**]. Here we will mainly provide just a summary of the material.

Some general properties of commuting graphs. We recall Definition 1.3.8 that $\Lambda^p_k(G)$ is defined to be the commuting graph on the set of elementary abelian p-subgroups of G of p-rank k. Of course *we assume in this section that* $m_p(G) \geq k$, so that the set of vertices of $\Lambda^p_k(G)$ is nonempty.

Historically the connection with uniqueness groups and k-generated p-cores was first made explicit by Aschbacher in the case $k = p = 2$. Specifically, we are interested in the case where the graph is disconnected. Since path-connection is an equivalence relation, *distinct connected components are disjoint*. And since G-conjugacy preserves the graph, it permutes the connected components. The first desirable property is transitivity on the set of components, which provides the following characterization:

PROPOSITION B.4.1. *Assume that G is transitive on the set of connected components of $\Lambda^p_k(G)$, with representative Λ_0. Then $\Lambda^p_k(G)$ is disconnected if and only if $N_G(\Lambda_0) < G$.*

PROOF. Using transitivity, the number of components is $|G : N_G(\Lambda_0)|$, while of course $\Lambda^p_k(G)$ is disconnected if and only if it has more than one component. □

This observation focuses attention on Λ_0 and $N_G(\Lambda_0)$. The next observation follows as the components partition the graph:

LEMMA B.4.2. *Let Λ_0 denote a connected component of $\Lambda_k^p(G)$.*
(1) For each $B \in \Lambda_0$, we have $N_G(B) \leq N_G(\Lambda_0)$.
(2) Hence as $B \leq N_G(B)$, we have $\langle \Lambda_0 \rangle \leq N_G(\Lambda_0)$. □

Part (1) suggests that $N_G(\Lambda_0)$ might be a candidate for a uniqueness subgroup, in the language of Example 1.3.1.

Our second desirable property is to replace $N_G(\Lambda_0)$ by a related uniqueness subgroup—the k-generated p-core $\Gamma_{k,S}(G)$ (in the language of Definition 1.3.6), for a suitable Sylow p-subgroup S of G. This core contains still more locals: the normalizers $N_G(C)$, for each subgroup C of S of rank at least k. In analogy with B.4.1, our goal will be to show that

(B.4.3) if $\Lambda_k^p(G)$ is disconnected, then $\Gamma_{k,S}(G) < G$ for $S \in Syl_2(G)$.

In fact we will see in B.4.7 that this holds when $k = 1$, since from B.4.5.2 we will get the equality $\Gamma_{k,S}(G) = N_G(\Lambda_0)$. But for larger k, we cannot necessarily expect even an inclusion relation between the two potential uniqueness subgroups, even when $p = 2$. However we do get an analogue of B.4.2.2:

LEMMA B.4.4. *Let Λ_0 be a connected component of $\Lambda_k^p(G)$, with $A \in \Lambda_0$ and $A \leq S \in Syl_p(G)$. Then $\langle \Lambda_0 \rangle \leq \Gamma_{k,S}(G)$.*

PROOF. Note $m_p(S) \geq k$, as we are assuming in this section that $m_p(G) \geq k$. Hence we get $S \leq N_G(S) \leq \Gamma_{k,S}(G) =: \Gamma$; so in particular, S is also Sylow in Γ.

Let $B \in \Lambda_0$ be on an edge with A. Then $B \leq C_G(A) \leq \Gamma_{k,S}(G) = \Gamma$. Furthermore BA is a p-group which is contained in Γ, and so BA lies in some Sylow p-subgroup of Γ—which must be of the form S^g for suitable $g \in \Gamma$. In particular, $B \leq S^g$, and we can repeat the previous argument, this time for C on an edge with B, to get $C \leq \Gamma_{k,S^g}(G) = \Gamma_{k,S}(G)^g = \Gamma^g = \Gamma$. Continuing in this way along paths through A in the graph, we conclude that $\langle \Lambda_0 \rangle \subseteq \Gamma$. □

Disconnectedness in $\Lambda_1^p(G)$ and strong p-embedding. The results for the simplest case where $k = 1$ are standard; see e.g. Peterfalvi [**Pet86**, Sec. I.1]. But we will be supplying some of the elementary details in our development below.

The vertices of $\Lambda_1^p(G)$ are the subgroups of order p, while the edges are given by the pairs $\{\langle x \rangle, \langle y \rangle\}$ such that $xy = yx$.

Our easy initial lemma describes a component and its normalizer. Of course transitivity in (1) gives the first desirable property appearing as the hypothesis of Proposition B.4.1.

LEMMA B.4.5.
(1) G is transitive on the connected components of $\Lambda_1^p(G)$ (say with representative Λ_0).
(2) For each $S \in Syl_p(N_G(\Lambda_0))$, we have $S \in Syl_p(G)$ and $\Gamma_{1,S}(G) = N_G(\Lambda_0)$.
(3) The vertices of Λ_0 are in fact the subgroups of order p of $N_G(\Lambda_0)$.

PROOF. Let Λ_0 denote a connected component of $\Lambda_1^p(G)$. Set $N := N_G(\Lambda_0)$ and let $\langle x \rangle \in \Lambda_0$. Then $\langle x \rangle$ is contained in some Sylow p-subgroup S of G. Since we have $1 \neq Z(S) \leq C_G(x)$, each subgroup $\langle y \rangle$ of $Z(S)$ of order p is on an edge with $\langle x \rangle$, and so lies in Λ_0; and B.4.2.1 shows that $Z(S) \leq N$. Then since $S \leq C_G(Z(S))$, we see similarly that all subgroups of S of order p lie in Λ_0, and $S \leq N$. As $S \leq N$, the initial statement in (2) follows. Next any vertex of $\Lambda_1^p(G)$ is G-conjugate to a subgroup of S and hence to a vertex of Λ_0, establishing the transitivity required for (1). Similarly each subgroup of N of order p is in an N-conjugate of S, and hence is a vertex of Λ_0. Thus (3) also holds.

It remains to establish the final equality in (2). Take some $1 \neq Q \leq S$. Then $N_G(Q)$ acts on the set of subgroups of Q of order p, which is a subset of the vertex set of Λ_0 by (3); hence we get $N_G(Q) \leq N_G(\Lambda_0) = N$. So we have shown that $\Gamma_{1,S}(G) \leq N$. For the reverse inclusion, set $H := \langle \Lambda_0 \rangle$. Then by B.4.4 and the inclusion just established, $H \leq \Gamma_{1,S}(G) \leq N = N_G(\Lambda_0)$, and hence H is in fact normal in N. Now $S \cap H \geq \Omega_1(S) \neq 1$, so by a Frattini argument (A.1.5.2) we have that $N = HN_N(S \cap H) \leq \Gamma_{1,S}(G)$. □

The notion of a strongly embedded subgroup in 2-local analysis (recall Definition 1.3.4) has an analogue for every prime p, and it is now easy to obtain the standard equivalence (cf. [**Pet86**, (5), p. 147]) of disconnectedness with the existence of strongly p-embedded subgroups.

DEFINITION B.4.6. A subgroup M of G is *strongly p-embedded in G* if and only if $M < G$, p divides $|M|$, and for every $g \in G \setminus M$, $M \cap M^g$ is a p'-group.

THEOREM B.4.7. *The following are equivalent:*
(1) $\Lambda_1^p(G)$ is disconnected.
(2) $N_G(\Lambda_0) < G$ for some (indeed for each) connected component Λ_0 of $\Lambda_1^p(G)$.
(3) $\Gamma_{1,S}(G) < G$ for some (indeed for each) $S \in Syl_p(G)$.
(4) G possesses a strongly p-embedded subgroup M.
(5) If we set $H := \langle C_G(u) : 1 \neq u \in S,\ u^p = 1 \rangle$, then H is not a normal subgroup of G.
Indeed, the strongly p-embedded subgroups of G are precisely the proper overgroups of $N_G(\Lambda_0)$ in G for some component Λ_0.

PROOF. By Lemma B.4.5.1, $\Lambda_1^p(G)$ satisfies the hypothesis of Proposition B.4.1, so that proposition says that (1) and (2) are equivalent. The equivalence of (2) and (3) follows from B.4.5.2.

Now assume (4). The strongly p-embedded subgroup M has order divisible by p, and so contains a subgroup $\langle x \rangle$ of order p; let Λ_0 denote the connected component containing the vertex $\langle x \rangle$. Any member $\langle y \rangle$ of Λ_0 on an edge with $\langle x \rangle$ lies in M because $x \in M \cap M^y$, so as Λ_0 is connected, we get $\langle \Lambda_0 \rangle \leq M$. Now if we take $g \in N_G(\Lambda_0)$, then $M \cap M^g$ contains $\langle \Lambda_0 \rangle$ and so has order divisible by p; hence $N_G(\Lambda_0) \leq M < G$. Thus (4) implies (2).

Next, assume (2), and set $M := N_G(\Lambda_0)$. Then $|M|$ is divisible by p, using B.4.5.2. Suppose for $g \in G$ that $|M \cap M^g|$ is divisible by p; then $M \cap M^g$ contains a subgroup $\langle x \rangle$ of order p. Hence by B.4.5.3, $\langle x \rangle$ lies in $\Lambda_0 \cap \Lambda_0^g$, so that $g \in N_G(\Lambda_0) = M$. Thus M is strongly p-embedded in G, so (2) implies (4).

Let H be as in (5). If (3) holds, then $N_G(H) \leq HN_G(H \cap S) \leq \Gamma_{1,S}(G) < G$, so (3) implies (5). Conversely H contains the centralizers of each of its elements of order p, so (5) implies (1)—unless H happens to contain all elements of G of order p. But in that case, all such elements are H-conjugate into S by Sylow's Theorem and so $H = \langle C_G(u) : 1 \neq u \in G, u^p = 1 \rangle \trianglelefteq G$, a contradiction completing the proof of the equivalences. The proof of the final statement is left to the reader. □

$\Lambda_2^2(G)$ **disconnected and the 2-generated 2-core** $\Gamma_{2,S}(G)$. If P is a p-group, we shall write $\Lambda_2(P)$ for $\Lambda_2^p(P)$. We note the following elementary fact.

LEMMA B.4.8. *For a p-group P with $m_p(P) \geq 3$, the graph $\Lambda_2(P)$ has a connected component $\Lambda_2^\circ(P)$ that contains each normal elementary subgroup of rank 2. Any other connected components consist of an isolated vertex: an E_{p^2}-subgroup V with $C_P(V) = \langle v \rangle \times K$, where $\Omega_1(K) = \Omega_1(Z(P)) =: \langle z \rangle$ has order p, and $V = \langle v, z \rangle$. In particular, $m_p\big(C_P(V)\big) = 2$.*

PROOF. By A.1.8, P has a normal elementary subgroup A of rank 2. Let Λ_0 denote the connected component of $\Lambda_2(P)$ containing the vertex A. Also we set $R := C_P(A)$. As $|\mathrm{Aut}(A)|_p = p$, we have $|P : R| \leq p$. If B is an E_{p^2} subgroup of R, then B is connected to A, and hence B is a vertex of Λ_0. Suppose that V is an E_{p^2} subgroup of P which is not a vertex of Λ_0. Then clearly $C_R(V)$ has rank 1 and $V = \langle z, v \rangle$, where $\langle z \rangle = \Omega_1\big(C_R(V)\big) = \Omega_1\big(Z(P)\big)$ and $C_P(V) = C_R(V) \times \langle v \rangle$. Thus V is an isolated vertex of the graph $\Lambda_2(P)$. Finally, if V is normal in P, then $P = C_P(V)A$ with $A \cap C_P(V) = \langle z \rangle$, and clearly $m_p(P) = 2$, a contradiction. □

We now specialize to the prime $p = 2$, and simplify our notation by writing $\Lambda_2(G) := \Lambda_2^2(G)$. Although the analysis of the 4-subgroups graph $\Lambda_2(G)$ is still fairly elementary, it is substantially more intricate than the discussion above for the graph $\Lambda_1^p(G)$. Thus, for the most part, we will just supply references for the analysis of $\Lambda_2(G)$.

Partial analogues for $\Lambda_2^p(G)$ of the results on $\Lambda_1^p(G)$. In order to obtain interesting results, it will now be important to assume certain further hypotheses—notably $m_2(G) \geq 3$, $G = O^2(G)$, and $Z^*(G) = O_{2'}(G)$.

First, under these restrictions, $\Lambda_2(G)$ has the desirable transitivity property indicated earlier, so the analogue of conclusion (1) from B.4.5 on $\Lambda_1(G)$ holds:

LEMMA B.4.9. *Assume that $G = O^2(G)$ with $m_2(G) \geq 3$. Then*

(1) G is transitive on the connected components of $\Lambda_2(G)$.

Let Λ_0 denote a connected component of $\Lambda_2(G)$. Then

(2) Λ_0 contains a vertex A which is normal in a Sylow 2-subgroup S of G; and

(3) $\Lambda_2(G)$ is disconnected if and only if $N_G(\Lambda_0) < G$.

PROOF. Let $S \in Syl_2(G)$, so that $m_2(S) \geq 3$. Then by B.4.8, $\Lambda_2(S)$ has a component $\Lambda_2^\circ(S)$ whose vertex set contains every normal 4-subgroup A of S. Choose Λ_0 to be the connected component of $\Lambda_2(G)$ containing $\Lambda_2^\circ(S)$, and further set $T := C_S(A)$, so that $|S : T| \leq 2$. By B.4.8, if V is a 4-subgroup of S which is not a vertex of Λ_0, then $V = \langle z, v \rangle$ with $z \in Z(S)$ the unique involution of $C_T(v)$. If we have $V = C_S(V)$, then S is of maximal class (see e.g. [**GLS96**, 10.24]), and so we get $m_2(S) = 2$, contrary to assumption. Hence $V < C_S(V)$, and it follows that $\langle z \rangle < C_T(v)$; then z is a square in $C_T(v)$, since it is the unique involution in that group. As $G = O^2(G)$, it follows by the Thompson Transfer Lemma B.2.9 that there exists $g \in G$ with $v^g \in T$ and with $C_S(v)^g \leq S$. In particular, z^g is a square in S, whence $z^g \in T$. Thus $V^g \leq T$ and so V^g is a vertex of Λ_0. Then V is a vertex of $\Lambda_0^{g^{-1}}$, whence every 4-subgroup of S is a vertex of some G-conjugate of Λ_0. Now (1) and (2) follow by Sylow's Theorem. Transitivity in (1) is in turn the hypothesis for Proposition B.4.1, whose conclusion gives (3). □

But then at the next step we get only a weak analogue of the result B.4.7 on $\Lambda_1(G)$, and the proof is more involved.

THEOREM B.4.10. *Assume that* $G = O^2(G)$, $Z^*(G) = O_{2'}(G)$, *and* $m_2(G) \geq 3$; *and let* Λ_0 *be a connected component of* $\Lambda_2(G)$.

If $\Lambda_2(G)$ *is disconnected, then some maximal subgroup of* G *contains* $N_G(\Lambda_0)$ *and* $\Gamma_{2,S}(G)$; *in particular,* $\Gamma_{2,S}(G) < G$.

PROOF. Let $S \in Syl_2(G)$, so that $m_2(S) \geq 3$. By B.4.9, we may assume that $A \in \Lambda_0$, with A normal in S. With this notation, [**GLS99**, 2.18.3.a] shows that $\Gamma^e_{2,S}(G) \leq N_G(\Lambda_0)$—where $\Gamma^e_{2,S}(G)$ is the subgroup of $\Gamma_{2,S}(G)$ generated by the normalizers of elementary subgroups E of rank at least 2 in S which satisfy $m_2(C_S(E)) \geq 3$; cf. [**GLS99**, Chap. 2,(18F)]. In particular, such vertices E are not isolated in $\Lambda_2(S)$.

Assume now that $\Lambda_2(G)$ is disconnected. Then by B.4.9.2 and the previous paragraph, we have $\Gamma^e_{2,S}(G) \leq N_G(\Lambda_0) < G$. But the difficulty now is that we do *not* know that $\Gamma_{2,S}(G) < G$ a priori, since there might be some 4-group B which is isolated in S such that $N_G(B)$ is not contained in $\Gamma^e_{2,S}(G)$ or $N_G(\Lambda_0)$.

This difficulty is handled by Theorem 2.18.4 in [**GLS99**]: if we now embed $N_G(\Lambda_0)$ in some maximal subgroup M of G, then that result shows that $\Gamma_{2,S}(G)$ also lies in M, completing the proof.

We mention that the proof of Theorem 2.18.4 in [**GLS99**] requires the hypotheses $G = O^2(G)$ and $Z^*(G) = O_{2'}(G)$. Moreover it depends crucially on the $\mathcal{K}$-group hypotheses 0.5.4. The intermediate result 2.18.9, showing that the Sylow 2-subgroup of S is of type J_2, is a version of the "Harada Reduction Theorem" of [**Har81**], which avoids $\mathcal{K}$-group hypotheses when G is of characteristic 2 type. Note that the hypotheses $G = O^2(G)$ and $Z^*(G) = O_{2'}(G)$ suffice—rather than the assumption that G is simple, which appears in [**GLS99**] and [**Har81**].[8] □

A reduction when G is not of component type. The following result involves the 2-generated core, and is quoted in our proof of the Dichotomy Theorem 0.3.10. The result appears to be well-known; but since we do not know of a reference, we sketch a proof. The final statement is a special case of the final statement in Aschbacher's Proper 2-Generated Core Theorem 1.5.10; but the proof here, when G is not of component type, is elementary.[9] (And this hypothesis also eliminates the case of $G \cong J_1$ which arises in Aschbacher's more general context.)

LEMMA B.4.11. *Assume* $m_2(G) \geq 3$, $G = O^2(G)$, *and* G *is not of component type. Then* $\Gamma_{2,S}(G) = \Gamma_{1,S}(G)$ *for* $S \in Syl_2(G)$.

Hence if $G = O^2(G)$, $Z^*(G) = O_{2'}(G)$, $m_2(G) \geq 3$, *and* G *is not of component type, then* $\Lambda^2_2(G)$ *disconnected implies that* $\Lambda^2_1(G)$ *is disconnected.*

[8] In the final contradiction given at the bottom of p. 98 in [**GLS99**], the reference to Burnside's argument [**Isa06**, 3.9] would require G simple. But an alternative contradiction can be obtained from the condition $|z^G| = 27$ there: this forces $z^G \subseteq N_G(U)$; and from the structure of the group $M = N_G(T) = N_G(U)$ in 2.19.5, we get $G = N_G(T) = M$—contrary to $G = O^2(G)$.

[9] The argument is parallel (and at one point identical) to the portion of our proof of the Dichotomy Theorem from just before B.5.3 through B.5.4.

PROOF. We begin by deriving the second statement as a corollary of the first: For the hypotheses of the second include those of the first, so that we can conclude $\Gamma_{1,S}(G) = \Gamma_{2,S}(G)$. Then the assumption that $\Lambda_2^2(G)$ is disconnected implies $\Gamma_{2,S}(G) < G$ using Theorem B.4.10 (where we saw in the proof that we could use Harada's argument to avoid $\mathcal{K}$-group hypotheses). Hence $\Gamma_{1,S}(G) < G$. Thus $\Lambda_1(G)$ is disconnected by B.4.7.

We turn to the proof of the first statement. Since $G = O^2(G)$ by hypothesis, we conclude from Lemma B.2.10 that $m_2(C_G(x)) \geq 3$ for all involutions $x \in G$, so that $m_2(C_S(t)) \geq 3$ for all extremal involutions $t \in S$.

So we must prove the equality in the first assertion under this hypothesis. As $\Gamma_{2,S}(G) \leq \Gamma_{1,S}(G)$, it remains to establish the reverse inclusion. Since any subgroup of S of 2-rank 1 has a unique involution by A.1.7, it will suffice to show for each involution x of S that $C_G(x) \leq \Gamma_{2,S}(G) =: \Gamma_2$.

First, x is fused to some extremal $t \in S$. By Corollary B.2.7, x is Γ_2-conjugate to t, so it now suffices to show that $C_t := C_G(t) \leq \Gamma_2$ for each extremal $t \in S$. As we saw earlier, $m_2(S_t) \geq 3$ for $S_t := C_S(t) \in Syl_2(C_t)$.

Now by coprime action 0.3.11.2, $O_t := O_{2'}(C_t)$ is generated by the various subgroups $C_{O_t}(A)$, for hyperplanes A of any nontrivial elementary 2-subgroup E of C_t; and as $m_2(S_t) \geq 3$, we may take $E \leq S_t$ with $m(E) \geq 3$. Then $m(A) \geq 2$, and hence $C_{O_t}(A) \leq \Gamma_2$. We conclude that $O_t \leq \Gamma_2$.

Next set $R := S \cap O_{2',2}(C_t)$, so that R is Sylow in the normal subgroup $O_{2',2}(C_t)$ of C_t by A.1.5.1, and also $t \in R$. Thus $O_{2',2}(C_t) = O_tR$; and by a Frattini argument (as in A.1.5.2), we then get that $C_t = (O_tR)N_G(R) = O_tN_G(R)$. We saw above that $O_t \leq \Gamma_2$, so it suffices to show that $N_G(R) \leq \Gamma_2$. And this holds by definition if $m_2(R) \geq 2$.

So we assume instead that $m_2(R) = 1$, and it remains to derive a contradiction.[10] As $N_G(R) \leq C_t$, which is 2-constrained as G is not of component type, we conclude from B.1.4 and B.1.9 that $N_G(R)$ is 2-constrained. As $m_2(G) \geq 3$, a Sylow 2-subgroup of G is not semidihedral. So by B.2.5, we have $m_2(R) \leq 2$, contrary to our assumption in this paragraph, completing the proof. □

Connectedness for J_2 and J_3. The 2-generated core theorem is also used to show that $\Lambda_2(G)$ is connected when $G = J_2$ and J_3—which is the final assertion of Theorem 1.5.9. This assertion seems to be only implicitly included in most discussions in the literature. The observation is almost established in the final paragraph of the Gorenstein-Lyons paper [**GL82**]. We can easily adapt their argument to give a more explicit proof of:

LEMMA B.4.12. *For $G = J_2$ or J_3, the graph $\Lambda_2(G)$ is connected.*

PROOF. Assume that $\Lambda_2(G)$ is disconnected. Then from B.4.10, there is a maximal subgroup M of G such that $N_G(A) \leq \Gamma_{2,S}(G) \leq M$ for some $S \in Syl_2(G)$ and normal 4-subgroup A of S.

For properties of the groups J_2 and J_3, we refer to Tables 5.3g and 5.3h in [**GLS98**]. In particular J_3 is not of component type, so the result holds in that case by B.4.11 and the Bender-Suzuki Theorem 1.3.5. In the J_2 case, the normal 4-subgroup A of S is unique, and $N_G(A)$ is maximal in G and solvable.

[10] This argument also eliminates $m_2(R) = 1$ in the next-to-last paragraph of the proof of B.5.4.

Consequently $M = N_G(A)$. But from the structure in [**GLS98**, Table 5.3g], $N_G(E) \cong E \times A_5$ for some four-group $E \leq S$. As $N_G(E) \leq \Gamma_{2,S}(G) \leq M$ and $N_G(E)$ is not solvable, this is a contradiction. □

A relation with the weak 2*-generated core.* Again we are not sure if the following facts have appeared explicitly in the literature, so we provide the elementary proofs.

LEMMA B.4.13. *Let G be a finite group with $m_p(G) \geq 2$, and let $S \in Syl_p(G)$. Then*

$$\Gamma_{2,S}(G) = N_G(S)\ \langle N_G(V) : \ V \text{ is an } E_{p^2}\text{-subgroup of } S\ \rangle.$$

PROOF. Let M denote the right-hand side; it is a group since $N_G(S)$ permutes the set of E_{p^2}-subgroups V of S, and hence normalizes the subgroup generated by their normalizers.

Now $M \leq \Gamma_{2,S}(G)$ by definition, so it remains to prove the reverse inclusion. Thus we may assume that there is some $D \leq S$ with $m_p(D) \geq 2$ and $N_G(D) \not\leq M$. We choose D such that $|D|$ is maximal subject to these restrictions. Then by maximality of $|D|$:
(B.4.14)

For all p-subgroups U of M with $m(U) > 1$ and $|U| > |D|$, $N_G(U) \leq M$.

By definition, $N_G(S) \leq M$, so $D < S$. Thus we get $D < U \in Syl_p(N_M(D))$. Now let $U \leq P \in Syl_p(N_G(D))$. By (B.4.14), $N_P(U) \leq M$, so:

(B.4.15) $$P = U \text{ and } N_G(P) \leq M.$$

We observe now that when $p = 2$, D cannot be either dihedral of order at least 8, or semidihedral: For otherwise, the outer automorphism group of D is a 2-group (e.g. A.1.9), and then we get $N_G(D) = C_G(D)P$. But D contains a 4-subgroup V of S, so by the definition of M, we have $C_G(D) \leq N_G(V) \leq M$, so it follows that $N_S(D) = C_G(D)P \leq M$ by (B.4.15), contradicting $N_G(D) \not\leq M$.

We return to the general case. Let $T \in Syl_p(N_G(D))$. Then using the restrictions on D from the previous paragraph, we may apply [**GLS96**, 10.11] with D, T in the roles of "Q, P" to see that D contains an E_{p^2}-subgroup V which is normal in T. Hence $T \leq N_G(V) \leq M$, so that $O^{p'}(N_G(D)) \leq M$. By a Frattini argument A.1.5.2, $N_G(D) = O^{p'}(N_G(D))N_{N_G(D)}(P)$, so $N_G(D) \leq M$ by (B.4.15). This final contradiction completes the proof. □

We use the above characterization to obtain:

LEMMA B.4.16. *Let G be a finite group with $m_p(G) \geq 3$, and let $S \in Syl_p(G)$.*

(a) Assume that $p = 2$, and that S has a normal elementary subgroup of rank at least 3. Then $\Lambda_2(S)$ has no isolated vertices; that is, each 4-subgroup V of S satisfies $m_2(C_S(V)) \geq 3$.

(b) Assume that $\Lambda_2(S)$ has no isolated vertices. Then $\Gamma^0_{2,S}(G) = \Gamma_{2,S}(G)$. Note by B.4.8, that $\Lambda_2(S)$ is connected if and only if $\Lambda_2(S)$ has no isolated vertices.

PROOF. To prove (a), we assume some 4-subgroup V of S has $m_2(C_S(V)) < 3$. Hence

(B.4.17) $$V = \Omega_1(C_S(V)) \text{ is of rank 2.}$$

Then $Z := \Omega_1(Z(S)) \leq \Omega_1(C_S(V)) = V$ by (B.4.17), so as $m_2(S) \geq 3$ by hypothesis, it follows from (B.4.17) that $Z < V$ and hence Z has order 2. By hypothesis, there is some normal elementary abelian subgroup E of S of rank at least 3; then $E \not\leq C_S(V)$ by (B.4.17), so that $V \not\leq E$. So as E is normal, $Z = V \cap E = C_E(V)$ again using (B.4.17). Hence we have $V = Z\langle v\rangle$ for some $v \in V \setminus Z$, with $C_E(V) = C_E(v)$. But $m(C_E(v)) \geq \lceil \frac{m(E)}{2} \rceil \geq 2$, and this is contrary to $C_E(v) = C_E(V) = Z$ of rank 1. This contradiction completes the proof of (a).

Assume the hypothesis of (b). Since $m_p(S) \geq 3$, while $m_p(C_S(V)) \geq 3$ for each E_{p^2}-subgroup V of S since $\Lambda_2(S)$ has no isolated vertices, we obtain that $\langle N_G(S), N_G(V)\rangle \leq \Gamma^0_{2,S}(G)$. We get $\Gamma_{2,S}(G) \leq \Gamma^0_{2,S}(G)$ from B.4.13, completing the proof. □

B.5. Application: A short elementary proof of the Dichotomy Theorem

This argument is essentially the proof of Theorem 2.96 of Gorenstein [**Gor83**], with a refinement by Ron Solomon—primarily[11] in order to replace several appeals to Aschbacher's Proper 2-Generated Core Theorem 1.5.10 by a more elementary argument based on Bender's Strongly Embedded Theorem 1.3.5. Gorenstein in turn attributes the underlying argument to Thompson, namely Theorem 10.7 in the N-group paper [**Tho71**]. In fact Thompson's argument (of about 4 pages, starting at p. 497 of [**Tho71**]) shows that for a minimal simple group, or more generally an N-group, if $m_2(G) \geq 3$ then the normalizer M of a maximal nontrivial signalizer for a Sylow 2-group contains the centralizers of the involutions in M. Our version of that argument culminates in a subgroup M which is strongly embedded in G.

First, we re-state the Dichotomy Theorem 0.3.10 in a version with an additional hypothesis (c), to avoid the small-rank subcase in the definition of groups of Gorenstein-Walter type:

THEOREM B.5.1 (Dichotomy Theorem). *Let G be a finite group satisfying*
(a) $O_{2'}(G) = 1$,
(b) $G = O^2(G)$,
(c) $m_2(G) \geq 3$.
Then G is either of component type or of characteristic 2 *type.*

We begin with some initial reductions. Let $S \in Syl_2(G)$. We may assume that G is not of component type, so for each involution x of G, $L_{2'}(C_G(x)) = 1$. In view of B.1.9, we can re-phrase this as:

(d) For each involution x of G, $C_G(x)$ is 2-constrained.

To show G is of characteristic 2 type, we must show that $F^*(H) = O_2(H)$ for all 2-locals H of G. By Lemma B.1.6, it suffices to show for all involutions x of G that $F^*(C_G(x)) = O_2(C_G(x))$. Therefore it suffices, in the light of (d) above, to show for all involutions x of G that $O_{2'}(C_G(x)) = 1$; so we assume by way of contradiction that:

(e) For some involution $y \in S$, $O_{2'}(C_G(y)) \neq 1$.

As a consequence of (d) (we saw this e.g. at 0.3.21), the subgroup functor defined by $\Theta(a) := O_{2'}(C_G(a))$, for $a^2 = 1 \neq a$, satisfies the balance condition in Definition B.3.2.4 (more specifically, (B.3.20)). So we are in the canonical situation to

[11] We mention also that this version does not require Gorenstein's hypothesis that G should be *fusion-simple*, which would require that $Z(G) = 1$.

apply signalizer functors, via Theorem B.3.25. Let Λ_0 be a connected component of $\Lambda_2(G)$, and for any 4-subgroup V of S, set

$$W_V := \Theta^+(V) = \langle \Theta(v) : v \in V^\# \rangle.$$

By B.4.9.1, G transitively permutes the connected components of $\Lambda_2(G)$. From this we draw two important conclusions. First, no V is an isolated vertex of $\Lambda_2(G)$, whence $m\big(C_G(V)\big) \geq 3$. Second, replacing y and Λ_0 by conjugates if necessary, we may assume that $y \in V$ for some $V \in \Lambda_0$, and Λ_0 contains some four-group $U \trianglelefteq S$ (by B.4.9.2). If possible we choose $U \leq Z(S)$.

Then Theorem B.3.25 yields that $N_G(\Lambda_0)$ normalizes the nontrivial odd-order subgroup W_V, and $W_{V'} = W_V$ for all $V' \in \Lambda_0$.

Set $M := N_G(W_V)$. Since $O_{2'}(G) = 1$ by hypothesis, we have

(f) $N_G(\Lambda_0) \leq M < G$.

In particular $\Lambda_2(G)$ is disconnected. Moreover M contains $N_G(V)$ for every vertex V in Λ_0, and hence for every V in the connected component $\Lambda_2(S)^\circ$ of $\Lambda_2(S)$ provided by Lemma B.4.8.

To show that $\Lambda_1(G)$ is disconnected, the main step is to show that

(B.5.2) M contains the centralizers of each of its involutions.

As a first step, we treat 2-central involutions:

LEMMA B.5.3. *We have:*
(1) $C_G(z) \leq M$ *for any* $z \in U^\# \cap Z(S)$.
(2) M *controls* G*-fusion in* S.
(3) $W_V = W_U$ *for all* 4*-subgroups* V *of* S.

PROOF. We first consider any involution u of U. Embed $T := C_S(U)$ in some $S^* \in Syl_2\big(C_G(u)\big)$, so that $|S^* : T| \leq 2$. Set $R := S^* \cap O_{2',2}\big(C_G(u)\big)$; so R is Sylow in $O_{2',2}\big(C_G(u)\big)$ by Lemma A.1.5.1. Claim $U \leq R$: If U centralizes the quotients of the $C_G(u)$-invariant chain $R > \langle u \rangle > 1$ then the claim follows from [**GLS96**, 12.5]). If not, then $R \not\leq T$, so $|S^* : T| = 2$. Thus $Z(S^*)$ is cyclic by the choice of U, and hence $\langle u \rangle = \Omega_1(Z(S^*))$. Further $|[R, U]| = 2$ and $[R, U]$ is invariant under $RT = S^*$, so $[R, U] = \langle u \rangle$, contrary to assumption. This completes the proof of the claim. By the claim, if Y is a U-invariant subgroup of odd order, then for all $u \in U^\#$, we have $[C_Y(u), U] \leq Y \cap O_{2',2}\big(C_G(u)\big) \leq O_{2'}\big(C_G(u)\big) \leq W_U$.

Now take $u = z \in U^\# \cap Z(S)$: then the roles of "S^*, R" in the previous paragraph are played by S and $Q := S \cap O_{2',2}\big(C_G(z)\big)$. We have $Q \leq S \leq M$, so Q normalizes W_U; hence Q also normalizes $Y := W_U^g$ for any $g \in N_G(Q)$. Set $\overline{C} := C_G(z)/O_{2'}(C_G(z))$. Then $[\overline{Q}, \overline{C_Y(z)}] \leq \overline{Q} \cap \overline{C_Y(z)} = 1$, so by (d) and B.1.9, we have $C_Y(z) \leq O_{2'}\big(C_G(z)\big)$. Then $C_Y(U) \leq C_Y(z) \leq O_{2'}\big(C_G(z)\big) \leq W_U$. Also by the previous paragraph, $[C_Y(u), U] \leq W_U$ for all $u \in U^\#$. So as $C_Y(U)$ and the subgroups $[C_Y(u), U]$ (for all $u \in U^\#$) generate Y by Coprime Action (e.g. 0.3.11.2 and [**GLS96**, 11.4]) we conclude that $Y \leq W_U$—so as $Y = W_U^g$ we get $Y = W_U$. Thus g normalizes W_U, so $N_G(Q) \leq N_G(W_U) = M$. Then by a Frattini argument, we get $C_G(z) = O_{2'}\big(C_G(z)\big)N_{C_G(z)}(Q) \leq M$, since $O_{2'}\big(C_G(z)\big) \leq W_U \leq M$. This establishes (1).

Next suppose that M does not control G-fusion in S. Then by the Alperin-Goldschmidt Fusion Theorem B.2.6, for some member D of the Alperin-Goldschmidt conjugation family (Definition B.2.4), $N_G(D) \not\leq M$. Set $V := \Omega_1\big(Z(D)\big)$. Then we get $\Omega_1(Z(S)) \leq C_S(D) \leq D$ by B.2.4.2, so $z \in V$. Since $N_G(D) \not\leq M$, while $C_G(z) \leq M$ by (1), we see that $\langle z \rangle$ is not characteristic in D; in particular $\langle z \rangle < V$. As $N_G(D) \leq N_G(V) \not\leq M$, it follows that V is an isolated vertex of $\Lambda_2(S)$. Hence, by B.4.8, $D \leq C_S(V) = \langle v \rangle \times K$ with $v \in V$ and with $\Omega_1(K) = \langle z \rangle$. As $\langle z \rangle$ is not characteristic in D, it follows that $K = \langle z \rangle$, whence $V = C_S(V)$. But then $m_2(S) = 2$, contrary to hypothesis. Thus (2) holds.

Let V be a 4-subgroup of S. If V is connected to U in $\Lambda_2(S)$, then $W_V = W_U$. So we may assume V is not connected to U. As $G = O^2(G)$ by hypothesis (b), it follows from (2) that $M = O^2(M)$. Then applying Lemma B.4.9.1 with M in the role of "G" there, we conclude that there is some $g \in M = N_G(W_U)$ with V^g connected to U. Therefore $W_{V^g} = W_U$, and hence $W_V = (W_U)^{g^{-1}} = W_U$, establishing (3). □

We are now in a position to prove (B.5.2) (recall Definition 1.3.4 of strong embedding):

LEMMA B.5.4. *G has a strongly embedded subgroup.*

PROOF. First, to show (B.5.2), it suffices in view of Lemma B.5.3.2 to show that $C_G(x) \leq M$ for all involutions x of S which are extremal in S. Thus $C_S(x)$ is Sylow in $C_G(x)$, so as $G = O^2(G)$ by (b), it follows from Lemma B.2.10 that we have $m_2(C_S(x)) \geq 3$.

By Lemma B.5.3.1 we may assume that $x \neq z$, so that $V := \langle x, z \rangle$ is a 4-subgroup of S. Hence $W_V = W_U$ using Lemma B.5.3.3, so we are able to conclude that $O_{2'}(C_G(x)) \leq W_V = W_U \leq M$.

Now set $R := S \cap O_{2',2}(C_G(x))$, so that R is Sylow in $O_{2',2}(C_G(x))$. Further a Sylow 2-subgroup of G is not semidihedral by our hypothesis (c) that $m_2(G) \geq 3$, and $C_G(x)$ is 2-constrained by our assumption that G is not of component type. So by B.2.5, $m_2(R) \geq 2$. Then $N_{C_G(x)}(R)$ permutes the subgroups W_V, as V ranges over the 4-subgroups of R; but by B.5.3.3, these are all equal to W_U, so that $N_{C_G(x)}(R) \leq N_G(W_U) = M$. Now by a Frattini argument, it follows that $C_G(x) = O_{2'}(C_G(x))N_{C_G(x)}(R)$. So $C_G(x) \leq M$, and (B.5.2) holds.

Set $H := \langle C_G(x) : x \text{ an involution of } S \rangle$. Thus $H \leq M$ and $S \in Syl_2(H)$. Indeed we get $N_G(S) \leq N_G(\Lambda_2(S)^o) \leq N_G(\Lambda_0) \leq M$, so by a Frattini argument $N_G(H) \leq HN_G(S) \leq M$. Thus H is not normal in G, and Theorem B.4.7 yields our lemma. □

Finally we may now apply the Bender-Suzuki Strongly Embedded Theorem, which we stated as 1.3.5 in the case of G simple. Indeed as we have $O_{2'}(G) = 1$ and $G = O^2(G)$ by hypotheses (a) and (b), and as $m_2(G) \geq 3$ by hypothesis (c), $O^{2'}(G)$ is indeed simple, and so is a rank-1 group of Lie type in characteristic 2. Then by 0.4.1.3, G is of characteristic 2 type, contrary to (e). This completes the proof of the Dichotomy Theorem B.5.1.

REMARK B.5.5. Our next three sections involve techniques discussed in Chapters B–E of [**AS04a**]. These methods are also covered in some of the other works we have been referencing. ◇

B.6. Failure of factorization

Modern sources for the theory of failure of factorization (FF) include Sections B.1 and B.2 of the quasithin work [**AS04a**], which in turn refer to [**Asc00**, Sec 32] and [**GLS96**, Sec 26]; also see [**Gor82**, Sec 4.12].

The theory originated in work of Thompson, for example [**Tho68**, 5.53]. We will give only a highly selective overview below.

We emphasize at the outset that any prime p may be used; but our exposition will take $p = 2$, since that is the case most frequently applied in the analysis of groups of characteristic 2 type.

Introduction: background on "general" factorizations. We begin with a common context for factorizations, which will allow us to compare the Thompson factorization from later in this section with the factorizations occurring in the more general theory of weak closure in Section B.8. (In particular, this initial discussion concerns factorizations which do not "fail".)

We assume that a nontrivial finite group $\overline{X}$ acts on an elementary abelian 2-group U, and regard U as an $\mathbb{F}_2$-module for $\overline{X}$. Ordinarily we assume that the action is faithful, and indeed that $O_2(\overline{X}) = 1$.

Most frequently this situation arises from an "internal module" U which is normal in some group X, so that $X/C_X(U)$ plays the role of the faithful group "$\overline{X}$". Furthermore U will typically be generated by the X-conjugates of $\Omega_1\big(Z(T)\big)$ (or of some suitable subgroup), where $T \in Syl_2(X)$—conditions that imply $O_2(\overline{X}) = 1$ (e.g. [**GLS96**, 26.3]).

Next $C_X(U)$ is also normal in X and $C_T(U)$ is Sylow in $C_X(U)$, so from the Frattini argument A.1.5.2, we obtain a factorization:

$$X = C_X(U)N_X\big(C_T(U)\big). \tag{B.6.1}$$

(Such elementary factorizations are frequently in the picture—if sometimes only implicitly—in the proof of the Uniqueness Case Theorem 8.0.2.) As $\Omega_1\big(Z(T)\big) \le U$, also $C_X(U) \le C_X(\Omega_1\big(Z(T)\big))$, so we have the factorization:

$$X = C_X(\Omega_1\big(Z(T)\big))N_X\big(C_T(U)\big), \tag{B.6.2}$$

in which the first term now might not be normal in X.

The theory of factorizations usually goes on to consider some subgroup W which is weakly closed in T, such as the Thompson subgroup of Definition B.6.4, or a generalized weak closure of U as in Definition B.8.1. If $W \le C_T(U)$, then W is also weakly closed in $C_T(U)$, so that we get the containment $N_X\big(C_T(U)\big) \le N_X(W)$. In this situation we can deduce from the previous factorization:

$$\text{If } W \le C_T(U), \text{ then } X = C_X(\Omega_1\big(Z(T)\big))N_X(W). \tag{B.6.3}$$

Finally in Aschbacher's more general theory of weak closure, we may wish to again adjust the first term—this time replacing $\Omega_1\big(Z(T)\big)$ by $C_T(W')$, for some weakly closed subgroup W' containing $\Omega_1(Z(T))$. As the inclusion $\Omega_1\big(Z(T)\big) \le C_T(W')$ is in the "wrong" direction, nontrivial arguments are required to produce the desired factorization, as discussed later in Section B.8.

The Thompson factorization, and its possible failure. In the Thompson factorization. the weakly closed subgroup "W" of the previous subsection is the subgroup generated by all maximal elementary abelian subgroups of T:

DEFINITION B.6.4 (The Thompson subgroup $J(T)$). For a 2-group T, the *Thompson subgroup* $J(T)$ is the subgroup generated by all elementary abelian subgroups A of T which satisfy $m(A) = m_2(T)$.

We now shift to slightly more customary notation, taking V, G in the roles of "U, X"; and usually we assume that T is Sylow in G.

Sometimes $J(T)$ will centralize V. In this case, as in (B.6.1)–(B.6.3) we get the *Thompson factorization* (e.g. [**AS04a**, B.2.7.1]):

$$G = C_G(V)N_G\big(J(T)\big), \tag{B.6.5}$$

and we say that the Thompson factorization holds.

But if (B.6.5) does not hold, i.e., Thompson factorization fails, then V is called a *failure of factorization module*, or *FF-module*, or *F-module*, for $\overline{G} = G/C_G(V)$. In this case $J(T) \not\leq C_G(V)$, so there is some maximal elementary subgroup A of T which does not centralize V. The maximality of A gives the useful inequality $m\big(V/C_V(A)\big) \leq m\big(A/C_A(V)\big)$. For example, this condition is achieved whenever $\overline{A}$ is generated by $\mathbb{F}_2$-transvections on V. More generally, one can try to determine all possible pairs $(\overline{G}, V)$, where V admits nontrivial action by some such "offending" elementary 2-subgroup $\overline{A}$ of $\overline{G}$.

REMARK B.6.6 (The parameter $q(\overline{G}, V)$). As part of the study of FF-modules (notably in our discussion of Chapter B of [**AS04a**], in Section 3.3), we can define $q(\overline{G}, V)$ as the minimum of the ratio

$$\frac{m\big(V/C_V(\overline{A})\big)}{m\big(\overline{A}/C_{\overline{A}}(V)\big)}, \tag{B.6.7}$$

as $\overline{A}$ varies over the subgroups of $\overline{G}$ which are "quadratic" on V, i.e. $[V, \overline{A}, \overline{A}] = 1$. The Thompson Replacement Lemma [**GLS96**, 25.2] shows that if this ratio is at most 1 for some $\overline{A}$, then it is also at most 1 for some replacement $\overline{A}^*$ which is quadratic. As the FF-modules are those satisfying $q(\overline{G}, V) \leq 1$, it is natural to study this condition.

In Chapter B of [**AS04a**] the parameter $\hat{q}(\overline{G}, V)$, is defined as the minimum of the same ratio, now taken over all subgroups $\overline{A}$ with "cubic" action: $[V, \overline{A}, \overline{A}, \overline{A}] = 1$. Notice in particular that $\hat{q}(\overline{G}, V) \leq q(\overline{G}, V)$. ◊

The determination of FF-modules. A fundamental problem is to determine the pairs $(\overline{G}, V)$ with $q(\overline{G}, V) \leq 1$, $\overline{G}$ a $\mathcal{K}$-group, which also satisfy $O_2(\overline{G}) = 1$ and perhaps certain further constraints. The problem has been solved by the combined work of a number of authors. The focus of attention is normally the group $F^*(\overline{G})\overline{A}$, where $\overline{A}$ is an offending subgroup, i.e. the ratio (B.6.7) is at most 1.

In the base case, $F^*(\overline{G})$ is quasisimple and V is irreducible; *in the examples, $F^*(\overline{G})$ is either of Lie type in characteristic 2, or an alternating group, and V is usually the natural module.* (This appears for example as case (1) in the more general Theorem 1 in [**Asc82**].)

In the most difficult subcase, $F^*(\overline{G}) \in Chev(2)$. In this case the original proof quoted unpublished work of Cooperstein–Mason [**CM80**] when $\overline{G} = F^*(\overline{G})$, and in the general case the appeal was to work of Mason's student McClurg in his unpublished thesis [**McC82**]. Guralnick and Malle have published a modern treatment in [**GM02, GM04**], with full proofs.

The cases where $F^*(\overline{G})$ is not in $Chev(2)$ were originally treated in [**Asc82**]. (Indeed Theorem 1 in [**Asc82**] even determines the larger class of F1-modules, where $m\big(V/C_V(\overline{A})\big) - 1 \leq m\big(\overline{A}/C_{\overline{A}}(V)\big)$, in the case $F^*(\overline{G})$ quasisimple and not in $Chev(2)$.) This work depends on various facts on subgroup structure and irreducible representations of simple $\mathcal{K}$-groups, discussed in Sections A.3 and A.4.

Another important result from [**Asc82**] shows that in an FF-module V for a group $\overline{G}$ the offending subgroups (more precisely, the minimal offenders in the set $\mathcal{P}(\overline{G}, V)$ of [**Asc82**]) normalize each component of $\overline{G}$—which effectively supplies a reduction to the case of $F^*(\overline{G})$ simple. There are simpler more recent proofs of this last result—including one due to Timmesfeld, reproduced as 26.24 in [**GLS96**].

We single out one of the early results on FF-modules, because of its frequent use: the solvable subcase.

Remark B.6.8 (Solvable failure of factorization). The treatment of the case where G is solvable (or even p-solvable), but Thompson factorization fails, is due to Glauberman [**Gla73**], and is reproduced in 32.5 in [**Asc00**]. The configuration arises only for $p = 2, 3$; and the normal closure of $J(T)$ in G is essentially a commuting product of groups, each of which is isomorphic to $SL_2(p)$ acting on its natural module. ♢

Other notions of "small" modules. Results related to failure of factorization are sometimes needed at various points in the treatment of groups of characteristic 2 type. For example, the weaker variant $\hat{q}(\overline{G}, V) \leq 2$ of the FF-condition is fundamental in the proof of the Quasithin Theorem 3.0.1. See in particular the discussion of the qrc-lemma in section D.1 of [**AS04a**]; we provide a brief discussion in Section 3.3. The determination of all pairs with $\hat{q}(\overline{G}, V) \leq 2$ was also obtained by Guralnick and Malle in [**GM02, GM04**].

Finally we mention that more detailed study of failure of factorization (as well as pushing-up and weak closure) often involves an analysis of "small" modules for known simple groups. We discuss results of this type in Section A.4.

B.7. Pushing-up, and the Local and Global $C(G,T)$ Theorems

We first indicate some general sources for the theory of pushing-up. There is a nice discussion of blocks and the Global $C(G,T)$ Theorem in the papers of Foote [**Foo80**] and Solomon [**Sol80**] in the proceedings of the 1979 AMS Summer Symposium at Santa Cruz. The quasithin case is developed largely independently in Chapter C of [**AS04a**]. See also Section 4.13 of [**Gor82**]. Below we will give a brief (and somewhat ad-hoc) introduction.

Pushing-up is often used to exploit the existence of uniqueness subgroups. We will begin our discussion with an important example of such a situation, where "$C(G,T) < G$". This occurs for example when a Sylow 2-subgroup T of G lies in a unique maximal 2-local subgroup M of G (that is, $M = !\mathcal{M}(T)$ in the language of Remark 8.0.4). For if C is a nontrivial characteristic subgroup of T, then we have $N_G(C) \leq M$ by the uniqueness of M.

More generally if H is a 2-local subgroup of G with $R := T \cap H \in Syl_2(H)$, and C is a nontrivial characteristic subgroup of R, then we can "push up" $N_H(C)$ to $N_G(C)$; and if $R < T$ then $R < N_T(C)$, so $N_G(C)$ has a larger 2-share than H.

Indeed as C varies over the nontrivial characteristic subgroups of R, we might hope that the locals $N_H(C)$ generate H. Thus we are led to define:

DEFINITION B.7.1. Let $R \leq T \in Syl_2(G)$. Set
$$C(G,R) := \langle N_G(C) : 1 \neq C \text{ char } R \rangle. \qquad \diamondsuit$$

The Local $C(G,T)$ Theorem. The groups G of characteristic 2 type which have the property $C(G,T) < G$ for $T \in Syl_2(G)$ can in fact be determined. The initial step is to describe the subcase in which $F^*(G) = O_2(G)$—for example, when G is a 2-local subgroup in a larger group of characteristic 2 type. This is accomplished in Aschbacher's Local $C(G,T)$ Theorem: we will sketch some history leading up to the result, which we will state below as Theorem B.7.3. (Cf. also Section C.1 in [**AS04a**].)

The theorem identifies certain "obstructions" to pushing-up:

DEFINITION B.7.2 ((Aschbacher) blocks). A group L is *short* if it satisfies:
$L = O^2(L)$;
$F^*(L) = O_2(L)$;
$W := [O_2(L), L] \leq \Omega_1(Z(O_2(L)))$;
$L/O_2(L)$ is quasisimple or of order 3; and
L is irreducible on $W/C_W(L)$.

A *block* of G is a short subgroup L which is subnormal in G.

A block L is a χ-*block* if $L/O_2(L)$ is $L_2(2^n)$ (for $n > 1$) or A_m for $m \geq 3$ odd, with $W/C_W(L)$ (respectively) the (irreducible) natural module or the irreducible core of the natural permutation module. (Correspondingly we use the terms $L_2(2^n)$-block and A_m-block.) Note that when L is an A_5-block, $[O_2(L), L]$ is the core of the standard permutation module for $L/O_2(L)$ regarded as A_5. On the other hand when L is an $L_2(4)$-block, setting $W := [O_2(L), L]$, then $W/C_W(L)$ is the standard $L_2(4)$-module for $L/O_2(L)$. $\diamondsuit$

Notice that W has a *unique* nontrivial composition factor under L. Indeed, the following properties follow easily:

If L is a solvable short subgroup, then $L \cong A_4$.

If L is a nonsolvable short subgroup, then L/W is quasisimple.

If L is a χ-block, then $O_2(L)$ is abelian with $|O_2(L)/W| \leq 2$.

Distinct blocks commute elementwise.

We can now state the Local $C(G,T)$ Theorem (where "M" might be a 2-local in a simple G).

THEOREM B.7.3 (The Local $C(G,T)$ Theorem [**Asc81a**]). *Let M be a finite group which satisfies $F^*(M) = O_2(M)$, and $C(M,T) < M$ for $T \in Syl_2(M)$.*

Then M is the product of $C(M,T)$ with one or more χ-blocks L.

As a corollary, we obtain the following statement, which is sometimes also referred to as the $C(G,T)$ Theorem. It is a re-statement of Theorems 2 and 3 in [**Asc81a**]:

THEOREM B.7.4. *Let G be a finite group of characteristic 2 type. Assume that $T \in Syl_2(G)$ and suppose that $C(G,T) < G$. Then one of the following holds:*

(1) $C(G,T)$ is a strongly embedded subgroup of G, and hence G is described by Bender's Strongly Embedded Theorem 1.3.5; or

(2) there is a maximal 2-local subgroup M of G and a χ-block L of M such that M is the unique maximal 2-local subgroup of G containing L.

Note that if (1) fails, then as $C(G,T)$ is not strongly embedded in G, there exists a maximal 2-local subgroup M of G (normalizing some nontrivial subgroup of T) which is not contained in $C(G,T)$. Choosing such an M so that a Sylow 2-subgroup S of $M \cap C(G,T)$ has maximal order, it then follows that $C(M,S) < M$ and $S \in Syl_2(M)$, and the Local $C(G,T)$ Theorem can be applied to M.

Historically, Thompson deduced, as an easy consequence of his factorization theorems, that if G is a finite group in which some Sylow p-subgroup P is a maximal subgroup, with p an odd prime, then $G = O_{p'}(G)P$. The analogous statement with $p = 2$ is false, as the symmetric group S_4 shows. Indeed, the group $G := L_2(17)$ is simple and has a maximal Sylow 2-subgroup T. Moreover, G is of characteristic 2 type and $T = C(G,T)$. The maximal 2-local subgroups of G isomorphic to S_4 are the obstructions to pushing-up in this case.

Bernd Baumann tackled the problem of describing such obstructions in [**Bau79**], and produced a theorem later refined and extended by Glauberman-Niles and Campbell. In particular, Baumann focused attention on the Baumann subgroup

$$\mathrm{Baum}(T) := C_T\big(\Omega_1(Z(J(T)))\big).$$

which has the following important property. (Recall that H is an abstract minimal parabolic over its Sylow 2-subgroup T if T is not normal in H, but T is contained in a unique maximal subgroup of H.)

THEOREM B.7.5 (Baumann's Lemma). *Suppose that L is an abstract minimal parabolic over its Sylow 2-group T, with $F^*(L) = O_2(L)$, and let $V := \langle \Omega_1(Z(T))^L \rangle$ and $L^* := L/C_L(V)$. Assume that $L^* \cong L_2(2^n)$ and $[V,L]/C_{[V,L]}(L)$ is a natural module for L^*. Finally, assume that $J(T)^* \neq 1$.*

Then Baum(T) is Sylow in its normal closure in L.

(A more detailed version of this lemma is proved as B.6.10 in [**AS04a**].) We can now state the Baumann-Glauberman-Niles Theorem, which takes care of an important minimal case of the $C(G,T)$ Theorem.

THEOREM B.7.6 (The Baumann-Glauberman-Niles Theorem). *Let S be a 2-group. Then S has nontrivial characteristic subgroups $C_1(S)$ and $C_2(S)$ with the following properties:*

(a) $C_1(S) \leq \Omega_1(Z(S))$;

(b) $C_2(S)$ is characteristic in $T :=$ Baum(S); and

(c) If L satisfies the hypotheses of Baumann's Lemma B.7.5 with $T \in Syl_2(L)$, then one of the following conclusions holds:

(1) $C_1(S) \leq Z(L)$;

(2) $C_2(S) \trianglelefteq L$; or

(3) $O^2(L)$ is a χ-block.

There is also a version of this result, with a proof due to Campbell in his thesis [**Cam79**], which gives an "explicit" definition of $C_1(T)$ and $C_2(T)$. Either proof of the Baumann-Glauberman-Niles Theorem depends on failure of factorization theory; indeed the hypotheses lead to W being an FF-module for L—see for example 2.c and 3.II of C.1.21, as well as C.1.29, in [**AS04a**].

We resume our informal discussion of the Local $C(G,T)$ Theorem. Since $J(T)$ normalizes each component of $M/O_2(M)$, we can focus attention on groups M in which $L := O^2(M)$ satisfies $L/O_2(L)$ quasisimple. Moreover, failure of Thompson factorization and the structure and representation theory of simple $\mathcal{K}$-groups (as in our Sections A.3 and A.4) reduces attention primarily to those cases where $L/O_2(L)$ is an alternating group or a group of Lie type in characteristic 2. The latter groups are, of course, generated by abstract minimal parabolics which are overgroups of T in which a certain "characteristic section" is a product of groups of Lie rank 1, or consist of a group of rank 2 on whose Dynkin diagram T acts nontrivially. Similarly abstract minimal parabolics over T generating alternating groups have characteristic sections which are products of symmetric groups S_{2^m+1}; indeed M is generated by $C(M,T)$ and abstract minimal parabolics with Sylow 2-subgroup Baum(T), whose characteristic sections are products of copies of $S_3 \cong L_2(2)$.

The primary technical tool for the further reduction of the Local $C(G,T)$ Theorem to the minimal parabolic setting is Glauberman's Argument,[12] a version of which is proved as C.1.21 in [**AS04a**]; but the underlying arguments needed are already in [**Asc81a**]. The argument assumes that M and L are as above, with $L/O_2(L)$ quasisimple and $T \in Syl_2(M)$. Set $S := Baum(T)$ and choose R with $O_2(M)S \leq R \leq T$; it considers a set $\mathcal{Y}(G,R)$ of R-invariant subgroups Y of the type studied by Glauberman-Niles. A typical conclusion is the following:

If $M = \langle N_G(S), Y\rangle$ for some $Y \in \mathcal{Y}(M,R)$, then either
(I) some nontrivial characteristic subgroup of S is normal in M; or
(II) L is a χ-block.

There is an alternative proof of the Local $C(G,T)$ Theorem by Gorenstein and Lyons in [**GL93**]; also see [**GLS96**, 27.5].

REMARK B.7.7 (Use of pushing-up for uniqueness theorems). Pushing-up is often used to establish uniqueness theorems, namely results of the form $M = !\mathcal{M}(X)$ for suitable subgroups M, X of G. Typically the argument centers on a proper subgroup R of $T \in Syl_2(G)$, which is Sylow in a 2-local $H \not\leq M$ having the property that $C(H,R) \leq H \cap M < H$. So the Local $C(G,T)$ Theorem B.7.3 with H and R in the roles of "G, T" shows that H contains blocks not in M. Then one can try to use the restricted structure of blocks to produce a contradiction to the original hypothesis.

An example of such an argument appears in our discussion of the proof of Theorem B.8.3. Many such arguments are made during the proofs of the Quasithin Theorem 3.0.1 and the Uniqueness Case Theorem 8.0.2. ◊

The Global $C(G,T)$ Theorem. The Global $C(G,T)$ Theorem completes the determination of simple groups G of characteristic 2 type in which $C(G,T) < G$. The point of departure is the existence of a maximal 2-local M having a χ-block, as guaranteed by the Local $C(G,T)$ Theorem (see B.7.4). The final result is the following theorem.

[12]The statement we are calling Glauberman's Argument evolved from Aschbacher's applications of ideas of Glauberman, as described on pages 7 and 10 of [**AS04a**].

THEOREM B.7.8 (Global $C(G,T)$-Theorem). *Let G be a finite simple group of characteristic 2 type with $T \in Syl_2(G)$. Suppose that $C(G,T) < G$. Then one of the following conclusions holds:*

(1) $G \cong A_6$, $L_3(2)$, $L_3(3)$, M_{11}, or $L_2(p)$ for p a Fermat or Mersenne prime;
(2) $G \cong M_{22}$ or M_{23}; or
(3) $G \cong L_2(2^n)$, $Sz(2^n)$, $U_3(2^n)$, $L_3(2^n)$, or $Sp(4,2^n)$, with $n \geq 2$.

As an immediate corollary, we obtain the following result.

COROLLARY B.7.9 (Sylow 2-Uniqueness). *Let G be a finite simple group of characteristic 2 type. Suppose that a Sylow 2-subgroup of G is contained in a unique maximal 2-local subgroup M of G. Then either G has a strongly embedded subgroup, or $G \cong L_3(3)$ or M_{11} or $L_2(p)$ for $p \geq 17$ a Fermat or Mersenne prime.*

PROOF. Suppose that $T \in Syl_2(G)$ is contained in a unique maximal 2-local subgroup M of G, but M is not strongly embedded in G. If $1 < U \trianglelefteq T$, then $N_G(U) \leq M$. In particular, this is true if U is characteristic in T, whence we have $C(G,T) \leq M < G$. Hence the hypotheses of the Global $C(G,T)$ Theorem hold. Moreover, if G is of Lie type in characteristic 2, then the uniqueness of M forces the Lie rank to be 1, in which case $M = N_G(T)$ is a strongly embedded subgroup of G. If $G = M_{22}$ or M_{23}, then T contains two normal E_{16} subgroups whose normalizers are distinct maximal 2-local overgroups of T. Hence conclusion (1) of the Global $C(G,T)$ Theorem holds; and as $A_6 \cong Sp(4,2)'$, the corollary follows. □

We remark that if $G \cong L_3(3)$, then T is contained in two different maximal subgroups of G, but these are maximal parabolics for the prime 3, and not 2-locals.

REMARK B.7.10 (The Thompson Strategy). Thus, for simple groups of characteristic 2 type, the Global $C(G,T)$ Theorem and its corollary lay the groundwork for implementing the Thompson Strategy (which we discuss at greater length in Section 3.1) by disposing of the case where a maximal 2-local overgroup M of a given Sylow 2-subgroup is unique. Thereafter, it is possible to study the completion $G_0 := \langle M, H \rangle$ of the amalgam (cf. 3.3.14) of M with a 2-local subgroup H not contained in M. Thus one has the crucial condition $O_2(G_0) = 1$. ◇

We now give a brief historical survey of the proof of the Global $C(G,T)$ Theorem in the spirit of [**Foo80**]. The crucial first step of proving a Standard Form Theorem for blocks was achieved by Richard Foote in [**Foo82**]. Previously Gilman [**Gil76**] had studied this for variants of blocks that he called "constrained components".

THEOREM B.7.11 (Standard form for blocks). *Let G be a finite group, M a maximal 2-local subgroup of G and J a nonsolvable block of M. Assume that M is the unique maximal 2-local subgroup of G containing J.*

Then either $M = G$ or J is normal in M.

The proof is facilitated by the property, which does not always hold for components, that if J is a nonsolvable block and x is an involution of $\mathrm{Aut}(J)$, then x centralizes an involution in $J \setminus Z(J)$. Indeed the analogous property is false for components J isomorphic to $SL(2,q)$, q odd.

REMARK B.7.12. The analogous statement is also false when the block J is isomorphic to A_4. However, under the additional hypotheses that $F^*(G)$ is simple and G is of characteristic 2 type, Solomon and S. K. Wong [**SW81**] (and independently Foote) were able to show that either $J \trianglelefteq M$ or $G \cong A_8$. ♢

Once standard form is established for a block J, it follows that $C := C_G(J)$ is a tightly embedded subgroup of G, i.e. $C \cap C^g$ has odd order for all $g \in G \setminus N_G(C)$. Thus Aschbacher's earlier work on tightly embedded subgroups [**Asc76b**], together with a refinement by Solomon and Timmesfeld in [**ST79**], may be applied to reduce to the case where $C_G(J)$ is either a cyclic 2-group or has an elementary abelian Sylow 2-subgroup.

The standard form problem for blocks J with $J/O_2(J) \cong A_{2n+1}$ was solved by Solomon in [**Sol81**]:

THEOREM B.7.13 (A_n Standard Form). *Let G be a finite simple group of characteristic 2 type. Assume that J is a short subgroup of G with $J/O_2(J) \cong A_{2n+1}$, $n \geq 1$. Assume further that $M := N_G(J)$ is the unique maximal 2-local subgroup of G containing J. Then one of the following conclusions holds:*

(1) $J \cong A_4$ and G has dihedral or semidihedral Sylow 2-subgroups, whence $G \cong A_6$, M_{11}, $L_3(3)$, or $L_2(p)$ for p a Fermat or Mersenne prime; or

(2) $J/O_2(J) \cong A_5$ and $G \cong U_4(2)$.

Note that $U_4(2) \cong \Omega(5,3)$ is the tip of an iceberg of examples, if the hypothesis of characteristic 2 type is dropped. Indeed the groups $M(n) \cong E_{2^n} \cdot S_n$ of monomial matrices in $O(n,q)$, q odd, provide examples of alternating blocks in groups G with $F^*(G)$ simple, for all $n \geq 5$. However, only $\Omega(5,3)$ passes the characteristic 2 type barrier. Indeed, the failure of characteristic 2 type in all other examples can be detected in $C_G(x)$ for most $x \in O_2(J)$, and this is key to the proof of the theorem. In fact, once it is shown that $C_G(x) \leq M$ for all involutions $x \in Q := O_2(M)$, it follows easily from the uniqueness of M that Q is a weakly closed TI-subgroup of G, at which point the theorem follows from work of Timmesfeld [**Tim75b**].

We illustrate these ideas with an elementary argument, provided to us by Ulrich Meierfrankenfeld, which covers an important special case.

LEMMA B.7.14. *Suppose that G is a finite simple group of characteristic 2 type, which does not have a strongly embedded subgroup. Suppose that a Sylow 2-subgroup S contained in a unique maximal 2-local subgroup M of G.*

Then not every χ-block J in 2-locals of G satisfies $J/O_2(J) \cong A_{2^n+1}$ with $n > 2$.

PROOF. Suppose false, and choose a 2-local L not lying in M with $|L \cap M|_2 > 1$ is maximal, and, subject to this, such that L is minimal. Let $T \in Syl_2(L \cap M)$ with $T < S$. By our maximal choice, $T \in Syl_2(L)$ and $C(L,T) < L$. Under our restrictive hypotheses on blocks, the Local $C(G,T)$ Theorem B.7.3 implies that $V := [O_2(L), O^2(L)]$ is a direct sum of irreducible cores of permutation modules for components of $O^2\big(L/O_2(L)\big)$, which is a direct product of copies of $A_{2^{n_i}+1}$ for various $n_i > 2$.

Set $Z := \Omega_1\big(Z(T)\big)$ and $H := N_G(Z)$. Since $N_S(Z) > T$, H is not contained in L. Moreover, by maximal choice of T, $H \leq N_G\big(O_2(H)\big) \leq M$. Now, for each block J of L, we have a unique 2-central involution in $O_2(L)$ whose stabilizer in $J/O_2(J)$ is isomorphic to $A_{2^{n_i}}$. Since $n_i > 2$, we get the crucial fact:

$$O_2(H \cap L) = O_2(L).$$

Since $L \le N_G(O_2(L))$, we see that $N_S(O_2(L)) = T$, since otherwise we would have $L \le N_G(O_2(L)) \le M$ by the maximal choice of T. Hence

$$N_{O_2(H)}(O_2(L)) \le O_2(H \cap L) = O_2(L).$$

It follows that $O_2(H) \le O_2(L)$. Hence, as $V \le Z(O_2(L))$, $[O_2(H), V] = 1$. Since G is of characteristic 2 type, it follows that $V \le O_2(H)$. But as $V = [O_2(L), O^2(L)]$ and $O_2(H) \le O_2(L)$, we conclude that

$$[O_2(H), O^2(L)] \le V \le O_2(H).$$

As $T \le H$, we deduce that $L = O^2(L)T \le N_G(O_2(H)) \le M$, a contradiction. □

The final step in the proof of the Global $C(G,T)$ Theorem was handled by Solomon and S. K. Wong [**SW81**] and, independently, by Aschbacher [**Asc81d**].

THEOREM B.7.15 ($L_2(2^n)$ Standard Form). *Let G be a finite simple group of characteristic 2 type. Assume L is a short subgroup of G with $L/O_2(L) \cong L_2(2^n)$, $n \ge 2$. Assume further that $M := N_G(L)$ is the unique maximal 2-local subgroup of G containing L.*

Then one of the following conclusions holds:
(1) $G \cong L_3(2^n)$ or $Sp(4, 2^n)$; or
(2) $n = 2$ and $G \cong J_3$, M_{22}, or M_{23}.

Note that in this case, there are no "shadows", in which $|G : M|_2$ is large, as there are with $O(n, q)$ in the alternating case. Indeed, if $T \in Syl_2(M)$ such that $T \le S \in Syl_2(G)$, then typically $J(T) = AB$, where A and B are the only two elementary abelian subgroups of T of maximum order, and we have $A = O_2(M)$. Hence $|N_S(T) : T| \le 2$ and one expects $J(N_S(T)) = J(T)$, whence $S = T$. Indeed Aschbacher proved in [**Asc78a**] that if $T < S$, then S either has class 2 or sectional 2-rank 4. (In the latter case, of course, $L/O_2(L) \cong L_2(4)$.) Then theorems of Gilman-Gorenstein [**GG75**] and Gorenstein-Harada [**GH74**] can be invoked to complete the proof. Solomon and Wong, on the other hand, quote Gomi [**Gom75**] for an alternate proof of the Gilman-Gorenstein theorem. For them an easy argument shows that either $M = O^2(M)C_M(L/O_2(L))$ or $G \cong M_{22}$ or M_{23}. Thus a tight configuration has been achieved in which, roughly, $O_2(M)$ is an indecomposable module for $M/O_2(M) \cong L_2(2^n)$. Since S is not of class 2, by assumption, it follows that $E := C_M(L) \ne 1$. By the uniqueness of M, E is a TI subgroup. A short fusion argument establishes that E is weakly closed, whence Timmesfeld's theorem [**Tim75b**] can once again be invoked to obtain a final contradiction, thereby completing the proof of the Global $C(G,T)$ Theorem.

Other pushing-up results. The proof in [**AS04b**] of the Quasithin Theorem 3.0.1 also involves pushing-up results related to the situation of $C(G,R) < G$, for a suitable proper subgroup R of a Sylow 2-subgroup T of G. Such results are included in our discussion in Section 3.3 of preliminary pushing-up results from Chapter C of [**AS04a**]; in particular see the discussion of the rank-2 case by Meierfrankenfeld and Stellmacher.

The proof of the $e(G) = 3$ Theorem 4.0.1 in [**Asc81b, Asc83a**] uses other pushing-up results (notably 3.16 in [**Asc83a**]), which appear in our discussion in Chapter 4.

Aschbacher's papers [**Asc83b, Asc83c**] treating the Uniqueness Case Theorem 8.0.2 also use some results from sections 8–10 of [**Asc81d**].

Finally we mention a result which is not a pushing-up result, but like the $C(G,T)$-Theorems is an improvement on, and application of, Bender's Strongly Embedded Theorem 1.3.5, in the viewpoint of the footnote just before Definition 1.3.4. It was established independently by Holt in [**Hol78**], and by F. Smith in [**Smi79a**]. We give the statement appearing as Theorem 4.32 in [**Gor82**]; a version also appears as Theorem SF on p. 89 of [**GLS94**].

THEOREM B.7.16 (Holt's Theorem). *Let G be a primitive permutation group. Assume further that $O_{2'}(G) = 1 = Z(G)$, and that some 2-central involution fixes exactly one point. Then either G has a strongly embedded subgroup, or $G \cong S_n$ or A_n for some odd n.*

In classification problems, this result was often used in the final stages; namely when a group H of odd index in G had been identified, in order to then show that $H = G$ by showing if $H < G$ that the representation of G on G/H satisfies Holt's condition—whereas G would be known not to be one of Holt's groups.

The hypothesis can be comparatively easy to verify in the final stages. It is equivalent to the existence of a maximal subgroup M and a 2-central involution z of G such that for all $g \in G$, $z^g \in M$ if and only if $g \in M$.

B.8. Weak closure

The basics for weak closure are developed in [**Asc81e**]; for a more recent treatment see Chapter E of [**AS04a**]. We will give only a brief overview here.

The theory of weak closure was developed by Aschbacher to be used in [**Asc83b, Asc83c**] to prove the Uniqueness Case Theorem 8.0.2; with some thought that it could also be useful in analyzing quasithin groups of characteristic 2 type—and later it was indeed used at numerous points in [**AS04b**] to prove the Quasithin Theorem 3.0.1.

The theory grew out of a study of Thompson's work on N-groups [**Tho68, Tho70, Tho71, Tho73**]; and can be seen in a primitive form in Aschbacher's work [**Asc78b**] on thin groups. It uses relatively little outside machinery: primarily Timmesfeld's work on root involutions in [**Tim75a**], his work on weakly closed TI-sets in [**Tim75b**], and some of Aschbacher's work on tightly embedded subgroups in the component paper [**Asc75a**] and in [**Asc76b**].

Generalized weak closures and related parameters. Weak closure arguments are designed to be used in a context like that of the Thompson strategy (mentioned in Section 3.1 for the Quasithin Theorem and in Section 8.2 for the Uniqueness Case Theorem), to exploit the interaction between two distinct 2-local subgroups M and H containing $T \in Syl_2(G)$.

Typically M is a maximal 2-local, and one seeks factorizations of H. In the usual Thompson strategy, $H \not\leq M$, and often H will be chosen minimal with respect to this and perhaps other properties. (But sometimes, as in Remark 8.5.23, H is instead contained in M.)

We also consider modules for these groups; so that in factorizations like those in Section B.6, there are two different pairs M, V and H, U which can play the role of the pair "X, U" of that section.

Weakly closed subgroups for factorizations. To get candidates for the weakly closed group "W" in a factorization like (B.6.3), we focus first on the module V for M. We can generalize the Thompson subgroup $J(T)$ of Definition B.6.4, which is generated by conjugates in T of maximal elementary subgroups A of T, by considering the conjugates in T of subgroups of V of small corank in V. (Cf. [**GLS96**, 26.13].)

DEFINITION B.8.1 (Generalized weak closures). Given an elementary abelian 2-subgroup V of T, for $i \geq 0$ we define

$$W_i(T,V) := \langle A : A \leq T \cap V^g \text{ for some } g \in G \text{ with } m(V^g/A) = i \rangle.$$

By construction this group is weakly closed in T; $W_0(T,V)$ is just the usual weak closure of V in T. Typically "V" is the module for M indicated above. We often use the abbreviation $W_i := W_i(T,V)$.

Sometimes we wish to establish a factorization of the form (B.6.3) for our second subgroup H in the role of "X", rather than for M. So from the discussion of (B.6.3) we want to show $W_i \leq C_T(U)$, for the module U for H, and replace $\Omega_1(Z(T))$ by:

$$C_j := C_T(W_j) .$$

In fact usually j is of the form $i + k$ with $k = n(H)$, where $n(H)$ is the parameter described in Remark B.8.4. ◇

Generalized weak closures (that is, cases with $i > 0$) appear already in Thompson's variant [**GLS96**, 26.16] of the usual Thompson factorization, and, in an analogous situation for p odd, in the Odd Order Paper, with $i = 1$.

Parameters relevant to factorizations. Various elementary arguments (e.g. E.3.17 in [**AS04a**]) show that obstructions to inclusions $W_i \leq C_T(U)$, and to factorizations like those indicated above, are associated with lower bounds on certain parameters defined for our groups and modules.

REMARK B.8.2 (Parameters associated to M and V). Assume that M is a maximal 2-local subgroup of G, with V a nontrivial elementary 2-subgroup normal in M. Set $\overline{M} := M/C_M(V)$ and define the parameter

$$r = r(G,V) := \min\{m(V/A) : A \leq V, C_G(A) \not\leq N_G(V) = M\}.$$

For example when the structure of $\overline{M}$ is dominated by a Lie type group in characteristic 2, this parameter r is related to the rank of a root space in the module V (and hence of a root group in M). For the role of r in establishing factorizations, see e.g. [**AS04a**, E.3.17] and subsequent results. In practice, we often have uniqueness theorems which will "control" suitable centralizers—typically guaranteeing that $C_G(A) \leq M$, for A of small corank in V. But we may also show $C_G(v) \leq M$ for suitable $v \in V^{\#}$. Such results provide a lower bound for r.

For Y a faithful $\mathbb{F}_2$-module for a group X, we define

$$m(X,Y) := \min\{m\big(Y/C_Y(t)\big) : t \text{ an involution of } X\}.$$

Most frequently we focus on $m := m(\overline{M}, V)$ where $\overline{M} := M/C_M(V)$. Lower bounds on the value of m guarantee that $\overline{M}$ induces no $\mathbb{F}_2$-transvections on V; and more generally give information about the rank of a root group in $\overline{M}$.

We can in effect combine these two parameters, by writing:

$$s = s(G,V) := \min\{\ r(G,V)\ ,\ m(\overline{M},V)\ \};$$

this notation is useful when we wish to exhibit a common lower bound on both parameters. $\diamondsuit$

Since these parameters are determined in local subgroups M and $C_G(A)$, we can hope to obtain bounds on their values by studying the properties of known groups; this is a major theme developed in [**Asc82**] primarily for application in the Uniqueness Case Theorem, and in [**AS04a**, Ch H] for application in the Quasithin Theorem (cf. our Section A.4). In particular, establishing "control of centralizers"—by proving $C_G(A) \leq M$ for large subgroups A of V—gives lower bounds for r, which is a first step in implementing weak closure arguments. This kind of control is one application of uniqueness theorems in the sense of Remark 8.0.4.

Some relations among the parameters. The parameters m and r are at least ostensibly independent; for example m is determined by the 2-local M, while r requires knowledge of the 2-locals $C_G(A)$. But there is one important situation in which a suitable bound on m guarantees the same bound on r; this result appears as 11.1 in [**Asc81e**]:

THEOREM B.8.3. *Assume that G is of characteristic 2 type; and for M, V as above that:*
(a) $M = !\mathcal{M}(N_M(C_T(V)))$, and
(b) $m(\overline{M}, V) > 2$.
Then $r(G, V) \geq m(\overline{M}, V)$; and so $s(G, V) = m(\overline{M}, V) > 2$.

The proof uses pushing-up: For the uniqueness hypothesis can be used to show that $Q := C_T(V)$ is Sylow in $H := C_G(U)$ when U is of corank $r < m$ in V and $C(H, Q) \leq H \cap M < H$—so that the Local $C(G, T)$ Theorem B.7.3 produces blocks of H not contained in M. But these block impose restrictions on internal modules for H, leading eventually to a contradiction to the bound $m > 2$ in the hypothesis.

This result is quoted e.g. in proving the basic weak closure result [**Asc83c**, 12.21] for the Uniqueness Case Theorem; and a version appears as E.6.3 in [**AS04a**] for the Quasithin Theorem.

In particular, the significance of the bound $s > 2$ is that 2 is just large enough to allow for the troublesome values $n(H) \leq 2$, in the factorizations of the form B.8.6 below—namely using $0, n(H)$ in the roles of "$i, i + k$", for the inequality $i + k < s$. So in these situations, an analysis of the parameter m suffices, and there is no need for an analysis of r.

Parameter bounds guaranteeing factorizations. We now return to the theme mentioned in Definition B.8.1, where we wish to obtain factorizations of H by showing $W_i \leq C_T(U)$, for the H-module U. Typically U will be generated by the H-conjugates of $\Omega_1(Z(T))$, but sometimes we may restrict to suitable subgroups of that normal closure. Such arguments require further notions recorded in:

REMARK B.8.4 (Parameters associated to H and U). Let $\hat{H} := H/C_H(U)$, so that $\hat{H}$ is faithful on U. We define $a(\hat{H}, U)$ to be the maximum k such that there is some elementary 2-subgroup A of $\hat{H}$ with $C_U(B) = C_U(A)$ for all $B \leq A$ with $m(A/B) < k$. When the structure of $\hat{H}$ is dominated by a group of Lie type in characteristic 2, $a(\hat{H}, U)$ is essentially the rank of a root space in U. Sometimes we instead write $a = a(H, U)$ for $a(\hat{H}, U)$.

The weak closure theory, as developed in [**Asc81e**] and in [**Asc82**] for use in the Uniqueness Case Theorem, involves certain families $\mathcal{E}_k$, $\mathcal{F}_k$ of groups, whose definition we will not repeat here. For the Quasithin Theorem, the minimum k with $H \in \mathcal{E}_k$ gives the value "$n(H) = k$" in Definition E.1.6 of [**AS04a**]. We omit the details, but comment that $n(H) \geq 1$ with equality when H involves no groups of Lie type over fields of even order $q > 2$ (for example when H is solvable). It will be convenient to use $n(H)$ in our fairly informal statements below of results from the weak closure literature. $\diamondsuit$

So we assume the setup involving M, H, etc. developed earlier in the section; We will discuss the fundamental result 6.11 in [**Asc81e**], which is stated for the module $U := \langle \Omega_1(Z(T))^H \rangle$.

Elementary arguments such as [**Asc81e**, 6.7], establish the following:

LEMMA B.8.5. *Assume $W_i > 1$ for some $i < s(G,V) - a(H,U)$. Then $W_i \leq C_G(U)$, and so $H = C_H(\Omega_1(Z(T)))N_H(W_i)$.*

We may use $n(H)$ in place of $a(H,U)$ in the statement of the lemma. We also want to replace $\Omega_1(Z(T))$ by $C_{i+n(H)}$. Possible obstructions at [**Asc81e**, 6.8–6.10] to such a factorization are ruled out under this bound, giving:

PROPOSITION B.8.6. *Set $k := n(H)$, and assume $W_i \neq 1$ for some i which satisfies $i + k < s(G,V)$. Then $H = \langle C_H(C_{i+k}), N_H(W_i) \rangle$. Similar hypotheses in* [**Asc81e**, 6.11.2] *lead to $H = C_H(C_{i+k})N_H(W_i)$.*

Versions of the above results appear as E.3.18 and E.3.19 in [**AS04a**]. And a variant essentially incorporating also the hypotheses of Theorem B.8.3 appears in (2) and (3) of 12.21 in [**Asc83c**]; while (1) of 12.21 is essentially Lemma B.8.5.

Further inequalities involving the parameters. The weak closure factorizations recorded above were used heavily in the proof of the Uniqueness Case Theorem 8.0.2 in [**Asc83b, Asc83c**]; and also to some extext in the proof of the Quasithin Theorem 3.0.1 in [**AS04b**].

However, the Quasithin work also extended the weak closure method, beginning in subsection E.3.3 in [**AS04a**]. We discussed some of that material in Section 3.3. Here we will merely summarize by saying that further inequalities are obtained among the parameters above, but now focusing on the relationship of $n(H)$ with a further parameter $w(G,V)$, defined to be the smallest w for which we have $W_w(T,V) \not\leq C_G(V)$.

B.9. Klinger-Mason analysis of bicharacteristic groups

The standard formulation of the Klinger-Mason Dichotomy, given below, follows easily from Theorem D in Klinger-Mason [**KM75**]. We are not sure if there is any derivation of it from Theorem D in the literature, so we supply one here.

THEOREM B.9.1 (Klinger-Mason Dichotomy). *Assume that G is a finite group of characteristic 2 type. If G is also of characteristic p type for some odd prime p, then $m_{2,p}(G) \leq 2$.*

In any case either G is quasithin, or there is an odd p such that G is not *of characteristic p type and $m_{2,p}(G) \geq 3$.*

PROOF. The second statement is an immediate consequence of the first. Theorem D proves that $m_{2,p}(G) \leq 2$ under our hypotheses and one additional hypothesis: that whenever P is a Sylow p-subgroup of G and A is a 2-subgroup of G normalized by P, then $A = 1$.

We complete the proof by verifying that either this extra hypothesis holds or $m_{2,p}(G) \leq 2$. Let $P \in Syl_p(G)$ and let B be an elementary abelian subgroup of P of maximal rank m. Then B contains every element of order p in $C_G(B)$, which is p-constrained as G is of characteristic p type by hypothesis. Let A be a B-invariant 2-subgroup of G. Then $C_A(b)$ is B-invariant for each $b \in B^\#$, so it lies in $O_{p'}(C_G(b))$ by the Bender-Thompson Lemma B.1.10. But $O_{p'}(C_G(b)) = 1$ as G is of characteristic p type, so $C_A(b) = 1$.

As b was arbitrary, either B is cyclic or $A = 1$ by coprime action B.1.8.2. So either $m_{2,p}(G) \leq 1$ or the extra hypothesis holds, completing the proof. □

We can easily prove a weaker result without Theorem D:

THEOREM B.9.2 (Weak Klinger-Mason Dichotomy). *Assume that G is of characteristic 2 type and also of characteristic p type for some odd prime p. Then either $m_{2,p}(G) \leq 1$ or G is of $GF(2)$ type.*

PROOF. Assume that $m_{2,p}(G) > 1$, and consider the set of $H \leq G$ which have $m_p(H) = m_{2,p}(G)$ and $F^*(H) = O_2(H)$. The set is nonempty since G has characteristic 2 type by hypothesis. Among such H, if possible further choose H with $Z(H) \neq 1$; and, subject to this constraint, choose H maximal. Let B be an elementary abelian p-subgroup of H of rank $m_{2,p}(G)$, and set $R := O_2(H) = F^*(H)$.

Suppose R is of symplectic type. Then $Z(R)$ is cyclic and $H \leq C_G(\Omega_1(Z(R)))$, hence equality holds by maximality in the choice of H. Thus G is of $GF(2)$ type in this case.

Hence, we may assume that R contains a noncyclic characteristic elementary abelian subgroup D. Then $\Omega_1(Z(H)) \leq \Omega_1(Z(R))D =: E$, and E is of rank at least 2 and normal in H. If B centralizes E, then consider $C := C_G(B)$ and $P := F^*(C)$. As G is of characteristic p type, P is a p-group (see Remark B.1.5). Since E acts faithfully on P and trivially on $B \leq Z(PE)$, the Thompson Dihedral Lemma B.1.7 implies that

$$PE \geq D_1 \times \cdots \times D_m \times B,$$

with each D_i dihedral of order $2p$ and with $m = m_2(E) \geq 2$. Let $\langle e_1 \rangle = E \cap D_1$. Then $C_G(e_1)$ contains $D_2 \times \cdots \times D_m \times B$. Hence

$$m_p(C_G(e_1)) \geq (m-1) + m_p(B) \geq 1 + m_p(B) > m_{2,p}(G),$$

a contradiction.

Hence B does not centralize E. By Maschke's Theorem (e.g. [**GLS96**, 9.5]), E is the sum of $C_E(B)$ and at least one nontrivial irreducible B-submodule. Let E_1 be one such submodule, and let $F := C_E(B) \oplus E_1$. Then $m_2(F) \geq 2$, with equality only if $p = 3$ and $C_E(B) = 1$. By Schur's Lemma (e.g. [**GLS96**, 9.2]) if we set $B_1 := C_B(E_1) = C_B(F)$, then B/B_1 is cyclic. Now consider $C_1 := C_G(B_1)$. As before, if $P_1 := O_p(C_1)$, then F acts faithfully on P_1, whence

$$P_1F \geq D_1 \times \cdots \times D_r \times B_1,$$

with each D_i dihedral of order $2p$ and with $r = m_2(F) \geq 2$. Set $F_1 := F \cap D_1$. Then

$$m_p\big(C_G(F_1)\big) \geq (r-1) + m_p(B_1) = (r-2) + m_p(B) = (r-2) + m_{2,p}(G);$$

so $r = 2$, and $m_p\big(C_G(F_1)\big) = m_{2,p}(G)$. We saw that $r = 2$ implies $p = 3$ and $C_E(B) = 1$. So $\Omega_1\big(Z(H)\big) \leq C_E(B) = 1$, and we get $Z(H) = 1$. On the other hand, we have $Z\big(C_G(F_1)\big) > 1$ and $m_p\big(C_G(F_1)\big) = m_{2,p}(G)$; this contradicts the choice of H with $Z(H) \neq 1$ if possible, and so completes the proof. □

REMARK B.9.3. First, the proof uses the characteristic p type hypothesis that $F^*(H) = O_p(H)$ only for the p-locals H given by $C_G(B)$ and $C_G(B_1)$, such that $B_1 \leq B$ and $m_p(B) = m_{2,p}(G)$. Lemma B.1.6 (with Lemma B.1.4) therefore implies that the Weak Klinger-Mason Dichotomy remains true if the characteristic p type assumption is replaced by the condition that for any $a \in G$ of order p,

$$\text{(B.9.4)} \qquad m_p\big(C_G(a)\big) \geq m_{2,p}(G) \text{ implies } F^*\big(C_G(a)\big) = O_p\big(C_G(a)\big).$$

Second, under the hypotheses of the Weak Klinger-Mason Dichotomy, Klinger and Mason continue the analysis in the case when G is of $GF(2)$ type. It is not difficult to show that $p \leq 5$, and indeed, that either $m_{2,p}(G) \leq 2$ or $p = 3$. With additional effort, they are able to eliminate the final case where $p = 3$, thereby obtaining Theorem D. There is an error in one subcase of their analysis, which was corrected by I. Korchagina in her Ph.D. thesis. ◇

Notice that even when $m_{2,p}(G) \geq 3$, we do not necessarily get a p-component in some p-local H, as desired in Step 3 of Strategy 1.1.1. It might be the case that $O_{p'}(H) \neq 1$ but $L_{p'}(H) = 1$ so that there are no p-components. But when we have $m_p(G) \geq 3$, Theorem B.3.25 then leads to a contradiction, unless $\Lambda_2^p(G)$ is disconnected, assuming that G is $\mathcal{K}$-proper. Hence this situation could be regarded as part of the uniqueness case in the sense of Step 1 of Strategy 1.1.1.

With this point of view, if we are not in the uniqueness case, we have the following trichotomy, which refines the basic Dichotomy Theorem 0.3.10, and is the point of departure for the deeper Gorenstein-Lyons Trichotomy 5.1.2.

THEOREM B.9.5 (Klinger-Mason Trichotomy Lemma). *Let G be a finite $\mathcal{K}$-proper nonabelian simple group which is not quasithin. Assume that p is an odd prime such that $m_{2,p}(G) \geq 3$, and $\Lambda_2^p(G)$ becomes connected when isolated vertices are removed. Then one of the following holds:*

(1) G is of GW type; or

(2) G is of $GF(2)$ type; or

(3) G is of characteristic 2 type and of p-component type.

PROOF. By the Dichotomy Theorem 0.3.10, either G is of characteristic 2 type, or G is of GW type, and in the latter case (1) holds.

Hence we may assume that G is of characteristic 2 type—but not quasithin. Thus $m_{2,p}(G) \geq 3$ for some odd prime p, and by hypothesis we may choose such a p so that the non-isolated vertices of $\Lambda_2^p(G)$ form a connected graph Λ.

Thus Λ is G-invariant and its vertices include all subgroups of order p^2 in any E_{p^3}-subgroup of G. Now the Signalizer Functor Theorem B.3.25 implies that $O_{p'}\big(C_G(a)\big) = 1$ for all $a \in G$ of order p such that $m_p\big(C_G(a)\big) \geq 3$. If for some such element a we have $E\big(C_G(a)\big) \neq 1$, then G is of p-component type, so that (3) holds.

Hence we may assume that $E(C_G(a)) = 1$ and thus $F^*(C_G(a)) = O_p(C_G(a))$ for all a of order p lying in some subgroup $A \cong E_{p^3}$ of G.

With Remark B.9.3, we see that the Weak Klinger-Mason Dichotomy B.9.2 holds. As by hypothesis we have $m_{2,p}(G) \geq 3$, we conclude that G is of $GF(2)$ type, so that (2) holds in this final case. $\square$

B.10. Some details of the proof of the Uniqueness Case Theorem

In this section, we collect some comments on certain details from the proof of the Uniqueness Case Theorem 8.0.2 which are perhaps obscure in [**Asc83b, Asc83c**], due for example to misprints, omission of details, results implicitly assumed from the literature, etc.

REMARK B.10.1 (Some details in the proof of 21.21, i.e. Lemma 8.5.15). Most of the conclusions of Lemma 8.5.15 are in fact provided by 13.6 in [**Asc83c**]. Here version (b) of the hypothesis of 13.6 holds because B is invariant under a Sylow p-subgroup P of the group $S(p)$ in 8.5.15.3. However, we also need to verify the Hypothesis 13.1 for Section 13.

We first check that 13.1.4 holds, namely $m_p(P) \leq 1$: for otherwise B would not be inverted in M by 8.5.14.4, whereas $P \in \mathcal{U}(M)$ by 6.6.1—so that B is inverted in M by 21.10. Next 13.1.4 implies $\sigma(M) = \sigma^1(M)$—so that it is not in fact necessary to *separately* assume this, as is done at the start of Section 13. To see this, note that $m_p(P) \leq 1$ and $m_p(M) \geq 3$ imply that $K(p) \neq 1$. In particular M is nonsolvable; thus we avoid the exceptional cases (2) and (3) in Definition 2.7.1, and hence $\sigma(M) = \sigma^1(M)$.

It is now fairly routine to use the restrictions on K in 8.5.14.2, together with various detailed properties of K in Section 21, to establish the remaining conditions in 13.1. For example, use 8.5.14.2 along with the list in 21.16 (and 21.9) to get 13.1.1. Use 8.5.14.2 (and also 8.5.14.1, applied in quoting 21.18) to get 13.1.2. In the situation of 13.1.3, 8.47 shows that $N_G(X) \not\leq M$; with $X \notin \mathcal{U}(M)$ by 21.1 since $X \leq N_G(T)$. Finally 21.19 gives 13.1.5. $\diamondsuit$

REMARK B.10.2 (Some details for the proof of 21.22, i.e. Lemma 8.5.17). We begin with some preliminary remarks (which are essentially implicit at the start of the proof in [**Asc83c**]): We may assume that:

$$H_0 \cap H_1 \not\leq M; \text{ in particular, each } H_i \not\leq M.$$

Now from [**AS04a**, A.1.6], and since $T \leq H_0 \cap H_1 \leq H_i$ with $F^*(H_i) = O_2(H_i)$, we obtain that also $F^*(H_0 \cap H_1) = O_2(H_0 \cap H_1)$. Therefore $H_0 \cap H_1 \in \mathcal{H}$. Thus we may assume that $L_0 \in \mathcal{L}_*$ lies in $H_0 \cap H_1$. (This choice, and that of the subgroup $B \in \mathcal{B}_1(L_0)$ in our main discussion in Chapter 8, will be required only fairly late in the proof.)

Next we deduce several consequences of the factorization of X in the hypothesis. In doing so, we are re-ordering part of the proof given in [**Asc83c**]. This is needed in order to establish the hypothesis for 7.7 prior to its use.

We restrict attention to the case where X is solvable. Let Y be a Hall $2'$-subgroup of X; and set $X_i := X \cap H_i$ and $Y_i := Y \cap H_i$. From $X = X_1X_2$ we get $Y = Y_1Y_2$, and further:

(B.10.3) $\qquad Y_i \notin \mathcal{U}(M)$ when Y_i is cyclic or E_{q^2} for some q.

For otherwise using 21.1 as before we would get $H_i \leq !\mathcal{M}(X_i) = M$, contrary to $H_i \not\leq M$.

Now suppose that some $p \in \sigma(M)$ also lies in $\pi(X)$. We obtain:

$$m_p(Y_i) \leq 1;$$

as otherwise $Y_i \in \mathcal{U}(M)$ using Theorem 8.1.3, contrary to (B.10.3). Since we have $m_p(M) \geq 3$, this implies that M is not solvable, as otherwise we could take M itself in the role of "X". In particular:

$$\sigma(M) = \sigma^1(M).$$

Therefore:

(B.10.4) Any solvable X is a $\sigma(M)'$-group.

For $m_p(Y_i) = 1$ forces $Y_i \in \mathcal{U}(M)$ as M is now strongly p-embedded, again contrary to (B.10.3).

We turn to the main part of the proof. First we will restrict the possibilities for $K \in \mathcal{C}(M)$—to the list excluded in part (2) of Hypothesis 8.5.14. This is accomplished by verifying that the other parts of that Hypothesis do hold, whereas the conclusions of Lemma 8.5.15 do not hold. We begin that verification.

The initial argument in [**Asc83c**] assumes that $\overline{M}^{\infty}$ does not centralize $F(\overline{M})$. Then $F(\overline{M})$ is not cyclic (Remark 8.5.13) so we may choose Y to be a noncyclic critical subgroup (cf. Lemma B.1.3) of a Sylow q-subgroup of $O_{2,F}(M)$ for some q. Notice $q \notin \sigma(M)$ by (B.10.4), and we saw that $\sigma(M) = \sigma^1(M)$ during its proof; these facts are needed to establish the hypothesis of 7.7, since its proof in turn quotes 6.9. Then 7.7.4 shows that $\Gamma_{1,Y}(G) \leq M$. Therefore $m_q(Y_i) \leq 1$—for we may mimic the argument in the case $m_p(Y_i) \leq 1$, since we can now replace Theorem 8.1.3 for p by Theorem 8.1.4 for q. Indeed we even get that Y_i has order q with $Y \cong E_{q^2}$. Since T normalizes both Y_i, $SL_2(q)$ is not a subgroup of $\mathrm{Aut}_M(Y)$, as in 6.10—so the hypothesis of that lemma must fail, giving some Y_i in $\mathcal{U}(M)$, contrary to (B.10.3).

Thus $[\overline{M}^{\infty}, F(\overline{M})] = 1$, as required for 8.5.14.1. This is essentially why 8.5.14.2 must fail. For note by Remark 8.5.13 that $TC_M(E(\overline{M}))$ is solvable, and hence is a $\sigma(M)'$-group by (B.10.4); thus the conditions on B in parts (3) and (4) of 8.5.14 are vacuously satisfied. On the other hand, neither of the conclusions in (3) of 8.5.15 can hold: since for $p \in \sigma(M)$, there is a T-invariant subgroup Y in either $S(p)$ or a $(2^m + 1)$-torus in $U_3(2^m)$ (in view of (4) there), contrary to (B.10.4).

So $K \in \mathcal{C}(M)$ must be one of the possibilities excluded in 8.5.14.2. The resulting possibilities for $I := \langle K^M \rangle$, which are mostly of Lie type and rank 2, are then successively eliminated; we sketch only certain parts of the proof: Recall that our $B \in \mathcal{B}_1(L_0)$ is contained in a Cartan subgroup of $L_0 \leq H_0 \cap H_1$, so that B permutes with T and $B \leq M$. Hence we may choose a Cartan subgroup Y of I (when $\overline{K} \cong J_1$, use the analogous subgroups of order 7), which is normalized by B. Then Y also normalizes L_0, since L_0 is a parabolic of each H_i, and $Y = Y_1 Y_2$ where $Y_i \leq H_i$ is T-invariant. Elementary arguments using these properties show that the action of B on $\overline{I}$ (which is faithful by the extension of 21.3 mentioned in a footnote in our discussion after Remark 8.5.11) is by *inner* automorphisms, so that B is isomorphic to a subgroup of C, and in particular $\pi(B) \subseteq \pi(C)$.

Suppose now that there is some odd divisor s of n—where we recall $n(L_0)$ and $m(K)$, with n dividing m since $\pi(B) \subseteq \pi(C)$ (cf. Remark 8.5.12). Let $r \in \pi(B)$

be a divisor of $2^s - 1$. Then earlier analysis in 21.8 and 21.15 (and in 21.17.3 for the case $\overline{I} \cong J_1 \times J_1$) shows that the r-part $\overline{B}_r$ lies in $O^{r'}(\overline{M}) = \overline{I}$, so that $B_r \leq Y$. But as $m_r(Y) = 2$ since $\overline{K}$ is of rank 2, we must then have $Y = B_rY_i$ for some i. So as $B_r \leq B \leq L_0 \leq H_0 \cap H_1$, we get $Y \leq H_i$. However $Y \in \mathcal{U}(M)$ again using 21.8, 21.15, and 21.17.3, so using 21.1 as before, we again obtain the contradiction that $H_i \leq !\mathcal{M}(Y) = M$.

Thus n is a power of 2, and so $3 \in \pi(B)$. Therefore $\overline{I} \cong L_2(2^m) \times L_2(2^m)$, with $n|m$. This final possibility is eliminated fairly easily, if a bit more technically: some arguments are similar to those in the previous paragraph, and there are further contradictions to 21.1 in the process; along with appeals to results from Sections 6 and 7, as well as to some specialized results from Section 8. ◊

REMARK B.10.5 (Some details for the proof of 23.3 in [**Asc83c**]). Recall from Remark 8.5.24 that $N \in \mathcal{M}(T) - \{M\}$, $N_1 = N \cap M$, and $U_1 = \langle Z^M \rangle$. We assume that $U_1 < U$. The irreducibility of N_1 on U_1 in Remark 8.5.24 gives irreducibility of M on U. We seek to establish the setup of Lemma 8.5.17, as described in Remark 8.5.16; so for $i := 0$ and 1, we want L_0T to normalize the generalized weak closures $W_i := W_i(T, U)$ of U in T.

We mention first that $U < W_0$—this is in fact required as a hypothesis for the applications below of 8.6 and 8.7 in [**Asc81e**]. To obtain $U < W_0$, use Timmesfeld's work [**Tim75b**] on weakly closed TI-sets: here U cannot be weakly closed—since as in (8.2.2) G is not on Timmesfeld's list.

Furthermore U is central in W_0 by (8.5.21); so W_0 is weakly closed in $C_T(U)$, and then from the usual Frattini factorization (8.3.3) we see that:

$$\overline{M} = \overline{N_M(W_0)}.$$

To get $W_0 \trianglelefteq L_0T$, we work with $E := \langle Z^{L_0} \rangle$. As L_0 is a minimal parabolic of L, E is a natural module for $L_0T/C_{L_0T}(E)$, so $a(L_0T, E) = n(L_0) = m(U_1) < m(U)$ as in 8.5.24. Suppose first that $U \leq O_2(L_0T)$; then we have the setup of 8.4 in [**Asc81e**]. Now in the second case of 8.4.1 in [**Asc81e**], $a(L_0T, E) \geq m(U)$; so instead the first case holds, giving $W_0 \leq C_T(E)$—which equals $O_2(L_0T)$ as in (8.5.19), so that W_0 is weakly closed in $O_2(L_0T)$, and hence is normal in L_0T. So assume that $U \not\leq O_2(L_0T)$. Then 8.6.1 of [**Asc81e**] establishes a "factorization" of the form $L_0T = \langle (L_0T \cap M), N_{L_0T}(W_0) \rangle$. Hence as $L_0T \cap M$ is contained in the Borel subgroup of L_0T, we conclude from (8.5.9) that W_0 is normal in L_0T here as well. So in either case, we get $L_0T \leq N_G(W_0)$, and then as $L_0T \not\leq M$, we have $N_G(W_0) \in \mathcal{H}$.

We next establish the bound $m(U) \geq 2n$. We can (using 20.5) embed L_0 in $L_1 \in \mathcal{L}(N_G(W_0))$, and then $L_1 \leq L_2 \in \mathcal{L}^*$. Now L_2 can also play the role of the "L" in the results established so far; in particular U_1 is determined by the subgroup L_0 in either L_2 or L. So applying 8.5.24 to $N_2 := !\mathcal{M}(L_2T)$ in the role of "N", we have:

$$U_1 = \langle Z^{N_2 \cap M} \rangle = \langle Z^{L_2 \cap M} \rangle = \langle Z^{L_1 \cap M} \rangle = \langle Z^{L_0 \cap M} \rangle = \langle Z^{L \cap M} \rangle = \langle Z^{N \cap M} \rangle;$$

where we recall that the Cartan subgroup in the Borel subgroup $L_0 \cap M$ induces the irreducible action of the maximal parabolic $L \cap M$ on the root space U_1 of rank n in V. Next, the argument on p. 540 of [**Asc83c**] based on "$N(WL_1)$" appears to involve a misprint, in assuming that L_1 is normal in $N_G(W_0)$—instead of just subnormal as in the definition of $L_1 \in \mathcal{L}(N_G(W_0))$. So in the remainder of this paragraph, we present an alternative argument provided by Aschbacher (in July 2008): We set $Y := N_M(W_0)$; we saw earlier that $\overline{M} = \overline{Y}$, so it follows that $U = \langle Z^Y \rangle$. Set $V_1 := \langle Z^{L_1} \rangle$, so that $U_1 = \langle Z^{L_0 \cap M} \rangle \leq U \cap V_1$. The module V_1 for the parabolic L_1 over L_0 in L_2 is natural, with root space U_1 (where $U_1 = V_1$ in case $L_0 = L_1$). Set $Y_1 := N_Y(L_1)$. Assume first there is $y \in Y_1 \setminus N_{Y_1}(V_1)$. Then the product $V_1 V_1^y$ is direct using Clifford's theorem (see A.1.21) applied to $Y_1 L_1$; hence U contains $U_1 \times U_1^y$ of rank at least $2n$, and we are done. So we may assume instead that Y_1 normalizes V_1. Since the natural module V_1 is the direct sum of L_1-conjugates of U_1, while conjugates under $L_1 \cap M$ lie in U_1 from the equations displayed above, we see, since U is a TI-set with $N_G(U) = M$, that $U_1 = U \cap V_1$. So as $N_{Y_1}(V_1)$ acts on $U \cap V_1$, we have $Y_1 = N_{Y_1}(V_1)$ acting on U_1. We saw that $U = \langle Z^Y \rangle$; so as $U_1 < U$, there is some $y \in Y \setminus N_Y(U_1)$. Now y acts on the group $M_1 := M \cap \langle L_1^Y \rangle \leq N_M(W_0) = Y$. Furthermore as $L_1 \in \mathcal{L}(N_G(W_0))$ is by definition a product of $2'$-components of $N_G(W_0)$, we see that $\langle L_1^Y \rangle$ is a product of conjugates of L_1 (commuting modulo a 2-group); so $M_1 \leq N_Y(L_1) = Y_1$, and hence M_1 normalizes U_1 by our remark above. And M_1 contains $L_0 \cap M$ which is irreducible on U_1, so that M_1 is also irreducible on U_1. Hence the product $U_1 U_1^y$ is direct by Clifford's theorem applied to $\langle y \rangle M_1$; so again U contains $U_1 \times U_1^y$ of rank at least $2n$, completing the proof in this remaining case.

We next prove $W_1 \trianglelefteq L_0 T$. Much as in the earlier argument for W_0, when we have $U \leq O_2(L_0 T)$, we use 8.4.1 of [**Asc81e**]: this time we have the inequalities we get are $a(L_0 T, E) \geq m(U) - 1 \geq 2n - 1 > n$—using the bound $m(U) \geq 2n$ from the previous paragraph along with $n > 2$ by (8.5.8). Therefore W_1 centralizes E, and then W_1 is normal in $L_0 T$. When $U \not\leq O_2(L_0 T)$, this time we apply 8.7[13] in [**Asc81e**], to get the analogue for W_1 of the earlier statement for W_0 from 8.6 of [**Asc83c**]; which similarly leads to W_1 normal in $L_0 T$.

The completion of the proof via Lemma 8.5.17 is now just a variant of the argument in the final paragraph of Remark 8.5.23: We have $L_0 T \leq H_0 \cap H_1$, where we now use $H_0 := N_G(W_0)$ along with $H_1 := N_G(Z(W_1)) \geq N_G(W_1)$. Again the conclusion of the Lemma gives $H_0 \cap H_1 \leq M$, contradicting $L_0 \not\leq M$.

So it remains to establish the hypothesis of the Lemma; namely we must show for $T \leq X \leq M$ with $n(X) = 1$ that $X = (X \cap H_0)(X \cap H_1)$. Again this would follow from the conclusion of Proposition B.8.6—but this time, we do not have the hypothesis $m(\mathrm{Aut}_M(U), U) \geq i + k = 0 + 1 = 1$ which is required there. So instead we assume that we have a minimal obstruction to our weak closure factorization, of the type denoted by $\mathcal{O}_1$ in [**Asc83c**] and [**Asc81e**]. (The proof on p. 540 of [**Asc83c**] refers just to those in 6.9 of [**Asc81e**]—but the underlying argument requires the reduction in 6.11.2 there, which in effect refers back to the argument at 4.5.2. Also (8.5.20) is implicitly used for the hypothesis of these arguments.) Indeed since U is a TI-set, properties such as 7.1 in [**Asc81e**] rather easily give the symmetry between M and some M^g required in such obstructions. As a result, our obstruction provides a hyperplane $U^g \cap M$ of U, which induces the full group of transvections on U with axis $U \cap M^g$. As M is irreducible on U with $m(U) \geq 2n > 2$, we conclude (use McLaughlin's Theorem A.6.1) that $\overline{M}$ is the full linear group $GL(U)$, which is simple. But we saw that $\overline{M} = \overline{N_M(W_0)}$, and we have $L_1 \in \mathcal{L}(N_G(W_0))$ subnormal[14] in $N_G(W_0)$ by definition; and so the subgroup $\overline{L_1 \cap M}$ is subnormal in $\overline{M}$, and is nontrivial since $L_0 \cap M$ is irreducible on U_1. This contradiction to simplicity of $\overline{M}$ completes the proof. ♢

[13]We mention this is misprinted as 8.6 in [**Asc83c**].

[14]This is misprinted as "normal" in [**Asc83c**].

REMARK B.10.6 (Some details for the proof of 23.4 in [**Asc83c**]). We assume that $Z < U$; then $S := C_T(U) < T$. Let $N_1 \in \mathcal{M}(T) - \{M\}$. Then $N_1 \in \mathcal{H}$; and setting $X := \langle \mathcal{L}(N_1) \rangle$, it follows from (8.5.4) that $N_1 = (N_1 \cap M)X$, and so $X \neq 1$.

Hence there is $L_1 \in \mathcal{L}(N_1)$. Then L_1 contains some $L_0 \in \mathcal{L}_*$, and lies in some $L \in \mathcal{L}^*$. So we may apply Remark 8.5.24 to $N := !\mathcal{M}(LT)$. We conclude since $U = U_1$ by (8.5.25) that $\mathrm{Aut}_N(U) = \overline{N \cap M}$ is cyclic of odd order and irreducible on U—with action induced by $L_0 \cap M \leq L_1 \leq N$. Therefore $\overline{L_1 \cap M}$, and also $A := \overline{X \cap M}$, are of odd order. In particular $T \cap X \leq C_T(U) = S$. Thus we get $S = T \cap X C_{N_1}(U)$; and as $C_{N_1}(U)$ is normal in $N_1 \cap M$, and $N_1 = (N_1 \cap M)X$, we conclude that $X C_{N_1}(U)$ is normal in N_1. It follows that S is strongly closed in T with respect to N_1. Then S is also strongly closed in T with respect to M, as $S = T \cap C_M(U)$ with $C_M(U)$ normal in M.

Next as A is cyclic of odd order and irreducible on U, by Schur's lemma we get $E := End_{GL(U)}(A)$ is contained in the odd-order multiplicative group of the field of size $|U|$, and $N_{GL(U)}(A)/E$ is a cyclic group of field automorphisms of order dividing $m(U)$. In particular since T normalizes A, T/S is cyclic. Then from the strong closure, it follows from a standard transfer result [**GLS96**, 15.15] that $T \cap O^2(N_1)$ and $T \cap O^2(M)$ are proper in T; indeed both must lie in the unique subgroup T_0 of T which is of index 2 and contains S.

To finish, the proof at this point quotes a result of McBride on control of fusion, in a preprint numbered [24] in [**Asc83b**]—which was never in fact published (and does not seem to survive). A modern treatment by Aschbacher (July 2008) is indicated below as Theorem B.10.7. To apply it, recall by (8.2.2) that G is not one of the known groups in conclusions (2)–(5) of that result. So conclusion (1) holds—and in particular, fusion in T is controlled by $\mathcal{M}(T)$. Hence $\mathcal{M}(T)$ also controls transfer. So since N_1 was arbitrary in $\mathcal{M}(T) - \{M\}$ in the previous paragraph, we conclude that $T \cap O^2(G) \leq T_0 < T$ and so $O^2(G) < G$, contrary to the simplicity of G. ♢

Aschbacher's Theorem 1 in [**Asc09**] is in fact phrased in the language of fusion systems; it has the following consequence (Theorem 2 there) for finite groups:

THEOREM B.10.7. *Assume G is of characteristic 2 type, and all 2-locals G are $\mathcal{K}$-groups. Fix some $T \in Syl_2(G)$. Then one of the following holds:*

(1) The subgroups $N_G(U)$, where $C_T(U) \leq U \trianglelefteq T$ and $U = O_2(N_G(U))$, control fusion in T.
(2) T is dihedral and $G \cong L_2(p)$, where p is a Fermat or Mersenne prime.
(3) T is semidihedral of order 16, and $G \cong M_{10}$, M_{11}, or $L_3(3)$.
(4) $|T| = 32$, and $G \cong Aut(L_3(3))$.
(5) $|T| = 2^7$, and $G \cong J_3$.

REMARK B.10.8 (Some details for the proof of 23.6 in [**Asc83c**]). For case (3) with $p = 3$ in Definition 2.7.1, we write $P = YJ(P)$ for a Sylow 3-subgroup of M, with Y of order 3. Further facts about M appear in 6.11 of [**Asc83c**]. The cyclic subgroup $O_{3'}(M) \cap C_T(Y)$ in 6.11.1 is described incorrectly as "$Z(C_T(Y))$", in the argument during the proof of 23.6 which shows that:

$$C_Z(Y) \text{ is cyclic.}$$

It follows that $[Z, Y] \neq 1$—since $m(Z) > 1$.

Next, the proof of the subsequent assertion "$N_{J(P)}(T) = 1$" seems to require some case analysis which is not included there; we now provide further details supplied by Aschbacher in July 2008. We may choose an element $b \in B \backslash C_B(Z)$ of some prime order r; then we obtain $C_Z(b) = 1$ from the action on the root space $U_1 = Z$ in Remark 8.5.24. Furthermore we get $C_{J(P)}(b) = 1$ from Lemma 8.5.28.1—since nontrivial subgroups of $J(P)$ are in $\mathcal{U}(M)$ by 6.11.3. In particular this shows we have $r > 3$, since $Z(P) \leq J(P)$. Now a Frattini argument shows that $M = C_M(Z)N_M(T)$; hence $3, r \in \pi(N_M(T))$, and we may take b in some Hall $\{3, r\}$-subgroup K of $N_M(T)$. Since b normalizes $C_K(Z)$, by another Frattini argument it acts on some Sylow 3-subgroup there, which we may take to lie in $J(P)$ since Y does not centralize Z; and then we see that b is fixed-point-free on $C_{J(P)\cap K}(Z)$. Suppose first that $C_{J(P)\cap K}(Z) \neq 1$. For $1 \leq d \leq 3 = m(J(P))$, a fixed-point-free element of odd order r in $GL_d(3)$ occurs only for $d = 3$ and $r = 13$; then $J(P) \leq C_K(Z)$, where in particular 12 divides $m(Z)$ from the fixed-point-free action of b. Furthermore Y cannot centralize b, since then b would normalize characteristic subgroups of P such as $Z(P)$ which are proper in $J(P)$, whereas b of order 13 is irreducible on $J(P)$. Hence using the structure in Remark 8.5.27, Y must act as a field automorphism on $\langle b \rangle$—so that $m(C_Z(Y))$ is divisible by $12/3 = 4$, contrary to $C_Z(Y)$ cyclic. So we may suppose instead that $C_{J(P)\cap K}(Z) = 1$, and hence $C_K(Z)$ is a $3'$-group. Now by 6.11.5, $[P, P]$ centralizes Z, so $1 = [P, P] \cap K = [P, P] \cap N_M(T)$. However, we saw that $N_P(T) \not\leq J(P)$; thus $N_{J(P)}(T) = 1$—since otherwise $[N_{J(T)}(T), N_P(T)]$ contains a nontrivial element of $[P, P] \cap N_M(T)$.

We can now finish the proof of this case as in 23.6: As $N_{J(P)}(T) = 1$, we may assume $Y = N_P(T)$, so $J(P)$ is Sylow in $C_M(Z)$ by the earlier Frattini factorization. Again as $\mathrm{Aut}_M(Z)$ is a group of semilinear transformations in 8.5.27 and $C_Z(Y)$ is cyclic, either $Z = [Z, Y]$ with n even, or $C_Z(Y)$ is of order 2 with $n = 3$. In the first case, $Y \leq B$—contrary to $r > 3$. In the other case where $n = 3$, B is of order 7—and so must centralize $J(P)$ of rank 3, contrary to $C_{J(P)}(b) = 1$. This completes the elimination of case (3) with $p = 3$ in Definition 2.7.1. (But also see the alternative argument at the end of this Remark.)

For case (2) with $p \geq 5$ in Definition 2.7.1, the elimination also involves contradictions to Lemma 8.5.28.1, but is easier: This time further information is given in 6.13. We have that $P_0 \times P_1$ is in a Sylow p-subgroup P; where P_0 is nontrivial on Z, from 6.13.1. We can then choose so that $b \in B \setminus C_G(Z)$ lies in P; and since P_1 is strongly closed[15] in P by 6.13.2, it is normal and so contains some $1 \neq X \leq Z(P)$—where $X \in \mathcal{U}(M)$ by 6.13.3, again contrary to Lemma 8.5.28.1.

As mentioned above, we also present a shorter treatment of case (3) with $p = 3$ in Definition 2.7.1, supplied by Aschbacher (July 2008): We have $M^* := M/C_M(Z)$ of odd order, with $1 \neq Y^* \leq P^*$; and we claim that $O^3(M^*) < M^*$. Suppose first that the quotient P^* is abelian. Then as $[P, P] = [J(P), Y]$ is a hyperplane of $J(P)$, P^* is elementary abelian of rank at most 2; so $\mathrm{Aut}(P^*)$ is a $\{2, 3\}$-group, and hence P^* is central in $N_{M^*}(P^*)$—so $O^3(M^*) < M^*$ by Burnside's

[15]We mention that P is even abelian in the argument of Gorenstein and Lyons which defines this case—see the statement of II.3.4.i (and (c)(i) in the definition of almost strongly p-embedded on p. 8) in [**GL83**]. But seemingly this fact is not assumed or used in Aschbacher's work in [**Asc83b, Asc83c**]—though it can easily be established independently there.

transfer theorem B.2.8. Thus we may suppose instead that P^* is nonabelian, so that $[P,P]^* \neq 1$. Now M^* is of odd order and in particular 3-solvable; so the normal subgroup $Q^* := P^* \cap O_{3',3}(M^*)$ of P^* must contain its centralizer in P^*, so that $Q^* \geq [P,P]^*$. If $Q^* = P^*$, then as $J(P)$ is strongly closed in P by 6.11.2, standard transfer (e.g. [**GLS96**, 15.15]; cf. also Remark B.2.3) shows that $Y^* \not\leq O^3(M^*)$. Otherwise $Q^* < P^*$; and then the (nontrivial) image of P^* in $M^*/O_{3',3}(M^*)$ is abelian, so we get $O^3(M^*) < M^*$ just as in the argument in the case P^* abelian—completing the proof of the claim. In particular this implies $O^3(M) < M$; but since $\Gamma_{2,P}(G) \leq M$, B.2.7 says that M controls 3-fusion in G—giving $O^3(G) < G$, contrary to the simplicity of G. $\diamondsuit$

References and Index

References used for both GW type and characteristic 2 type

[AS76a] Michael Aschbacher and Gary M. Seitz, *Involutions in Chevalley groups over fields of even order*, Nagoya Math. J. **63** (1976), 1–91. MR MR0422401 (54 #10391)

[AS76b] ———, *On groups with a standard component of known type*, Osaka J. Math. **13** (1976), no. 3, 439–482. MR MR0435200 (55 #8161)

[AS78] M. Aschbacher and G. Seitz, *Correction to: "Involutions in Chevalley groups over fields of even order" (Nagoya Math. J.* **63** *(1976), 1–91)*, Nagoya Math. J. **72** (1978), 135–136. MR MR514895 (80b:20058)

[AS81] Michael Aschbacher and Gary M. Seitz, *On groups with a standard component of known type. II*, Osaka J. Math. **18** (1981), no. 3, 703–723. MR MR635729 (83a:20018)

[Asc74] Michael Aschbacher, *Finite groups with a proper 2-generated core*, Trans. Amer. Math. Soc. **197** (1974), 87–112. MR MR0364427 (51 #681)

[Asc93] ———, *Simple connectivity of p-group complexes*, Israel J. Math. **82** (1993), no. 1-3, 1–43. MR MR1239044 (94j:20012)

[Ben71] Helmut Bender, *Transitive Gruppen gerader Ordnung, in denen jede Involution genau einen Punkt festläßt*, J. Algebra **17** (1971), 527–554. MR MR0288172 (44 #5370)

[Ben75] ———, *Goldschmidt's 2-signalizer functor theorem*, Israel J. Math. **22** (1975), no. 3-4, 208–213. MR MR0390056 (52 #10882)

[BS04] Curtis D. Bennett and Sergey Shpectorov, *A new proof of a theorem of Phan*, J. Group Theory **7** (2004), no. 3, 287–310. MR MR2062999 (2005k:57004)

[Cur65] Charles W. Curtis, *Central extensions of groups of Lie type*, J. Reine Angew. Math. **220** (1965), 174–185. MR MR0188299 (32 #5738)

[FT63] Walter Feit and John G. Thompson, *Solvability of groups of odd order*, Pacific J. Math. **13** (1963), 775–1029. MR MR0166261 (29 #3538)

[GH69] Daniel Gorenstein and Koichiro Harada, *A characterization of Janko's two new simple groups*, J. Fac. Sci. Univ. Tokyo Sect. I **16** (1969), 331–406 (1970). MR MR0283075 (44 #308)

[GH74] ———, *Finite groups whose 2-subgroups are generated by at most 4 elements*, American Mathematical Society, Providence, R.I., 1974, Memoirs of the American Mathematical Society, No. 147. MR MR0367048 (51 #3290)

[Gil76] Robert Gilman, *Components of finite groups*, Comm. Algebra **4** (1976), no. 12, 1133–1198. MR MR0430053 (55 #3060)

[GL82] Daniel Gorenstein and Richard Lyons, *Signalizer functors, proper 2-generated cores, and nonconnected groups*, J. Algebra **75** (1982), no. 1, 10–22. MR MR650406 (83i:20017)

[Gla66] George Glauberman, *Central elements in core-free groups*, J. Algebra **4** (1966), 403–420. MR MR0202822 (34 #2681)

[Gla76] ———, *On solvable signalizer functors in finite groups*, Proc. London Math. Soc. (3) **33** (1976), no. 1, 1–27. MR MR0417284 (54 #5341)

[Gol72a] David M. Goldschmidt, *2-signalizer functors on finite groups*, J. Algebra **21** (1972), 321–340. MR MR0323904 (48 #2257)

[Gol72b] ———, *Solvable signalizer functors on finite groups*, J. Algebra **21** (1972), 137–148. MR MR0297861 (45 #6913)

[Gor69a] Daniel Gorenstein, *On finite simple groups of characteristic 2 type*, Inst. Hautes Études Sci. Publ. Math. (1969), no. 36, 5–13. MR MR0260864 (41 #5484)

[Gri72] Robert L. Griess, Jr., *Schur multipliers of the known finite simple groups*, Bull. Amer. Math. Soc. **78** (1972), 68–71. MR MR0289635 (44 #6823)

[Gri73] ———, *Schur multipliers of finite simple groups of Lie type*, Trans. Amer. Math. Soc. **183** (1973), 355–421. MR MR0338148 (49 #2914)

[Gri74] ———, *Schur multipliers of some sporadic simple groups*, J. Algebra **32** (1974), no. 3, 445–466. MR MR0382426 (52 #3310)

[Gri80] ———, *Schur multipliers of the known finite simple groups. II*, The Santa Cruz Conference on Finite Groups (Univ. California, Santa Cruz, Calif., 1979), Proc. Sympos. Pure Math., vol. 37, Amer. Math. Soc., Providence, R.I., 1980, pp. 279–282. MR MR604594 (82g:20025)

[Gri85] R. L. Griess, Jr., *Schur multipliers of the known finite simple groups. III*, Proceedings of the Rutgers group theory year, 1983–1984 (New Brunswick, N.J., 1983–1984) (Cambridge), Cambridge Univ. Press, 1985, pp. 69–80. MR MR817238 (87g:20027)

[GW75] Daniel Gorenstein and John H. Walter, *Balance and generation in finite groups*, J. Algebra **33** (1975), 224–287. MR MR0357583 (50 #10051)

[Har76] Koichiro Harada, *On the simple group F of order* $2^{14} \cdot 3^6 \cdot 5^6 \cdot 7 \cdot 11 \cdot 19$, Proceedings of the Conference on Finite Groups (Univ. Utah, Park City, Utah, 1975) (New York), Academic Press, 1976, pp. 119–276. MR MR0401904 (53 #5730)

[Jan69] Zvonimir Janko, *Some new simple groups of finite order. I*, Symposia Mathematica (INDAM, Rome, 1967/68), Vol. 1, Academic Press, London, 1969, pp. 25–64. MR MR0244371 (39 #5686)

[Suz62] Michio Suzuki, *On a class of doubly transitive groups*, Ann. of Math. (2) **75** (1962), 105–145. MR MR0136646 (25 #112)

[Suz64] ———, *On a class of doubly transitive groups. II*, Ann. of Math. (2) **79** (1964), 514–589. MR MR0162840 (29 #144)

[Tho68] John G. Thompson, *Nonsolvable finite groups all of whose local subgroups are solvable*, Bull. Amer. Math. Soc. **74** (1968), 383–437. MR MR0230809 (37 #6367)

[Tho70] ———, *Nonsolvable finite groups all of whose local subgroups are solvable. II*, Pacific J. Math. **33** (1970), 451–536. MR MR0276325 (43 #2072)

[Tho71] ———, *Nonsolvable finite groups all of whose local subgroups are solvable. III*, Pacific J. Math. **39** (1971), 483–534. MR MR0313378 (47 #1933)

[Tho73] ———, *Nonsolvable finite groups all of whose local subgroups are solvable. IV, V, VI*, Pacific J. Math. **48** (1973), 511–592, ibid. 50 (1974), 215–297; ibid. 51(1974), 573–630. MR MR0369512 (51 #5745)

[Tit62] J. Tits, *Groupes semi-simples isotropes*, Colloq. Théorie des Groupes Algébriques (Bruxelles, 1962), Librairie Universitaire, Louvain, 1962, pp. 137–147. MR MR0148667 (26 #6174)

References mainly for *GW* type (see [Gor82][Gor83] for full list)

[ABG70] J. L. Alperin, Richard Brauer, and Daniel Gorenstein, *Finite groups with quasi-dihedral and wreathed Sylow 2-subgroups.*, Trans. Amer. Math. Soc. **151** (1970), 1–261. MR MR0284499 (44 #1724)

[ABG73a] ———, *The extended* ZJ*-theorem*, Finite groups '72 (Proc. Gainesville Conf., Univ. Florida, Gainesville, Fla., 1972), North-Holland, Amsterdam, 1973, pp. 6–7. North-Holland Math. Studies, Vol. 7. MR MR0354855 (50 #7332)

[ABG73b] ———, *Finite simple groups of* 2*-rank two*, Scripta Math. **29** (1973), no. 3-4, 191–214, Collection of articles dedicated to the memory of Abraham Adrian Albert. MR MR0401902 (53 #5728)

[Alw79] L. Alward, *Standard components of type* $O^-(8,2)$, Ph.D. thesis, U. Oregon, 1979.

[Asc75a] Michael Aschbacher, *On finite groups of component type*, Illinois J. Math. **19** (1975), 87–115. MR MR0376843 (51 #13018)

[Asc77a] ———, *A characterization of Chevalley groups over fields of odd order*, Ann. of Math. (2) **106** (1977), no. 2, 353–398. MR MR0498828 (58 #16865a)

[Asc77b] ———, *A characterization of Chevalley groups over fields of odd order. II*, Ann. of Math. (2) **106** (1977), no. 3, 399–468. MR MR0498829 (58 #16865b)

[Asc80a] ———, *Correction to: "A characterization of Chevalley groups over fields of odd order. I, II" [Ann. of Math. (2)* **106** *(1977), no. 2, 353–398; MR* **58** *#16865a; ibid. (2)* **106** *(1977), no. 3, 399–468; MR* **58** *#16865b]*, Ann. of Math. (2) **111** (1980), no. 2, 411–414. MR MR569077 (81j:20026)

[Asc82a] ———, *A characterization of some finite groups of characteristic* 3, J. Algebra **76** (1982), no. 2, 400–441. MR MR661863 (83h:20022)

[Asc82b] ———, *The Tits group as a standard subgroup*, Math. Z. **181** (1982), no. 2, 229–252. MR MR674275 (84a:20025)

[Asc97] ———, *3-transposition groups*, Cambridge Tracts in Mathematics, vol. 124, Cambridge University Press, Cambridge, 1997. MR MR1423599 (98h:20024)

[Asc08] ———, *Standard components of alternating type centralized by a* 4*-group*, J. Algebra **319** (2008), 595–615.

[Ben70] Helmut Bender, *On the uniqueness theorem*, Illinois J. Math. **14** (1970), 376–384. MR MR0262351 (41 #6959)

[BS59] Richard Brauer and Michio Suzuki, *On finite groups of even order whose* 2*-Sylow group is a quaternion group*, Proc. Nat. Acad. Sci. U.S.A. **45** (1959), 1757–1759. MR MR0109846 (22 #731)

[BSW58] R. Brauer, Michio Suzuki, and G. E. Wall, *A characterization of the one-dimensional unimodular projective groups over finite fields*, Illinois J. Math. **2** (1958), 718–745. MR MR0104734 (21 #3487)

[Bur77] N. Burgoyne, *Finite groups with Chevalley-type components*, Pacific J. Math. **72** (1977), no. 2, 341–350. MR MR0457550 (56 #15755)

[CH81] Kai Nah Cheng and Dieter Held, *Finite groups with a standard component of type* $L_3(4)$*. I*, Rend. Sem. Mat. Univ. Padova **65** (1981), 59–75 (1982). MR MR653283 (83h:20020)

[CH85] ———, *Finite groups with a standard-component of type* $L_3(4)$*. II*, Rend. Sem. Mat. Univ. Padova **73** (1985), 147–167. MR MR799904 (87c:20037)

[DS81] Stephen L. Davis and Ronald Solomon, *Some sporadic characterizations*, Comm. Algebra **9** (1981), no. 17, 1725–1742. MR MR631885 (83a:20024)

[Ega80] Y. Egawa, *Standard components of type M_{24} and $\Omega^+(8,2)$*, Ph.D. thesis, Ohio State U., 1980, New MS to be submitted soon for publication.

[Ega81] Yoshimi Egawa, *Standard components of type M_{24}*, Comm. Algebra **9** (1981), no. 5, 451–476. MR MR606647 (82c:20028)

[EY82] Yoshimi Egawa and Tomoyuki Yoshida, *Standard subgroups of type* $2\Omega^+(8, 2)$, Hokkaido Math. J. **11** (1982), no. 3, 279–285. MR MR680816 (84b:20014)

[Fin75] Larry Finkelstein, *Finite groups with a standard component of type Janko-Ree*, J. Algebra **36** (1975), no. 3, 416–426. MR MR0379653 (52 #558)

[Fin76a] ———, *Finite groups with a standard component isomorphic to M_{23}*, J. Algebra **40** (1976), no. 2, 541–555. MR MR0414700 (54 #2795)

[Fin76b] ———, *Finite groups with a standard component isomorphic to* HJ *or* HJM, J. Algebra **43** (1976), no. 1, 61–114. MR MR0427450 (55 #482)

[Fin77a] ———, *Finite groups with a standard component isomorphic to M_{22}*, J. Algebra **44** (1977), no. 2, 558–572. MR MR0427451 (55 #483)

[Fin77b] ———, *Finite groups with a standard component of type J_4*, Pacific J. Math. **71** (1977), no. 1, 41–56. MR MR0444764 (56 #3112)

[Fin77c] ———, *Finite groups with a standard component whose centralizer has cyclic Sylow 2-subgroups*, Proc. Amer. Math. Soc. **62** (1977), no. 2, 237–241. MR MR0439928 (55 #12809)

[Foo78] Richard Foote, *Finite groups with maximal 2-components of type $L_2(q)$, q odd*, Proc. London Math. Soc. (3) **37** (1978), no. 3, 422–458. MR MR512020 (82k:20023)

[Fri77a] Franz J. Fritz, *On centralizers of involutions with components of 2-rank two. I*, J. Algebra **47** (1977), no. 2, 323–374. MR MR0450391 (56 #8686a)

[Fri77b] ———, *On centralizers of involutions with components of 2-rank two. II*, J. Algebra **47** (1977), no. 2, 375–399. MR MR0450392 (56 #8686b)

[FS79a] Larry Finkelstein and Ronald M. Solomon, *Standard components of type M_{12} and $\cdot 3$*, Osaka J. Math. **16** (1979), no. 3, 759–774. MR MR551587 (81g:20033)

[GHNS05] R. Gramlich, C. Hoffman, W. Nickel, and S. Shpectorov, *Even-dimensional orthogonal groups as amalgams of unitary groups*, J. Algebra **284** (2005), no. 1, 141–173. MR MR2115009 (2006f:20058)

[Gla68] George Glauberman, *A characteristic subgroup of a p-stable group*, Canad. J. Math. **20** (1968), 1101–1135. MR MR0230807 (37 #6365)

[Gla74] ———, *On groups with a quaternion Sylow 2-subgroup*, Illinois J. Math. **18** (1974), 60–65. MR MR0332969 (48 #11294)

[GMS78] Robert L. Griess, Jr., David R. Mason, and Gary M. Seitz, *Bender groups as standard subgroups*, Trans. Amer. Math. Soc. **238** (1978), 179–211. MR MR0466300 (57 #6180)

[Gol74] David M. Goldschmidt, *2-fusion in finite groups*, Ann. of Math. (2) **99** (1974), 70–117. MR MR0335627 (49 #407)

[Gol75] ———, *Strongly closed 2-subgroups of finite groups*, Ann. of Math. (2) **102** (1975), no. 3, 475–489. MR MR0393223 (52 #14033)

[Gom78a] Kensaku Gomi, *Finite groups with a standard subgroup isomorphic to* PSU(4, 2), Pacific J. Math. **79** (1978), no. 2, 399–462. MR MR531327 (82k:20024)

[Gom78b] ———, *Finite groups with a standard subgroup isomorphic to* $\mathrm{Sp}(4, 2^n)$, Japan. J. Math. (N.S.) **4** (1978), no. 1, 1–76. MR MR528868 (80h:20024)

[Gom80] ———, *Standard subgroups of type* $\mathrm{Sp}_6(2)$. *I, II*, J. Fac. Sci. Univ. Tokyo Sect. IA Math. **27** (1980), no. 1, 87–107, 109–156. MR MR573334 (81j:20024)

[Gri] Robert L. Griess, Jr., *Finite groups with standard component a 4-fold cover of M_{22}*, unpublished U Michigan preprint; now covered by [**HS08**].

[GS79a] Robert Gilman and Ronald Solomon, *Finite groups with small unbalancing 2-components*, Pacific J. Math. **83** (1979), no. 1, 55–106. MR MR555039 (81g:20028)

[GS79b] Robert L. Griess, Jr. and Ronald Solomon, *Finite groups with unbalancing 2-components of $\{\hat{L}_3(4), \mathrm{He}\}$-type*, J. Algebra **60** (1979), no. 1, 96–125. MR MR549100 (80k:20013)

[GW65a] Daniel Gorenstein and John H. Walter, *The characterization of finite groups with dihedral Sylow 2-subgroups. I*, J. Algebra **2** (1965), 85–151. MR MR0177032 (31 #1297a)

[GW65b] ———, *The characterization of finite groups with dihedral Sylow 2-subgroups. II*, J. Algebra **2** (1965), 218–270. MR MR0177033 (31 #1297b)

[GW65c] ———, *The characterization of finite groups with dihedral Sylow* 2*-subgroups. III*, J. Algebra **2** (1965), 354–393. MR MR0190220 (32 #7634)

[Har] Koichiro Harada, *Finite groups with standard component of type* M_{22}, unpublished Ohio State U preprint circa 1979.

[Har75] ———, *On finite groups having self-centralizing* 2*-subgroups of small order*, J. Algebra **33** (1975), 144–160. MR MR0354857 (50 #7334)

[Har77] Morton E. Harris, *Finite groups having an involution centralizer with a* 2*-component of dihedral type. II*, Illinois J. Math. **21** (1977), no. 3, 621–647. MR MR0480720 (58 #873b)

[Har80a] ———, *Finite groups having an involution centralizer with a* 2*-component of type* PSL(3, 3), Pacific J. Math. **87** (1980), no. 1, 69–74. MR MR590868 (81k:20024)

[Har81a] ———, *Finite groups having an involution centralizer with a* PSU(3, 3) *component*, J. Algebra **72** (1981), no. 2, 426–455. MR MR641335 (84d:20018)

[Har81b] ———, PSL(2, q) *type* 2*-components and the unbalanced group conjecture*, J. Algebra **68** (1981), no. 1, 190–235. MR MR604303 (83a:20021)

[HS77] Morton E. Harris and Ronald Solomon, *Finite groups having an involution centralizer with a* 2*-component of dihedral type. I*, Illinois J. Math. **21** (1977), no. 3, 575–620. MR MR0480719 (58 #873a)

[HS08] Koichiro Harada and Ronald Solomon, *Finite groups having a standard component* L *of type* $\hat{M}_{12}$ *or* $\hat{M}_{22}$, J. Algebra **319** (2008), 621–628.

[Hun72] David C. Hunt, *A characterization of the finite simple group* $M(22)$, J. Algebra **21** (1972), 103–112. MR MR0299673 (45 #8721)

[Hun73] ———, *A characterization of the finite simple group* $M(23)$, J. Algebra **26** (1973), 431–439. MR MR0322046 (48 #410)

[Lyo72] Richard Lyons, *A characterization of the group* $\mathrm{U}_3(4)$, Trans. Amer. Math. Soc. **164** (1972), 371–387. MR MR0286881 (44 #4088)

[Mac70] Anne R. MacWilliams, *On* 2*-groups with no normal abelian subgroups of rank* 3*, and their occurrence as Sylow* 2*-subgroups of finite simple groups*, Trans. Amer. Math. Soc. **150** (1970), 345–408. MR MR0276324 (43 #2071)

[Man79] J. Manferdelli, *Standard components of type* ·2, Ph.D. thesis, U. Calif. Berkeley, 1979.

[Miy79] Izumi Miyamoto, *Finite groups with a standard subgroup isomorphic to* $U_4(2^n)$, Japan. J. Math. (N.S.) **5** (1979), no. 1, 209–244. MR MR614698 (82i:20027)

[Miy80] ———, *Finite groups with a standard subgroup of type* $U_5(2^n)$, $n > 1$, J. Algebra **64** (1980), no. 2, 430–459. MR MR579070 (81g:20031)

[Miy82] ———, *Standard subgroups isomorphic to* ${}^2F_4(2^n)$, J. Algebra **77** (1982), no. 1, 261–273. MR MR665177 (83k:20019)

[Sei77] Gary M. Seitz, *Standard subgroups of type* $L_n(2^a)$, J. Algebra **48** (1977), no. 2, 417–438. MR MR0450412 (56 #8706)

[Sei79a] ———, *Chevalley groups as standard subgroups. I*, Illinois J. Math. **23** (1979), no. 1, 36–57. MR MR516569 (82j:20038a)

[Sei79b] ———, *Chevalley groups as standard subgroups. II, III*, Illinois J. Math. **23** (1979), no. 4, 516–553, 554–578. MR MR540399 (82j:20038b)

[Sei81] ———, *Some standard groups*, J. Algebra **70** (1981), no. 1, 299–302. MR MR618396 (82g:20037)

[Sol74] Ronald Solomon, *Finite groups with Sylow* 2*-subgroups of type* 3, J. Algebra **28** (1974), 182–198. MR MR0344338 (49 #9077)

[Sol75] ———, *Finite groups with intrinsic* 2*-components of type* $\hat{A}_n$, J. Algebra **33** (1975), 498–522. MR MR0372034 (51 #8251)

[Sol76a] ———, *Maximal* 2*-components in finite groups*, Comm. Algebra **4** (1976), no. 6, 561–594. MR MR0422415 (54 #10404)

[Sol76b] ———, *Standard components of alternating type. I*, J. Algebra **41** (1976), no. 2, 496–514. MR MR0422405 (54 #10394)

[Sol77] ———, *Standard components of alternating type. II*, J. Algebra **47** (1977), no. 1, 162–179. MR MR0444767 (56 #3115)

[Sol78a] ———, 2*-signalizers in finite groups of alternating type*, Comm. Algebra **6** (1978), no. 6, 529–549. MR MR0466303 (57 #6183)

[Sol78b] ———, *Some standard components of sporadic type*, J. Algebra **53** (1978), no. 1, 93–124. MR MR498549 (81h:20018)

[Str81] G. Stroth, *On standard subgroups of type* ${}^2E_6(2)$, Proc. Amer. Math. Soc. **81** (1981), no. 3, 365–368. MR MR597641 (82j:20039)

[Sys81] S. A. Syskin, *Standard components of type* F_3, Algebra i Logika **20** (1981), no. 4, 465–482, 485, English translation: Algebra and Logic 20 (1981), no. 4, 308–318 (1982). MR MR663062 (84a:20026)

[Yam79a] Hiromichi Yamada, *Finite groups with a standard subgroup isomorphic to* $G_2(2^n)$, J. Fac. Sci. Univ. Tokyo Sect. IA Math. **26** (1979), no. 1, 1–52. MR MR539772 (80g:20028)

[Yam79b] ———, *Finite groups with a standard subgroup isomorphic to* ${}^3D_4(2^{3n})$, J. Fac. Sci. Univ. Tokyo Sect. IA Math. **26** (1979), no. 2, 255–278. MR MR550686 (80k:20016)

[Yam79c] ———, *Standard subgroups isomorphic to* PSU(5, 2), J. Algebra **58** (1979), no. 2, 527–562. MR MR540657 (82k:20030)

[Yam79d] ———, *Standard subgroups isomorphic to* PSU(6, 2) *or* SU(6, 2), J. Algebra **61** (1979), no. 1, 82–111. MR MR554853 (81f:20025)

[Yam85] ———, *A remark on the standard form problem for* ${}^2F_4(2^{2n+1})$, $n \geq 1$, Hitotsubashi J. Arts Sci. **26** (1985), no. 1, 47–60. MR MR839725 (87f:20024)

[Yos74] Tomoyuki Yoshida, *A characterization of Conway's group* C_3, Hokkaido Math. J. **3** (1974), 232–242. MR MR0369513 (51 #5746)

References used primarily for characteristic 2 type

[AGL81] Michael Aschbacher, Daniel Gorenstein, and Richard Lyons, *The embedding of 2-locals in finite groups of characteristic 2-type.*, Ann. of Math. (2) **114** (1981), no. 2, 335–456. MR MR632843 (83i:20016a)

[And80] S. Andrilli, *On the uniqueness of O'Nan's simple group*, Ph.D. thesis, Rutgers U., 1980.

[AS91] Michael Aschbacher and Yoav Segev, *The uniqueness of groups of type J_4*, Invent. Math. **105** (1991), no. 3, 589–607. MR MR1117152 (92g:20026)

[AS04a] Michael Aschbacher and Stephen D. Smith, *The classification of quasithin groups. I*, Mathematical Surveys and Monographs, vol. 111, American Mathematical Society, Providence, RI, 2004, Structure of strongly quasithin K-groups. MR MR2097623 (2005m:20038a)

[AS04b] ———, *The classification of quasithin groups. II*, Mathematical Surveys and Monographs, vol. 112, American Mathematical Society, Providence, RI, 2004, Main theorems: the classification of simple QTKE-groups. MR MR2097624 (2005m:20038b)

[Asc72] Michael Aschbacher, *On finite groups generated by odd transpositions. I*, Math. Z. **127** (1972), 45–56. MR MR0310058 (46 #9161)

[Asc73] ———, *A condition for the existence of a strongly embedded subgroup*, Proc. Amer. Math. Soc. **38** (1973), 509–511. MR MR0318308 (47 #6855)

[Asc75b] ———, *2-components in finite groups*, Comm. Algebra **3** (1975), no. 10, 901–911. MR MR0414686 (54 #2781)

[Asc76a] ———, *Finite groups in which the generalized Fitting group of the centralizer of some involution is symplectic but not extraspecial*, Comm. Algebra **4** (1976), no. 7, 595–616. MR MR0407144 (53 #10927)

[Asc76b] ———, *Tightly embedded subgroups of finite groups*, J. Algebra **42** (1976), no. 1, 85–101. MR MR0422400 (54 #10390)

[Asc77] ———, *On finite groups in which the generalized Fitting group of the centralizer of some involution is extraspecial*, Illinois J. Math. **21** (1977), no. 2, 347–364. MR MR0442089 (56 #477)

[Asc78a] ———, *A pushing up theorem for characteristic 2 type groups*, Illinois J. Math. **22** (1978), no. 1, 108–125. MR MR0470063 (57 #9830)

[Asc78b] ———, *Thin finite simple groups*, J. Algebra **54** (1978), no. 1, 50–152. MR MR511458 (82j:20032)

[Asc80b] ———, *On finite groups of Lie type and odd characteristic*, J. Algebra **66** (1980), no. 2, 400–424. MR MR593602 (81k:20023)

[Asc81a] ———, *A factorization theorem for 2-constrained groups*, Proc. London Math. Soc. (3) **43** (1981), no. 3, 450–477. MR MR635565 (83a:20019b)

[Asc81b] ———, *Finite groups of rank 3. I*, Invent. Math. **63** (1981), no. 3, 357–402. MR MR620676 (82j:20033)

[Asc81c] ———, *On the failure of the Thompson factorization in 2-constrained groups*, Proc. London Math. Soc. (3) **43** (1981), no. 3, 425–449. MR MR635564 (83a:20019a)

[Asc81d] ———, *Some results on pushing up in finite groups*, Math. Z. **177** (1981), no. 1, 61–80. MR MR611470 (82e:20016)

[Asc81e] ———, *Weak closure in finite groups of even characteristic*, J. Algebra **70** (1981), no. 2, 561–627. MR MR623826 (82j:20034)

[Asc82] ———, GF(2)*-representations of finite groups*, Amer. J. Math. **104** (1982), no. 4, 683–771. MR MR667532 (83j:20025)

[Asc83a] ———, *Finite groups of rank 3. II*, Invent. Math. **71** (1983), no. 1, 51–163. MR MR688262 (84h:20008)

[Asc83b] ———, *The uniqueness case for finite groups. I*, Ann. of Math. (2) **117** (1983), no. 2, 383–454. MR MR690850 (84g:20021a)

[Asc83c] ———, *The uniqueness case for finite groups. II*, Ann. of Math. (2) **117** (1983), no. 3, 455–551. MR MR701255 (84g:20021b)

[Asc94] ———, *Sporadic groups*, Cambridge Tracts in Mathematics, vol. 104, Cambridge University Press, Cambridge, 1994. MR MR1269103 (96e:20020)

[Asc02a] ———, *Characterizing $U_3(3)$ by counting involutions*, J. Ramanujan Math. Soc. **17** (2002), no. 1, 35–49. MR MR1906419 (2003b:20021)

[Asc02b] ———, *Finite groups of $G_2(3)$-type*, J. Algebra **257** (2002), no. 2, 197–214. MR MR1947319 (2004b:20025)

[Asc03a] ———, *A 2-local characterization of M_{12}*, Illinois J. Math. **47** (2003), no. 1-2, 31–47, Special issue in honor of Reinhold Baer (1902–1979). MR MR2031304 (2004i:20024)

[Asc03b] ———, *A 2-local characterization of the Higman-Sims group*, J. Algebra **260** (2003), no. 1, 16–31, Special issue celebrating the 80th birthday of Robert Steinberg. MR MR1973574 (2004e:20022)

[Asc09] ———, *Generation of fusion systems of characteristic* 2 *type*, ?? **??** (2009?), ??–??, submitted for publication.

[Bau79] Bernd Baumann, *Über endliche Gruppen mit einer zu $L_2(2^n)$ isomorphen Faktorgruppe*, Proc. Amer. Math. Soc. **74** (1979), no. 2, 215–222. MR MR524288 (80g:20024)

[Bei77] Bert Beisiegel, *Semi-extraspezielle p-Gruppen*, Math. Z. **156** (1977), no. 3, 247–254. MR MR0473004 (57 #12683)

[Ben67] Helmut Bender, *Über den grössten p′-Normalteiler in p-auflösbaren Gruppen*, Arch. Math. (Basel) **18** (1967), 15–16. MR MR0213439 (35 #4303)

[BGL77] N. Burgoyne, R. Griess, and R. Lyons, *Maximal subgroups and automorphisms of Chevalley groups*, Pacific J. Math. **71** (1977), no. 2, 365–403. MR MR0444795 (56 #3142)

[Blo67] David M. Bloom, *The subgroups of* PSL(3, *q*) *for odd q*, Trans. Amer. Math. Soc. **127** (1967), 150–178. MR MR0214671 (35 #5520)

[Bou68] N. Bourbaki, *Éléments de mathématique. Fasc. XXXIV. Groupes et algèbres de Lie. Chapitre IV: Groupes de Coxeter et systèmes de Tits. Chapitre V: Groupes engendrés par des réflexions. Chapitre VI: systèmes de racines*, Actualités Scientifiques et Industrielles, No. 1337, Hermann, Paris, 1968. MR MR0240238 (39 #1590)

[BT71] A. Borel and J. Tits, *Éléments unipotents et sous-groupes paraboliques de groupes réductifs. I*, Invent. Math. **12** (1971), 95–104. MR MR0294349 (45 #3419)

[Bur83] N. Burgoyne, *Elements of order 3 in Chevalley groups of characteristic 2*, Appendix (Sec. 34) to Gorenstein-Lyons: The local structure of finite groups of characteristic 2 type, vol. 42, Mem. Amer. Math. Soc., no. 276, American Mathematical Society, Providence, R.I., 1983, pp. 425–431.

[BW77] N. Burgoyne and C. Williamson, *Semi-simple classes in Chevalley type groups*, Pacific J. Math. **70** (1977), no. 1, 83–100. MR MR0507018 (58 #22322)

[Cam79] N. Campbell, *Pushing up in finite groups*, Ph.D. thesis, Caltech, 1979.

[Car89] Roger W. Carter, *Simple groups of Lie type*, Wiley Classics Library, John Wiley & Sons Inc., New York, 1989, Reprint of the 1972 original, A Wiley-Interscience Publication. MR MR1013112 (90g:20001)

[CKS76] Charles W. Curtis, William M. Kantor, and Gary M. Seitz, *The* 2*-transitive permutation representations of the finite Chevalley groups*, Trans. Amer. Math. Soc. **218** (1976), 1–59. MR MR0422440 (54 #10429)

[Col73] Michael J. Collins, *The characterization of finite groups whose Sylow* 2*-subgroups are of type $L_3(q)$, q even*, J. Algebra **25** (1973), 490–512; erratum, ibid. 27 (1973), 199–203. MR MR0320139 (47 #8680)

[Con69] J. H. Conway, *A group of order* $8,315,553,613,086,720,000$, Bull. London Math. Soc. **1** (1969), 79–88. MR MR0248216 (40 #1470)

[Con71] ———, *Three lectures on exceptional groups*, Finite simple groups (Proc. Instructional Conf., Oxford, 1969), Academic Press, London, 1971, pp. 215–247. MR MR0338152 (49 #2918)

[CW73] J. H. Conway and D. B. Wales, *Construction of the Rudvalis group of order* 145, 926, 144, 000, J. Algebra **27** (1973), 538–548. MR MR0335620 (49 #400)

[Dem73a] U. Dempwolff, *On the second cohomology of* GL(n, 2), J. Austral. Math. Soc. **16** (1973), 207–209, Collection of articles dedicated to the memory of Hanna Neumann, II. MR MR0357639 (50 #10107)

[Dem73b] Ulrich Dempwolff, *Zentralisatoren zentraler Involutionen in* $L_n(2)$, Illinois J. Math. **17** (1973), 465–497. MR MR0320135 (47 #8676)

[Dem74] ———, *A characterization of the Rudvalis simple group of order* $2^{14}\cdot 3^3\cdot 5^3\cdot 7\cdot 13\cdot 29$ *by the centralizers of noncentral involutions*, J. Algebra **32** (1974), 53–88. MR MR0352242 (50 #4729)

[DGS85] A. Delgado, D. Goldschmidt, and B. Stellmacher, *Groups and graphs: new results and methods*, DMV Seminar, vol. 6, Birkhäuser Verlag, Basel, 1985, With a preface by the authors and Bernd Fischer. MR MR862622 (88a:05076)

[Die55] Jean Dieudonné, *La géométrie des groupes classiques*, Ergebnisse der Mathematik und ihrer Grenzgebiete (N.F.), Heft 5, Springer-Verlag, Berlin, 1955. MR MR0072144 (17,236a)

[DW77a] Ulrich Dempwolff and S. K. Wong, *Another characterization of* $L_n(2)$, Arch. Math. (Basel) **28** (1977), no. 1, 41–44. MR MR0430092 (55 #3099)

[DW77b] ———, *On finite groups whose centralizer of an involution has normal extra special and abelian subgroups. I*, J. Algebra **45** (1977), no. 1, 247–253. MR MR0427447 (55 #479)

[DW78] ———, *On finite groups whose centralizer of an involution has normal extraspecial and abelian subgroups. II*, J. Algebra **52** (1978), no. 1, 210–217. MR MR0486106 (58 #5894)

[Fen73] Daniel Fendel, *A characterization of Conway's group. 3*, J. Algebra **24** (1973), 159–196. MR MR0314962 (47 #3511)

[FF79] Larry Finkelstein and Daniel Frohardt, *A* 3*-local characterization of* $L_7(2)$, Trans. Amer. Math. Soc. **250** (1979), 181–194. MR MR530049 (80f:20015)

[FF81a] ———, *Simple groups with a standard* 3*-component of type* $A_n(2)$, *with* $n \geq 5$, Proc. London Math. Soc. (3) **43** (1981), no. 3, 385–424. MR MR635563 (83f:20012)

[FF81b] ———, *Standard* 3*-components of type* Sp(6, 2), Trans. Amer. Math. Soc. **266** (1981), no. 1, 71–92. MR MR613786 (82f:20033)

[FF84] ———, *Standard* 3*-components of type* SL(4, 2), J. Algebra **86** (1984), no. 2, 407–421. MR MR732257 (85k:20046)

[Fin73] Larry Finkelstein, *The maximal subgroups of Conway's group* C_3 *and McLaughlin's group*, J. Algebra **25** (1973), 58–89. MR MR0346046 (49 #10772)

[Foo82] Richard Foote, *Component type theorems for finite groups in characteristic* 2, Illinois J. Math. **26** (1982), no. 1, 62–111. MR MR638555 (83a:20020)

[FR73] L. Finkelstein and A. Rudvalis, *Maximal subgroups of the Hall-Janko-Wales group*, J. Algebra **24** (1973), 486–493. MR MR0323889 (48 #2242)

[FR74] ———, *The maximal subgroups of Janko's simple group of order* 50, 232, 960, J. Algebra **30** (1974), 122–143. MR MR0354846 (50 #7323)

[Fra51] J. S. Frame, *The classes and representations of the groups of* 27 *lines and* 28 *bitangents*, Ann. Mat. Pura Appl. (4) **32** (1951), 83–119. MR MR0047038 (13,817i)

[Fro83] Daniel Frohardt, *A trilinear form for the third Janko group*, J. Algebra **83** (1983), no. 2, 349–379. MR MR714249 (84m:20023)

[FS79b] Larry Finkelstein and Ronald Solomon, *Finite simple groups with a standard* 3*-component of type* Sp($2n$, 2), $n \geq 4$, J. Algebra **59** (1979), no. 2, 466–480. MR MR543264 (81a:20021)

[FSS77] L. R. Fletcher, B. Stellmacher, and W. B. Stewart, *Endliche Gruppen, die kein Element der Ordnung* 6 *enthalten*, Quart. J. Math. Oxford Ser. (2) **28** (1977), no. 110, 143–154. MR MR0447394 (56 #5706)

[FT62] Walter Feit and John G. Thompson, *Finite groups which contain a self-centralizing subgroup of order 3.*, Nagoya Math. J. **21** (1962), 185–197. MR MR0142623 (26 #192)

[FW69] Paul Fong and W. J. Wong, *A characterization of the finite simple groups* PSp(4, q), $G_2(q)$, ${D_4}^2(q)$*. I*, Nagoya Math. J. **36** (1969), 143–184. MR MR0255666 (41 #326)

[GG75] Robert Gilman and Daniel Gorenstein, *Finite groups with Sylow* 2*-subgroups of class two. I, II*, Trans. Amer. Math. Soc. **207** (1975), 1–101; ibid. 207 (1975), 103–126. MR MR0379662 (52 #567)

[GG83] Robert H. Gilman and Robert L. Griess, Jr., *Finite groups with standard components of Lie type over fields of characteristic two*, J. Algebra **80** (1983), no. 2, 383–516. MR MR691810 (84g:20024)

[GH71a] Daniel Gorenstein and Koichiro Harada, *Finite simple groups of low 2-rank and the families* $G_2(q)$, $D_4^2(q)$, q *odd*, Bull. Amer. Math. Soc. **77** (1971), 829–862. MR MR0306301 (46 #5427)

[GH71b] ———, *On finite groups with Sylow 2-subgroups of type* $\hat{A}_n$, $n = 8, 9, 10,$ *and* 11, J. Algebra **19** (1971), 185–227. MR MR0285599 (44 #2817)

[GL75] Robert L. Griess, Jr. and Richard Lyons, *The automorphism group of the Tits simple group* ${}^2F_4(2)'$, Proc. Amer. Math. Soc. **52** (1975), 75–78. MR MR0390054 (52 #10880)

[GL77] Daniel Gorenstein and Richard Lyons, *Non-solvable signalizer functors on finite groups*, Proc. London Math. Soc. (3) **35** (1977), no. 1, 1–33. MR MR0450396 (56 #8690)

[GL83] ———, *The local structure of finite groups of characteristic* 2 *type*, Mem. Amer. Math. Soc. **42** (1983), no. 276, vii+731. MR MR690900 (84g:20025)

[GL93] ———, *On Aschbacher's local* $C(G;T)$ *theorem*, Israel J. Math. **82** (1993), no. 1-3, 227–279. MR MR1239050 (94k:20023)

[Gla70] George Glauberman, *Prime-power factor groups of finite groups. II*, Math. Z. **117** (1970), 46–56. MR MR0294483 (45 #3553)

[Gla71] G. Glauberman, *Global and local properties of finite groups*, Finite simple groups (Proc. Instructional Conf., Oxford, 1969), Academic Press, London, 1971, pp. 1–64. MR MR0352241 (50 #4728)

[Gla73] George Glauberman, *Failure of factorization in p-solvable groups*, Quart. J. Math. Oxford Ser. (2) **24** (1973), 71–77. MR MR0318301 (47 #6848)

[Gla77] G. Glauberman, *Factorizations in local subgroups of finite groups*, American Mathematical Society, Providence, R.I., 1977, Regional Conference Series in Mathematics, No. 33. MR MR0470072 (57 #9839)

[GM02] Robert M. Guralnick and Gunter Malle, *Classification of* $2F$*-modules. I*, J. Algebra **257** (2002), no. 2, 348–372. MR MR1947326 (2003m:20008)

[GM04] ———, *Classification of* $2F$*-modules. II*, Finite groups 2003, Walter de Gruyter GmbH & Co. KG, Berlin, 2004, pp. 117–183. MR MR2125071 (2006b:20062)

[GMS89] Robert L. Griess, Jr., Ulrich Meierfrankenfeld, and Yoav Segev, *A uniqueness proof for the Monster*, Ann. of Math. (2) **130** (1989), no. 3, 567–602. MR MR1025167 (90j:20030)

[GN83] George Glauberman and Richard Niles, *A pair of characteristic subgroups for pushing-up in finite groups*, Proc. London Math. Soc. (3) **46** (1983), no. 3, 411–453. MR MR699096 (84g:20042)

[Gol72] David M. Goldschmidt, *Weakly embedded 2-local subgroups of finite groups*, J. Algebra **21** (1972), 341–351. MR MR0323896 (48 #2249)

[Gol80] ———, *Automorphisms of trivalent graphs*, Ann. of Math. (2) **111** (1980), no. 2, 377–406. MR MR569075 (82a:05052)

[Gom75] Kensaku Gomi, *Finite groups all of whose non-2-closed 2-local subgroups have Sylow 2-subgroups of class* 2, J. Algebra **35** (1975), 214–223. MR MR0374265 (51 #10465)

[Gor71] D. Gorenstein, *Centralizers of involutions in finite simple groups*, Finite simple groups (Proc. Instructional Conf., Oxford, 1969), Academic Press, London, 1971, pp. 65–133. MR MR0335622 (49 #402)

[Gri76] Robert L. Griess, Jr., *The structure of the "monster" simple group*, Proceedings of the Conference on Finite Groups (Univ. Utah, Park City, Utah, 1975) (New York), Academic Press, 1976, pp. 113–118. MR MR0399248 (53 #3099)

[Gri82] ———, *The friendly giant*, Invent. Math. **69** (1982), no. 1, 1–102. MR MR671653 (84m:20024)

[Gut79] Martin M. Guterman, *A characterization of* $F_4(4^n)$ *as a group with standard* 3*-component* $B_3(4^n)$, Comm. Algebra **7** (1979), no. 10, 1079–1102. MR MR533205 (82k:20025)

[GW71] Daniel Gorenstein and John H. Walter, *The* π*-layer of a finite group*, Illinois J. Math. **15** (1971), 555–564. MR MR0289624 (44 #6812)

[Hal82] J. I. Hall, *Certain 2-local blocks with alternating sections*, Comm. Algebra **10** (1982), no. 16, 1721–1747. MR MR673821 (84b:20015)

[Har68] Koichiro Harada, *Finite simple groups with short chains of subgroups*, J. Math. Soc. Japan **20** (1968), 655–672. MR MR0230811 (37 #6369)

[Har78] ______, *On Yoshida's transfer*, Osaka J. Math. **15** (1978), no. 3, 637–646. MR MR510501 (80c:20017)

[Har80b] ______, *On finite simple groups possessing 2-local blocks of orthogonal type*, Comm. Algebra **8** (1980), no. 5, 441–449. MR MR561539 (81b:20016)

[Hel69] Dieter Held, *The simple groups related to* M_{24}, J. Algebra **13** (1969), 253–296. MR MR0249500 (40 #2745)

[HH56] P. Hall and Graham Higman, *On the p-length of p-soluble groups and reduction theorems for Burnside's problem*, Proc. London Math. Soc. (3) **6** (1956), 1–42. MR MR0072872 (17,344b)

[Hig57] Graham Higman, *Groups and rings having automorphisms without non-trivial fixed elements*, J. London Math. Soc. **32** (1957), 321–334. MR MR0089204 (19,633c)

[HM69] Graham Higman and J. McKay, *On Janko's simple group of order* 50, 232, 960, Theory of Finite Groups (Symposium, Harvard Univ., Cambridge, Mass., 1968), Benjamin, New York, 1969, pp. 65–77. MR MR0245668 (39 #6974)

[Hol78] D. F. Holt, *Transitive permutation groups in which an involution central in a Sylow 2-subgroup fixes a unique point*, Proc. London Math. Soc. (3) **37** (1978), no. 1, 165–192. MR MR0575516 (58 #28153)

[HS68] Donald G. Higman and Charles C. Sims, *A simple group of order* 44,352,000, Math. Z. **105** (1968), 110–113. MR MR0227269 (37 #2854)

[HW68] Marshall Hall, Jr. and David Wales, *The simple group of order* 604,800, J. Algebra **9** (1968), 417–450. MR MR0240192 (39 #1544)

[Jam73] G. D. James, *The modular characters of the Mathieu groups*, J. Algebra **27** (1973), 57–111. MR MR0330277 (48 #8614)

[Jan66] Zvonimir Janko, *A new finite simple group with abelian Sylow 2-subgroups and its characterization*, J. Algebra **3** (1966), 147–186. MR MR0193138 (33 #1359)

[Jan76] ______, *A new finite simple group of order* $86 \cdot 775 \cdot 571 \cdot 046 \cdot 077 \cdot 562 \cdot 880$ *which possesses* M_{24} *and the full covering group of* M_{22} *as subgroups*, J. Algebra **42** (1976), no. 2, 564–596. MR MR0432751 (55 #5734)

[JP76] Wayne Jones and Brian Parshall, *On the 1-cohomology of finite groups of Lie type*, Proceedings of the Conference on Finite Groups (Univ. Utah, Park City, Utah, 1975) (New York), Academic Press, 1976, pp. 313–328. MR MR0404470 (53 #8272)

[JW69] Zvonimir Janko and S. K. Wong, *A characterization of the Higman-Sims simple group*, J. Algebra **13** (1969), 517–534. MR MR0260866 (41 #5486)

[KL06] Inna Korchagina and Richard Lyons, *A* $\{2,3\}$*-local characterization of the groups* $A_{12}, \mathrm{Sp}_8(2), F_4(2)$ *and* F_5, J. Algebra **300** (2006), no. 2, 590–646. MR MR2228213 (2007c:20032)

[Kon71] Marc W. Konvisser, *Embedding of abelian subgroups in p-groups*, Trans. Amer. Math. Soc. **153** (1971), 469–481. MR MR0271228 (42 #6111)

[Ku97] Chao Ku, *A characterization of* M_{22}, J. Algebra **187** (1997), no. 1, 295–303. MR MR1425573 (97m:20025)

[LS77] Jeffrey S. Leon and Charles C. Sims, *The existence and uniqueness of a simple group generated by* $\{3,4\}$*-transpositions*, Bull. Amer. Math. Soc. **83** (1977), no. 5, 1039–1040. MR MR0444765 (56 #3113)

[LW76] J. Richard Lundgren and S. K. Wong, *On finite simple groups in which the centralizer* M *of an involution is solvable and* $O_2(M)$ *is extraspecial*, J. Algebra **41** (1976), no. 1, 1–15. MR MR0442077 (56 #465)

[Lyo72] Richard Lyons, *Evidence for a new finite simple group*, J. Algebra **20** (1972), 540–569. MR MR0299674 (45 #8722)

[Mag71] Spyros S. Magliveras, *The subgroup structure of the Higman-Sims simple group*, Bull. Amer. Math. Soc. **77** (1971), 535–539. MR MR0283077 (44 #310)

[Mar72] R. P. Martineau, *On 2-modular representations of the Suzuki groups*, Amer. J. Math. **94** (1972), 55–72. MR MR0360777 (50 #13224)

[McB82a] Patrick P. McBride, *Nonsolvable signalizer functors on finite groups*, J. Algebra **78** (1982), no. 1, 215–238. MR MR677718 (84g:20049b)

[McB82b] Patrick Paschal McBride, *Near solvable signalizer functors on finite groups*, J. Algebra **78** (1982), no. 1, 181–214. MR MR677717 (84g:20049a)

[McL69a] Jack McLaughlin, *A simple group of order* 898, 128, 000, Theory of Finite Groups (Symposium, Harvard Univ., Cambridge, Mass., 1968), Benjamin, New York, 1969, pp. 109–111. MR MR0242941 (39 #4268)

[McL69b] ———, *Some subgroups of* $\mathrm{SL}_n(\mathbf{F}_2)$, Illinois J. Math. **13** (1969), 108–115. MR MR0237660 (38 #5941)

[Mit14] Howard H. Mitchell, *The subgroups of the quaternary abelian linear group*, Trans. Amer. Math. Soc. **15** (1914), no. 4, 379–396. MR MR1500986

[MS93] Ulrich Meierfrankenfeld and Bernd Stellmacher, *Pushing up weak BN-pairs of rank two*, Comm. Algebra **21** (1993), no. 3, 825–934. MR MR1204756 (94d:20025)

[Nor80] Simon Norton, *The construction of* J_4, The Santa Cruz Conference on Finite Groups (Univ. California, Santa Cruz, Calif., 1979), Proc. Sympos. Pure Math., vol. 37, Amer. Math. Soc., Providence, R.I., 1980, pp. 271–277. MR MR604593 (82a:20030)

[O'N76a] Michael E. O'Nan, *Some characterizations by centralizers of elements of order three*, Proceedings of the Conference on Finite Groups (Univ. Utah, Park City, Utah, 1975) (New York), Academic Press, 1976, pp. 79–84. MR MR0404428 (53 #8230)

[O'N76b] ———, *Some evidence for the existence of a new simple group*, Proc. London Math. Soc. (3) **32** (1976), no. 3, 421–479. MR MR0401905 (53 #5731)

[Pag69] D. Page, *Some problems in the theory of finite groups*, Ph.D. thesis, Oxford U., 1969.

[Par76] David Parrott, *A characterization of the Rudvalis simple group*, Proc. London Math. Soc. (3) **32** (1976), no. 1, 25–51. MR MR0390043 (52 #10869)

[Par77] ———, *On Thompson's simple group*, J. Algebra **46** (1977), no. 2, 389–404. MR MR0447396 (56 #5708)

[Par81] ———, *Characterizations of the Fischer groups. I, II, III*, Trans. Amer. Math. Soc. **265** (1981), no. 2, 303–347. MR MR610952 (82d:20018)

[Pat72] N. Patterson, *On Conway's group* .0 *and some subgroups*, Ph.D. thesis, U. Cambridge, 1972.

[Pol71] Harriet Pollatsek, *First cohomology groups of some linear groups over fields of characteristic two*, Illinois J. Math. **15** (1971), 393–417. MR MR0280596 (43 #6316)

[Pri77] A. R. Prince, *On 2-groups admitting* A_5 *or* A_6 *with an element of order* 5 *acting fixed point freely*, J. Algebra **49** (1977), no. 2, 374–386. MR MR0457565 (56 #15770)

[PW76] N. J. Patterson and S. K. Wong, *A characterization of the Suzuki sporadic simple group of order* 448, 345, 497, 600, J. Algebra **39** (1976), no. 1, 277–286. MR MR0399253 (53 #3104)

[Rei76] Arthur Reifart, *A characterization of Thompson's sporadic simple group*, J. Algebra **38** (1976), no. 1, 192–200. MR MR0393217 (52 #14027)

[Rei78a] ———, *A general characterization of the Steinberg simple group* $D_4^2(2^3)$, J. Algebra **50** (1978), no. 1, 63–68. MR MR0463285 (57 #3238)

[Rei78b] ———, *On finite simple groups with large extraspecial subgroups. I*, J. Algebra **53** (1978), no. 2, 452–470. MR MR502644 (80j:20019a)

[Rei78c] ———, *On finite simple groups with large extraspecial subgroups. II*, J. Algebra **54** (1978), no. 1, 273–289. MR MR511465 (80j:20019b)

[Rob76] Derek J. S. Robinson, *The vanishing of certain homology and cohomology groups*, J. Pure Appl. Algebra **7** (1976), no. 2, 145–167. MR MR0404478 (53 #8280)

[Sch04] I. Schur, *Über die darstellung der endliche gruppen durch gebrochene lineare substitutionen*, J. Reine angew. Math. **27** (1904), 20–50.

[Sei82] Gary M. Seitz, *Generation of finite groups of Lie type*, Trans. Amer. Math. Soc. **271** (1982), no. 2, 351–407. MR MR654839 (83h:20021)

[Sim73] Charles C. Sims, *The existence and uniqueness of Lyons' group*, Finite groups '72 (Proc. Gainesville Conf., Univ. Florida, Gainesville, Fla., 1972), North-Holland, Amsterdam, 1973, pp. 138–141. North–Holland Math. Studies, Vol. 7. MR MR0354881 (50 #7358)

[Smi] Fredrick Smith, *On blocks as uniqueness subgroups*, unpublished U. Illinois-Chicago preprint, 1978.

[Smi74] Fredrick L. Smith, *A characterization of the* .2 *Conway simple group*, J. Algebra **31** (1974), 91–116. MR MR0349832 (50 #2325)

[Smi76a] Fredrick Smith, *On the centralizers of involutions in finite fusion-simple groups*, J. Algebra **38** (1976), no. 2, 268–273. MR MR0396738 (53 #599)

[Smi76b] P. E. Smith, *A simple subgroup of* $M?$ *and* $E_8(3)$, Bull. London Math. Soc. **8** (1976), no. 2, 161–165. MR MR0409630 (53 #13382)

[Smi77a] Fredrick Smith, *On a result of M. Aschbacher*, Comm. Algebra **5** (1977), no. 3, 279–288. MR MR0430057 (55 #3064)

[Smi77b] ———, *On finite groups with large extra-special* 2*-subgroups*, J. Algebra **44** (1977), no. 2, 477–487. MR MR0427448 (55 #480)

[Smi77c] ———, *On groups with an involution z such that the generalized Fitting subgroup E of $C(z)$ is extra-special and $C(z)/E \supseteq$* $(\mathrm{Out}(E))'$, Comm. Algebra **5** (1977), no. 3, 267–277. MR MR0437639 (55 #10563)

[Smi79a] ———, *On transitive permutation groups in which a* 2*-central involution fixes a unique point*, Comm. Algebra **7** (1979), no. 2, 203–218. MR MR515456 (80d:20004)

[Smi79b] Stephen D. Smith, *Large extraspecial subgroups of widths* 4 *and* 6, J. Algebra **58** (1979), no. 2, 251–281. MR MR540638 (80g:20030)

[Smi80a] ———, *A characterization of finite Chevalley and twisted groups of type E over* GF(2), J. Algebra **62** (1980), no. 1, 101–117. MR MR561119 (81b:20018)

[Smi80b] ———, *A characterization of orthogonal groups over* GF(2), J. Algebra **62** (1980), no. 1, 39–60. MR MR561116 (81b:20017)

[Smi81] ———, *A characterization of some Chevalley groups in characteristic two*, J. Algebra **68** (1981), no. 2, 390–425. MR MR608542 (82d:20021)

[Sol81] Ronald Solomon, *On certain* 2*-local blocks*, Proc. London Math. Soc. (3) **43** (1981), no. 3, 478–498. MR MR635566 (83g:20018)

[SS70] T. A. Springer and R. Steinberg, *Conjugacy classes*, Seminar on Algebraic Groups and Related Finite Groups (The Institute for Advanced Study, Princeton, N.J., 1968/69), Lecture Notes in Mathematics, Vol. 131, Springer, Berlin, 1970, pp. 167–266. MR MR0268192 (42 #3091)

[ST79] Ronald Solomon and F. G. Timmesfeld, *A note on tightly embedded subgroups*, Arch. Math. (Basel) **31** (1978/79), no. 3, 217–223. MR MR521473 (80b:20017)

[Ste62] Robert Steinberg, *Générateurs, relations et revêtements de groupes algébriques*, Colloq. Théorie des Groupes Algébriques (Bruxelles, 1962), Librairie Universitaire, Louvain, 1962, pp. 113–127. MR MR0153677 (27 #3638)

[Ste63] ———, *Representations of algebraic groups*, Nagoya Math. J. **22** (1963), 33–56. MR MR0155937 (27 #5870)

[Ste68a] ———, *Endomorphisms of linear algebraic groups*, Memoirs of the American Mathematical Society, No. 80, American Mathematical Society, Providence, R.I., 1968. MR MR0230728 (37 #6288)

[Ste68b] ———, *Lectures on Chevalley groups*, Yale University, New Haven, Conn., 1968, Notes prepared by John Faulkner and Robert Wilson. MR MR0466335 (57 #6215)

[Str76] G. Stroth, *A characterization of Fischer's sporadic simple group of the order* $2^{41} \cdot 3^{13} \cdot 5^6 \cdot 7^2 \cdot 11 \cdot 13 \cdot 17 \cdot 19 \cdot 23 \cdot 31 \cdot 47$, J. Algebra **40** (1976), no. 2, 499–531. MR MR0417277 (54 #5334)

[Str78] Gernot Stroth, *Einige Gruppen vom Charakteristik* 2*-Typ*, J. Algebra **51** (1978), no. 1, 107–143. MR MR487469 (81g:20032)

[Str80] ———, *Endliche Gruppen, die eine maximale* 2*-lokale Untergruppe besitzen, so daß $Z(F^*(M))$ eine TI-Menge in G ist*, J. Algebra **64** (1980), no. 2, 460–528. MR MR579071 (81j:20025)

[Suz69a] Michio Suzuki, *Characterizations of linear groups*, Bull. Amer. Math. Soc. **75** (1969), 1043–1091. MR MR0260889 (41 #5509)

[Suz69b] ———, *A simple group of order* 448, 345, 497, 600, Theory of Finite Groups (Symposium, Harvard Univ., Cambridge, Mass., 1968), Benjamin, New York, 1969, pp. 113–119. MR MR0241527 (39 #2867)

[SW81] Ronald Solomon and S. K. Wong, *On $L_2(2^n)$-blocks*, Proc. London Math. Soc. (3) **43** (1981), no. 3, 499–519. MR MR635567 (83g:20019)

[Thm70] Gomer Thomas, *A characterization of the unitary groups* $\mathrm{PSU}_5(q^2)$, $q = 2^n$, J. Algebra **14** (1970), 245–259. MR MR0252506 (40 #5726)

[Tim73] F. G. Timmesfeld, *A characterization of the Chevalley- and Steinberg-groups over F_2*, Geometriae Dedicata **1** (1973), no. 3, 269–321. MR MR0330279 (48 #8616)

[Tim75a] ———, *Groups generated by root-involutions. I, II*, J. Algebra **33** (1975), 75–134; ibid. 35 (1975), 367–441. MR MR0372019 (51 #8236)

[Tim75b] Franz Georg Timmesfeld, *Groups with weakly closed TI-subgroups*, Math. Z. **143** (1975), no. 3, 243–278. MR MR0407143 (53 #10926)

[Tim77] F. G. Timmesfeld, *On elementary abelian TI-subgroups*, J. Algebra **44** (1977), no. 2, 457–476. MR MR0427455 (55 #487)

[Tim78a] Franz Timmesfeld, *Finite simple groups in which the generalized Fitting group of the centralizer of some involution is extraspecial*, Ann. of Math. (2) **107** (1978), no. 2, 297–369. MR MR486255 (81i:20016a)

[Tim78b] ———, *On the structure of 2-local subgroups in finite groups*, Math. Z. **161** (1978), no. 2, 119–136. MR MR504481 (80d:20020)

[Tim79a] F. G. Timmesfeld, *A condition for the existence of a weakly closed TI-set*, J. Algebra **60** (1979), no. 2, 472–484. MR MR549941 (82j:20040)

[Tim79b] Franz Timmesfeld, *Correction to: "Finite simple groups in which the generalized Fitting group of the centralizer of some involution is extraspecial"*, Ann. of Math. (2) **109** (1979), no. 2, 413–414. MR MR528971 (81i:20016b)

[Tim79c] ———, *A note on 2-groups of* $\mathrm{GF}(2^n)$*-type*, Arch. Math. (Basel) **32** (1979), no. 2, 101–108. MR MR534918 (80f:20020)

[Tim81] ———, *On finite groups in which a maximal abelian normal subgroup of some maximal 2-local subgroup is a TI-set*, Proc. London Math. Soc. (3) **43** (1981), no. 1, 1–45. MR MR623717 (82h:20020)

[Tit63] J. Tits, *Groupes simples et géométries associées*, Proc. Internat. Congr. Mathematicians (Stockholm, 1962), Inst. Mittag-Leffler, Djursholm, 1963, pp. 197–221. MR MR0175903 (31 #179)

[TW02] Jacques Tits and Richard M. Weiss, *Moufang polygons*, Springer Monographs in Mathematics, Springer-Verlag, Berlin, 2002. MR MR1938841 (2003m:51008)

[Wal69a] David B. Wales, *Generators of the Hall-Janko group as a subgroup of* $G_2(4)$, J. Algebra **13** (1969), 513–516. MR MR0251133 (40 #4364)

[Wal69b] John H. Walter, *The characterization of finite groups with abelian Sylow 2-subgroups.*, Ann. of Math. (2) **89** (1969), 405–514. MR MR0249504 (40 #2749)

[Wei55] A. J. Weir, *Sylow p-subgroups of the classical groups over finite fields with characteristic prime to p*, Proc. Amer. Math. Soc. **6** (1955), 529–533. MR MR0072143 (17,235e)

[Yam69] Hiroyoshi Yamaki, *A characterization of the simple group* Sp(6, 2), J. Math. Soc. Japan **21** (1969), 334–356. MR MR0255670 (41 #330)

[Yos78] Tomoyuki Yoshida, *Character-theoretic transfer*, J. Algebra **52** (1978), no. 1, 1–38. MR MR0491920 (58 #11095)

Expository references mentioned

[AG67] J. L. Alperin and Daniel Gorenstein, *Transfer and fusion in finite groups*, J. Algebra **6** (1967), 242–255. MR MR0215914 (35 #6749)

[Asc00] M. Aschbacher, *Finite group theory*, second ed., Cambridge Studies in Advanced Mathematics, vol. 10, Cambridge University Press, Cambridge, 2000. MR MR1777008 (2001c:20001)

[Asc04] Michael Aschbacher, *The status of the classification of the finite simple groups*, Notices Amer. Math. Soc. **51** (2004), no. 7, 736–740. MR MR2072045

[Bau76] Bernd Baumann, *Endliche nichtauflösbare Gruppen mit einer nilpotenten maximalen Untergruppe*, J. Algebra **38** (1976), no. 1, 119–135. MR MR0409639 (53 #13391)

[BCC+70] A. Borel, R. Carter, C. W. Curtis, N. Iwahori, T. A. Springer, and R. Steinberg, *Seminar on Algebraic Groups and Related Finite Groups. (Held at The Institute for Advanced Study, Princeton, N. J., 1968/69)*, Lecture Notes in Mathematics, Vol. 131, Springer-Verlag, Berlin, 1970. MR MR0258840 (41 #3486)

[Ben81] Helmut Bender, *Finite groups with dihedral Sylow 2-subgroups*, J. Algebra **70** (1981), no. 1, 216–228. MR MR618389 (83c:20011b)

[BF55] Richard Brauer and K. A. Fowler, *On groups of even order*, Ann. of Math. (2) **62** (1955), 565–583. MR MR0074414 (17,580e)

[BG81] Helmut Bender and George Glauberman, *Characters of finite groups with dihedral Sylow 2-subgroups*, J. Algebra **70** (1981), no. 1, 200–215. MR MR618388 (83c:20011a)

[BG94] ______, *Local analysis for the odd order theorem*, London Mathematical Society Lecture Note Series, vol. 188, Cambridge University Press, Cambridge, 1994, With the assistance of Walter Carlip. MR MR1311244 (96h:20036)

[Bra57] Richard Brauer, *On the structure of groups of finite order*, Proceedings of the International Congress of Mathematicians, Amsterdam, 1954, Vol. 1, Erven P. Noordhoff N.V., Groningen, 1957, pp. 209–217. MR MR0095203 (20 #1709)

[Bur55] W. Burnside, *Theory of groups of finite order*, Dover Publications Inc., New York, 1955, 2d ed. MR MR0069818 (16,1086c)

[BW76] N. Burgoyne and C. Williamson, *On a theorem of Borel and Tits for finite Chevalley groups*, Arch. Math. (Basel) **27** (1976), no. 5, 489–491. MR MR0430091 (55 #3098)

[CCN+85] J. H. Conway, R. T. Curtis, S. P. Norton, R. A. Parker, and R. A. Wilson, *Atlas of finite groups*, Oxford University Press, Eynsham, 1985, Maximal subgroups and ordinary characters for simple groups, With computational assistance from J. G. Thackray. MR MR827219 (88g:20025)

[CM80] B. Cooperstein and G. Mason, *Some questions concerning the representations of Chevalley groups in characteristic two*, preprint, U. Calif. Santa Cruz, 93 pages, 1980.

[Cur71] C. W. Curtis, *Chevalley groups and related topics*, Finite simple groups (Proc. Instructional Conf., Oxford, 1969), Academic Press, London, 1971, pp. 135–189. MR MR0340442 (49 #5197)

[Dic58] Leonard Eugene Dickson, *Linear groups: With an exposition of the Galois field theory*, with an introduction by W. Magnus, Dover Publications Inc., New York, 1958. MR MR0104735 (21 #3488)

[Fis71] Bernd Fischer, *Finite groups generated by 3-transpositions. I*, Invent. Math. **13** (1971), 232–246. MR MR0294487 (45 #3557)

[Foo80] Richard Foote, *Aschbacher blocks*, The Santa Cruz Conference on Finite Groups (Univ. California, Santa Cruz, Calif., 1979), Proc. Sympos. Pure Math., vol. 37, Amer. Math. Soc., Providence, R.I., 1980, pp. 37–42. MR MR604554 (84g:20023a)

[FS73] Paul Fong and Gary M. Seitz, *Groups with a* (B, N)*-pair of rank* 2. *I, II*, Invent. Math. **21** (1973), 1–57; ibid. 24 (1974), 191–239. MR MR0354858 (50 #7335)

[GLS94] Daniel Gorenstein, Richard Lyons, and Ronald Solomon, *The classification of the finite simple groups*, Mathematical Surveys and Monographs, vol. 40, American Mathematical Society, Providence, RI, 1994. MR MR1303592 (95m:20014)

[GLS96] ———, *The classification of the finite simple groups. Number 2. Part I. Chapter G*, Mathematical Surveys and Monographs, vol. 40, American Mathematical Society, Providence, RI, 1996, General group theory. MR MR1358135 (96h:20032)

[GLS98] ———, *The classification of the finite simple groups. Number 3. Part I. Chapter A*, Mathematical Surveys and Monographs, vol. 40, American Mathematical Society, Providence, RI, 1998, Almost simple K-groups. MR MR1490581 (98j:20011)

[GLS99] ———, *The classification of the finite simple groups. Number 4. Part II. Chapters 1–4*, Mathematical Surveys and Monographs, vol. 40, American Mathematical Society, Providence, RI, 1999, Uniqueness theorems, With errata: *The classification of the finite simple groups. Number 3. Part I. Chapter A* [Amer. Math. Soc., Providence, RI, 1998; MR1490581 (98j:20011)]. MR MR1675976 (2000c:20028)

[GLS02] ———, *The classification of the finite simple groups. Number 5. Part III. Chapters 1–6*, Mathematical Surveys and Monographs, vol. 40, American Mathematical Society, Providence, RI, 2002, The generic case, stages 1–3a. MR MR1923000 (2003h:20028)

[GLS05] ———, *The classification of the finite simple groups. Number 6. Part IV*, Mathematical Surveys and Monographs, vol. 40, American Mathematical Society, Providence, RI, 2005, The special odd case. MR MR2104668 (2005m:20039)

[Gor69b] Daniel Gorenstein, *On the centralizers of involutions in finite groups*, J. Algebra **11** (1969), 243–277. MR MR0240188 (39 #1540)

[Gor80a] ———, *Finite groups*, second ed., Chelsea Publishing Co., New York, 1980. MR MR569209 (81b:20002)

[Gor80b] ———, *An outline of the classification of finite simple groups*, The Santa Cruz Conference on Finite Groups (Univ. California, Santa Cruz, Calif., 1979), Proc. Sympos. Pure Math., vol. 37, Amer. Math. Soc., Providence, R.I., 1980, pp. 3–28. MR MR604552 (82c:20030)

[Gor82] ———, *Finite simple groups*, University Series in Mathematics, Plenum Publishing Corp., New York, 1982, An introduction to their classification. MR MR698782 (84j:20002)

[Gor83] ———, *The classification of finite simple groups. Vol. 1*, The University Series in Mathematics, Plenum Press, New York, 1983, Groups of noncharacteristic 2 type. MR MR746470 (86i:20024)

[Gra04] R. Gramlich, *Phan theory*, Habilitationsschrift, TU Darmstadt, 2004.

[Hal76] Marshall Hall, Jr., *The theory of groups*, Chelsea Publishing Co., New York, 1976, Reprinting of the 1968 edition. MR MR0414669 (54 #2765)

[Har81] Koichiro Harada, *Groups with nonconnected Sylow 2-subgroups revisited*, J. Algebra **70** (1981), no. 2, 339–349. MR MR623812 (82i:20023)

[Hig68] G. Higman, *Odd characterizations of finite simple groups*, U. Michigan lecture notes, 77 pages, 1968.

[Hig72] ———, *Some p-local conditions for odd p*, Mimeographed lecture notes, Oxford U., 1972.

[Hup67] B. Huppert, *Endliche Gruppen. I*, Die Grundlehren der Mathematischen Wissenschaften, Band 134, Springer-Verlag, Berlin, 1967. MR MR0224703 (37 #302)

[Isa06] I. Martin Isaacs, *Character theory of finite groups*, AMS Chelsea Publishing, Providence, RI, 2006, Corrected reprint of the 1976 original [Academic Press, New York; MR0460423]. MR MR2270898

[Jan72] Zvonimir Janko, *Nonsolvable finite groups all of whose 2-local subgroups are solvable. I*, J. Algebra **21** (1972), 458–517. MR MR0357584 (50 #10052)

[KM75] Kenneth Klinger and Geoffrey Mason, *Centralizers of p-groups in groups of characteristic 2, p-type*, J. Algebra **37** (1975), no. 2, 362–375. MR MR0390047 (52 #10873)

[LS74] Vicente Landazuri and Gary M. Seitz, *On the minimal degrees of projective representations of the finite Chevalley groups*, J. Algebra **32** (1974), 418–443. MR MR0360852 (50 #13299)

[Lun73] J. Richard Lundgren, *On finite simple groups all of whose 2-local subgroups are solvable*, J. Algebra **27** (1973), 491–515. MR MR0332953 (48 #11278)

[Mas] G. Mason, *The classification of finite quasithin groups*, typescript, U. California Santa Cruz, ca. 1981; about 800 pages,.

[McC82] P. McClurg, *roughly, on FF-modules for almost-simple groups*, Ph.D. thesis, U. Calif. Santa Cruz, about 1982.

[MSS03] Ulrich Meierfrankenfeld, Bernd Stellmacher, and Gernot Stroth, *Finite groups of local characteristic p: an overview*, Groups, combinatorics & geometry (Durham, 2001), World Sci. Publ., River Edge, NJ, 2003, pp. 155–192. MR MR1994966 (2004i:20020)

[Pet84] Thomas Peterfalvi, *Simplification du chapitre VI de l'article de Feit et Thompson sur les groupes d'ordre impair*, C. R. Acad. Sci. Paris Sér. I Math. **299** (1984), no. 12, 531–534. MR MR770439 (86d:20020)

[Pet86] ———, *Le théorème de Bender-Suzuki. I, II*, Astérisque (1986), no. 142-143, 141–233, 235–295, 296, Révision dans les groupes finis. Groupes du type de Lie de rang 1. MR MR873959 (88h:20018)

[Pet00] ———, *Character theory for the odd order theorem*, London Mathematical Society Lecture Note Series, vol. 272, Cambridge University Press, Cambridge, 2000, Translated from the 1986 French original by Robert Sandling and revised by the author. MR MR1747393 (2001a:20016)

[Pha77a] Kok Wee Phan, *On groups generated by three-dimensional special unitary groups. I*, J. Austral. Math. Soc. Ser. A **23** (1977), no. 1, 67–77. MR MR0435247 (55 #8207)

[Pha77b] Kok-wee Phan, *On groups generated by three-dimensional special unitary groups. II*, J. Austral. Math. Soc. Ser. A **23** (1977), no. 2, 129–146. MR MR0447427 (56 #5739)

[Smi75] Fredrick Smith, *Finite simple groups all of whose 2-local subgroups are solvable*, J. Algebra **34** (1975), 481–520. MR MR0376847 (51 #13022)

[Smi80] Stephen D. Smith, *The classification of finite groups with large extraspecial 2-subgroups*, The Santa Cruz Conference on Finite Groups (Univ. California, Santa Cruz, Calif., 1979), Proc. Sympos. Pure Math., vol. 37, Amer. Math. Soc., Providence, R.I., 1980, pp. 111–120. MR MR604567 (82c:20034)

[Sol80] Ronald Solomon, *Some results on standard blocks*, The Santa Cruz Conference on Finite Groups (Univ. California, Santa Cruz, Calif., 1979), Proc. Sympos. Pure Math., vol. 37, Amer. Math. Soc., Providence, R.I., 1980, pp. 43–46. MR MR604555 (82i:20028)

[Str96] G. Stroth, *The uniqueness case*, Groups, difference sets, and the Monster (Columbus, OH, 1993), Ohio State Univ. Math. Res. Inst. Publ., vol. 4, de Gruyter, Berlin, 1996, pp. 117–126. MR MR1400413 (97k:20027)

[Tit81] J. Tits, *A local approach to buildings*, The geometric vein, Springer, New York, 1981, pp. 519–547. MR MR661801 (83k:51014)

[Wal86] John H. Walter, *The B-conjecture; characterization of Chevalley groups*, Mem. Amer. Math. Soc. **61** (1986), no. 345, iv+196. MR MR840833 (87h:20035)

Index

Page locations for definitions, as well as for references which are particularly fundamental, are indicated in **boldface**.